Design and Investment of High Voltage NanoDielectrics

Ahmed Thabet Mohamed
Aswan University, Egypt & Qassim University, Saudi Arabia

IGI Global
PUBLISHER of TIMELY KNOWLEDGE

A volume in the Advances in Computer and Electrical Engineering (ACEE) Book Series

Published in the United States of America by
IGI Global
Engineering Science Reference (an imprint of IGI Global)
701 E. Chocolate Avenue
Hershey PA, USA 17033
Tel: 717-533-8845
Fax: 717-533-8661
E-mail: cust@igi-global.com
Web site: http://www.igi-global.com

Library of Congress Cataloging-in-Publication Data

Names: Mohamed, Ahmed Thabet, 1974- author.
Title: Design and investment of high voltage nanodielectrics / by Ahmed Thabet Mohamed.
Description: Hershey, PA : Engineering Science Reference, an imprint of IGI Global, [2021] | Includes bibliographical references and index. | Summary: "This book examines the principle properties to the principle nanoparticles and its impacts on the fundamental polymers' properties"-- Provided by publisher.
Identifiers: LCCN 2020006903 (print) | LCCN 2020006904 (ebook) | ISBN 9781799838296 (hardcover) | ISBN 9781799854821 (paperback) | ISBN 9781799838302 (ebook)
Subjects: LCSH: Nanostructured materials--Electric properties. | Dielectrics.
Classification: LCC TA418.9.N35 M6315 2021 (print) | LCC TA418.9.N35 (ebook) | DDC 620.1/95--dc23
LC record available at https://lccn.loc.gov/2020006903
LC ebook record available at https://lccn.loc.gov/2020006904

This book is published in the IGI Global book series Advances in Computer and Electrical Engineering (ACEE) (ISSN: 2327-039X; eISSN: 2327-0403)

British Cataloguing in Publication Data
A Cataloguing in Publication record for this book is available from the British Library.

Advances in Computer and Electrical Engineering (ACEE) Book Series

Srikanta Patnaik
SOA University, India

ISSN:2327-039X
EISSN:2327-0403

Mission

The fields of computer engineering and electrical engineering encompass a broad range of interdisciplinary topics allowing for expansive research developments across multiple fields. Research in these areas continues to develop and become increasingly important as computer and electrical systems have become an integral part of everyday life.

The **Advances in Computer and Electrical Engineering (ACEE) Book Series** aims to publish research on diverse topics pertaining to computer engineering and electrical engineering. **ACEE** encourages scholarly discourse on the latest applications, tools, and methodologies being implemented in the field for the design and development of computer and electrical systems.

Coverage

- Qualitative Methods
- Computer Science
- Circuit Analysis
- Computer Hardware
- Power Electronics
- Programming
- Applied Electromagnetics
- Chip Design
- Digital Electronics
- VLSI Fabrication

IGI Global is currently accepting manuscripts for publication within this series. To submit a proposal for a volume in this series, please contact our Acquisition Editors at acquisitions@igi-global.com or visit: https://www.igi-global.com/publish/.

The Advances in Computer and Electrical Engineering (ACEE) Book Series (ISSN 2327-039X) is published by IGI Global, 701 E. Chocolate Avenue, Hershey, PA 17033-1240, USA, www.igi-global.com. This series is composed of titles available for purchase individually; each title is edited to be contextually exclusive from any other title within the series. For pricing and ordering information please visit https://www.igi-global.com/book-series/advances-computer-electrical-engineering/73675. Postmaster: Send all address changes to above address.

Titles in this Series

For a list of additional titles in this series, please visit: https://www.igi-global.com/book-series/advances-computer-electrical-engineering/73675

Research Advancements in Smart Technology, Optimization, and Renewable Energy
Pandian Vasant (University of Technology Petronas, Malaysia) Gerhard Weber (Poznan University of Technology, Poland) and Wonsiri Punurai (Mahidol University, Thailand)
Engineering Science Reference • © 2021 • 407pp • H/C (ISBN: 9781799839705) • US $225.00

HCI Solutions for Achieving Sustainable Development Goals
Fariza Hanis Abdul Razak (Universiti Teknologi MARA (UiTM), Malaysia) Masitah Ghazali (Universiti Teknologi Malaysia (UTM), Malaysia) Murni Mahmud (International Islamic University Malaysia, Malaysia) Chui Yin Wong (Multimedia University, Malaysia) and Muhammad Haziq Lim Abdullah (Universiti Teknikal Malaysia, Melaka, Malaysia)
Engineering Science Reference • © 2021 • 300pp • H/C (ISBN: 9781799849360) • US $195.00

Innovations in the Industrial Internet of Things (IIoT) and Smart Factory
Sam Goundar (The University of the South Pacific, Fiji) J. Avanija (Sree Vidyanikethan Engineering College, India) Gurram Sunitha (Sree Vidyanikethan Engineering College, India) K Reddy Madhavi (Sree Vidyanikethan Engineering College, India) and S. Bharath Bhushan (Sree Vidyanikethan Engineering College, India)
Engineering Science Reference • © 2020 • 300pp • H/C (ISBN: 9781799833758) • US $225.00

Advancements in the Design and Implementation of Smart Grid Technology
Ravi Samikannu (Botswana International University of Science and Technology, Botswana) Karthikrajan Senthilnathan (VIT University, India) Balamurugan Shanmugam (Quants IS & CS, India) Iyswarya Annapoorani (VIT University, India) and Bakary Diarra (Institute of Applied Sciences University of Sciences, Techniques and Technologies of Bamako, Mali)
Engineering Science Reference • © 2020 • 300pp • H/C (ISBN: 9781799836575) • US $205.00

Industrial Internet of Things and Cyber-Physical Systems Transforming the Conventional to Digital
Pardeep Kumar (Quaid-e-Awam University of Engineering, Science, and Technology, Pakistan) Vasaki Ponnusamy (Universiti Tunku Abdul Rahman, Malaysia) and Vishal Jain (Bharati Vidyapeeth's Institute of Computer Applications and Management (BVICAM), India)
Engineering Science Reference • © 2020 • 431pp • H/C (ISBN: 9781799828037) • US $225.00

IGI Global
PUBLISHER of TIMELY KNOWLEDGE

701 East Chocolate Avenue, Hershey, PA 17033, USA
Tel: 717-533-8845 x100 • Fax: 717-533-8661
E-Mail: cust@igi-global.com • www.igi-global.com

Table of Contents

Preface

Welcome to the amazing nano world! God willing, you will find in this book scientific and laboratory applications of nanosciences in the field of industrial applications and in the scope of high-voltage engineering. This book encourages researchers and investors to follow up the applications received for scientific and laboratory use in all environmental fields and calls for modern education's interaction with nanosciences and nanotechnologies. Hence, it is necessary to train new generations of researchers, engineers, and technologists in the fields of nanoscience. As a contribution from us in the education of Nano scopes, we present this book which offers a framework that gathers the basic ideas necessary to understand modern developments based on nanoscience and nanotechnology in the fields of electrical engineering.

Nanotechnology is the science of modifying molecules or atoms to make new products. The new nanotechnology revolution aims to develop a new type of industrial electronic applications used in various environmental fields, where scientists turn to engineers designing wires and equipment at the atomic level. In such system, the fields of physics are intertwined and overlap. Moreover, chemistry, biology, electricity, electronics, and mechanics are forcefully brought together, and this interconnectedness and overlap can promise many new innovations.

The future will be the Nanotech industry, which could lead the world into a new industrial revolution. Nano is a billionth of a meter, and the diameter of a human hair is 80,000 nanometers. At this level of technology, all normal physical and chemical laws do not exist. For example, nano-carbon tubes are 100 times stronger than steel and six times lighter than it. Therefore, this technology is distinguished in the areas of its human use, since its products are smaller, cheaper, lighter and better able to perform the functions assigned to it.

Nanotechnology is a very new, active and very fast research field. Many scientists working in this field confirm that nanotechnology will bring about a new industrial breakthrough in various areas of life, causing fundamental transformations in economics and technology.

Nanotechnology applications attract the attention of many scientists, industrialists and funders, and the assistance of governments and developed countries to nanotechnology research has increased in recent years.

Nanotechnology can give developing countries the opportunity, like what happened in China, to invest in research and development, support university programs aimed at finding scientists and engineers specializing in nanotechnology and send them to study in nanotechnology institutes.

CHALLENGES

Nanotechnology leads to new industrial revolution in the world, and the scientific nature of this technology carries with it many benefits for mankind through the fruits of scientific research, provision of capabilities and the desire towards development in order to benefit from this technology and invest in it to serve humanity.

The goal of nanotechnology focus on the economic side is to encourage the development of an industrial development plan that aims to localize and transfer nanotechnology, based on the importance of this field at various scientific, industrial and economic levels globally. Our ultimate goal is to introduce a nanotechnology-based research on industrial products with economic returns that contribute to supporting the economy.

Control of research and development, as well as acquisition of intellectual property for any new technology and innovation are the keys to successful marketing, sweeping markets and achieving the best investment. On the commercial level, this means the need to assess the importance and potential of new innovative technologies for industries. Then governments or companies decide the best strategy to exploit that great potential, and this strategy relies on choosing how to join such research and its authors, patents and inventors. Definitely, rich countries of the world and their subsidiaries are the ones that control the emerging technologies revolving around wealth and power. Since nanotechnology is a new page, overlap between it and the industries and services sector is necessary. The establishment of nanotechnology companies often starts from research that has worked in academic institutes and colleges, and this is especially true in the United States of America. Subsequently, the scientific experiences of the founders may be invaluable.

In industry, small sciences and technologies have stormed the door to multiple and varied applications that include various scientific, industrial and engineering fields. These sciences and technologies are concerned with nanoscale dimensions characterized by new mechanical, electronic and electrical properties due to the high ratio of their surface to their size. In this part, this book will look at the applications of nanotechnology in the industry, which have begun to spread widely and receive great acceptance due to its quality and accuracy, and it is noted that the nanotechnology applications in the industry are many and cannot be limited. In this book, we will attempt to address the most important high voltage industrial applications. Industry is considered in all its forms the primary engine for growth because it increases production and the standard of living and improves health, education and economic situation, if we preserve the environment as much as possible.

In recent years, mankind has entered the era of nanotechnology-based consumer products, since nanotechnology has become a profitable business for international companies. As nano is the focus of science today, there is hope for increasing interest in it across all industrial applications. The use of nanomaterials improves technological performance in production and gives new functions to products such as solutions related to light weight. In addition, it reaps benefits and enhances electrical performance, thermal insulation, high performance and quality.

Nanomaterials are among the most important modern concepts. When economic units introduce them into production, they offer products with characteristics and advantages that exceed those produced in traditional ways. Given the scientific progress and the huge technological development that has led to increasing competition intensity, economic units should be directed to applying concepts that help in the introduction of materials and the structuring of modern industrial product costs. It further aims to review the impact of nanomaterials on cost accounting and delivery of research. Thus, the costs of the modern industrial product will shift from direct materials, direct wages and industrial costs to industrial costs that are not direct only. The use of new nanomaterials in production contributes to improving control and reaping benefits as a result of their use in the industry, in addition to increasing the level of durability of these industrial processes, as well as improving and evaluating the performance of production lines in terms of productivity and cost-efficiency to improve product performance functions.

The introduction of nanomaterials into production lines contributes to accelerating the market absorption of products that are produced using nanoparticles. An improvement in the existing manufacturing processes can be observed by incorporating the nanomaterials which lead to the best efficiency in resource use, safety, sustainability, recycling, as well as quality and standards. Moreover, technical knowledge is improved by integrating nanomaterial manufacturing processes in terms of productivity, environmental performance and cost-effectiveness. All of this, in turn, contributes to the business plans that encourage investment in the private sector and the effect this has on the growth of business in the future.

SEARCHING FOR A SOLUTION

Dielectrics and liquids play a main role in the effectiveness and reliability of electric power. This is concluded by the present work by illustrating the importance of the new nanotechnology for solid dielectric materials in our life, presenting theoretical and experimental studies for using nanoparticles in industry, specifying the characterization and applications of dielectrics, studying the nanodielectrics fabrication, focusing on nanodielectric theories and the new technologies for filling nanoparticles inside dielectrics, presenting the realistic investment of nanodielectrics, the degradation of nanodielectrics and space charge in nanodielectrics surfaces and barriers. Nanotechnology is a name coming to the world of scientific discoveries, inventions and amazing scientific innovations at lightning speed and has broken into all engineering, medical, environmental and industrial fields. Consequently, we cannot underestimate the contribution of electrical engineering and huge effort to this discovery from nanoscience. The focus of this book will revolve on topics of nanodielectrics and its applications in industry, the importance of nanoparticles and their applications in the industry, clarification of the applications of electrical nanodielectrics, their design, theory and manufacture, as well as all possible techniques for filling nanoparticles in the industrial applications of nanodielectrics. Additionally, this book discusses the manufacture of nanodielectrics, the importance of electrical nanomaterials and explains the effects of nanoparticles on the characterization of space charging and Nano films. This book also presents the breakdown of electrophoresis of nanoscale insulators, the surfaces of nanodielectrics and nanoparticles, as well as investment techniques in nanotechnology. Finally, the book also clarifies the investment techniques in nanotechnology. Nanotechnology has tremendous potential, which makes it capable of contributing to amazing progress in the welfare of human life and completely changing the direction of our next world for the better, if used properly.

In this book, an attempt is made to provide easy, simple and rich information about nanotechnology in insulators to make the reader, the specialist and the non-specialist in direct contact with this technology, following up its development and benefitting from it. It is also useful for those who want to find more information and discussion of the details of this technique and who can find in its rich list of references a view on recent research in this field. The future will be for nanotechnology in all fields generally and engineering industrial fields particularly. This technology and the breakthrough it witnesses will lead the world to a new industrial revolution.

This book aims to link nanostructures in electrical engineering and high effort with nanotechnology for industry and industrial investments, focus on the economic side and encourage and develop economic plans. Such plans aim at localizing and transferring nanotechnology in various fields, especially the industrial engineering field to reach industrial products with an economic return that contributes to supporting the economy based on new scientific research that renders it more powerful and efficient. The introduction of nanostructures into production will leave its mark on the costs of the modern industrial product.

Scientific and technological progress resulted in nanomaterials with their advanced and developable properties and super durability. Because of their significant effects on the characteristics, features and capabilities of products, it allows them to achieve a competitive advantage in the industry. The tremendous scientific progress and the great technological development have contributed to highlighting materials with advanced specifications that help the industrial fields enhance their products and their competitive position in terms of quality, efficiency and accuracy, as the costs will turn into indirect industrial costs. The dependence of industrial units on nanomaterials poses the economic problem represented by its call for natural resources because nanotechnology enables access and misses the required properties as long as the properties of the material and its specifications are determined within the economic unit.

This book explains the importance of electrical nanomaterials to nanodielectrics manufacture and sheds light on the effects of nanoparticles on the characterization of space charging in nano films. Furthermore, this book displays the breakdown of electrophoresis of nanoscale insulators, the surfaces of nanodielectrics and nanoparticles, as well as investment techniques in nanotechnology. Finally, the investment techniques in investment technology are highlighted. This is a brief overview of the contents of the book and its main topics. In conclusion, the aim of this book is to develop the uses of dielectrics in high voltage and to develop, design and manufacture new samples with excellent durability. In addition, it seeks to investigate the relationship between nanostructures in electrical engineering and its applications in industry.

ORGANIZATION OF THE BOOK

Nanotechnology is a name emerging in the world of scientific discoveries, inventions and amazing scientific innovations at lightning speed and has broken into all engineering, medical, environmental and industrial fields. Hence, the share of electrical engineering and the huge efforts in this discovery from the nanoscience cannot be overlooked and will be the focus of this book.

This book tackles the topics of nanodielectrics and their applications in industry, the importance of nanoparticles and their applications in the industry, the applications of electrical nanodielectrics, their design, theory and manufacture, as well as techniques for filling nanoparticles in the industry. Moreover, it shows the design, theories, and applications of nanoscale insulators. Thus, this book explains NEW vision in ten chapters which are summarized as follows:

Chapter 1 describes the importance of the new nanotechnology for solid dielectric materials in our life, the main definitions of nanotechnology, historical background of nanotechnology materials, concepts of nanofluid development and the purpose of the present work.

Chapter 2 presents the characterization and applications of dielectrics that have handled the dielectrics forms, classification of dielectrics, thermoplastic dielectrics, thermosetting dielectrics and elastomers dielectrics.

Chapter 3 offers a thorough study for using nanoparticles in industry, which discusses the importance of nanoparticles, nanoparticles identifications and particulates of variant nanoparticles.

Chapter 4 introduces recent nanodielectric theories which discuss the nanodielectric composites, nanodielectric interphase prediction models, interphase power law IPL model, inhomogeneous interphase and multi-nanoparticles technique.

Chapter 5 discusses the fabrication of nanodielectrics, which addresses the synthesis of co-dielectrics, synthesis of organic nanoparticles/Dielectrics, synthesis of inorganic nanoparticles/ dielectrics, synthesis of metal nanoparticles/Dielectrics, synthesis of multi-nanodielectrics, synthesis of nanodielectrics with coating nanoparticles, synthesis of thin films nanodielectrics and preparation of membranes.

Chapter 6 reviews the new technologies for filling nanoparticles inside dielectrics by explaining the computational solid-state physics of nanodielectrics, modeling and simulation techniques, bare spherical nanoparticles, non-spherical nanoparticles, physical process analysis, nanodielectrics technology and fillers in commercial dielectrics.

Chapter 7 offers realistic characterization of nanodielectrics which tackles the polyethylene nanodielectrics characterization, polypropylene nanodielectrics, polyvinyl chloride nanodielectrics and focus on new multi-nanocomposites insulation materials.

Chapter 8 reviews the space charge degradation of nanodielectrics, which explains the space charge measurements, thin films nanodielectric materials, characterization of thin films nanodielectrics, effects of nanoparticles on space charge characterization forecasting and recommendations for nanodielectric degradation.

Chapter 9 presents nanodielectric surfaces and barriers, thermal interface materials (TIM), wetting nanodielectric surfaces, packaging and battery applications and the proposed investment procedures in the future.

Chapter 10 tackles the degradation of nanodielectrics by reviewing electrical degradation and lifetime, nanodielectric thin films, degradation of nanodielectrics under ac electric fields, degradation of nanodielectrics under dc electric fields, thermal stability analysis and recommendations for investment nanodielectric technologies.

Ahmed Thabet Mohamed

Acknowledgment

In the Name of Allah, Most Gracious, Most Merciful

First and foremost, praises and thanks to Allah, the Almighty, for His showers of blessings throughout my research work to complete the book successfully. All deepest thanks are due to Almighty Allah, the Merciful and the Compassionate for the uncountable gifts granted to me.

I am extremely grateful to my parents for their love, prayers, caring and sacrifices for educating and preparing me for my future. I am very much thankful to my lovely wife for her love, understanding, prayers and continuing support to complete this book. My sincere appreciation goes to my parents and my wife for their generous advice, insightful suggestions and support during the whole scope of this work. Without their help, love and care, this work would never have been accomplished. Furthermore, I express my special thanks to all the research teamwork of Nanotechnology Research Center at Aswan University.

The present work was supported by Nanotechnology Research Center at Aswan University that is established with the aid of Science and Technology Development Fund (STDF), Egypt, Grant No: Project ID 505, 2009-2011.

Ahmed Thabet Mohamed

Introduction

Nanotechnology is the technology of the future. The one who holds the reins of this technology will hold the key to promising future economies, since technological progress in recent times passes steadily to other scientific horizons, especially in nanotechnology by enriching the scientific, research, social and economic life with everything. This book discusses the topics of nanodielectrics and their applications in industry, as well as the importance of nanoparticles and their applications in the industry. This book also clarifies the applications of electrical nanodielectrics, their design, theory and manufacture, as well as techniques for filling nanoparticles in the industry through displaying the design, theories and applications of nanoscale insulators. At the end of this research, this book concludes that nanotechnology is one of the most important technologies nowadays and in the future, and it has become at the forefront of the most important fields of science. This is due to its importance for improvement of products, mothers' treatment and human service in all areas of life. The future will be for nanotechnology and its products. The significance of this modern technology lies in its being inexpensive compared to the rest of the technologies, and its economic returns are very high. In all fields of science and applied science, physical, biological, chemical, electronic, mechanical and information sciences are intertwined and interrelated. The industries in which nanotechnology will be introduced, will erase from the market all industries produced with traditional technology because they will be better, cheaper, less consuming of energy and materials and cleaner as they do not pollute the environment and are more robust and durable than them. The role of nanotechnology is the reclassification of the costs of the modern industrial product.

In our present life, there is an overwhelming demand for this emerging and new technology, so that we can catch up with the scientific process and unleash the scientific energies in people's minds in the country to demonstrate their merit and competence. In light of the aforementioned, we are driven by hope in presenting the concept of nanotechnology, its importance currently and in the future and its impact on various scientific, practical, industrial, commercial and engineering fields. Hence, this will be a starting point for activating nanotechnology in our scientific and practical life. Great hope lies for future scientific revolution in physics, chemistry, biology, engineering and others. Therefore, it is necessary to work on utilizing the distinctive properties of nanomaterials in creating innovation. Inventions benefit humanity in the areas of peace, accelerate and facilitate life and substitute malicious earth mother whose science has not yet reached a radical cure and in many other services. This book also highlights the importance of nanotechnology in the fields of electrical engineering and high voltage insulators, as well as its applications in detail. It also stresses the importance of scientific research and its role in achieving scientific and industrial development and breakthrough. Accordingly, scientific research must take advantage of these examples and the distinguished scientific experiments of nanomaterials to bring about innovations and inventions that benefit nations.

Ahmed Thabet Mohamed

Section 1
Design of High Voltage NanoDielectrics

In this section, NanoDielectrics theories have been described to know how to design high voltage nanodielectrics and so define the fabrication procedures.

Chapter 1
Introduction

ABSTRACT

Nanotechnology development was initially "pushed" by fundamental knowledge (nanoscience and nano engineering) and the long-term promise of its transformative power. For this reason, we have done the preparation and governance of nanotechnology differently. This chapter describes the importance of the new nanotechnology for solid dielectric materials in our life. The main definitions of nanotechnology and historical background of nanotechnology materials are obvious. For nano, research policies have been motivated by long-term vision rather than short-term economic and political decisions. So, this chapter contains the concepts of nanodielectrics and nanofluids development. The purpose of present work is presented in this chapter.

The field of nanotechnology is a standout amongst the vast majority of prevalent zones to present innovative work on the whole specialized foul orders. This clearly incorporates dielectric science and engineering. Furthermore, in this field, the investigations disguise an expansive extent about topics. This might incorporate microelectronics that are concerned with illustration of the discriminating measurement scale, for up to date gadgets are right away beneath 100 nm. Other territories incorporate dielectric-based biomaterials, nanoparticle pill delivery, mini emulsion particles, power module cathode dielectric bound catalysts, layer-by-layer self-amassed dielectric films, electro spun nanofibers, engraving lithography and dielectric blends.

Nanocomposites category is the materials to which nanoparticles are added during the manufacture of those materials, and as a result, the nanomaterials show great improvement in its properties. For example, adding carbon nanotubes will change the properties of the electrical and thermal conductivity of the material. The addition of other types of nanoparticles may improve the optical and electrical insulation properties, as well as mechanical properties, such as hardness and strength. The volume ratio of the added nanoparticles must be very low (in the range of 0.5% to 5%) due to the fact that the ratio between the surface area to the volume of the nanoparticles is high. Furthermore, considerably in the field for nanocomposites, numerous different topics exist including composite reinforcement, obstruction properties, fire resistance, electro-optical properties, cosmic applications and bactericidal properties.

DOI: 10.4018/978-1-7998-3829-6.ch001

Nanotechnology is profoundly inserted in the configuration for propelled gadgets to electronic as optoelectronic provisions. The dimensional scale for electronic units needs to currently enter the nano-range. The utility from claiming dielectric-based nanocomposites over these zones is a very much different directing, including a significant number of possibility provisions and sorts for nanocomposites. Person-particular nanocomposite sort which has received significant enthusiasm includes conjugated dielectrics and carbon nanotubes. A later survey about this region has indicated a reiteration for possibility requisitions including photovoltaic (PV) phones and photodiodes, super-capacitors, sensors, printable conductors, light emitting diodes (LEDs), in addition to field impact transistors (Baibarac & Gomez-Romero, 2006). Additionally, the vast majority publicized requisition for dielectric nanocomposites might have been used for claiming exfoliated clay support from claiming TPO (thermoplastic polyolefin). Obviously, a lot of people have used more provisions directing, including exfoliated clay reinforcement, however, a number of these have not been publicized (Baughman, 2002; Moniruzzaman & Winey, 2006).

1.1 NANOMATERIALS AND APPLICATIONS

Nanomaterials take many forms; each with a structure, properties and a measure of its diameter and length, and each of them has distinctive uses as well. Nanomaterials can be classified according to the form of Quantum Dots category. This category is a three-dimensional semiconductor nanostructure ranging between 2 and 10 nanometers, and this corresponds to 10 - 50 atoms in a single diameter and 100 - 100,000 atoms in a single quantum point size, when the diameter of the quantum point is equal to 10 nanometers. Fullerene category was discovered in 1985 and it is another strange nanoscale of carbon, where a-60-carbon atom molecule has a symbol of 60C. A fullerene globular molecule is like spotted soccer, as shown in the figure below. Various applications have appeared for each of these compounds, including 60C3K and 60C2RbCs which have shown super conductivity. Moreover, other forms have been discovered, such as conical, tubular and spherical fluorines. Nano balls category is one of the most important carbon nanotubes that belongs to the 60C class of fluorines, but they slightly differ in composition, as they are multi-shells and they are also empty of center. The nanoballs have no gaps on their surface, and because of their composition, they look like onions. Scientists have called them "onions", and the diameter of a single ball may reach 500 nanometers or more. Nanoparticles (nanoparticles) category is new to use. These particles were in manufactured or natural materials since ancient times. Nanoparticles can be defined as microscopic atomic or molecular assemblies which range from few atoms (a molecule) to million atoms, and they are almost spherically bound together with radius less than 100 nanometers. When the size of the nanoparticle reaches the nanoscale in one dimension, it is called a quantum well, but when its nanoscale is in two dimensions, it is called a quantum wire. When the quantum wire has 3 dimensions, it is called Quantum dots. Here, it is indicated that the change in the nanoscale dimensions of the three aforementioned structures will affect their electronic properties, which leads to a significant change in the optical properties of the nanostructures. Recently, nanoparticles have been made from ferrite, dielectric, semiconductor and hybrid formulations (such as encapsulated nanoparticles), as well as from models of nanoparticles of a semi-solid nature. Copper nanoparticles (less than 50 nm) are of high rigidity and are not malleable or withdrawable, unlike ordinary copper particles which can be folded, knocked and pulled. Nanotubes category is a slice that folds in a cylindrical shape; the end of the tube is often open and the other is closed in a semicircle. They are made from organic materials (carbon) or inorganic materials (metal oxides, such as vanadium oxide and manganese). These tubes have strength, rigidity and electrical conductivity, but the metal oxides are heavier and weaker than carbon tubes. The diameter of the nanotube is between 1 nm and 100 nm, and its length is 100 μm to form nanowires. The nanotubes have several forms, as they may be straight, spiral, winding, resonant, conical, and so on. Nano fibers category of materials has received recent attention for their industrial importance. It takes many forms, such as hexagonal and spiral fibers, and wheat-like fibers. Nanofibers are distinguished by the fact that the surface area to their size is large because the number of surface peaks is large in relation to the total number, and this is what gives them distinctive mechanical properties, such as hardness, tensile strength, etc., However, it suffers from the difficulty of controlling its continuity, integrity and pavement. These fibers are used in medicine and transplantation, such as joints, wound healing and the transmission of medications in the body. They are also used in military fields, such as reducing air resistance. Nanowires category includes the nanowires that may be less than one nanometers in diameter and are of different lengths, i.e. a length-to-width ratio greater than 1000 times. Hence, they are attached to one-dimensional materials and they are superior to conventional ordinary wires because the electrons in them are quantified to one side. This makes them occupy specific energy levels that differ from those broad levels present in the physical material. These wires are not present in nature, but are prepared in the laboratory in many ways, including chemical curettage of a large wire or extrusion of a large wire by high-energy particles. They take many forms, including spiral or five-way symmetric, and when prepared, they are suspended from the upper end or deposited on another surface. Nanoscale wires have many future uses, such as linking electronic components within a small circuit and building electric and electronic circuits.

1D Nanomaterials are explained as the category that includes materials with a dimension of less than 311 nm. This category includes one-dimensional nanomaterials (that is, after only one nanoscale). Thin layer films or films such as nanomaterials are used in coating work. Surface Nano coating is the same as that used for coating surfaces of oral products by imposing rust erosion, or the thin films used in thin films. Dilute food products come with a hand that prevents them from pollution. Likewise, materials are made of semiconductors. Different conductors, such as the silicon chips, are used in the manufacture of solar cells. **2D nanomaterials** require in this group of nanomaterials that a dimension of two dimensions is less than 311 dimensions. Nanotubes or carbon nanotubes are taken into consideration. Nanowires are useful models for the class of materials. It is not surprising that the carbon nanotubes were filtered because they were used as support and fortifying molds. The mouth is used to raise the values of its salads and improve its mechanical properties, and in particular, blindness raises its resistance to collapse. In addition, it combines other unique features, such as the superior ability to conduct thermal and electrical conductivity. In addition to the distinguished chemical properties, pipes and nanowires are expected to be used in the manufacture of solar cell components, electronic slides, sensors and electronic devices. **3D Nanomaterials** require in this group of nanomaterials that spheres represent nanoparticles, such as nanoparticles and oily powders. The ultra-fine cyclic materials are similar to this category of valuable technological materials that have been attributed to Bania trilogy. Look at the three-dimensional nanomaterials on the three axes X, Y and Z of less than 311 nm. It is worth noting that this category of three-dimensional nanomaterials, whether blind, ultra-fine powders or powders come on top of the global nanomaterial production list in general. This is because of its multiple uses in modern technological fields and applications. Hence, as an example of blind nanomaterials on the market, nanoparticles of mouth oxides are economically available. It is very important for mouth oxides to enter the oil, such as silicon oxide (SiO_2), titanium oxides (TiO_2), aluminum oxide (Al_2O_3) and iron oxides ($Fe_3\,O_4$).

The mechanical properties of blindness material come as a list of properties benefiting from small size granules and the presence of huge numbers of substances on the outer surfaces. For example, increasing hardness values for oral and alloy materials increase their strength. Blinds of blind loads are located by reducing the particle size and control measurements. Reducing the measurements of granules of cycloids will give them more durability and this characteristic is not found in the ceramic materials known for their use in bombs and their resistance to formation. Nanotech Research aims at developing cyclic materials and raising the values of durability and ability to form and bear shock aids to dampen new types of enabling materials. For example, blind granules, such as titanium carbide, are used in the manufacture of cutting and engraving tools used to cut high-strength objects. Likewise, they are used in accessing oil reservoirs and groundwater circulation, through dealing with rocks, since geological layers are very hard. Thus, they are used instead of expensive black diamonds whose properties are less than those of these new nanomaterials. In this book, it has been found that nanomaterials have high hardness and durability, such as aluminum oxide granules and zirconium oxide, as an application area. Furthermore, it is also used to laminate the inner surfaces of the engine cylinders in order to increase life. The agglomerations are composed of oral nano granules that are combined with granules of other materials.

Melting Point acts as the values of the material fusion temperature degrees are affected by the reduction of the particle size measurements. For example, blindness occurs at the degree of heat at which the transformation of the pure gold symbol turns from the solid to the toxic state and this is known as the melting point at 3194 ° C. This value changes with the change of positions and the order of the gold symbol resulting from the reduction of measures, driving away granules and increasing the surface area of the outer surface. The value of the points of melody of the gold symbol is varied according to the dimensions of the granules of the diagonals. The diameters are different from the chemical composition and are free from impurities. The values of the material fusion points decrease, as the particle measurements decrease to an increase in areas. Due to the external surfaces, the different positions and the arrangement of time, the gold symbol is that of the blind.

Optical properties explain some changes in the characteristics of the distinguished nanomaterials. They are different from the granulated material that is large. It is interesting to see how the effect of grain size extends to changing the optical properties of the dispersal material or optical cracking of the surface of matter. An example of this is the known tincture of grains of pure gold. The diagonals which are more than 211 nanometers are the yellow gold moons that we know, but if they are minimized, these granules are transparent and have increased reduction. Optical cracking is combined between narrow-range spectral emission and wide-range excitation spectrum. The field of electronics and optics is one of the most applied fields of nanomaterials. It can be seen in the manufacture of high-resolution screens of ultra-high contrast and purity of color, such as monitors and modern computers.

Magnetic properties of nanomaterials, like the strength of the magnet, depend quantitatively on the dimensions of the granules of the material made from them. In the magnet, the smaller the grains, the greater the external surface area and the presence of the grains. If the surfaces are managed, the magnet becomes stronger, more effective and strengthened. Accordingly, nanomaterials have their characteristics. Magnetism is the source of the materials used in the production of the super-strong magnets used in huge generators. Nanoparticles are also used in magnetic materials in the manufacture of high-precision loading devices, as well as in the manufacture of photography devices and general medical diagnosis.

The effect of the nanoparticle size of nanomaterials and the density of granular boundary positively counter affect the effect of the electrical properties of the superconducting power. This is recently shown in the manufacture of micro-sensors and electronic chips at the various modern briefs, since such are used in the manufacture of mobile phone components and computers. This has enabled these industrial sectors to produce lightweight, high-tech specifications with low cost at the same time. The quantitative marking of the nanoparticles enables them to improve and enhance the properties and attribute, according to the theory of quantum mechanics that came to correct Newton's classical laws. Varying properties of nanomaterials go back to the following reasons:

- Particles size: The properties of materials, such as conduction and color, do not change by changing the size, except when their size reaches the nanometer scale, their properties change. For example, silicon in normal size is considered an opaque material that does not radiate, but when it is at the size of 1 nm, it radiates in blue, and when it is at the size of 3 nm, it radiates in red.
- Particles Shape: The properties of the nanoparticle depend on the shape that is spherical, tubular, hexagonal, or other.
- Particles Structure: It means what kind of self or particles the nanoparticle is made of and how many.
- Aggregation Degree: Some nanoparticles are the molecules or the particles in which they are spaced, others - the molecules or their atoms are agglomerated adjacent to each other, and the difference in the degree of aggregation of the molecules from one particle to another causes a change in properties.
- Particles Distribution: The distribution of particles or particles within the particle may be uniform or irregular and may be - stable or unstable. For example, silicon particles are distributed regularly in the solution and the entire solution is radiated, but after leaving it for several days, the distribution becomes irregular and goes down to the bottom. The solution is not completely radiated.
- Quantitative Restriction: Some materials are limited to two dimensions, so the movement of electrons will be in one direction, and after the materials are limited to one dimension, then the movement of electrons will be in two directions.

Today, nanotechnology has become the locomotive that pushes nations to achieve their major industrial revolutions and industrial boom events, and nanotechnology research and applications are characterized as a process of design and application with great economic returns. Nanotechnology is, in my opinion, the technology capable of achieving the highest degree of accuracy in the functions, shapes, sizes of materials, their components and their properties. This, in turn, helps to control the functions of the tools used in all fields, especially engineering fields. Therefore, preparing nanomaterials plays a very important role as they acquire unique properties in electrical, magnetic and optical properties as a result of the new arrangement that the atoms take. The increased interest in nanotechnology is due to the distinct new features and properties that materials acquire when they are very young.

1.2 NANOPARTICLES IN INDUSTRY

Nano is derived from the Greek nanos, meaning dwarf or infinitesimal. Nanoscience is the science that studies and synthesizes chemical, physical, and mechanical properties with the study of these phenomena arising from the reduction or enlargement of their sizes. Nanotechnology is the technology that deals with atoms and individual molecules consisting of materials that are concerned with the design, manufacture and installation of new materials that have the same properties but are very small and withstand more. This is to create and produce new and useful means, innovations and inventions. Nanomaterials are the cornerstone of nanoscience science and technology. They are the advanced materials that can be produced so that the dimensions of their dimensions or the dimensions of their internal granules range from (1nm - 100nm). Their smallness and sizes have led to the availability of very distinct characteristics, such as efficiency and quality. Nanomaterials are the industry materials for the twenty-first century and its basic building blocks. They are the main and important pillar of the technology of the century, which is considered a standard for the progress of civilizations of nations and keeping pace with the development and nanoscale industrial boom. The fineness of the small size of the granules of the nanomaterials and the intensity of the granular boundary positively affect their electrical properties, which are their superior ability to conduct electrical current.

Science is a broad sea and the wheel of science is in continuous progress and never stands. Every day, we find new things in various scientific fields, and there is no doubt that nanotechnology has become the subject of modern science and the focus of its attention and has become at the forefront of the most important fields in physics, chemistry, biology and others. The origin of the word "nano" is derived from the Greek word (nanos) which means dwarf, meaning everything that is small, and nontechnology means nanomaterial technology or micro-microscopic technology. Nanoscience is a study of the basic principles of molecules. Vehicles that measure less than 100 nanometers are nanometers, which are unit of measurement equal to 6-10 mm or 9-10 meters. The principle of this technique is based on capturing and manipulating the nano-atoms of any material, moving it from its original positions to other places and then merging it with the atoms of other materials to form a crystalline network in order to obtain nanoscale materials with high-performance properties. Therefore, nanoparticles gain scientific significance as they lie between the large volumetric composition of matter and the atomic and molecular structure.

But how can we benefit from this technology in the future, can it meet all the needs of man in all fields and will it become more popular when it comes to serving man in life matters??

It can be said that nanomaterials are those distinct class of advanced materials that can be produced so that the dimensions of their dimensions or the dimensions of their internal granules are between 1 nm and 100 nm. The smallness of these materials has caused their characteristics to differ from the larger materials (greater than 100 nm). These materials are the building materials for the twenty-first century and an important pillar of the technologies of this century. Nanomaterials vary in source and vary in their proportions, such as being organic or inorganic, natural or synthetic (manufactured).

Recently, it has become possible to manufacture nanoparticles from metals, insulators, semiconductors and hybrid formulations (such as encapsulated nanoparticles) and to manufacture models of nanoparticles of a semi-solid nature, which are liposomes. Other images of nanoparticles are semiconducting quantum dots and nanocrystals. Copper nanoparticles that are less than 50 nanometers in size are of high hardness and are not malleable or withdrawable, and this is the opposite of what happens to ordinary copper, where it can be easily bent, knocked and pulled.

The manufacture and study of artificial nanoparticles in the form of metallic granules dates back to the 1960s. These materials consist of metal particles of nanoscale sizes embedded in a metallic, ceramic, or semiconducting matrix that is not miscible and since then an explosion has occurred in the number and types of systems that have been developed and investigated, which involve A wide variety of treatment methods.

In fact, a wide range of nanocomposites has been prepared showing a spectrum of striking functional and structural properties from a technical point of view. Processing methods have evolved enabling very accurate control of the microstructures, which in turn enables a careful fibrosis of these properties. Several nanocomposites have already been produced and they are currently used in commercial and industrial applications. Thus, the fillers bring a key part in the improvement in claiming business utilization for dielectrics. Originally, they were basically seen similarly as shabby diluents, like the sake filler. However, their capacity should beneficially change. A large number of properties before long will start to be recognized and they will be utilized for numerous purposes nowadays. The term utilitarian filler will be regularly utilized on portray materials that would more than gatherings give cost decrease. Cases of practical fillers include carbon dark, precipitated silica reinforcements do sort treads, aluminum and magnesium hydroxide fire retardant additives to numerous dielectric types. Furthermore, calcined clay and wollastonite support few thermoplastics. Traditionally, fillers were acknowledged as additives. Because of their unfavorable geometrical features, surface range or surface compound composition, they might best reasonably progress the qualities of the dielectric. Their significant commitment might have been in bringing down the cost in materials towards trading a greater amount of exorbitant dielectric. Other things could reasonably be expected; investment favorable circumstances have become quicker forming cycles. Concerning illustration, there is an aftereffect about expanded warm conductivity. Contingent upon the sort of filler, different dielectric properties might cause effects. For example, melt viscosity might be altogether expanded through the consolidation from claiming stringy materials.

On the other hand, shape shrinkage or even warm extension might have a chance to be reduced; a basic impact of practically inorganic fillers. Likewise, the requisition from claiming dielectric influences decision from claiming filler. For example, in order to make conductive materials ready, exceptional fillers must get the obliged properties. In addition, the system for transformation imposes certain imperatives on the decision and medication of the filler before its utilization. For example, dielectrics transformed towards high engineering require fillers, which don't hold numerous dampness. This influences both the decision of the filler or its pretreatment. The decision of additives used to move forward the consolidation of the filler relies on the requisition and the properties required from an item, provided that it is additionally a dead set. Eventually, Tom's perusing the transforming system. For example, extraordinary lubricating operators lessen the viscosity of a melt, while the viscosity for filler dispersions may be regulated by the surface medication for filler. For some cases, the request about expansion may be critical, alternately extraordinary filler pretreatment will be used to accomplish the wanted effects (Ajayan et al., 2003; Harsha & Tewari, 2002; Kim et al., 2008; Landgraeber et al., 2008; Lenz et al., 2009; Siegmeth et al., 2009).

1.3 DIELECTRICS

Dielectric is a large molecule (macromolecule) composed of repeating structural units typically connected by covalent chemical bonds. While dielectric in popular usage suggests plastic, the term actually refers to a large class of natural and synthetic materials with a variety of properties. Due to the extraordinary range of properties accessible in dielectric materials (Painter & Coleman, 1997), they have come to play an essential and ubiquitous role in everyday life (McCrum et al., 1997), from plastics and elastomers on the one hand to natural bio-dielectrics, such as DNA and proteins that are essential for life on the other. Emerging nanotechnologies hold great promise for creating new means of detecting high dielectrics and cleaning waste dielectrics. The use of nanoparticles can absorb and trap organic contaminants in dielectrics. If testing continually succeeds, the process can also be very effective in preparing new safe industrial insulations and non-toxic dielectrics in the world.

Dielectrics are insulators, their charges tend not to move easily in nonmetallic solids. It's possible to have "islands" of charge in glass, ceramics and plastics, as long as the material that provides safe passage for electric charges is a *conductor*. The plastic coating on an electrical cord is an insulator. The glass or ceramic plates used to support power lines and keep them from shorting out to the ground are insulators. Whenever a nonmetallic solid is used in an electrical device, it's called an insulator. Perhaps the only time the word dielectric is used is in reference to the non-conducting layer of a capacitor. Dielectrics in capacitors serve three purposes to keep the conducting plates from coming in contact, allowing for smaller plate separations and therefore higher capacitances. Dielectrics tend to increase the effective capacitance by reducing the electric field strength, which means you get the same charge at a lower voltage. It also tends to reduce the possibility of shorting out by sparking (more formally known as "dielectric breakdown") during operation at high voltage.

When a metal is placed in an electric field, the free electrons flow against the field until they run out of conducting material. In no time at all, we'll have the excess electrons on one side and a deficit on the other. One side of the conductor has become negatively charged and the other positively charged. Release the field and the electrons on the negatively charged side now find themselves too close for comfort. Like charges repel and the electrons run away from each other as fast as they can until they're distributed uniformly throughout; one electron for every proton on average in the space surrounding every atom. A conducting electron in a metal is like a racing dog fenced in a pasture. They are free to roam around as much as they want and can run the entire length, width and depth of the metal on a whim. Life is much more restrictive for an electron in an insulator. By definition, charges in an insulator are not *free* to move. This is not the same thing as saying they *can't* move. An electron in an insulator is like a guard dog tied to a tree, free to move around, but within limits. Placing the electrons of an insulator in the presence of an electric field is like placing a tied dog in the presence of a mailman. The electrons will strain against the field as far as they can in much the same way that our hypothetical dog will strain against its leash as far as it can. Electrons on the atomic scale are more cloudlike than doglike, however. The electron is effectively spread out over the whole volume of an atom and isn't concentrated in any one location. A good atomic dog wouldn't be named Spot, I suppose.

When the atoms or molecules of a dielectric are placed in an external electric field, the nuclei are pushed with the field resulting in an increased positive charge on one side, while the electron clouds are pulled against it resulting in an increased negative charge on the other side. This process is known as polarization, and a dielectric material in such state is said to be polarized. There are two principal methods by which a dielectric can be polarized: stretching and rotation.

Stretching an atom or molecule results in an induced dipole moment added to every atom or molecule. Rotation occurs only in polar molecules; those with a permanent dipole moment like the water molecule shown in the diagram below. Polar molecules generally polarize more strongly than nonpolar molecules. Water (a polar molecule) has a dielectric strength 80 times higher than that of nitrogen (a nonpolar molecule that is the major component of air). This happens for two reasons, one of which is usually trivial. First, all molecules stretch in an electric field whether they rotate or not. Nonpolar molecules and atoms stretch, while polar molecules stretch and rotate. However, this combination of actions only has a tiny effect on the overall degree to which a substance will polarize. What's more important is that polar molecules are already strongly stretched naturally. The way the hydrogen atoms sit themselves on the arms of an oxygen atom's electron clouds distorts the molecule into a dipole. All of this takes place on an interatomic or molecular scale. At such tiny separations, the strength of the electric field is relatively huge for what would otherwise be an unremarkable voltage (13.6 V for an electron in a hydrogen atom, for example). Stretching and rotation are not the end of the story when it comes to polarization. They are just the simplest methods to describe to the casual observer. In general, the polarization of a dielectric material is microscopic electrostatic strain in response to a macroscopic electrostatic stress. An external field applied to a dielectric can't make charges move macroscopically, but it can stretch and distort them microscopically. It can push them into uncomfortable positions, and when released, it allows them to fall back into a relaxed state. The thing that makes the polarizing in an insulator different from stretching an elastic body like a spring is that eliminating the stress doesn't necessarily release the strain. Some insulators will remain in their polarized state for hours, days, years or even centuries. The longest characteristic times have to be extrapolated from incomplete observations about more reasonable duration. No one is going to sit around and wait two thousand years to see the polarization of a chunk of plastic dwindle away to zero. It isn't worth the wait.

Finally, it's somewhat important to keep in mind that the charges "stored" in a dielectric layer aren't available as a pool of free charges. To extract them, you still need metal plates. It's important to remember that the only reason anyone seems to care about this phenomenon is that it helps us to make better capacitors. I think that's where this discussion should wrap up.

1.4 NANODIELECTRICS

Nanodielectrics are the second resins in the encasing engineering, which comprise claiming dielectrics loaded for an extensive amount of micron-sized inorganic fillers. Recently, destined dielectric nanocomposites need characterized aid. Likewise, a substance from claiming dielectrics is filled for a little sum from claiming nanoparticles. Nano-fillers, regardless of how little they are (few wt.%), bring in content an enormously expansive surface concerning illustration contrasted with micro-fillers. Therefore, it may be arguable that it is imperative that research should address how nanoparticles might cooperate with dielectric matrices for essential understanding of qualities that develop because of nano-structuration.

Dielectric nanocomposites are recently rising propelled materials with possibility total provision in transportation and electricity. More fields include hardware engineering, nourishment packaging, building and different commercial enterprises (Tanaka et al., 2004). They might be utilized in illustration of high utilitarian materials, such as barrier-functional materials, covering materials, fire resistant materials and foamed materials. It ought to be noted that materials would be recorded in the request from claiming assessed business request to secondary execution materials (Chujo, 2003). Polyolefin will be recognized in a large portion anticipated in the market. What's more, polypropylene in particular is the real material in the polyolefin business. Barrier-functional materials, as speculated, should represent the biggest advertise stake.

Nanodielectrics are characterized concerning illustration dielectrics by a little measure from claiming nanoparticles. They ought to be called additives by nature instead of fillers, yet the expressions "fillers" will be utilized here. Illustration could be effectively acknowledged in the encasing field. The nanoparticles would be: 1 equals 100 nm done size, 1 should be 10%wt previously in content, and ought further to bolster a chance to be homogeneously scattered in the dielectric grid. The camwood is made towards regulating the blending method, the inter capitulars vein method, and the sol-gel strategy that concentrates on dividing claiming durable nanoparticles from one another (Such division could be directed to forcing shear energy toward the middle of neighboring nanoparticles or toward expanding compound. The more physical closeness between the nanoparticles, the more their contacting dielectric grid becomes.

Nanotechnology provisions are applicable on power transmission for wires. Furthermore, cables might help enhance the effectiveness for power transmission wires. The constituent materials are artificially inactive with respect to one another. Furthermore, they could withstand amazing temperatures without synthetic responses alternately at whatever calculable reduction in quality. The material utilized within the center of the link replaces the steel utilized within accepted cables.

Different electrical transmission foundation nanotechnology requisitions might assist in enhancing different segments of the electric transmission infrastructure, thereby conceivably diminishing natural effects. For instance, regarding transformers, substations, sensors, transformers, liquids holding nano-materials might provide more proficient coolants and transformers, potentially decreasing the footprints, or considerably the number, from claiming transformers. Nanoparticles expand high temperature transfer, and robust nanoparticles behavior that is heat superior to fluid. Nanoparticles sit tight suspended for fluids more extended over bigger particles. In addition, they bring a significantly more amazing surface area, which will be the place at which high temperature exchange takes place. Utilizing nanoparticles in the advancement from claiming HTS transformers might bring about conservative units with no combustible liquids, which might help build siting adaptability. Substations batteries are essential to load-leveling top shaving. Giving uninterruptible supplies from claiming power will control substation switchgear and initiate reinforcement control frameworks. In brief, the more productive batteries can lessen the footprints for substations and potentially the number for substations inside a column.

Likewise, nano-electronics have the possibility to change sensors and power-control gadgets. Nanotechnology-enabled sensors might introduce self-calibrating, as well as self-diagnosing. They might put issue calls to specialists at whatever point issues are predicted or encountered. Such sensors might also take into consideration remote screening about frameworks for an ongoing premise. Smaller than expected, sensors deployed all around a whole transmission organization might furnish entry with information and the majority of data that was inaccessible before. The ongoing energized status from claiming appropriation feeders might speed blackout restoration and period adjusting. Furthermore, accordance passing can be simpler to manage, making a difference in enhancing the general operation of the appropriation feeder system.

REFERENCES

Ajayan, P. M., Schadler, L. S., & Braun, P. V. (2003). *Nanocomposite Science and Technology*. Wiley-VCH. doi:10.1002/3527602127

Baibarac, & Gomez-Romero. (2006). Nanocomposite Materials Based on Conducting Polymers and Carbon Nanotubes. From Fancy Materials to Applications. *Journal of Nanoscience and Nanotechnology*, *6*, 1–14. doi:10.1166/jnn.2006.903

Baughman, R. H. (2002). Carbon Nanotubes—The Route Toward Applications. *Science*, *297*(5582), 787–792. doi:10.1126cience.1060928 PMID:12161643

Chujo, K. (2003). *Polymer Nanocomposites*. Kogyo Chosakai Japan.

Harsha, A. P., & Tewari, U. S. (2002). Tribo Performance of Polyaryletherketone Composites. *Polymer Testing*, *21*(6), 697–709. doi:10.1016/S0142-9418(01)00145-3

Kim, W. Y., Greidanus, N. V., Duncan, C. P., Masri, B. A., & Garbuz, D. S. (2008). Porous Tantalum Uncemented Acetabular Shells in Revision Total Hip Replacement: Two to Four Year Clinical and Radiographic Results. *Hip International*, *18*(1), 17–22. doi:10.1177/112070000801800104 PMID:18645969

Landgraeber, S., von Knoch, M., Loer, F., Wegner, A., Tsokos, M., Hussmann, B., & Totsch, M. (2008). Extrinsic and Intrinsic Pathways of Apoptosis in Aseptic Loosening After Total Hip Replacement. *Biomaterials*, *29*(24-25), 3444–3450. doi:10.1016/j.biomaterials.2008.04.044 PMID:18490052

Lenz, R., Mittelmeier, W., Hansmann, D., Brem, R., Diehl, P., Fritsche, A., & Bader, R. (2009). Response of Human Osteoblasts Exposed to Wear Particles Generated at the Interface of Total Hip Stems and Bone Cement. *Journal of Biomedical Materials Research*, *89*(2), 370–378. doi:10.1002/jbm.a.31996 PMID:18431768

McCrum, N. G., Buckley, C. P., & Bucknall, C. B. (1997). *Principles of Polymer Engineering*. Oxford University Press.

Moniruzzaman, M., & Winey, K. I. (2006). Polymer Nanocomposites Containing Carbon Nanotubes. *Macromolecules*, *39*(16), 5194–5205. doi:10.1021/ma060733p

Painter, P. C., & Coleman, M. M. (1997). *Fundamentals of Polymer Science: an Introductory Text*. CRC Press.

Siegmeth, A., Duncan, C. P., Masri, B. A., Kim, W. Y., & Garbuz, D. S. (2009). Modular Tantalum Augments for Acetabular Defects in Revision Hip Arthroplasty. *Clinical Orthopaedics and Related Research*, *467*(1), 199–205. doi:10.100711999-008-0549-0 PMID:18923882

Tanaka, T., Montanari, G. C., & Mülhaupt, R. (2004). Polymer Nanocomposites as Dielectrics and Electrical Insulation –Perspectives for Processing Technologies, Material Characterization and Future Applications. *IEEE Transactions on Dielectrics and Electrical Insulation*, *11*(5), 763–784. doi:10.1109/TDEI.2004.1349782

Chapter 2
Characterization and Applications of Dielectrics

ABSTRACT

A polymer is an expansive atom (macromolecule) created about rehashing decimal structural units regularly joined by covalent compound bonds. At the same time, polymer is prevalent in use and is recommended over plastic. This chapter contains the characterization and applications of polymer. It handled also the polymer forms, classification of polymers. It contains also thermoplastic polymers and thermosetting polymers, elastomers polymers to obtain the required applications.

The polymer really alludes to an expansive population of common and manufactured materials for an assortment of properties. Because of the uncommon extend of receptive properties clinched alongside polymeric materials (Painter & Coleman, 1997), they bring and assume a key and universal part for ordinary term (McCrum et al., 1997), resining with plastics and elastomers in this particular case. Moreover, common biopolymers, such as DNA and proteins that are vital to existence are different. The report card investigates and examines the recorded sorts about polymers and its physical, mechanical, and electrical properties. The advancement about vulcanization later in the nineteenth century enhanced the sturdiness of the regular polymer rubber, signifying the resin with popularized semi-synthetic polymer. In the 20th century, the resin has been made with totally manufactured polymer, Bakelite, towards reacting phenol, as well as formaldehyde at unequivocally regulated temperature and weight.

2.1 POLYMERS FORMS

Polymers come in numerous structures including plastics, rubber and fibers. Plastics would stiffer over rubber but have diminished low-temperature properties. Generally, plastic differs from rubbery materials because of the area for its glass move temperature (Tg). A plastic has a Tg over room temperature. At the same time, an elastic will have Tg beneath room temperature. Tg will be practically and obviously characterized toward evaluating the excellent association of versatile modulus with temperature

DOI: 10.4018/978-1-7998-3829-6.ch002

for polymers. Plastics are additionally divided under thermoplastics and thermosets. A thermoplastic material is a secondary sub-atomic weight polymer that is not cross-linked. A thermoplastic material could exist in a straight or spread structure. Upon warming a thermoplastic, an exceedingly viscous fluid may be structured so that they can make molded utilizing plastics transforming gear. A thermoset has the sum of the chains tied together with covalent bonds to a system (cross-linked). A thermoset doesn't have a chance to be reprocessed. When cross-linked, thermoplastic material makes reprocessing towards warming the suitable temperature.

Regardless of the huge developments for union and characterization of polymers, a right understanding about polymer atomic structure didn't develop until the 1920s. At that time, researchers accepted that polymers were groups of little particles (called colloids) without positive sub-atomic weights, held together by an obscure force, an idea known as cooperation hypothesis. A paramount commitment should have engineered polymer science which might produce polymers incorporating the energy about step-growth polymerization and for expansion of polymerization, chain transfer, excluded volume, the Flory-Huggins theory, and the Flory gathering.

Manufactured polymer materials, such as nylon, polyethylene, Teflon, and silicone have framed the foundation for a burgeoning polymer business. During a considerable length of time, these have additionally demonstrated noteworthy developments for normal polymer union. For a greater part, industrially essential polymers today are actually engineered and generated on high volume and once suitably scaled natural engineered systems. Engineered polymers today find requisition in about each industry and region from claiming an aggregation. Likewise, Polymers have broadly aided the utilization of adhesives and lubricants and structural parts for results extending from youngsters' toys like airplanes. They have been utilized over an assortment for biomedical provisions extending resinting with implantable units for controlled medication regardless of conveyance. Polymers, such as poly (methylmethacrylate) find requisition as a photograph of opposing materials utilized within semiconductor manufacturing and low-k dielectrics for utilization over high-octane microprocessors. Recently, polymers have also been utilized like adaptable substrates in the advancement of natural light-emitting diodes to electronic shows.

There are three classes of polymers: Thermoplastic polymers ordinarily called Thermoplastics, thermosetting polymers known as Thermosets, and Elastomers ordinarily known as rubbers. Thermoplastics are long-chain straight atoms that effortlessly shape. Thermoset polymers expect a lasting shape alternately set when heated, although a portion will be situated in room temperature. The thermosets resin is similar to powders of alternately fluids which would react with a second material, or which through catalyzed polymerization brings about another result whose properties are contrary to resining with individuals about the possibility of beginning material. Elastomers are polymeric materials whose extents could make a drastic change, eventually perusing the application of a generally humble force, as a return on their first qualities at the point when the compel is discharged. Characteristic polymeric materials, such as shellac, amber and common elastic have been used for a long time. Biopolymers, such as proteins and nucleic acids assume essential parts for living forms. An assortment for different regular polymers exists, such as cellulose, which is the fundamental constituent for wood and paper. The rundown from claiming engineered polymers incorporates manufactured rubber, Bakelite, neoprene, nylon, PVC, polystyrene, polyethylene, polypropylene, polyacrylonitrile, PVB, silicone and a lot more. Polymers have received focus in fields like claiming polymer chemistry, polymer physics, and polymer science.

3.1 THERMOPLASTIC POLYMERS

A thermoplastic is a polymer that turns into a fluid at a point of warming and freezes to a glassy state if sufficiently cooled. The greater number of thermoplastics are high-molecular-weight polymers whose chains copartner through powerless Van der Waals strengths (polyethylene); stronger dipole-dipole collaborations and hydrogen holding (nylon); actually stacking fragrant rings (polystyrene). Thermoplastic polymers vary from thermosetting polymers (Bakelite) since they can, dissimilar to thermosetting polymers, be melted and remolded. Numerous thermoplastic materials are expansion polymers; e.g., vinyl chain-growth polymers, such as polyethylene and polypropylene. The types of thermoplastic polymers will be illustrated below:

3.1.1 Acrylonitrlle Butadlene-Styrene (ABS)

ABS polymers are determined from acrylonitrile, butadiene, and styrene and have the accompanying structure:

Acrylonitrlle Butadlene-Styrene (ABS)

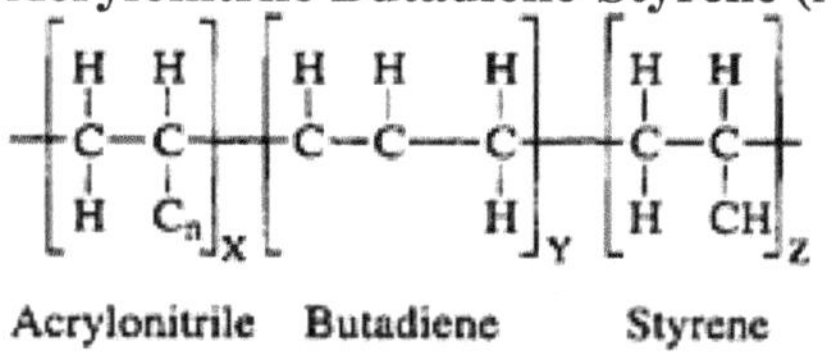

ABS (acrylonitrile butadiene styrene) is made with elastic particles in place, expanding the sturdiness. The particles go about as furor stabilizers, thereby delaying split resin and expanding the strain on disappointment (Socrate et al., (2001) ; Seelig & Van Der Giessen, (2002). The properties for ABS polymers cause modification, eventually perusing the relative sums of acrylonitrile, butadiene, and styrene. has a chance to acquire higher strength, preferred toughness, more excellent dimensional solidness and other properties during the liability of other trademark. ABS plastics are utilized generally for mechanical purposes; they additionally bring handy electrical properties that are reasonably steady through an extensive variety from claiming frequencies. These properties are minimally influenced by temperature and climatic stickiness in the worthy working extend from claiming temperatures (Harper, 1975).

Physical and mechanical Properties: sway imperviousness and sturdiness are the most remarkable mechanical Properties. Sway safety doesn't tumble off quickly at temperatures brought down. Dependability under constrained load is fantastic. The point when sway disappointment can take place is when disappointment may be flexible instead of fragile. Dampness has minimal impact on the physical properties for ABS, which aides in administering dimensional solidness for ABS items. The physical and mechanical properties for ABS are strengthened with glass fibers. This support builds the pliable strength, ductile modulus, flexural strength and compressive strength. At the same time, the warm extension may be diminished. The degree for progress may be reliant upon the rate of glass fiber support.

ABS will be alloyed alternately mixed for polycarbonate. Hence, some of the best qualities for both materials come about. Thermoplastics become simpler with process, have secondary heat and sway resistance, and may be additionally prudent over polycarbonate alone. The ABS/PC compound has an effect quality from claiming 10.7 ft-lb/in. This score is great and exceeds the high effect quality about building thermoplastics. However, this is not secondary concerning the illustration of the effect quality for pc. Polycarbonate has incredulous thickness for admiration to effect quality which the ABS/PC compound doesn't have. The scored sway esteem of the compound drops toward main 2-4ft-lb/in, at the 1/8 to 1/4-inch range. The ABS/PC compound has a flexural quality of give or take that is 15% more terrific over that of the polycarbonate alone. The compound stays all the more inflexible over polycarbonate alone up to more or less 200°F (93°C). ABS could additionally make alloyed for other thermoplastics in place for enhancing sure properties, as well as physical and mechanical properties from claiming ABS/Nylon compound and the ABS/PVC fire redesign compound.

Erosion imperviousness properties: immaculate ABS polymer will be assaulted towards oxidizing agents, solid acids and will cause anxiety split in the vicinity about certain natural mixes. It will be safe on aliphatic hydrocarbons, will not so on fragrant and chlorinated hydrocarbons. ABS will be corrupted by ultraviolet light unless protective additives are consolidated under the detailing. Ordinary Applications: ABS polymers are used to process benefits of the business machine, as well as Polaroid housings, blowers, bearings, gears, pump impellers, concoction tanks, seethe hoods, ducts, piping and electrical channel.

3.1.2 Acrylics

Acrylics are dependent upon polymethyl methacrylate. They also bring a concoction structure as shown below:

Acrylics

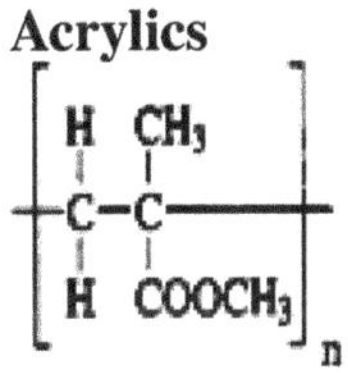

Acrylic polymers are exacerbating prepared eventually by perusing and holding acrylic corrosive or acrylic ester with different atoms. Acrylic polymers use the alluring aspects for acrylic corrosive and acrylic ester (Slone, 2010). Due to their compound structure, acrylic resins are naturally safe with staining and passing from claiming light transmission. Physical and mechanical Properties: parts formed resining with acrylic powders in their immaculate state might have a chance to be clear and almost optically immaculate. The dampness level on acrylics may be reliant upon the relative moistness of the earth. As the relative moistness of air increases, acrylics will absorb moisture, which will bring about a slight dimensional extension. Acrylics might additionally make alloyed with polycarbonate. Erosion imperviousness Properties: Acrylics show remarkable climate capability. They are assaulted towards solid solvents, gasoline, acetone and other comparable liquids. Acrylics are utilized to lenses, airplanes, buildings, glazing, lighting fixtures, coatings, material fibers, fluorescent road lights, open air signs and pontoon windshields.

3.1.3 Chlorotrifluoroethylene (CTFE)

It is a fluorocarbon for the Emulating concoction structure:

Chlorotrifluoroethylene (CTFE)

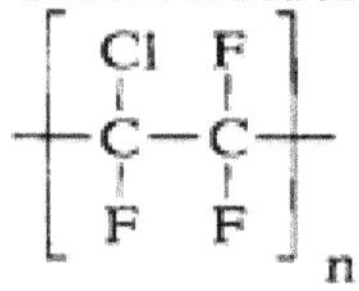

Physical and Mechanical Properties: CTFE, alternately chlorotrifluoroethylene, may be a chlorofluorocarbon with concoction recipe CF2CCIF. It will be regularly utilized like a refrigerant to cryogenic requisitions (Lide, 1998). CTFE has more amazing pliable and compressive quality over PTFE inside its administration temperature. The electrical properties about CTFE are excellent. Although dielectric misfortunes are higher over the individuals of PTFE, CTFE doesn't bring about low friction or bearing properties of PTFE. Erosion imperviousness Properties: In spite of the fact that CTFE has an extensive variety of compound resistance, it mostly has an imperviousness over PTFE, FEP and PFA. It is liable with swelling to exactly chlorinated solvents in raised temperatures and is assaulted by the same chemicals that ambush PTFE.

3.1.4 Ethylenetetrafluorethylene (ETFE)

ETFE is an alternating copolymer of ethylene and tetrafluorethylene sold and has the following structural formula:

Ethylenetetrafluorethylene (ETFE)

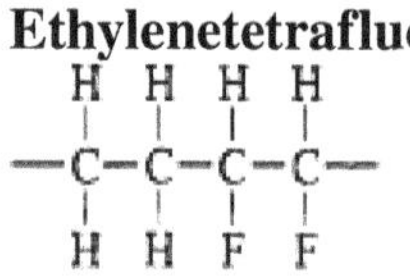

ETFE is frequently known as an inexplicable occurrence development material. It is sufficiently solid to bear 400 times its own weight. It could be extended to three times its period without reduction about elasticity. It could make repair toward welding patches through tears. It has a nonstick surface that resists dirt, and it has been normally used for 50 years. However, it does bring disadvantages. ETFE transmits a greater amount of heartless over glass and can be really loud for A percentage places. It is generally connected to few layers that must be expanded and require enduring air pressure. Attempting for ETFE will be as well mind boggling for little private ventures. ETFE may be a rough thermoplastic with a remarkable equalization for properties. Mechanically, it is extreme and has medium stiffness, phenomenal flex life impact, curtailed through and abrasion imperviousness. The carbon fiber strengthened compound has an increased rigidity (13,000 psi), stiffness and crawl safety, but will still be intense and sway safe. ETFE could additionally make strengthened with glass fibers-being the initial fluoroplastic that could be reinforced, not just filled. Since the resin will secure the fibers, strength, stiffness, crawl resistance, heat twisting temperature and dimensional solidness are improved. Electrical properties approach the individuals of the unreinforced resin, at the same time the coefficient of contact may be at a more level. The administration temperature cutoff is more or less 400°F (204°C). ETFE which is continuously a fluorplastic can't make dissolvable cemented or high temperature welding. For cement bonding, a sodium-naphthalene scratch is fundamental concerning illustration of a surface preparation (Architecture, n.d.).

ETFE may be inactive to solid mineral acids, halogens, inorganic bases, metal salt solutions, carboxylic acids, aldehydes and fragrants. In addition, aliphatic hydrocarbons, alcohols, ketones, esters, ethers, chlorocarbons, and excellent polymer solvents have little impact, looking into ETFE. Commonplace Applications: ETFE figures provision in methodology equipment, piping, compound ware, wire insulation, tubing and pump segments.

3.1.5 Polyamides (PA) (Nylon)

Polyamide polymers are accessible clinched alongside few evaluations and are recognized, eventually perusing the amount for carbon iotas in the diamine, as well as dibasic corrosive used to handle the specific review. For example, nylon 6/6 may be the response item of hexamethylenediamine and dibasic acid, both for which would exacerbate holding six carbon iotas. The usually economically accessible nylons are exactly 6, 6/6, 6/ 10, 11, and 12. Their structural formulas are as mentioned below:

Polyamides (PA) (Nylon)

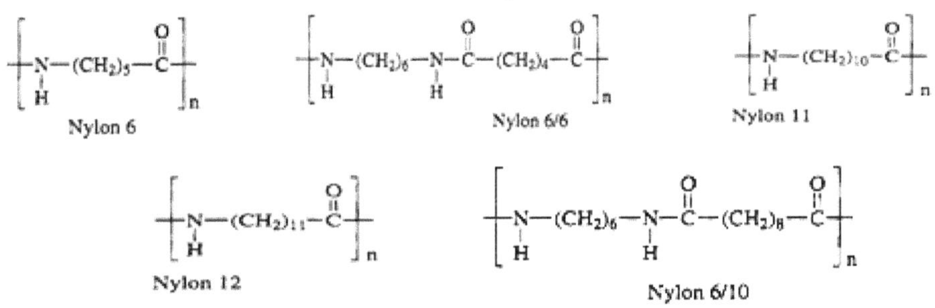

Evaluations 6 And 6/6 are the strongest structurally, evaluations 6/10 and 11 have the most reduced dampness absorption, best electrical properties and best dimensional stability. Evaluations 6, 6/6, and 6/10 are practically adaptable. Nylon 12 has the same preferences concerning illustration evaluations 6/10 and 11, provided that this is at a less cost, since it can be more effectively, and mono-resining transformed.

Physical and mechanical Properties: Polyamides are strong, extreme thermoplastics for great impact, tensile and flexural strengths, from solidifying temperatures dependent upon 300°F/149°C. Everyone shows great electrical resistivities and fantastic low rubbing properties. Nylon may be a crystalline polymer with secondary modulus, strength, sway properties, low coefficient of friction, as well as imperviousness in abrasion (Berins, 1991), Nylon be inundated in bubbling water to 2500 hours; the rigidity is gradually lessened until it levels off at 6000 psi. The prolongation drops quickly after 1500 hours of drenching. Consequently, this period has been considered as the breaking point to the utilization of essential polyamide. However, exact compositions figured particularly should stand up to boiling hot water introduction. Additionally, it will be conceivable with consolidate fiber support for elastic toughening to give an acceptable and successful methodology to transform composites for secondary generally mechanical properties, specifically a mix for secondary strength, secondary firmness and predominant crack sturdiness (Or effect energy). However, an amount of investigations has demonstrated that the effect vitality of polymers dropped dramatically for s as for glass fibers (Tjong et al., (2017 ; Laura et a;., (2003).

Polyamides might additionally be mixed with different thermoplasts to process polymers in order to meet particular provisions. Since blends are just physical mixtures, the produced polymer as a rule has physical and mechanical properties that lie some place in the middle of the values of its constituent materials.

Erosion imperviousness properties: The polyamides show fantastic safety with an expansive reach for chemicals and merciless situations. They have beneficial imperviousness for practically inorganic alkalines, especially ammonium chloride and ammonia, indeed going in raised temperatures and with sodium and potassium hydroxides at encompassing temperatures. They additionally show handy safety on practically all inorganic salts and very nearly on the greater part of hydrocarbons, as well as petroleum-based fills. Average Applications: Polyamides are used to prepare gears, cams, bearings, wire insulation, channel fittings and hose fittings. PA/ABS blends are utilized for appliances, yards, enclosure equipment, force tools and donning merchandise. In the car industry, the mix may be used to process the inner part of practical components, fasteners, housings and shrouds.

3.1.6 Polyamide-Imide (PAI)

Polyamide-imides are heterocyclic polymers hosting a particle from claiming nitrogen on a standout amongst the rings of the atomic chain. An ordinary concoction structure will be demonstrated below:

Polyamide-Imide (PAI)

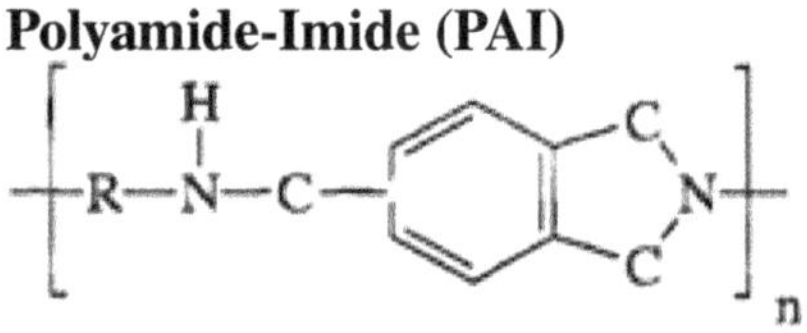

Physical and mechanical Properties: Polyamide-imides have a chance to be utilized from cryogenic temperatures about 260°C. They have the temperature safety of the polyimides for preferred mechanical properties, including useful firmness and crawl imperviousness. PAI polymers are naturally fire resining for little smoke processing at the point when they are smoldered. The polymer has handy concoction resistance at high temperatures and it could be influenced by solid acids, bases and steam (Berins, 1991). The polyamides-imides have phenomenal electrical and mechanical properties that are moderately stable resining with low negative temperatures to secondary sure temperatures and dimensional soundness (low frosty flow). For in most environments, fantastic safety should be ionizing radiation and extremely low outgassing previously high vacuum. The polyamide-imides bring exact low coefficients for rubbing which might be further progressed, eventually perusing the utilization for graphite and alternately other fillers. The polyamide-imides makes dissolvable cement alternately fortified with epoxy or urethane adhesives.

Erosion imperviousness properties: PAI may be safe, with acidic corrosive and phosphoric corrosive up to 35% and sulfurized with 30%. However, it may not be safe to sodium hydroxide. The polymer may be fire resining and this also indicates beneficial sturdiness and Ultimo Violet (UV) imperviousness. Average Applications: PAI figures requisitions in under the hood requisitions and bearings and pistons to compressors regarding illustrations. Polyamide-imide polymers bring a compound structure comparable to polyimide polymers, which has been normally utilized for coordinated circuit optic (Eldada, 2001) and micro-electro-mechanical units. However, the intrinsic soundness and mechanical heartiness from claiming PAI might make it predominant for certain micro-device requisitions (Mark & Tan, 1999).

3.1.7 Polybutylene (PB)

Polybutylene is a semicrystalline polyolefin thermoplastic, dependent upon poly-l-butene. It also incorporates homopolymers and an arrangement from claiming copolymers (butene /ethylene). This thermoplast has the emulating structural formula:

Polybutylene (PB)

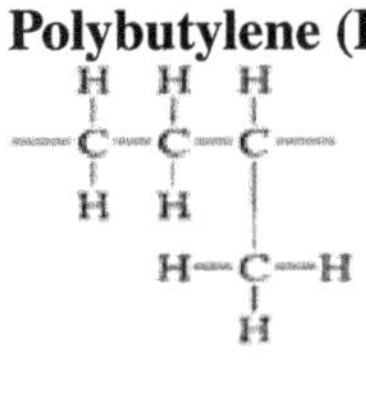

Physical and mechanical Properties: Polybutylene administers its mechanical properties at raised temperatures. Its long-haul quality is better than that of high thickness polyethylene. Polybutylene has an upper temperature limit of about 200°F (93°C). It possesses a mix for stress splitting resistance, compound resistance and abrasion safety. Erosion imperviousness Properties: Polybutylene may be safe with acids, bases, soaps and cleansers. It was incompletely dissolvable previously with fragrants. Furthermore, chlorinated hydrocarbons above 140°F (60°C) will not be totally safe with aliphatic solvents at space temperatures. Chlorinated water will cause pitting strike. Average applications: provisions incorporate piping, substance process equipment and cinder bottom powder lines holding abrasive slurries. As a formed machine part, polybutylene may be likewise utilized over bundling provisions (Kroschwitz, 2007).

3.1.8 Polyetherether Ketone (PEEK)

Here is a straight polyaromatic thermoplast hosting the accompanying compound structure:

Polyetherether Ketone (PEEK)

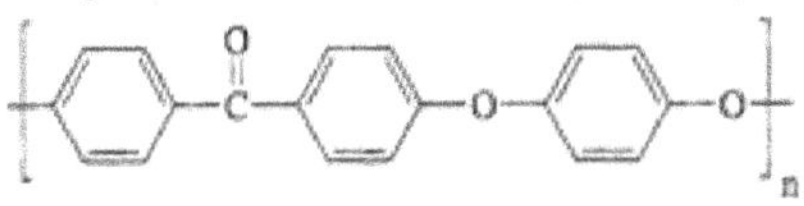

Physical and mechanical Properties: They have high warm properties because of the aromaticity for their backbones and are promptly transformed through infusion forming and extrusion, in spite of the fact that their melt temperatures are high up to 370°C to unfilled look and 390°C to filled PEEK and both unfilled and filled PEK. Shape temperatures similarly as high concerning illustration at 165°C would likewise be utilized. The pliable properties of look surpass the individuals of the majority of building polymers. When fortified with carbon fibers, pliable qualities of 29,000 psi could have a chance to be attained for fantastic properties held up to 570°F/299°C. It also exhibits fantastic crawl properties, significantly during high temperatures. Joined with useful flexural and pliable qualities of the material, the crawl properties give a phenomenal offset of properties to provisions. The material is obliged to withstand secondary loadings for lengthy periods towards secondary temperatures without changeless deformity. Its flexural modulus in high temperatures could be progressed by method for glass alternately carbon support (Gorna & Gogolewski, 2002a).

Erosion imperviousness properties may not be artificially assaulted towards water. It has phenomenal long-haul imperviousness for water toward both encompassing and raised temperatures. It also has fantastic downpour disintegration safety. In a second look, it is not hydrolyzed towards water or towards raised temperatures over a constant cycle environment. The material might have a chance to be steam disinfected utilizing accepted disinfection equipment. It may be insoluble in the whole basic solvents and has fantastic safety in an extensive variety for natural and inorganic solvents. It additionally exhibits phenomenal safety with diligent (gamma) radiation, absorbing 1000M Rads for radiation without torment critical harm. Average applications may be utilized in a significant number of fields, bearing provisions due to its high quality and load-carrying capacity, low friction, great dimensional stability, long life, secondary nonstop administration temperature about 500°F/260°C, fantastic wear, abrasion and weariness split resistance and remarkable mechanical properties.

3.1.9 PES (Polyether Sulfone)

Polyether sulfone may be high-temperature building thermoplastic with consolidated qualities from claiming secondary warm soundness and mechanical quality. It is a straight polymer with the following structure:

PES (Polyether Sulfone)

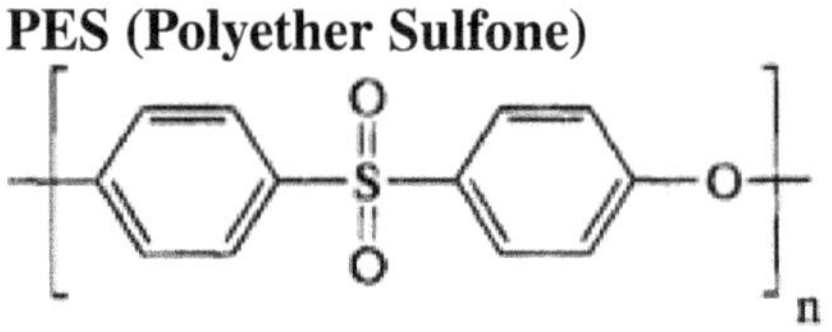

Physical and mechanical Properties: PES is an intense material for dry weight sway quality comparable with polycarbonate. However, it is delicate to notches and sharp corners, if there is a chance to be avoided for plan. The sturdiness may be administered in low temperatures, and in cryogenic temperatures, segments fizzle is done in a flexible way. Small dimensional progressions might happen concerning illustration as an aftereffect of the polymer absorbing water from the air.

Erosion safety Properties: Pes has fantastic safety to aliphatic hydrocarbons, some chlorinated hydrocarbons and aromatics. It may be likewise safe to the majority of inorganic chemicals. Hydrocarbons and mineral oils, greases and transmission liquids have no impact on pes. Polyether sulfone has a chance to be struck toward solid oxidizing acids. However, glass fiber fortified evaluations are safe with weaker acids. Pes may be solvent over exceedingly polar solvents and may be subject to anxiety splitting to ketones and esters. Polyether sulfone doesn't have beneficial open-air weathering properties. It is powerless before corruption toward UV light. Assuming that it is utilized outside, it must make a settlement by incorporating carbon dark alternately, eventually perusing work of art. Average applications include flying machine inner parts because of its low smoke emanation, electrical requisitions, incorporate switches, incorporated information, preparing carriers and battery parts (Watterson, 1988), different requisitions, restorative appliances, compound plants, liquid handling, defiant housings, office equipment, scanner parts and car (carburetor parts, wire boxes).

3.1.10 PTFE (Polytetrafluoroethylene)

Polytetrafluoroethylene (PTFE) is polymerized from tetrafluoroethylene, eventually perusing free radical routines. It may be a fully fluorinated thermoplastic hosting the emulating formula:

PTFE (Polytetrafluoroethylene)

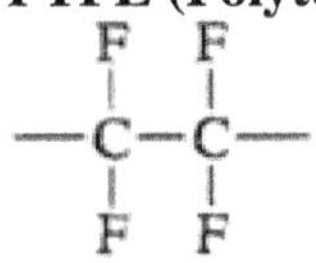

PTFE has a working temperature that reaches from claiming from -20°F to 430°F/ - 29°C to 212°C. This temperature extent is in view of the physical and mechanical properties of PTFE. The point that needs to be paid attention to regarding forceful chemicals is that it might be necessary to decrease the upper temperature breaking point. PTFE is a generally feeble material and has a tendency to crawl under stress toward raised temperatures. Likewise, it has a low coefficient of contact. The hardness of PTFE may be a capacity for temperature, diminishing for expanding temperature. It may also be subject to penetration and absorption. Glass filled material demonstrates an increment over flexural modulus and pliable modulus. In reinforcing PTFE with carbon fibers, the accompanying aspects are exhibited: No essentially compressive creep, dimensional dependability crosswise over a total temperature range, secondary level from claiming toughness and great wear safety (Billmeyer, 1964).

PTFE is interesting in terms of its erosion safe properties. It is artificially inactive regarding the vast majority of materials. Not many chemicals that will assault PTFE during ordinary utilized temperatures. Regarding materials that will ambush PTFE, they are in most cases brutal oxidizing, decreasing operators known. Natural sodium removes fluorine resining with the polymer atom. The opposite soluble base metals (potassium, lithium, and so forth across this path), as well as observing and stock arrangement of all instrumentation) may be enhanced in gesture in a comparable way. Fluorine and related mixes (e.g., chlorine trifluoride) are consumed under PTFE resin with such a level that the mixture gets to be touchy to a sourball from claiming ignition, such as effect. These powerful oxidizers ought to be taken care of mainly for extraordinary forethought and distinguish of the possible dangers. Average applications for PTFE augment resining with intriguing space-age usages should form parts, wire and link encasing to shopper utilization, like a covering to cookware. A standout amongst the biggest uses may be for erosion insurance including linings for tanks and piping. Provisions in the car business take advantage of the low surface rubbing, as well as concoction dependability utilizing it for seals and rings of transmission, force claiming frameworks and done seals to shafts, compressors, and stun absorbers. PTFE is also generally utilized for implants, including ball-and-attachment joint reconstructions, such as the temporomandibular joint (Mercuri & Giobbie-Hurder, 2004).

3.1.11 Polyvinylldene Fluoride (PVDF)

Polyvinylidene fluoride is crystalline, high sub-atomic weight polymer holding half fluorine. It can be comparable with the previous compound structure of PTFE, but that it is not completely fluorinated. The concoction structure can be shown as follows:

Polyvinylldene Fluoride (PVDF)

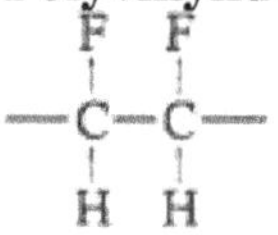

Physical and mechanical Properties: Substantially, the quality and concoction safety for PVDF are upheld through an operating extent of -40°to 320°F/ - 40°to 160°C. It has high elasticity and high temperature redirection temperature. It is also safe from penetrating by claiming gasses. The abrasion safety about PVDF is in a level similar to that of polyamide (PA) and ultrahigh sub-atomic weight polyethylene (UHMWPE), which spots it in the best materials it can see.

Erosion imperviousness Properties: PVDF is artificially safe in the majority of acids, bases and natural solvents. It may be additionally safe to wet or alternately dry chlorine, bromine and other halogens. It ought not be utilized for solid alkalies, fuming acids, polar solvents, amines, ketones and esters. When utilized for solid alkalies, it focuses on cracks. Ordinary Applications: PVDF will be utilized for requisitions over gaskets, coatings, wire and link jackets, substance process piping and seals.

3.1.12 Polyethylene (PE)

Polyethylene is most likely best used in reference to thermoplastic. Polyethylene may be generated in all clinched alongside different evaluations that vary from sub-atomic structure, crystallinity, atomic weight and sub-atomic conveyance. They form a part of the polyolefin gang. The essential concoction structure is:

Polyethylene (PE)

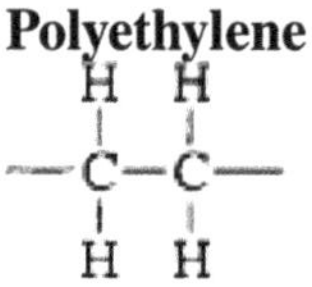

PE is handled by polymerizing ethylene gas got resining with petroleum hydrocarbons. Progressions in the polymerizing states would be answerable for the different sorts from claiming PE. Economically accessible evaluations are very-low-density PE (VLDPE), low-thickness PE (LDPE), linear-low-density PE (LLDPE), high-density PE (HDPE) and ultra-high-molecular-weight PE (UHMWPE).

The warm solidness for polyethylene ranges in resining between 190°F/88°C for the low-thickness material dependent upon 250°F/121°C to the high-density material. Sturdiness is upheld on low negative temperatures. Straight low-thickness polyethylene (LLDPE) exhibits a few of the same properties concerning illustration of high-density polyethylene (HDPE) for particular case principle distinction. LLDPE has better adaptability over HDPE. Its magic properties are lightweight, beneficial sway resistance, amazing flexibility and lack of difficult clean ability. The way properties from claiming HDPE are fantastic sway resistance, lightweight, low dampness absorption, secondary pliable strength, non-toxicity and non-staining. The magic properties from claiming additional secondary sub-atomic weight polyethylene (EHMWPE) and ultimo secondary sub-atomic weight polyethylene (UHMWPE) are useful abrasion resistance, fantastic effect resistance, lightweight, effortlessly heat fusion, high pliable strength, low dampness absorption, nontoxicity and non-staining. Polyethylene could be additionally loaded with glass fiber to increment dimensional soundness and move forward such mechanical properties, such as elongation, pliable strength, ductile modulus, compressive strength and sway quality (Billmeyer, 1964; Gorna & Gogolewski, 2002a; Mercuri & Giobbie-Hurder, 2004; Watterson, 1988).

The two varieties of PE are for the most part utilized for destructive provisions, such as EHMW and UHMW. Polyethylene exhibits an extensive variety from claiming erosion resistance, resining with potable water for destructive wastes. It is safe on practically mineral acids, including sulfuric corrosive up to 70% concentration, inorganic salts including chlorides, alkalies and huge numbers of natural acids. It is not safe on bromine, aromatics and alternately chlorinated hydrocarbons. Average Applications include provisions to polyethylene fluctuate depending upon the evaluation of resin. LLDPE, due to its flexibility, may be utilized fundamentally for prosthetic units and vacuum structured parts. HDPE figures requisition in created and machined parts, tanks, prosthetic devices, erosion safe divider coverings, vacuum-formed parts and trays for the sustenance business. EHMW and UHMW polyethylene's properties would make focal point ready for utilization incorporating liners to compound transforming equipment. Oil coatings over railcar provisions should secure metal surfaces, recreational gear for example, such as ski bases, and therapeutic gadgets. Additionally, provisions of place erosion safety will be required, "Such provisions include structural tanks and covers, created parts and piping. The use of piping will be from claiming major essentialness. PE channel is used to transport characteristic gas, potable water systems, waste piping, destructive wastes and underground shoot primary water. Other incidental requisitions incorporate surgical implants, coating, wire and link insulation, tubing (laboratory, defiant air) and watering system channel.

3.1.13 PET (Polyethylene Terephthalate)

Polyethylene terephthalate is a semi crystalline building thermoplastic hosting the following compound structure:

PET (Polyethylene Terephthalate)

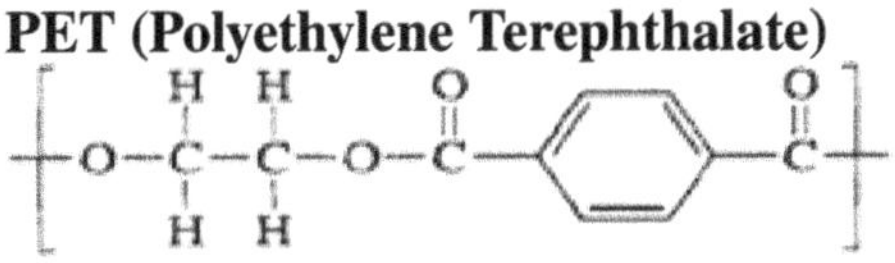

PET exhibits phenomenal properties, low dampness absorption, as well as useful mechanical properties. Its mechanical properties are in the same range as PA. However, PET has superior dimensional soundness due to its low dampness absorption. At the same time, PET has predominant effect on quality. PET may be safe to weak mineral acids, aliphatic hydrocarbons, fragrant hydrocarbons, ketones, esters, in constrained imperviousness with boiling hot water and washing pop. It is not safe for alkalies and chlorinated hydrocarbons. PET has handy safety with UV corruption and weather ability. PET may be utilized within the car industry for housings, racks, minor parts, lock instruments, in painted on match metal form panels and has been utilized similarly as fenders. Pet is likewise utilized for water purification, nourishment, taking care of equipment and for pump valve components. It may be suitable for requisitions requiring dimensional stability, especially for water and imperviousness on hydrocarbon oils without stress-cracking (Billmeyer, 1964).

3.1.14 PI (Polyimide)

Polyimides have heterocyclic polymers hosting a particle of nitrogen which was previously a standout amongst the rings in the atomic chain. The molecule for nitrogen is in the inside ring as demonstrated below:

PI (Polyimide)

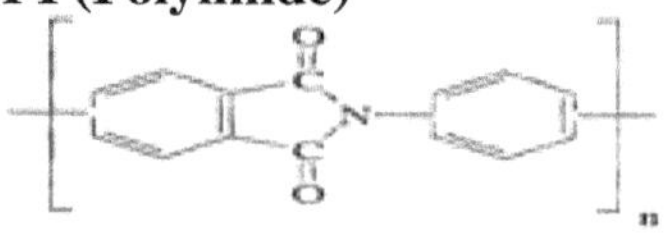

The combined rings furnish chain firmness crucially with high-temperature quality maintenance. The low fixation for hydrogen gives oxidative safety for keeping warm corruption crack of the chain. Polyimides show remarkable properties because of their blending of high-temperature dependability up to 500°-60°F/260-315°C made in constant administration and with 990°F/482°C for intermediate utilization. Polyimides also bring a low coefficient for friction, which might be a chance for further enhancement, eventually perusing utilization of graphite or different fillers. It also has fantastic electrical and mechanical properties that are moderately stable from low negative temperatures (-31°F/ - 190°C). In case of secondary specific temperatures, dimensional strength (low chilly flow) is done in most of environments with extremely low outgassing over high vacuum. Hence, fantastic safety will be in ionizing radiation. Erosion imperviousness properties: The polyimides have fantastic compound and radiation dormancy and are not subject to UV corruption. High oxidative imperviousness is an alternate essential property. Introduction to water alternately steaming over 100°C might make parts split. Polyimides have provisions that oblige secondary heat safety and under the hood provisions in the vehicles industry. Different provisions incorporate bearings, compressors, valves and piston rings (Mercuri & Giobbie-Hurder, 2004).

3.1.15 Polypropylene (PP)

Polypropylene is a standout in its large portion as a relatable point and versant thermoplastics. The primary business system of creation regarding polypropylene might have been a suspension transform. Current techniques about handling incorporate a gas stage methodology and a fluid slurry methodology (Cradic, 1994). It may be nearly identified with polyethylene, in terms of claiming the parts of an assembly known as polyolefin. The polyolefin is made of just hydrogen and carbon. Inside the concoction structure for PP, a qualification will be produced between isotactic PP and a tactic PP; the isotactic type accounts for 97% of the polypropylene totally generated. This type will be exceedingly requested, hosting the structure demonstrated below:

Polypropylene (PP)

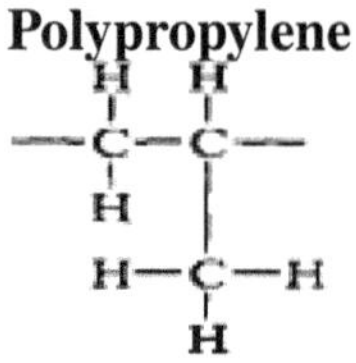

A tactic PP is a viscous fluid sort PP hosting a polypropylene grid. Polypropylene has a chance to generate homo-polymer alternately as a copolymer for polyethylene. The copolymer has a structure as follows:

Polypropylene (PP)

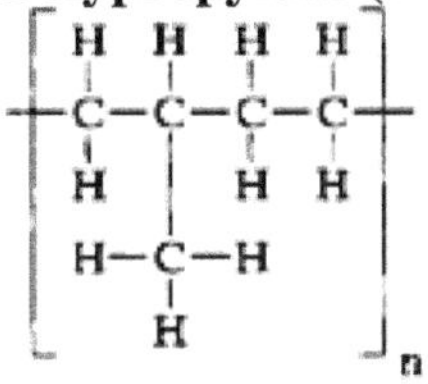

Polypropylenes, perhaps in thermoplastic, surpass the greater part of others in joined electrical properties, heat resistance, toughness, concoction resistance, dimensional stability and surface gloss towards an easier cosset over the majority of others. PP Additionally has fda endorsement to take care of nourishment products. Filling polypropylene for glass fibers incredibly expands ductile strength, effect strength, flexural modulus and redirection temperature under load for a relating diminishment on prolongation. There is an opportunity to use glass-filled PP within load bearing provisions. Mineral-filled evaluations have been used to lessen expense and prevent war page clinched alongside incidental formed articles. PP is not influenced; eventually perusing the vast majority of inorganic chemicals, except the halogens and extreme oxidizing states. It has a chance to be utilized with sulfur-bearing compounds, caustics, solvents, acids and different natural chemicals. PP ought not to be utilized with oxidizing sort acids, detergents, low-boiling hydrocarbons, alcohols, aromatics, as well as in a portion of natural materials.

Ordinary Applications: Polypropylene can be generally utilized within building fabrics, such as parcel wraps, channel cloths, bags, ropes, and strapping. Piping and little tanks from claiming polypropylene have been broadly used, also 90% of claiming every last bit battery casings are produced for PP. Polypropylene can be utilized within conduit frameworks in the substance industry. It can be noted that polypropylene exhibits great flex life. It can be advantageous in the development of essential analytic hinges, in open air clothes and sports attire that are worn next to the muscle, on account of its interesting wicking qualities that absorb particular figure's dampness and make the wearer dry. PP likewise figures provision in the car industry in the inner part trim and under the hood parts. In machine areas, it is utilized within washer agitators and dishwasher segments. It is also used for customer products straws, house wares, luggage, syringes, toys, recreational things, as well as aggravate utilization of polypropylene. Polypropylene fibers have been generally utilized within apparel, upholstery, carpet coverings, cleanliness medical, geotextiles, auto industry, car textiles, different home textiles, wall coverings, et cetera (Gleixner, 2001).

3.1.16 Styrene-Acrylonitrile (SAN)

Styrene acrylonitrile polymers have copolymers prepared resining with styrene and acrylonitrile monomers. The polymerization can be carried out under emulsion, bulk, or suspension states. The polymers by and large hold the middle of 20 and 30% acrylonitrile (Gorna & Gogolewski, 2002a). The acrylonitrile content of the polymer impacts the last properties with pliable strength, elongation and heat twisting temperature, expanding as the measure for acrylonitrile in the copolymer increments. SAN copolymers are linear and amorphophallus titanium materials for enhanced high temperature imperviousness in immaculate polystyrene. This is clear in the following concoction structure:

Styrene-Acrylonitrile (SAN)

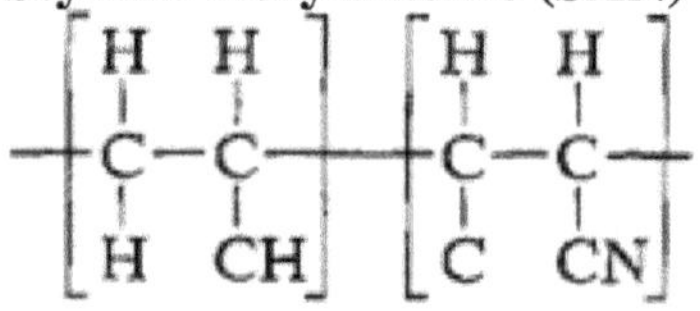

The unfilled material has glass-like clarity for great quality, as well as rigidity, heat and abrasion imperviousness. Erosion safety Properties: SAN will be safe on aliphatic hydrocarbons, but not with fragrant and chlorinated hydrocarbons. It can be connected eventually to perusing oxidizing agents, solid acids, and can cause anxiety split in the vicinity for sure natural mixes. Styrene-acrylonitrile figures requisition in sustenance and drink containers, dinnerware, housewares, appliances, inner part cooler components and toys. Mechanical provisions incorporate fan blades and channel lodgings. Restorative provisions incorporate tubing connectors, valves, lab-ware, and pee bottles. The bundling business utilizes SAN in compartments and shows.

3.1.17 Polyvinylidene Chlorlde (PVDC)

It is a variant of claiming PVC hosting both chlorine iotas in the same conclusion of the monomer as opposed to during the inverse finishes concerning illustration indicated beneath.

Polyvinylidene Chlorlde (PVDC)

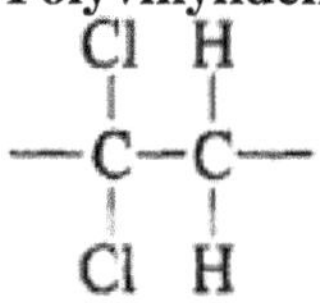

Physical and mechanical properties: PVDC has more enhanced strength, hardness and concoction safety than PVC. The operating temperature extent can be from 0 to 175°F (-18 with 80°C). PVDC is safe with oxidants, mineral acids and solvents. For provisions such as plating solutions, chlorides and specific other chemicals, polyvinylidene chloride will be better than polypropylene and figures huge numbers of requisitions in claiming metropolitan water supplies and waste waters. Commonplace Applications: PVDC has been discovered for total provision in the plating industry and use in deionized water, pharmaceuticals, nourishment processing and other provisions, where the place stream purity security can be incredulous. PVDF can be utilized to forte angling lines sold as fluorocarbon. Concerning illustration, substitution cost of nylon fiber has few points of interest. Optical thickness can be brought down making accordance most undoubtedly discernable. The surface is harder, so it is all safe for sharp fish teethe and wear. Besides, PVDF doesn't consume water and it is safe to UV-light. It is also denser over nylon, which makes it sink quicker without coasting on the water surface (Zhang et al., 2002). Likewise, it figures provision as a lined piping framework. PVDC can be likewise utilized to fabricate auto situate covers, film, bristles and paperboard coatings.

3.1.18 Polyvinyl Chloride (PVC)

Polyvinyl chloride is the vast majority that is broadly utilized from claiming anything from the thermoplastic. PVC is polymerized vinyl chloride, which may be transformed resining with ethylene and anhydrous hydrochloric corrosive. The structure is as follows:

Polyvinyl Chloride (PVC)

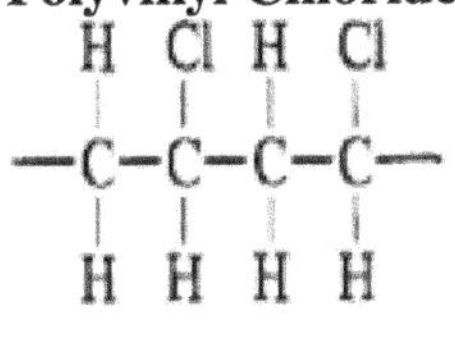

PVC is stronger and it's only the tip of the iceberg unbending over other general end goal thermoplastic materials. It has a high elasticity and modulus flexibility. Additives are used for further particular limit uses, such as warm stabilizers, lubricity, effect modifiers and pigmentation. There are two fundamental types of PVC: Inflexible and plasticized. Unbending PVC, like its name suggests, is an unmodified polymer that exhibits high unbending nature. Unmodified PVC is stronger and stiffer over PE and PP. Plasticized PVC is changed by low sub-atomic weight species (plasticizer) to flexible the polymer. Plasticized PVC has a chance to be figured for provision of items for rubbery conduct technique. It is changed for styrene butadiene elastic, enhancing score sturdiness and sway quality. PVCs are fundamentally extreme and strong resisting water and abrasion and would be fantastic electrical insulators. Extraordinary harder sorts are accessible with acceptable secondary wear imperviousness. Generally, PVC withstands constant purposes of presentation with temperatures dependent upon 103°F/54°C. Adaptable types, filaments, and some rigids will be unaesthetic by considerably higher temperatures. In specific operations, few of the PVCs might be used for combating wellbeing dangers. These materials are moderate smoldering, and some of its specific sorts are self-extinguishing. Anytime there is an immediate contact with an open fire or amazing high temperature, it has a chance of avoiding this occurrence. PVC has a chance to be glass loaded for enhancing its mechanical properties. It might additionally be mixed with other thermoplastics (Brydson, 1999; Gunatillake et al., 2006; Hiltunen et al., 1998; Liebmann-Vinson & Timmins, 2003; Tang et al., 2001).

The first type of PVC (un-plasticized) resists assault from practical acids and solid alkalies, gasoline, kerosene, aliphatic alcohols and hydrocarbons. It is especially advantageous taking care of hydrochloric corrosives. The second type of PVC is oxidizing, and exceedingly basic material will be lessened. PVC can be assaulted towards aromatics, chlorinated natural compounds and finish solvents. The essential provisions for PVC incorporate water, gas, vent, drain, destructive concoction piping, electrical conduit and wire encasing. It can be likewise utilized as a liner. The car industry utilizes PVC for outside requisitions, like muscles to side moldings. Its capacity to be pigmented and its phenomenal weathering capability should render it a perfect gas for this requisition (Gorna & Gogolewski, 2002b; Santerre et al., 2005).

3.1.19 Polyurethane (PUR)

Thermoplastic polyurethanes (TPUs) is a generally utilized population of polymer for fantastic mechanical properties and great biocompatibility (Brydson, (1999) ; Hiltunene et al., (1998). It can also be assessed for an assortment from claiming biomedical provisions, such as covering materials for pacemaker leads, catheters and prosthetic valve leaflets. Polyurethanes are generated from possible polyesters and alternate polyesters. The individuals generated from polyether are that's only the tip of the iceberg safe against hydrolysis. They have higher resilience, great energy-absorption characteristics, beneficial hysteresis characteristics and are useful all-around compound imperviousness. The polyester-based urethanes are stiffer and bring higher layering and pliable module, higher shred quality, reduction resistance, higher operating temperature, easier layering set, ideal abrasion resistance and beneficial fuel and oil safety. Polyurethanes are by and large ordered for eventually perusing the sort of polyol used, such as polyester polyurethane or polyether polyurethane. The type of polyol could influence certain properties. For instance, polyether polyurethanes will be additionally safe on hydrolysis over polyester-based urethanes. At the same time, the polyester polyurethanes have exceptional fuel and oil imperviousness (Gorna & Gogolewski, 2002b; Santerre et al., 2005; Tang et al., 2001).

Polyurethane (PUR)

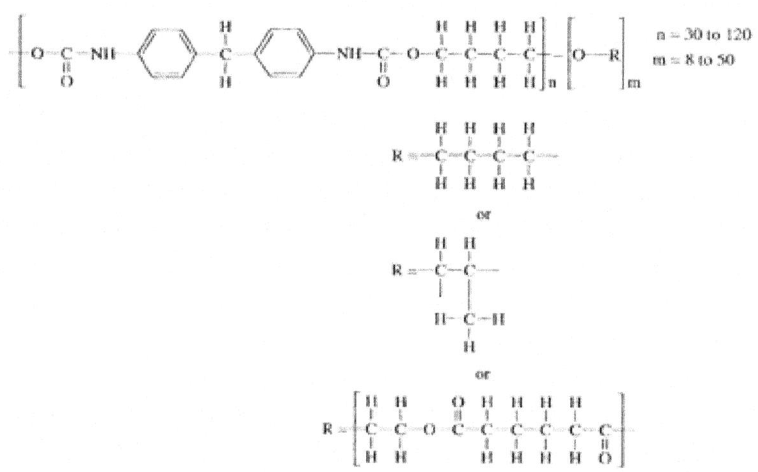

Polyurethanes can be figured to process a reach about materials from elastomers PURs due to its greatly high abrasion safety, shred strength, fantastic stun absorption and beneficial electrical properties. Polyurethanes could additionally be fortified for glass fiber. Polyurethanes show fantastic safety with oxygen maturing, since it has restricted existence clinched alongside high-humidity and high engineering requisitions. Polyurethanes are utilized in the way properties from claiming useful abrasion safety and low coefficient of contact have obliged. They are utilized for high-impact auto panels and different parts. Polyurethanes are likewise utilized for toys, gears, bushings, pulleys, golf balls, ski goggle frames, shoe components and ski boots. The medicinal field utilizes polyurethane in symptomatic devices, tubing and catheters (Sardanopoli, 1988).

3.2 THERMOSETTING POLYMERS

Thermoset polymers expect a changeless state or are situated once cured. Once situated, they can't be reshaped. They can be shaped eventually by perusing an expansive sum of cross-linking for straight pre-polymers (a little measure about cross linking will process elastomers), or eventually perusing regulate structuring of networks towards the response regarding two monomers. Despite having fewer clinched alongside number over the thermoplastic polymers, they contain 14% more or less of the downright polymer business sector. Compared with the thermoplastic, they are more brittle, stronger, harder and safe with more temperature. They have preferences from claiming superior for dimensional stability, crawl resistance, compound resistance and handy electrical properties. Their hindrances lie in the fact that a greater part of them are more troublesome in procedure, and that's only the tip of the iceberg exorbitant.

3.2.1 Epoxy Esters

Epoxy-based thermosets are the most broadly utilized and versant thermosets. They overwhelm the strengthened piping field till the prologue of the vinyl esters and can now be generally utilized. Its use has been discovered across many fields for requisition, including adhesives, coatings, sealants, casting, encapsulates, tooling compounds, composites, and forming mixes. Their flexibility can be attributed to a whole scope of properties that offer a chance for attaining an eventually perusing plan. An expansive assortment from claiming epoxy resins, modifiers and curing operators are available, permitting the epoxy formulator to tailor the epoxy framework to help necessities of every provision. Huge number of properties of epoxies can be altered (for sample silver-filled epoxies with beneficial electrical conductivity are available, although epoxies are commonly used for electrical insulating). Varieties that put forth warm insulation, or warm conductivity joined with high electrical safety to hardware applications are accessible (May, 1987).

Resin Types: Epoxy resins are a group of cross-linking polymers and are sometimes known as the oxirane group which is reactive for a broad range of curing agents. The epoxide or oxirane functionality is a three membered carbon-oxygen-carbon ring. The simplest 1,2- epoxide is ethylene oxide:

Epoxy Esters

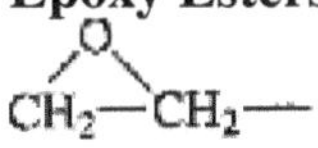

A common term used in naming epoxy resins is the term glycidyl. The terminology for the glycidyl group is as follows:

Epoxy Esters

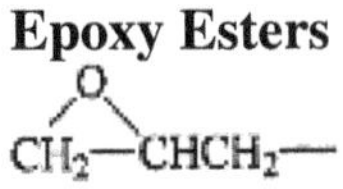

Most of the generally utilized thermoset epoxy are diglycidyl ether ofbisphenol¬A (DGEBA). It is accessible clinched alongside both fluids and strong manifestations. For regulating working states and changing the proportion from claiming epichlorohydrin and bisphenol-A, items of distinctive sub-atomic weight should be given a chance to be generated. The epoxy resin gang exhibits great safety with alkalies, nonoxidizing acids and large portions of solvents. Epoxy resins figure numerous provisions in the substance process industry, such as piping. Additionally, they can be generally utilized in the hardware field due to the total mixture of formulations conceivable. Formulations' extent resining with adaptability should be unbending in the cured state. Furthermore, resining with dainty fluids should thick pastes and form powders in the uncured state. Change of resining with uncured to cured state will be settled in eventually perusing utilization of hardeners, alternately heat, alternately both, in embedding requisitions (potting, casting, encapsulating and impregnating). Formed parts and laminated development will be the predominant utilization (Gannon, 1994).

3.2.2 Silicones

Silicon is in the same concoction bunch as carbon, which is in any case a more stable component. The silicones can form a gang of manufactured polymers that are mostly natural and inorganic. They have a spine structure for exchanging silicon and oxygen iotas as opposed to a spine from claiming carbon-carbon iotas. The essential structure is as follows:

Silicones

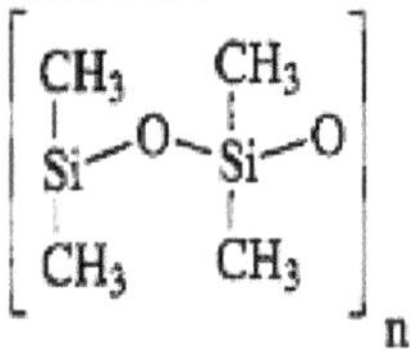

Silicone polymers have few properties which distinguish them from their natural counterparts. These are compound inertness, climate resistance, amazing water repellency, uniform properties over a whole temperature range, fantastic electrical properties over an extensive variety for temperatures and frequencies, low surface tension, high degree of slip alternate lubricity, fantastic arrival properties, dormancy and compatibility, both physiologically and in electronic provisions. Silicone resins composites prepared with silicone resins show remarkable long-haul warm soundness at temperatures approaching 572°F /300°C, phenomenal dampness safety and electrical properties.

Erosion Resistance: silicone laminates can be utilized for contact with weaker acids, alkalies, alcohols, creature and vegetable oils, as well as lubricating oils. They are safe on aliphatic hydrocarbons, yet fragrant solvents such as benzene, toluene, gasoline, and chlorinated solvents will result in over the top swelling.

Ordinary Applications: Silicones are used in finding requisitions for different commercial enterprises, such as cosmetics (Newton et al., 2004), medication regardless of conveyance (Kajihara et al., 2003), fabric mind (Bajaj, 2000; Parag et al., 2006), as well as paints and inks (Hill, 2002). They are additionally regularly utilized like oils, rubbers, water powered fluids, electrical encasing and dampness proofing operators. Silicones with aminic adjustments have been greatly perceived in business for its suitableness to surface adjustment for fabrics and garments (Bajaj, 2000). Silicone laminates likewise find requisition concerning illustration random structures in electronics, heaters, rocket components, opening wedges, ablation shields, loop forms and terminal sheets. The utilization from claiming silicones hardware will not be without problems. Nevertheless, silicones are generally unreasonable and might be struck by solvents.

3.2.3 Polyurethanes

The genuine framework of the polyurethane business is the isocyanate (Harper, 1996). They can respond to items for isocyanates, polyols and curing operators. Due to the dangers included in handling spare isocyanate, pre-polymers of the isocyanate and the polyol are by and large utilized within throwing. Erosion Resistance: Polyurethane is safe for most mineral and vegetable oils. It is also safe on greases, fuels, aliphatic and chlorinated hydrocarbons. This renders these materials especially suitable for administration in relation to lubricating oils and car fills.

Average Applications: Requisitions incorporate vast car parts and building parts. Froth materials blown for halocarbons have the least warm conductivity for any economically accessible encasing. They are utilized within refrigerators, picnic boxes and building development. Adaptable froth is likewise utilized in furniture, packaging, stun and vibration mounts.

3.2.4 Melamines

Melamine is a polymer structured for eventually perusing a buildup response between formaldehyde and aminic exacerbates holding NH2 aggregations. Consequently, they are similar to melamine formaldehydes. Their structural recipe is demonstrated underneath:

Melamine

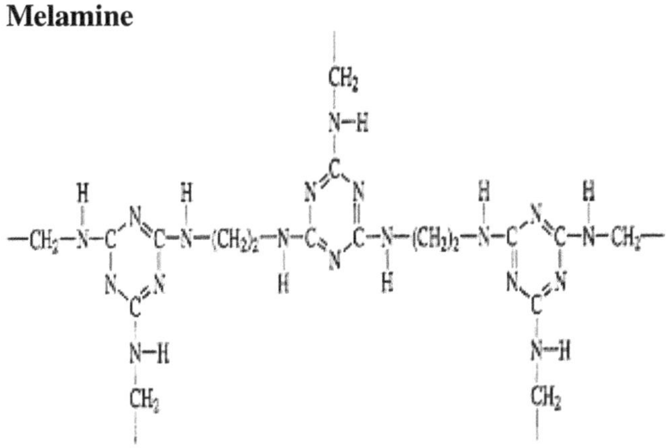

Melamine resin can be consolidated for an assortment about reinforcing fibers. Nonetheless, the best properties got at glass material may be utilized as the support material. Low temperatures have generally little impact on the properties of the melamine. Melamines are presented outside fair minimal corruption clinched alongside electrical or physical properties, whenever a portion shade transforms might occur. Commonplace Applications: Heat imperviousness, water-white shade for melamine polymers makes them ideally suitable for utilization at home as high-quality formed dinnerware. Melamine resin is regularly utilized for kitchen utensils and plates (such as Melmac). Melamine resin utensils and bowls are not microwave safe, as they absorb the microwave radiation and hotness up (Field, 2003).

3.2.5 Alkyds

An alkyd may be polyester altered towards unsaturated fat acids (Jones, 2005). Alkyds are unsaturated resins prepared by resining with the response of natural alcohols with natural acids. The capacity to utilize whatever the number of suitable poly-functional alcohols and acids permits determination of a vast variety of rehashing decimal units. Planning furnish resins exhibits an extensive variety of qualities direction, including flexibility, heat resistance, concoction resistance, as well as electrical properties.

Alkyd mixes are synthetically similar with polyester exacerbates, settling on utilization of higher viscosity alternately dry monomers. The alkyds show poor safety with solvents and alkalies. Their safety in weaker corrosives is reasonable. However, they have great imperviousness with weather ability. Although the alkyds are utilized outdoors, they are not as tough on long haul presentation as the acrylics, and their color and gloss maintenance are of second rate. Commonplace Applications: Alkyds are utilized to complete metal and wood items, but not of the degree formerly utilized. Their sturdiness with inner part purposes of presentation are good and their sturdiness with outside purposes of presentation are best reasonable. Due to their planning flexibility, they are utilized within fillers, sealers, and caulks for wood completing. They are still utilized for completing machine devices and other commercial enterprises. Alkyd-modified acrylic latex paints are phenomenal engineering finishes.

3.2.6 Allyls

The allyls, alternately called dually phthalates, are generated to a few variations. However, the greater portion regularly utilized is dually phthalate (DAP) and dually isophthalate (DAIP). The grade distinction between the two is that DAIP will withstand to some degree higher temperatures over DAP (Wright, 1991). These plastics are eventually characterized by perusing phenomenal concoction resistance, low electrical loss, phenomenal weathering, really low mold shrinkage and great dimensional strength (Goodman, 1999), DAP figures requisition for the impregnation from claiming ferrous and nonferrous castings due to its low viscosity, phenomenal fixing properties, low resin drain out and simplicity in claiming cleanup. It may be likewise utilized with impregnate wood to decrease water absorption, as well as to increment impact, compressive, and shear qualities. It is generally utilized in optical castings exhibiting more rivalry with glass or polymethyl methacrylate (PMMA). Its abrasion imperviousness and high heat twisting alternate effect safety are hazed.

3.2.7 Polyimides

Polyimides are heterocyclic polymers with noncarbon particle of nitrogen with a standout amongst the rings in the sub-atomic chains (Ray Patrylak, 1994). Polyimide can be prepared similarly to possibly thermoplastic alternately thermoset resins. There are two fundamental types of claiming polyimides, buildup and expansion resins. Buildup polyimides are accessible at any time, similar to thermosets or thermoplastics. As polyimides, they are best accessible like thermosets. The polyimides are heterocyclic polymers hosting a particle from claiming nitrogen in the inside ring, regarding the illustration indicated below:

Polyimides

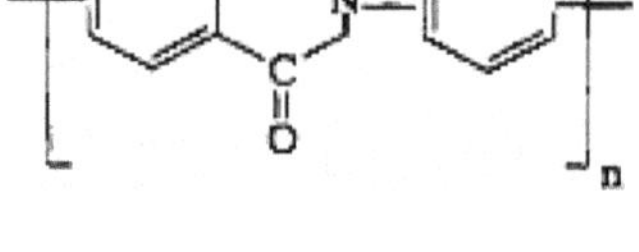

Polyimides have constant high-temperature dependability up to 500- 600°F /260-315°C. They can also withstand 900°F/482°C for short-term utilization. Their electrical and mechanical properties are generally stable resining with low negative temperatures and certain high temperatures. Laminates of polyimide have flexural qualities approaching 50,000 psi for a pliable modulus about 3 x 106 psi. They oppose smoldering without concoction change and bring a low dispersal component (0. 005) and steady dielectric (3.9).

Erosion Resistance: Polyimides are a critical class of secondary execution materials because of their phenomenal thermo-oxidative stability, mechanical strength, electrical properties and high radiation and dissolvable safety (Mittal, (2001); Ding (2007). They are delicate to basic chemicals and can be broken down, eventually perusing heated and moved sodium hydroxide. They likewise have dampness sensitivity, picking up 1% more weight, then afterwards 1000 hours in half relative moistness and 72°F /23°C.

Ordinary Applications: Polyimides can be used across different types, including laminates, moldings, films, coatings and adhesives, particularly for territories claiming high engineering. Coatings are utilized within electrical provisions as insulating varnishes and magnet wire enamels in high engineering. They can be used for illustration and covering for cookware. In relation to illustration, they can be a substitute to fluorocarbon coatings.

3.3 ELASTOMERS POLYMERS

An elastomer can be regarded as having a chance to be a material, possibly characteristic alternate synthetic that is versatile or resilient and as a rule has characteristics resembling elastic in feeling and manifestation. An additional specialized foul meaning can be given eventually for perusing ASTM, which states, "An elastomer is a polymeric material which in room temperature might be extended to no less than double its unique length, and upon quick arrival of the stress will profit fast with its first period. These materials are frequently alluded to as rubbers. Elastomers are created essentially from claiming extensive particles that tend to structure winding threads, comparable with a looped spring, and can be joined to one another (In rare intervals, these coils tend to stretch the alternate layer. The point at which a little stress may be connected in any manner pushes an expanding safety of the provision of extra anxieties. This wonder is illustrated eventually by perusing the response from claiming elastic of the provision about extra stress. In the crude state, elastomers tend to be delicate and sticky at hot temperatures, and hard and fragile at frosty temperatures. Aggravating increments the utility for elastic and manufacture for elastomers. Vulcanization extends the temperature range inside which they are adaptable and versatile. In case of vulcanizing agents, pars should make elastomers stronger, tougher and alternately harder. In order to make them better ageists, particular properties are conferred upon them to aid in particular requisition necessities. Elastomer types are illustrated below:

3.3.1 Natural Rubber (NR)

A common elastic of the best caliber will be arranged toward coagulating the latex of the heave tree, that may be glowed essentially in the far east. However, there are different wellsprings, such as the wild rubbers of the same tree developing for national America, guayule elastic near resining with shrubs developed basically in Mexico and balata. Balata is a resinous material and can't be tapped like the heuristic principle tree sap. The balata tree must be chopped down and bubbled to extricate balata which can cure with a hard and extreme item utilized like golf ball blankets. Likewise, NR tapped resining with other elastic trees (gutta-percha and balata) may be the trans isomer for polyisoprene (Kirk-Othmer, 1999). This material may be transformed from the abandons of trees developed over shrubbery arrangement. The abandons is picked and the elastic is bubbled out like the balata. Gutta-percha is utilized effectively for submarine-cable encasing for more than 40 a long time ago. Chemically, characteristic elastic is a polymer from claiming methyl butadiene (isoprene):

Natural Rubber (NR)

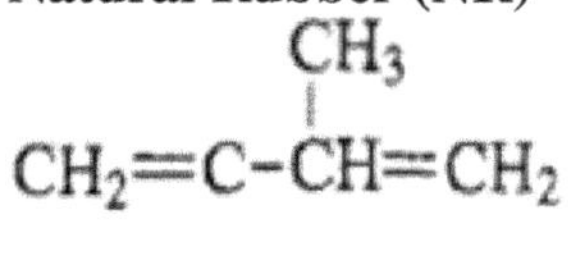

Purified crude elastic gets sticky clinched alongside hot climate and in fragile done cool climate. Its profitable properties are following vulcanization. Vulcanization is the majority paramount NR substance response. Depending upon the level of curing, characteristic elastic is arranged as soft, semi hard or hard elastic. Main delicate elastic meets the ASTM from claiming an elastomer, and in this manner the majority of the data taken after that relate main with delicate elastic. The properties of semi tough and hard elastic contrast somehow, especially in the region for claiming erosion imperviousness.

Vulcanization has the best impact on the mechanical properties of characteristic elastic. Vulcanized elastic has a chance to be extended to approximately ten times its period, and toward this side of the point, it will bear a load about 10 tons/in. ‘. can be compacted with one-third from claiming its thickness many times without damage. The point when a large portion sorts about vulcanized rubbers are stretched, their safety builds to more stupendous extent over their development. Indeed, being extended just short of the purpose of rupture, they recoup very nearly and constantly on their unique extents with respect to being discharged et cetera bit by bit, recuperating a part of the remaining twisting. The remarkable property from claiming regular elastic in examination of the engineered rubbers may be its flexibility. It has phenomenal bounce back properties, whether heated or chilly and it is safe to the sun, weather and Ozone: Icy water preserves characteristic rubber. If used in air, especially during sunlight, elastic has a tendency to be turned tricky and fragile. It has the best reasonable imperviousness on ozone. Dissimilar to the engineered elastomers, common elastic softens and reverts for agincourt in daylight. In general, it has moderately poor weathering and agincourt properties (Kirk-Othmer, 1999).

Average Applications: Characteristic elastic figures are significantly used in assembling from claiming pneumatic tires, tubes, control transmission belts, conveyer belts, gaskets, mountings, hose, concoction tank linings, printing press platens, callous or stun absorbers, as well as seals against air, moisture, sound, and soil. Elastic has been utilized for a lot of people for quite some time, regarding illustration, as a lining material to steel tanks; especially to security against erosion against inorganic salt solutions, particularly brine, alkalies and non-oxidizing acids. These linings have the point of claiming, continuously prompting repairer set up. Regular elastic is likewise utilized for lining pipelines used to pass on these sorts from claiming materials. From claiming these requisitions, it has been supplanted by engineered rubbers that have been created again at quite the same time.

3.3.2 Isoprene Rubber (IR)

Chemically, regular elastic may be common cis-polyisoprene. The engineered type of common rubber, engineered cis-polyisoprene, is known as isoprene elastic. The physical and mechanical properties of isoprene elastic are similar to the physical and mechanical properties of characteristic rubber. The one significant distinction continuously is that isoprene doesn't have a smell. This characteristic permits the utilization of isoprene elastic for specific food-handling requisitions. Isoprene elastic can be compounded, processed, and utilized in the same way as common elastic. Other than the absence of odor, isoprene elastic has no points of interest in common elastic. Depolymerized polyisoprene fluid is utilized like a sensitive plasticizer for cement tapes, heated melts, brake linings, grinding wheels, and wire and link sealants (Buchoff, 1996).

3.3.3 Neoprene (CR)

Neoprene is a standout amongst the most seasoned and a large portion versant of the engineered rubbers. Synthetically, it is polychloroprene. Polychloroprene is processed toward free-radical emulsion polymerization for principally trans-2-chloro-2-butenylene moieties. 23 chloroprene elastic possesses direct oil resistance, exact beneficial climate and oil resistance, as well as handy safety with oxidative chemicals. Its fundamental unit is a chlorinated butadiene whose recipe is:

Neoprene (CR)

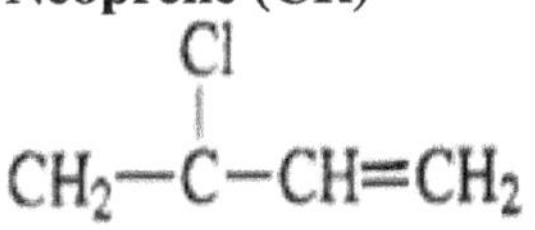

The crude material is acetylene, which makes this item additionally unreasonable over exactly different elastomeric materials. Neoprene might have been acquainted commercially, eventually perusing DuPont over 1932, as an oil resistant substitute to regular elastic. Its progressive properties are fundamentally the same with the individuals from claiming characteristic rubber. However, its tendency for compound safety defeats a significant number of the shortcomings of common elastic. Neoprene is also accessible in an assortment of claiming types. In neoprene latex that is comparable with characteristic elastic latex, neoprene will be processed in a "fluid" structure like a possibly exacerbated latex scattering or a dissolvable result. When these materials are hardened and alternately cured, they have the same physical and concoction properties, like the robust or cell division structures. A neoprene compound can be transformed into gatherings that give any properties needed. The point is that when the hardness of neoprene is over 55 shore A, its flexibility surpasses that of common elastic, eventually perusing nearly 5%. At hardness beneath 50 shore A, its flexibility may not be as beneficial concerning illustration of a regular rubber, despite the fact that its flexibility will be measured at 75%, which is a high esteem. Due to its secondary resilience, neoprene items have low hysteresis, as well as the least high temperature development throughout element operations. Contrasted with regular rubber, neoprene is moderately impermeable with gasses. Due to this impermeability, neoprene could be used to seal against Freon blowing agents, propane, butane and different gasses. Neoprene is utilized within a number of electrical applications, in spite of its dielectric qualities. Cutoff is utilized as encasing with low voltage (600V) and low recurrence (60Hz). Due to its secondary degree of imperviousness with indoor and open-air maturing and its safety on weathering, neoprene may be regularly utilized as a protective external coat on encasing in the least voltages. It can be likewise safe with high-voltage crown release impacts that cause extreme surface cutting for large portion sorts for elastomers (Meier, 1996). In the most extreme working temperature of claiming 200°F (93°C), neoprene proceeds will administer beneficial physical properties and has fantastic safety to long haul high temperature corruption. Dissimilar to different elastomers, neoprene doesn't mellow or melt if heated, in any case of the high temperature level. High temperature disappointment comes from solidifying of the result and absence of flexibility. Neoprene manifests a greatly solid mechanical bond for cotton fabric. Whenever suitable medicines of additives are provided, it can additionally be produced to stick on such man-consuming shark fibers. Similar to glass, nylon, rayon, acrylic and polyester fibers, it might additionally be formed to contact with metals, especially carbon and compound steels, stainless steels, aluminum and aluminum alloys, brass and copper, utilizing any of the economically accessible holding operators.

Closed-cell neoprene is a versatile intricate from claiming individual, non-connecting phones that confer the included point from claiming non-absorbency. This property makes closed-cell neoprene particularly suitable for fixing provisions at the place liquid contact is expected. The results include, for example, wet suits for divers, shoe soles, car deck cover seals and for other applications, the place at which the compressible nonabsorbent weather-resistant material may be hazed. Froth neoprene is similar to open-cell neoprene in that it is a compressible material with interconnecting units. Its primary zone of provision may be for cushioning, for example, in mattresses, seating, as well as rug underlay. Due to the great high temperature and oil safety from claiming neoprene, it has likewise discovered provision similar to a railroad auto lubricator. This absorbent open-cell structure gives a wicking activity to convey oil of the diary bearings.

3.3.4 Nitrile Rubber (NBR, BUNA-N)

Nitrile rubbers is an outgrowth of German Buna-N alternately Perbunan. They are copolymers of butadiene and acrylonitrile (CH2 =CH -C=N) and have a standout amongst the four. The greater part wildly utilized elastomers, XNBR, are carboxylic-acrylonitrile-buta¬diene-nitrile elastic with a progressed abrasion imperviousness to a certain degree over that of the standard NBR nitrile rubbers. The primary favorable circumstances of nitrile rubbers are at their low cost, useful oil and abrasion resistance, beneficial low-temperature and swell aspects. Their more terrific imperviousness with oils, fuel and solvents contrasted with that of neoprene is their grade playing point. As with other elastomers, suitable aggravating can move forward certain properties.

The physical and mechanical properties from claiming nitrile rubbers are fundamentally the same to the individuals from claiming characteristic elastic. These rubbers have the remarkable capability to hold both their quality and versatility at greatly low temperatures. It is this property that makes them profitable to use as hose utilized within operating the water powered controls in airplanes. Buna-N doesn't have remarkable heat safety. It has the greatest working temperature from claiming 250°F (120°C), as it has a propensity to solidify during high temperatures. Nitrile rubbers back burning and blaze. Their electrical properties are generally poor. Therefore, they don't figure total utilization in the electrical field, since there are some problems. There is a large number of different elastomeric materials for significantly predominant electrical properties. Due to their safety to sun, weather, and ozone, nitrile rubbers have poor safety with daylight. Moreover, Ozone's weathering qualities are not handy (Toth, 1996). Nitrile elastic provisions are belting, sheeting, link jacketing, hose to fuel lines and air conditioners, sponge, gaskets, arctic/aviation elastomeric o-rings, seals, precision dynamic abrasion seals and shoe soles. Nitrile rubbers are coextruded similarly as the internal tube with chlorinated polyethylene external tube for car provisions. Due to its remarkable imperviousness for will fills and water powered fluids, Buna-N's significant zone for provision is in the fabrication of air ship hose, gas and oil hose and settling towards oneself fuel tanks. Different requisitions incorporate carburetor diaphragms, gaskets, cables, printing rolls and hardware mountings.

3.3.5 Butyl Rubber (IIR) and Chlorobutyl Rubber (CIIR)

Butyl elastic (IIR) is an isobutylene-based elastic which incorporates copolymers of isobutylene and isoprene, halogenated butyl rubbers and isobutylene/p-methylstyrene/bromo-p-methylstyrene terpolymers. IIR has a chance to be slurry polymerized resining with isobutylene copolymerized with little sums of isoprene for methyl chloride diluent at -130 to148°F (-90 to -100°C).

Butyl Rubber (IIR) and Chlorobutyl Rubber (CIIR)

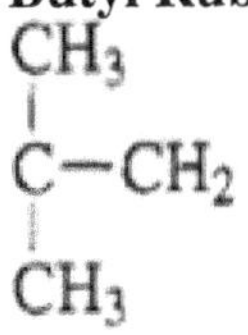

Likewise, its guardian material included little proportions of butadiene or isoprene. Business butyl elastic might hold numerous 5% butadiene, similar to a copolymer. It may be a universally useful engineered elastic whose remarkable physical properties are low permeability corrosive with air (approximately one-fifth that of regular rubber) and high vitality absorption. Chlorobutyl elastic may be chlorinated isobutylene-isoprene. It has the same general properties of butyl elastic for somewhat higher suitable working temperatures (Rogers & Waddell, 1999).

At room temperature, the strength about butyl elastic may be poor and in case the temperature increases, the strength increments. In raised temperatures, butyl elastic exhibits handy strength. Its abrasion resistance, shred resistance, pliable strength and bond with fabrics and metals will be beneficial. Butyl elastic has a greatest constant administration temperature of 250-350°F (120-150°C) for handy safety on high temperature agincourt. Its electrical properties are for the most part great, yet not remarkable over any other classification. The fire imperviousness about butyl elastic is poor. Two economically accessible halogenated butyl elastic subsidiaries are bromobutyl (BUR) and chlorobutyl (CUR). The halogen iotas are consolidated under the polymer on the isoprene units. These mixes have compound properties for tolerance of the polymers that are covulcanized for other elastomers more promptly over UR. The end-use properties and requisitions are comparable to the individuals of UR. An ordinary provision will be the spread of ventilating hoses. Chlorobutyl (CUR) rubbers bring the greatest working temperature for 300°F (150°C). It may also work as low as low as -30°F (-34°C). The opposite physical and mechanical properties are similar to the individuals from claiming butyl elastic. Regarding imperviousness to sun, weather and ozone, butyl elastic has phenomenal imperviousness to sun, weather and ozone. Its weathering qualities are outstanding, similar to its safety on water absorption. Due to its impermeability, butyl elastic figures have numerous utilizations in the assembling of inflatable things, such as existence jackets, aggregation boats, balloons and inward tubes. The fantastic imperviousness it exhibits in the vicinity of water and steam makes it suitable to hoses and diaphragms. Provisions are additionally found concerning illustration, adaptable electrical insulation, stun and vibration absorbers, curing bags for tire vulcanization and forming.

3.3.6 Chlorosulfonated Polyethylene Rubber (HYPALON)

Chlorosulfonated polyethylene manufactured elastic (CSM) is made by DuPont under the exchange name Hypalon. It is an immersed chlorohydrocarbon elastic handled from Cl_2, SO_2 and an amount of polyethylenes, and it holds something like 20 with 40% chlorine and 1 with 2% sulfur microorganisms similar to sulfonyl chloride. It can be compared to neoprene, yet it does have more favorable circumstances than neoprene specific sorts from claiming administration. It has preferred heat, ozone resistance, preferred electrical properties, exceptional color stability and finer compound safety. Hypalon, when legitimately compounded, also exhibits beneficial imperviousness to wear and abrasion, beneficial flex life, secondary sway resistance, and great imperviousness that is a lasting deformity under overwhelming stacking (Kirk-Othmer, 1999).

Due to the polymer's common ozone resistance, it is not essential to include antiozonates during aggravating. The antiozonates are solid discoloring operators, and when included with elastomers, they have foundation shades on blur and turn flimsy. Hypalon smolders to real fire circumstances and may be ordered as self-smothering. If the fire is removed, the elastomer quits blazing. This advantage is because of its chlorine content, which makes it safer on smoldering than only hydrocarbon polymers. Hypalon's safety for abrasion is better than that of regular elastic and many different elastomers, eventually perusing to the extent of 2 to 1. It also possesses high imperviousness on weariness splitting and cut development resining with steady flexing. These final properties aggravate Hypalon's suitableness for obtaining exceptional results for element operation. Useful safety with impact, crushing, cutting, gouging and different sorts of physical ill-use are available for elastic parts prepared from this elastomer. The electrical business utilizes Hypalon in blanket car ignition loop and essential wire, atomic force station cable, control cable and welding link. Likewise, it has an additional security from storms in sea, control and lighting link once seaward oil platforms are sheathed for Hypalon, due to its high temperature and radiation imperviousness.

3.3.7 Ethylene-Acrylic (EA) Rubber

Ethylene-acrylic elastic is generated from ethylene and acrylic corrosive. Likewise, for other engineered elastomers, the properties of the EA rubbers can be modified towards aggravating. Essentially, EA is a cost-effective, hot-oil-resistant elastic for handy low-temperature properties. Ethylene-acrylic elastomers bring useful shred quality and rigidity and secondary prolongation in break. Furthermore, addition of exceptionally low layering situated values is an included advantage, making the result suitable for a significant number of hoses, as well as sealing and reduction gasket provisions. EA elastomers have amazingly useful imperviousness to sun, weather and ozone. Long term exposures bring no impact with respect to these rubbers. Ethylene-acrylic elastic is utilized within such results concerning illustration gaskets, hoses, seals, boots, damping components, low-smoke carpet tiling, link jackets to seaward oil platforms, ships and building plenum installations. Ethylene-acrylic elastic over motor parts gives handy imperviousness against heat, fluids, and wear and in addition useful low-temperature fixing capability (Kirk-Othmer, 1999; Rogers & Waddell, 1999; Toth, 1996).

3.3.8 Acrylate-Butadiene Rubber (ABR) and Acrylic Ester-Acrylic Halide (ACM) Rubbers

Acrylate-butadiene and acrylic ester-acrylic halide rubbers are similar to ethylene-acrylic rubbers. Physical and mechanical Properties: the ABRs and ACM rubbers show great flexibility and shred safety at poor sway imperviousness. Abrasion imperviousness and layering situated are handy. The most extreme temperatures rating is 340°F (170°C), the same with respect to the EA rubbers. Acrylate-butadiene and acrylic ester-acrylic halide rubbers show useful resistance to sun, weather, and ozone. Applications: These rubbers are utilized for safety, in cases of climatic states and heat (Kirk-Othmer, 1999).

3.3.9 Ethylene-Propylene Rubbers (EPDM and EPT)

Ethylene-propylene elastic is an engineered hydrocarbon-based elastic aggravated, possibly resining with ethylene-propylene diene monomer or alternately ethylene-propylene terpolymer. EPDM molecule size is a noteworthy parameter to Santoprene's mechanical properties, with more little particles revealing higher quality and prolongation (Margolis, 1996). Ethylene-propylene elastic possesses many properties better than the characteristic elastic and traditional universally useful elastomers. Clinched alongside portion applications, it performs superior to other materials, and at the same time over different provisions. It is recently more drawn out, alternately requiring more support and might indeed expense in most. EPDM has remarkable heat resistance, having the ability to work during temperatures for 300-350°F (148-176°C) and at the same time discovering requisition at temperatures as low as -70°F (-56°C). Physical and mechanical properties are reasonably expected to intensify EPDM to give whichever high flexibility alternately with secondary damping. The point when it is exacerbated will give high resilience. The items are comparable with regular elastic vivacity, or at least hysteresis values. The vitality absorbing mixes have low flexibility values approaching the individuals from claiming fame elastomers utilized for stun and vibration damping. If the compound has been exacerbated to flexibility or secondary damping, its properties remain moderately consistent through a whole temperature extent. The electrical properties of EPDM are excellent, especially for high voltage encasing. The properties are likewise stable, after long periods from claiming drenching done water. Fantastic imperviousness is likewise shown against cutting brought on by high voltage crown release. Ethylene-propylene elastic is especially safe with sun, weather and ozone strike. Phenomenal climate imperviousness is acquired if the material is formulated for color, whiteness or bootleg. The elastomer remains free of surface crazing and retains a high rate of its properties and quite some of its purposes of presentation. Ozone safety is inalienable in the polymer. Regarding its useful purposes, it might be viewed as resistant to ozone ambush. It is not necessary to include any uncommon aggravating parts to process this imperviousness.

Ethylene-propylene elastic figures requisition in the electrical business and in the assembling for electrical gear. A standout amongst the elemenresiny provisions is the protecting tape. It is utilized for medium-voltage (up on 35 kV) and auxiliary organized control cable, coverings to line and multiplex circulation wire, jacketing and encasing to sorts encountered with urban decay because of deindustrialization, innovation development, government lodging, SJ adaptable cords and encasing to car ignition loop link. Accessories, for instance formed terminal covers, plugs, transformer connectors, line-tap and exchanging devices, splices, and insulating and semiconductor tape are likewise generated for all the resining with BPOM. Medium and high-voltage underground force appropriation link insulated with BPOM offers a significant number of favorable circumstances. It gives phenomenal imperviousness for tearing down and disappointment brought on towards high voltage contaminants and anxiety. Its fantastic electrical properties make it suitable for high-voltage link encasing. It withstands overwhelming crown release without causing harm.

3.3.10 Polysulfide Rubbers (ST and FA)

Polysulfide rubbers are made by joining together ethylene ($CH_2=CH_2$) for a basic polysulfide, the sulfur microbe's manifestations and only the polymerized atom. Otherwise, they would have been called Thiokol rubbers. As a rule, these elastomers don't have great elasticity, yet they have useful safety with high temperature and are safe to the vast majority of solvents. Contrasted with nitrile rubber, they have poor pliable strength, a pungent odor, poor rebound, high crawl under strain and poor abrasion safety. Changed natural polysulfides are aggravated, eventually perusing and substituting different unsaturated exacerbates to ethylene, which brings about mixes that have minimal objectionable smell. High proportions of ST are NBR or neoprene decline swelling from aromatics, fuels, ketones and esters. The low-temperature adaptability can be Additionally progressed. Higher proportions of NBR or neoprene on ST bring about physical properties improvement, and compressive situated safety, following heat maturing. The electrical insulating properties for polysulfide ST are poor, as its fire safety. FA polysulfide elastic exacerbates presentation phenomenal imperviousness to ozone, weathering and introduction to ultraviolet light. Their imperviousness is better than that from claiming ST polysulfide rubbers. In high focuses of ozone, the utilization of 0.5 and only nickel dibutyldithiocarbamate (NBC) for every 100 parts from claiming fa polysulfide elastic will move forward the ozone imperviousness. ST polysulfide elastic exacerbated with carbon dark will be safe with ultraviolet light and daylight. Its imperviousness with ozone may be great. However, it can be progressed from claiming NBC, but this expansion corrupts the material's situated layering. ST polysulfide elastic additionally possesses acceptable climate imperviousness. FA polysulfide elastic is a standout amongst the elastomeric materials regularly utilized to the creation from claiming elastic rollers for printing and covering gear. The significant purpose behind this is its secondary level for claiming imperviousness from a huge number of claiming solvents, including ketones, esters, fragrant hydrocarbons and plasticizers that are similarly utilized as vehicles for different printing inks and coatings. Additionally, because of its handy climate imperviousness, it may be advantageous in outside caulking mixes (Meier, 1996).

3.3.11 Polyamides

Polyamide TPEs are normally polyester-amides, polyetheresteramide square copolymers and alternately polyether piece amides (PEBA) (PEBA piece copolymer sub-atomic building design can be compared with average piece copolymers as shown below:

Polyamides

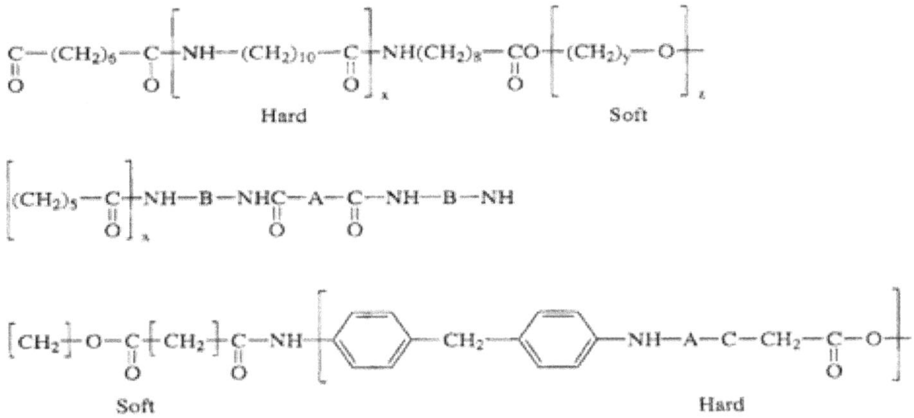

Where:

$A = C_{19}$ to C_{21} dicarboxylic acid

$B = \left[CH_2\right]_3 O \left[\left[CH_2\right]_4 O\right]_b \left[CH_2\right]_3$

The polyamide is the difficult (thermoplastic) segment. At the same time, the polyester, polyetherester, and polyether segments are the delicate (elastomeric) fragment. Polyamide materials bring a surprising blending for secondary ductile strength, ductility and sturdiness. The withstands great secondary impact, especially in very low temperatures. As stated by ASTM test 0-746, the chilly weakness temperature is -94°F (-70°C). Nylons 11 and 12 have a very high versatile memory that permits them to withstand rehashed extending and flexing through lengthy periods from claiming occasion. They additionally show handy abrasion safety. Their absorption of dampness may be extremely low, which implies that parts transformed from these materials will bring beneficial dimensional dependability in case of the stickiness of the earth. Dampness likewise has little impact on the mechanical properties, especially the modulus of flexibility. The electrical encasing properties of the polyamides evaluation 11 and evaluation 12 are useful. They have stable volume resistivity and offer phenomenal imperviousness against the following circular segment. These materials are self-extinguishing. Imperviousness against sun, weather, and ozone: Polyamides are safe with sun, weather and ozone. A number of metals are covered with polyamide in gatherings to give security from merciless climate (Rader, 1996).

Polyamides discover a significant number of different requisitions coming about because of a huge number of advantageous properties. An extensive variety of adaptability permits material that is delicate enough to high-quality bike seats to be handled, as well as different materials whose quality and unbending nature are tantamount to the individuals for a significant number of metals. Super flexible evaluations are likewise accessible and utilized for shoe soles, gaskets, diaphragms and seals. Due to the high versatile memory about polyamides, these parts of withstand repeater extending and flexing through lengthy period of the long haul. Since polyamides can meet determination SAE J844 for air-brake hose, looped tubing may be generated for this motivation for utilization around trucks. Looped air-brake hose generated from this material has been utilized with respect to trucks that have voyaged over 2 million miles without an absolute account for disappointment. High-fueled hose and fuel lines can also handle resining with this material. It can also be handled with corrosion-resistant and wear-resistant coverings for airplane control cables, car cables and electrical wires.

3.3.12 Silicone (SI) and Fluorosilicone (FSI) Rubbers:

Silicone rubbers, also called polysiloxanes, are an arrangement from claiming exacerbates whose polymer structure comprises silicon and oxygen iotas as opposed to the carbon structures of the vast majority of different elastomers. The silicones have subsidiaries for silica, SiO_2 or O=Si=O. The point at which the iotas are joined together in this way, the twofold linkages will break. Moreover, as Methyl assemblies enter the linkages, silicone elastic may be produced:

Silicone (SI) and Fluorosilicone (FSI) Rubbers

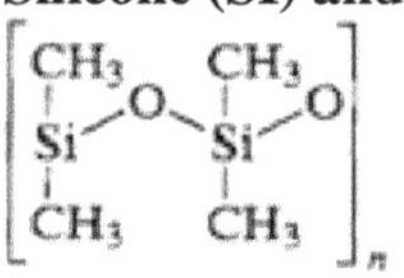

Silicon is in the same concoction bunch as carbon, since it is to a great extent a stable element, and subsequently more stable mixes are handled. The essential structure can be changed for vinyl or fluoride groups, which enhances properties, such as shred resistance, oil resistance and concoction imperviousness. This brings about a crew of silicones that blanket an extensive variety of physical and natural necessities. Silicones are few of the large portion of accessible heat-resistant elastomers and the vast majority which is adaptable in low temperatures. Silicone's property mark is combined of the following: high-temperature imperviousness [-500°F (260°C)], great adaptability toward -100°F (-73°C), beneficial electrical properties, beneficial layering set, and shred imperviousness and solidness against a total temperature go. Their viable operating temperature go resining with -60 to 450°F (-51 to 232°C). They show fantastic properties during the most reduced temperatures. Fluorosilicones have a compelling operating temperature that reach -100 on 375°F (-73 to 190°C). Silicone rubbers have remarkable electrical properties, better than a large portion of elastomers. The decay item of carbon-based elastomers is conductive carbon black, which might be brilliant and in this way leaves nothing to insulation, while the decay result of the silicone rubbers is the insulating silicone dioxide. This property is preferable in encasing electric motors. The polysiloxanes have poor abrasion resistance, pliable strength and shred resistance, but they show useful layering set safety and bounce back properties in both frosty and hot situations. Their imperviousness with fire is handy. Fluorosilicones (FSIs) have basically the same physical and mechanical properties as Silicones. However, it slightly changes in the bond with metals and in impermeability. Safety with sun, weather and ozone: silicone and fluorosilicone rubbers presentat phenomenal safety with sun, weathering and ozone. Their properties are essentially unaesthetic by long haul introduction (Meier, 1996; Rader, 1996).

Due to the exceptional warm soundness or their insulating values, silicone rubbers figure large portion employments in the electrical industries, principally over appliance, heaters, furnaces, aviation devices and car parts. Their phenomenal weathering qualities and total temperature reach has additionally brought about their job, like caulking exacerbates. If silicone or fluorosilicone rubbers are infused for high-density conductive filler, an electric way is made. These conductive elastomers can be likewise utilized and only in EMI/RFI/EMP protecting transform clinched alongside manifestations, such as O-rings and gaskets to furnish protection to regulation, while preventing the escape of EMI internally created towards the device. Protection from avoidance should keep the interruption from claiming EMI/RF /EMP made eventually perusing outside sources under the secured device. Avoidance or regulation, in addition to weight and alternately vacuum fixing to gatherings give both EMI/EMP weakening and weight regulation or weatherproofing, grounding and contacting to furnish a trustworthy low-impedance association. In the behavior of electric vitality with ground, regularly utilizing the place of mechanical mating may be blemished or illogical. The emulating gear can be fit to generate EMI or defenseless EMI, flying machine and aviation electronics, simple instrumentation, car electronics, business machines, correspondence systems, advanced instrumentation, radio-frequency defiant and radar restorative electronics, miliresiny and marine electronics, as well as security frameworks (miliresiny and commercial).

3.3.13 Vinylidene Fluoride (HFP, PVDF)

Polyvinylidene fluoride (PVDF) is a homo-polymer from claiming l,l-difluoroethene for exchanging CH2 and CF2 assemblies along the polymer chain. These aggregations confer an interesting polarity that impacts its solvency and electrical properties. The polymer has the trademark soundness for fluoropolymers. This point when laid open, will cause hostile thermal, chemical, and ultrasuede states. In general, PVDF is a standout amongst the easiest fluoropolymers with transform and might be effortlessly reused without influencing its physical and concoction properties. Likewise, with other elastomeric materials, aggravating can be used to move forward certain particular properties. Cross-linking of the polymer chain and control of the sub-atomic weight are also finished to enhance specific properties.

Physical and mechanical properties: PVDF elastomers bring secondary ductile and effect qualities. The encompassing temperature elasticity in yield for 4,000--7,000 psi (28-48 Mpa) and the unnotched sway strength of l5-S0 ft-lb/in. (800-4270 kJ /m) demonstrate that evaluations from claiming this polymer will be solid and intense. These properties are held over a whole temperature range. Fantastic safety will crawl, and weariness will likewise be exhibited towards PVDF. At the same time, for slim areas, films, filament and tubing would be adaptable and transparent. PVDF wire encasing has fantastic safety to cut-through. The place load bearing is important, these polymers are unbending and safe to creep under mechanical load. Imperviousness with deformity under load is amazingly useful over the temperature range from claiming -112 to 302°F (-SO on 150°C). Applications: PVDF figures huge numbers of provisions which can be seen in properties from claiming erosion resistance, totally suitable working temperature range, mechanical quality and toughness, secondary abrasion resistance, secondary dielectric strength and imperviousness to weathering, ultraviolet light, radiation and growths which are all helpful. In the electrical and hardware fields, PVDF is utilized for multi-wire jacketing, plenum cables, heat-shrinkable tubing, anode lead wire, workstation wiring and link binds. Due to its acknowledgement in the field of pharmaceuticals, exchange houses are lined with PVDF. Its erosion safety is additionally an element in these provisions. In over fluid-handling systems, PVDF figures provisions as jacketing material, valve diaphragms and membranes to microporous filters and ultrafiltration. Likewise, an aftereffect from claiming its imperviousness to growths and its remarkable erosion resistance, it is utilized in the encasing of material to underground anode couch installations (Meier, 1996; Rader, 1996).

3.3.14 Ethylene-Tetrafluoroethylene (ETFE) Elastomer

This elastomer can be sold under the exchange name Tefzel towards DuPont. Ethylene¬tetrafluoroethylene (ETFE) is an altered incompletely fluorinated copolymer of ethylene and polytetrafluoroethylene (PTFE). Since it holds more than 75% TFE towards weight, it has finer imperviousness with abrasion and cut-through over TFE. At the same time, it has the majority of erosion imperviousness properties.

Compared with glass, ETFE film is 1% of the weight. It transmits more light and costs 24% to 70% in most cases. It's also versatile (able to bear 400 times its own weight, self-cleaning (due to its nonstick surface) and recyclable. In the great holders, it may be inclined to punctures eventually perusing sharp edges; subsequently it is most accioli utilized for roofs (ETFE, n.d.). Ethylene-tetrafluoroethylene has phenomenal mechanical strength, stiffness and abrasion resistance, with an administration temperature reach from claiming -370 to 300°F (-223 on 149°C). It also exhibits useful shred safety and useful electrical properties. However, its remarkable property is its safety with an extensive variety of corrodents. Safety with sun, weather, and ozone: Ethylene-tetrafluoroethylene has remarkable imperviousness to sunlight, ozone and climate. This feature, coupled with its extensive variety of erosion resistance, makes the material especially suitable to open air requisitions subject to climatic erosion. The central requisitions of ETFE are found clinched alongside such items, like gaskets, packings and seals (O-rings, lip and X-rings) in ranges, in case erosion is an issue. The material is likewise utilized in sleeve, part curled and push bearings, as well as bearing pads for channel and supplies backing the place development. Furthermore, withdrawal or development might happen.

3.3.15 Ethvlene-Chlorotrifluoroethylene (ECTFE) Elastomer

Ethylene-chlorotrifluoroethylene (ECTFE) elastomer is a 1: 1 exchanging copolymer of ethylene and chlorotrifluoroethylene. This compound structure provides for the polymer an interesting consolidation for its properties. It possesses fantastic concoction resistance, beneficial electrical properties and broad-use temperature reach [from cryogenic to 340°F (171°C)]. It also meets the prerequisites of the UL-94V-O verthandi fire test clinched alongside thicknesses as low as 7 mils. ECTFE is an extreme material for a fantastic sway quality over its whole operating temperature extent. About every last one of fluoropolymers, ECTFE ranks around the best for abrasion imperviousness. Most of systems utilized within transforming polyethylene have a chance to be used to procedure ECTFE. It gives a chance to be extruded, infusion molded, rotomolded and connected by conventional fluidized couch for electrostatic coating systems. Ethylene-chlorotrifluoroethylene possesses profitable electrical properties. It has high resistivity and low passing. The dispersal element varies to some extent for frequency, especially over 1 kHz. The ac misfortune properties of ECTFE are better than the individuals of vinylidene fluoride and approach the individuals of PTFE. The dielectric steady is stable crosswise over an expansive temperature and recurrence go. Other paramount physical properties incorporate a low coefficient of friction, fantastic machinability and the capacity with a chance to be pigmented. In thicknesses as low as 7 mils, ECTFE has an UL-94-V-0 rating. The oxygen list (ASTM D2863) is 60 around An -h-in, thick example and 48, looking into a 0. 0005-in fiber yarn. ECTFE is a strong, exceptionally impact-resistant material that retains handy properties again for an extensive variety of claiming temperatures. Safety to sun, weather, and ozone: Ethylene-chlorotrifluoroethylene is greatly safe with sun, weather and ozone ambush. Its physical properties experience next to no transform after long exposures. This elastomer figures huge numbers of provisions in the electrical business, concerning illustration wire, link encasing and jacketing, plenum link insulation, oil well wire and link insulation, logging wire jacketing and jacketing for cathodic protection, aircraft, impostor transit and car wire, connectors, loop forms, resistor sleeves, wire tie wraps, tapes, tubing, adaptable printed meandering and even link. Provisions are also found clinched alongside different commercial enterprises, such as diaphragms, adaptable tubing, closures, seals, gaskets, convoluted tubing and hose; especially in the chemical, cryogenic and aviation commercial enterprises. Materials of ECTFE are likewise utilized for lining vessels, pumps and different supplies (ETFE, n.d.; Margolis, 1996; Rader, 1996).

3.3.16 Epichlorhydrin Rubber

The epichlorhydrin polymer aggregation incorporates the homopolymer (CO), the copoly¬mer for ethylene oxide (ECO) and terpolymers (GECO). Physical and Mechanical properties: In general, Epichlorhydrin elastic possesses a mix of the handy properties of neoprene and nitrile elastic. Epichlorhydrin rubbers bring fantastic warm solidness for most extreme working temperatures to co-elastic from claiming 300°F/150°C and to ECO elastic from claiming 275°F/135°C. The different polymers generally show separate execution levels with respect to flexibility and low-temperature adaptability. The ECO form has progressive properties over common elastic. It has useful maturing properties. Their secondary chlorine content imparts greatly on reasonable fire safety. Imperviousness to sun, weather and ozone: These rubbers have fantastic safety with sun, weathering and ozone.

Applications: Epichlorhydrin elastic provisions incorporate seals and tubes for air-conditioning and fuel frameworks. Epichlorohydrin may be a versant forerunner in the union for a significant number of natural mixes. For example, it can be changed over to glycidyl nitrate, a vivacious folio utilized within hazardous and propellant compositions (Gould, 1966).

REFERENCES

Architecture. (n.d.). http://architecture.about.com/od/construction/g/ETFE.htm

Bajaj, P. (2000). Finishing of textile materials. *Journal of Applied Polymer Science*, *83*(3), 631–659. doi:10.1002/app.2262

Berins. (1991). *Plastics Engineering Handbook* (5th ed.). Springer.

Billmeyer. (1964). *Polymer Science*. Interscience Publishers.

Brydson. (1999). *Plastics Materials* (7th ed.). Butterworth-Heinemann.

Buchoff, L. (1996). Liquid and Low-Pressure Resin Systems. In *Handbook of Plastics, Elastomers, and Composites* (3rd ed.). McGraw-Hill.

Cradic, G. W. (1994). *Modern Plastics Encyclopedia Handbook*. McGraw-Hill.

Ding, M. (2007). Isomeric polyimides. *Progress in Polymer Science*, *32*(6), 623–668. doi:10.1016/j.progpolymsci.2007.01.007

Eldada, L. (2001). Advances in telecom and datacom optical components. *Optical Engineering (Redondo Beach, Calif.)*, *40*(7), 1165. doi:10.1117/1.1372703

ETFE. (n.d.). http://www.todoarquitectura.com/revista/40/en04_ETFE.asp

Field. (2003). Melamine Plastic. Home Maintenance and Repair. Michigan State University Extension. http://web1.msue.msu.edu/imp/mod02/01500096.html

Gannon, J. (1994). Plastics Handbook, Modern. Plastics Magazine.

Gleixner, G. (2001). Flame reresindant PP fibres-lateat developments. *Chem Fibers Int, 51*, 422–424.

Goodman, S. H. (1999). *Handbook of Thermoset Plastics* (2nd ed.).

Gorna, K., & Gogolewski, S. (2002a). Biodegradable polyurethanes for implants. II. In vitro degradation and calcification of materials from poly(ecaprolactone)- poly(ethylene oxide) diols and various chain extenders. *Journal of Biomedical Materials Research, 60*(4), 592–606. doi:10.1002/jbm.10100 PMID:11948518

Gorna, K., & Gogolewski, S. (2002b). In-vitro degradation of novel medical biodegradable aliphatic polyurethanes based on ε-caprolactone and Pluronics® with various hydrophilicities. *Polym Deg. STAB, 75*(1), 113–122. doi:10.1016/S0141-3910(01)00210-5

Gould, R.F. (1966). Advanced Propellant Chemistry. *ACS Chemistry Series, 54.*

Gunatillake, P., Mayadunne, R., & Adhikari, R. (2006). Recent developments in biodegradable polymers. *Biotechnology Annual Review, 12*, 1387–2656. PMID:17045198

Harper, C. A. (1975). *Handbook of plastic and elastomers*. McGraw-Hill.

Harper, C. A. (1996). *Handbook of Plastics Elastomers, and Composites*. McGraw-Hill.

Hill, R. M. (2002). *Silicone (siloxane) surfactants* (3rd ed., Vol. 14). Academic Press.

Hiltunen, K., Tuominen, J., & Seppälä, J. V. (1998). Hydrolysis of lactic acid-based poly(ester-urethane) s. *Polymer International, 47*(2), 186–192. doi:10.1002/(SICI)1097-0126(1998100)47:2<186::AID-PI47>3.0.CO;2-E

Jones, F. (2005). *Alkyd Resins in Ullmann's Encyclopedia of Industrial Chemistry*. Wiley-VCH.

Kajihara, M., Sugie, T., Maeda, H., Sano, A., Fujioka, K., Urabe, Y., Tanihara, M., & Imanishi, Y. (2003). Novel drug delivery device using silicone: Controlled release of insoluble drugs or two kinds of water-soluble drugs. *Chemical & Pharmaceutical Bulletin, 51*(1), 15–19. doi:10.1248/cpb.51.15 PMID:12520121

Kirk-Othmer. (1999). *Concise Encyclopedia of Chemical Technology*. John Wiley & Sons.

Kroschwitz. (2007). *Polymer Science and Engineering*. Wiley-Interscience.

Laura, D. M., Keskkula, H., Barlow, J. W., & Paul, D. R. (2002). Effect of glass fiber surface chemistry on the mechanical properties of glass fiber reinforced, rubber [hyphen]toughened nylon 6. *Polymer, 43*(17), 4673–4687. doi:10.1016/S0032-3861(02)00302-6

Laura, D. M., Keskkula, H., Barlow, J. W., & Paul, D. R. (2003). Article. *Polymer*, 44.

Lide, D. R. (1998). *Handbook of Chemistry and Physics* (87th ed.). CRC Press.

Liebmann-Vinson, A., & Timmins, M. (2003). *Biodegradable polymers: degradation mechanisms, Part 2* (Vol. 2). PBM Series.

Margolis. (1996). *Thermoplastic Elastomer Markets in 2000.* Multiclient Report.

May, C. A. (1987). *Epoxy Resins: Chemistry and Technology (2nd ed.).* Marcel Dekker Inc.

McCrum, N. G., Buckley, C. P., & Bucknall, C. B. (1997). *Principles of Polymer Engineering*. Oxford University Press.

Meier, J. (1996). *Fundamentals of Plastics and Elastomers. In Handbook of Plastics, Elastomers, and Composites* (3rd ed.). McGraw-Hill.

Mercuri, L. G., & Giobbie-Hurder, A. (2004). Long-term outcomes after total alloplastic temporomandibular joint reconstruction following exposure to failed materials. *Journal of Oral and Maxillofacial Surgery, 62*(9), 1088–1096. doi:10.1016/j.joms.2003.10.012 PMID:15346359

Mittal, K. L. (Ed.). (2001). *Polyimides and other high temperature polymers: synthesis, characterization and applications* (Vol. 1). VSP.

Newton, J., Stoller, C., & Sresinch, M. (2004). Silicone technologies as delivery systems via physical associations. *Cosmet Toiletries, 119*(5), 69–70, 72–74, 76, 78.

Painter, P. C., & Coleman, M. M. (1997). *Fundamentals of Polymer Science: an Introductory Text*. CRC Press.

Parag, Somasundaran, & Kulkarni. (2006). Study of properties of modified silicones at solid–liquid interface: Fabric-silicone interactions. *J Colloid Interface Sci, 298*(2), 987–990.

Rader, C. (1996). *Thermoplastic Elastomers. In Handbook of Plastics, Elastomers, and Composites* (3rd ed.). McGraw-Hill.

Ramsteiner, F., Heckmann, W., Mckee, G. E., & Breulmann, M. (2002). Influence of void formation on impact toughness in rubber modified styrenic-polymers. *Polymer, 43*(22), 5995–6003. doi:10.1016/S0032-3861(02)00485-8

Ray Patrylak, R. (1994). *Plastics Handbook, Modern Plastics Magazine*. McGraw-Hill.

Rogers & Waddell. (1999, Feb.). A Review of Isobutylene-Based Elastomers Used in Automotive Applications. *Rubber World.*

Santerre, J. P., Woodhouse, K., Laroche, G., & Labow, R. S. (2005). Understanding the biodegradation of polyurethanes: From classical implants to tissue engineering materials. *Biomaterials, 26*(35), 7457–7470. doi:10.1016/j.biomaterials.2005.05.079 PMID:16024077

Sardanopoli, A. A. (1988). *Thermoplastic Polyurethanes. In Engineering Plastics* (Vol. 2). ASM International.

Seelig, T., & Van Der Giessen, E. (2002). Localized plastic deformation in terny polymer blends. *International Journal of Solids and Structures, 39*(13-14), 3505–3522. doi:10.1016/S0020-7683(02)00161-0

7. Slone. (2010). *Acrylic Ester Polymers in Encyclopedia Of Polymer Science and Technology*. John Wiley & Sons, Inc.

Socrate, S., Boyce, M. C., & Lazzeri, A. (2001). A micromechanical model for multiple crazing in high impact polystyrene. *Mechanics of Materials, 33*(3), 155–175. doi:10.1016/S0167-6636(00)00068-5

Tan, L.-S. (1999). *Polymer Data Handbook Poly(amide-imide)* (J. E. Mark, Ed.; 1st ed.). Oxford University Press.

Tang, Y. W., Labow, R. S., & Santerre, J. P. (2001). Enzyme-induced biodegradation of polycarbonate polyurethanes: dependence on hard-segment concentration. *J Biomed Mater Res., 56*, 516–528.

Tjong, S. C., Xu, S. A., Li, R. K. Y., & Mai, Y. W. (2017). *Composites Science Technology*. Academic Press.

Toth, R. (1996). *Elastomers and Engineering Thermoplastics for Automotive Applications. In Handbook of Plastics, Elastomers, and Composites* (3rd ed.). McGraw-Hill.

Watterson, E. C. (1988). *Polyether Sulfones in Engineering Plastics* (Vol. 2). ASM International.

Wright, R. E. (1991). *Molded Thermosets*. Hanser Verlag.

Zhang, Q. M., Bharti, V., Kavarnos, G., & Schwartz, M. (Eds.). (2002). *Poly(Vinylidene Fluoride) (PVDF) and its Copolymers* (Vol. 1-2). John Wiley & Sons. doi:10.1002/0471216275.esm063

Chapter 3 Nanoparticles in Industry

ABSTRACT

This chapter presents the audit and investigation of the principle properties of the principle nanoparticles and its impacts on the fundamental polymers' properties. Researching different variables have an influence on the electrical properties of the polymers insulating materials and the ideal properties which will be granted to the novel nano-science materials. Moreover, this report card examines the desire that starts with nanoparticles properties and particulate nanoparticles. Thus, this chapter presents a deep study for using nanoparticles in industry which had handled the importance of nanoparticles. This chapter contains also nanoparticles identifications and details about particulate of variant nanoparticles.

The principal notice of a portion of the recognizing ideas clinched alongside nanotechnology might have been over 1867 by James Clerk Maxwell when he suggested as a thought test a little substance known for its capacity to handle individual atoms. The 1st perceptions and measure estimations about nanoparticles may have emerged during the initial decades of the twentieth century. They are basically constructed around a nitty gritty investigation for gold sols and other nanomaterials with sizes down to 10 μm. Ultramarine blue that employs the dull field technique to see particles of sizes considerably short of light wavelength has been utilized. Nanometer has been utilized unequivocally to characterize molecule extent and set it at 1/1,000,000 of millimeter.

3.1 IMPORTANCE OF NANOPARTICLES

The use of nanoparticles as materials has been initially driven by a desire for more level costs. Nanoparticles were inexpensive, accordingly utilizing them might make the material less expensive. Cost decrease is not the only reason for its utilization. Nanoparticles have also received attention in planning composite materials. A percentage of the principle purposes behind utilizing nanoparticles are examined. Cost diminishment relies on the relative cost of the polymer and the nanoparticle. Nanoparticle costs depend incredibly on the molecule extent. In the rundown below, nanoparticles are separated under extensive molecule span materials (up on 100mm; e.g., ground $CaCO_3$), medium molecule measure (around 10

DOI: 10.4018/978-1-7998-3829-6.ch003

mm; e.g., clay), little molecule span (around 1 mm; e.g., TiO_2 or precipitated $CaCO_3$), and really little molecule size (below 0.1 mm; e.g., raged silica). Nanoparticles can be utilized to expansion, alternately for the decline of the thickness of a result. The thickness of a nanoparticle can be chance to be as secondary as 10 g/cm^3 or as low as 0. 03 g/cm^3. There might be an expansive contrast in the middle of the thickness of the nanoparticle and the polymer. Along these lines, an expansive extent of item densities can be reached. There are high thickness results (above 3 g/cm^3) for materials utilized within appliances alternately casings for electronic units. The more regular are densities beneath 2 g/cm^3 and glass fiber filled composites continuously forming an ordinary sample. The successful thickness of the polymer may be a chance to be diminished towards filling a froth for empty polymer spheres. In this example, the thickness of a material could be a level over 0. 1 g/cm^3. Optical properties of exacerbated materials rely on the physical aspects of the nanoparticle and the other major parts including the polymer. Nanoparticles every now and then cause issues in shade matching and must be accounted for in item shade plan. Many types of nanoparticle bring a unique color which will be advantageous in material coloring. As of late, metal powders have been utilized within mix for pigments to aggravate the composite show up metal. For hundreds of years, sticky surfaces have been dusted with powder (e.g., talc) which keep them divided. Talc is comprehensively utilized within link and profile expulsion to acquire a smooth birch surface. Similarly, infusion molding and the requisition from claiming aluminum trihydroxide provide for a preferred complete surface. Talc, $CaCO_3$ and diatomite furnishes anti-blocking properties. Graphite and other nanoparticles diminish the coefficient of contact for materials. PTFE, graphite and MoS_2 permit the preparation of self-lubricating parts. Here, PTFE, a polymer made in powder form, acts in terms of illustration as a nanoparticle to other polymers. Matte surfaced paint will be used for silica nanoparticles.

Nanoparticles lessen shrinkage for polymer foams. Mica and glass fiber decrease warpage and build the high temperature twisting temperature. Similarly, intumescent nanoparticles expand to volume quickly as they corrupt, thermally extending the material and blocking shoot spread. Nanoparticles might be diminishing warm conductivity. The best encasing properties of composites are obtained with empty circular particles concerning Illustration of a nanoparticle. Conversely, metals, powders and other thermally conductive materials considerably increment the dispersal of warm vitality. Volume resistivity, static dispersal and other electrical properties might be impacted towards the decisions from claiming nanoparticle. Conductive nanoparticles for powder or fiber form, metal covered plastics and metal covered pottery increment the conductivity. A significant number of nanoparticles expand the electric resistivity. These are utilized within electric link insulations. Ionic conductivity can be altered eventually on perusing silica nanoparticles. Gas and fluid per manganic corrosive are impacted eventually on perusing the decision of nanoparticle. The platelet structure of mica, or alternately talc as a nanoparticle over paints and plastics declines the transmissions of gasses and fluids. All mechanical properties are influenced towards nanoparticles. Nanoparticle combinations if chosen, will streamline an assortment about mechanical properties. Nanoparticles strengthen and give acceptable abrasion imperviousness.

3.2 NANOPARTICLES IDENTIFICATIONS

The knowledge of execution qualities of nanoparticles leads us to an ID number for nanoparticle properties which permit distinctive nanoparticles to be compared and assessed. Every last bit of materials talked about is solid. However, they may be accessibly clinched alongside a pre-dispersed state. Nanoparticles may be inorganic or natural and from claiming a created compound arrangement. They may additionally be an absolute element, common products, mixtures from claiming separate materials for obscure proportions (waste and reused materials), or materials of a proprietary creation, spherical, cubical, irregular, block, plate, flake, fiber or mixtures for diverse shapes. Nanoparticles extent can be from couple nanometers to many millimeters (nanocomposites with pavements alternately textured coatings). They can be monodispersing outlined mixture from claiming sizes, gaussian distribution or unpredictable conveyance. From 10 to 400m2/g, contingent upon the particular surface zone particles, bring separate levels for porosity from totally non-porous and smooth birch to exact porous with a range of pore sizes, arranged from 2 (carbon black) on 12 (calcium hydroxide). Whole varieties are workable between non-conductive and conductive, as well as attractive and non-magnetic properties. These different properties about nanoparticles are used to portray distinctive results. The possibility provisions of nanoparticles have a dead set, eventually, perusing its set from claiming properties recorded over, but often different aspects must be known to select the ideal nanoparticles or alternate nanoparticles to particular requisition.

3.3 PARTICULATE NANOPARTICLES

3.3.1 Aluminum Flakes and Powder

The engineering organization of processing of aluminum powders and flakes started when a sheltered transform of fabricate has been produced in one of the corridors of Columbia college. This strategy is at present utilized today for making the majority of pigments. The guidelines of fabricate are in of view of wet ball processing aluminum in the vicinity of an ointment and mineral spirits (Tavman, 1996; Yang & Schruben, 1994).

3.3.2 Aluminum Borate Whiskers

It has been discovered that the aluminum borate whisker is a standout amongst the least expensive ceramic whiskers. Therefore, it is used to strengthen aluminum grid composites (Zhu & Iizuka, 2003). When aluminum borate whiskers are consolidated, the whiskers because of acid/based association will do well to natural inclination with a motivation behind being finished, dependent upon over home in polyamide stage (Persson & Bertilsson, 1996; Zhu & Iizuka, 2003).

3.3.3 Aluminum Oxide

Recalcitrant evaluations bring extensive molecule sizes in the extent of 5-25 μm and precisely low surface zone toward 0. 3-1 m^2/g. Their particular gravity will be secondary at 3. 95 g/cm^3 as previously demonstrated. Calcinated alumina are prepared by the layer calcination procedure from aluminum trihydroxide in rotating kilns. Throughout the process, water is uprooted, and a stable α-alumina structure is acquired. The molecule size for calcinated evaluations is comparative to recalcitrant evaluations, unless they are processed. More diminutive molecule span evaluations have a particular surface zone of 3-10 m^2/g. Actuated alumina bring molecule sizes in the range from claiming 6-80 μm, yet altogether substantial particular surface zones in the extent of 220-325 m^2/g. They could promptly absorb water to harmony through 18-22%.

The evaluations generated towards nano-phase advances Cor have aided in uprooting an engineering manner towards vanishing the metal and its ensuing oxidation. This method produces standard round particles as indicated in Figure 1 (Ford, 1998). These materials have properties which can't be duplicated, eventually perusing routine evaluations for alumina taken from minerals and alternately perusing compound amalgamation. The nanoparticles upgrade mechanical execution of plastic materials (tensile, hardness, wear, and so forth across this way, observing and stock arranging of all instrumentation that can be enhanced.). The hardness from claiming compacted pottery increments as the molecule span abates, and it there may be time permits on acquiring materials which allow for a respectable light transmission. These materials are in the business currently. Furthermore, huge numbers of high tea requisitions will be discovered (Grohens et al., 1997).

Figure 1. SEM of NanoTec aluminum oxide (Ford, 1998)

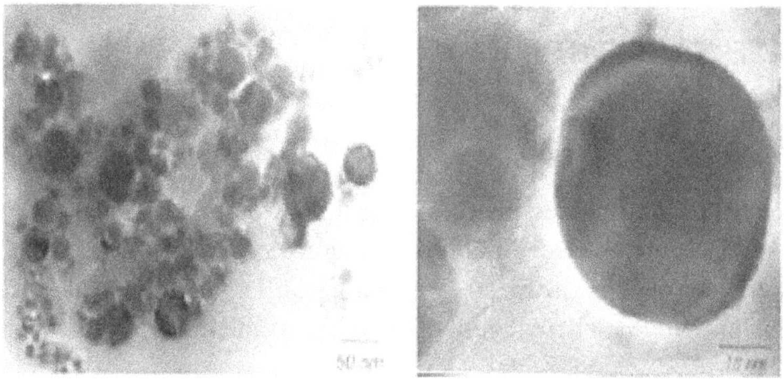

3.3.4 Aluminum Trihydroxide

Properties about aluminum Trihydroxide are demonstrated. The preparations transform to aluminum trihydroxide can be viewed as a spinoff of aluminum metal generation in the main phase, in which the metallurgical review about aluminum trihydroxide is processed (Dayer & Mead, 1998). At the same time, this evaluation holds various impurities and obliges purification. Nanoparticle review generation is separate from the preparation of metallurgical review and yields an immaculate aluminum trihydroxide. Two popular properties make aluminum trihydroxide thick: Its fire retarding abilities and its low absorption of ultimo Violet(UV).

The low absorption from claiming UV renders it a suitable material to provisions done to UV reparable materials. Its fire retarding action may be because of cooling, obstruction layer formation and weakening. The cooling ability of aluminum trihydroxide arises from its capability with arrival water at raised temperatures with top discharge of around 300·C. The response toward itself is endothermic, and in addition, water must have a chance to be dissipated to expend extra heat vitality. Aluminum trihydroxide which is afterwards decomposed, is a type of boundary which slows the stream from claiming oxygen and framing for gasses. Expansive amounts (e.g., 150 phr) about nanoparticle must be used to get fire retarding properties (dilution factor). This allows fire retardancy to influence the mechanical and rheological properties of materials. Since the sums of nanoparticle cannot be altogether reduced, additives, such as exacerbates for zinc are utilized for considering a portion diminishment in all $(OH)^3$ fixation.

Mechanical properties are enhanced, eventually perusing the morphological tenet and surface covering of the nanoparticle. Evaluations are accessible and can be utilized for large portions of plastics without a dread of debasing their mechanical execution. The issue about rheology for materials throughout preparing and use tends to eventually peruse the change of the morphological tenet for particles and additives which assist to decrease viscosity. Figures 2 indicates how morphological tenet can be customized to enhance viscosity. Figure 2(a) indicates a precipitated review which is created from claiming blocky round particles. The watchful determination of a proper molecule span appropriation for these morphologically separate species brings about a low viscosity material. Figure 2(b) demonstrates another review which has platy particles that provide for a higher viscosity (as could be expected).

Figure 2. SEM of aluminum trihydroxide (Dayer & Mead, 1998)

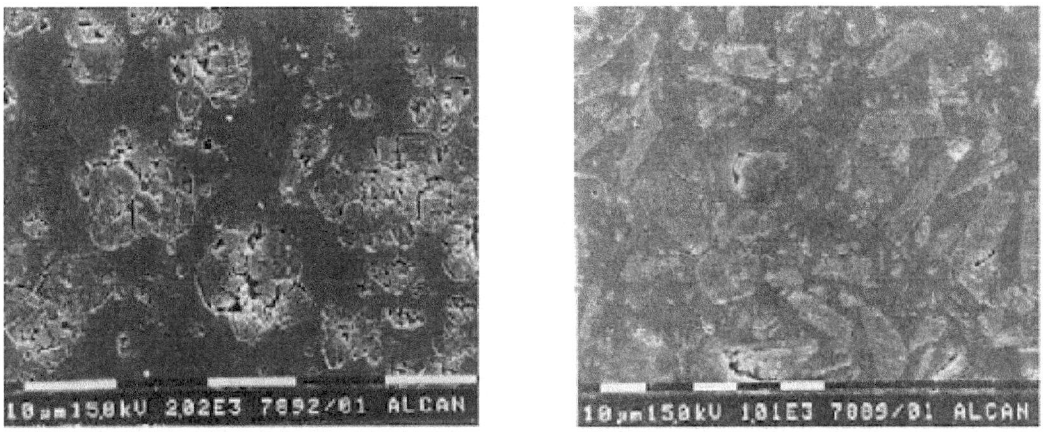

(a) Decreasing viscosity. (b) Increasing viscosity.

3.3.5 Anthracite

Anthracite abounds are minerals and can be cost-effectively mined and grounded. It has been discovered (Kretzschmar, 1996) that materials holding it have enhanced strength, stiffness, natural stress cracking, heat redirection temperature, antistatic properties, weathering resistance and concoction safety regardless of loading for generous amounts of anthracite (up will 60 wt.%). The hindrances have color, flow ability for melt and expanded dampness absorption. Person real playing point causes developing enthusiasm. The vast majority of nanoparticles utilized today are non-combustible and they remain as powder when plastic materials are incinerated towards the limit for a few reusing operations. Anthracite is a low powder substance and gives calorific quality.

3.3.6 Sodium Antimonate

Properties of sodium antimonate allow it to be utilized with halogen, holding mixes to it to go about as viable shoot retardant. The hotspot from claiming chlorine might hail from polymer (e.g., PVC, chlorinated rubber, and so forth through this way, observing and stock arranging of all instrumentation may be enhanced), or other chlorinated alternately brominated material. The profits of utilizing sodium antimonate through antimony oxide incorporate its low tinting quality and the corrosive searching proficiency. For these reasons, it can be utilized within semi-opaque or dull shaded materials and over polymers, such as polyesters and polycarbonates which are corrosive delicate.

3.3.7 Antimony Pentoxide

Antimony pentoxide is an alternative to antimony trioxide. It figures provisions done with semi-transparent materials and dim shades due to its low tinting quality. Similarly, as for antimony trioxide, antimony pentoxide can be utilized together with halogen-containing exacerbates in capacity, concerning illustration as a fire retardant. The opposite points of interest of antimony pentoxide incorporate its refractive file which is closer to practical materials, its exceptionally little molecule size, its secondary particular surface area and its considerably bring down thickness. Due to its little molecule size, it is utilized a lot within the material industry, since its expansion has a better impact for shade, alternately ahead of mechanical properties. Preparation of fine-denier fibers requires a stable scattering and a little molecule extent nanoparticle. The fire retardancy of laminates is enhanced with antimony pentoxide. As a result, little particles are less demanding to fuse in the interfiber spaces the sum of what has been demonstrated. Antimony pentoxide, as added substance for plastic materials, such as polyolefins and ABS, will be handled in a pre-dispersed structure holding halogen exacerbates and a polymeric folio which has a low liquefying list to help consolidation.

3.3.8 Antimony Trioxide

Antimony oxide is normally generated starting with stibnite (antimony sulfide) or eventually perusing oxidizing antimony metal. Numerous hypotheses endeavor to clarify the component from claiming fire retardancy. The fire retarding movement may be thought of as occurring in the vapor stage over the smoldering surface. For antimony oxide to work, the halogen and antimony oxide must create a vapor period which will occur in temperatures over 315°C. At these temperatures, antimony halides and oxy-halides are structured and go about as fire smothering moieties, eventually perusing quenching radicals concerning illustration of this type. The tinting quality relies on molecule measure. Whether molecule sizes are beneath 300 μm, they fall underneath unmistakable reach. Over this value, tint quality declines as the molecule size expands. The high tint quality review generally has molecule sizes reaching 1. 1-1. 8 μm and the low tint quality evaluation has molecule sizes over an extent of 1. 8-3 μm. The tint quality might additionally cause influence, eventually perusing crystalline manifestation. The orthorhombic type declines tint quality. Antimony oxide might be profitably consolidated for huntite/hydro- magnesite nanoparticles which offer phenomenal fire retarding properties. (Toure, Lopez Cuesta, Gaudon et al, 1996; Toure, Lopez Cuesta, Longerey et al, 1996) Furthermore, zinc borate can be used to lessen the measure about antimony trioxide. Other execution enhancing additives incorporate zinc stannate and ammonitic octamolybdate (Herbert, 1996).

3.3.9 Apatite

Properties and provisions from claiming apatite are indicated in (Nagieb & El-Sakr, 1997). Due to its fantastic biocompatibility and bone holding capacity (Labella et al., 1994), apatite (main HA) has not just been utilized as a bone substitute material, but also a nanoparticle about composite biomaterials with natural polymer grid such as polylactide (PLA), chitosan, collagen, polyhydroxybutyrate (PHB) (Yu et al., 2005), polytheretherketone (PEEK) (Ni & Wang, 2002) et cetera concerning illustration.

3.3.10 Ash, Fly

Fly cinder has ended up being extensively utilized as a modest nanoparticle. It is not utilized in extensive amounts during the available time. Examination investigations (Rebeiz et al., 1996; Wong & Truss, 1994) show that materials could move forward when fly cinder is utilized as a nanoparticle. The real obstruction is wellbeing and security, since fly cinder holds crystalline silica.

3.3.11 Barium Metaborate

Barium metaborate is a multifunctional added substance which inhibits corrosion, increments UV stability, inhibits shape growth and has fire retarding properties when utilized in consolidation with halogenated materials. The business items in Buckman Laboratories is an altered result which holds 90% for animated elements.

3.3.12 Barium Sulfate

Barites are the most common normal barium minerals, discovered clinched alongside immaculate manifestation as well as together with many different minerals. Most of the incessant substitution cost of barium is that of strontium and alternately radium. Barium sulfate, generally utilized within the industry and in medicinal applications, starts from regular barites and manufactured materials.

The personal satisfaction of the nanoparticle relies on the purity of the material utilized for handling and the system for preparing. The simplest system for preparing incorporates grinding and dry arrangement. Better results are achieved by concentration, wet grinding, bleaching and order. The result from claiming most elevated caliber is blanc fixe (permanent white). It is transformed starting with the response between barium carbonate and sulfurized. Since the main different response results are water and carbon dioxide, item purity relies on the caliber for crude materials utilized. Figure 4 demonstrates at present the better review, formed eventually perusing which has a molecule measure comparable to titanium dioxide (0. 35 μm). This may be an exactly uncommon nanoparticle which has a center settled on crazy manufactured barium sulfate (an insulator) covered with a semi-conducting layer for antimony doped for SnO_2. This material has high brightness, electric conductivity and light transparency over slim coatings. The material can be used to dispense with static charges starting with plastics and painted surfaces. During give or take, 19% PVC material has a permeation edge and surface resistivity that drops quickly, eventually perusing 8 requests from the claiming extent. Sachtoperse is still more diminutive, clinched alongside molecule size starting with 0. 2 μm and beneath 0. 1 μm, relying upon evaluation. This is utilized as a nucleating added substance to polymers, such as pet. It abates cycle occasion and lessens the transforming temperature, builds crystallization rate and keeps flocculation of pigments. Figure 4 demonstrates the component of which Sachtoperse keeps pigment flocculation. Pigment particles (lighter particles) stick to Sachtoperse (smaller darker particles) which go about as a spacer. This methodology brings about brighter shades and moved forward gloss. When pictures for engineered evaluations are compared to picture about ground barites as shown in Figure 5, the morphologic contrasts become clear.

These contrasts are not basically for molecule size and conveyance, as well as fit as a fiddle of particles (Bijwe, 1997).

Figure 3. TEM micrograph of Sacon.

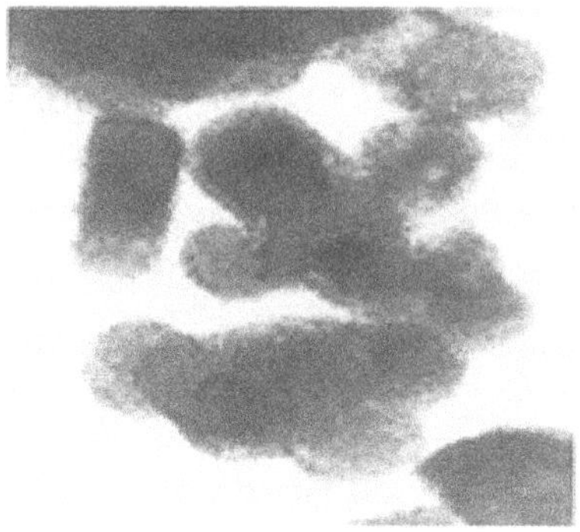

Figure 4. Anti-flocculating action of Sachtoperse, Huberbrite.

Figure 5. SEM micrograph of milled barite, HU.

Chemical composition is an alternate component which determines quality, especially to compound and therapeutic requisitions as well as in paints and coatings and in turn influences brilliance. Barium is profoundly lethal in any case of water dissolvable salt; therefore, for each requisition, the water solvent barium must be regulated. Other ordinary admixtures hold iron, copper, manganese and lead, rely upon application and their fixation may be additionally confined. Characteristic results hold 94-99% BaSO4, while blanc fixe holds start with 97. 5 to over 99%.

For a few applications, a refractive file is imperative to match the middle of the molecule span of a portion. Barium evaluations and the refractive file from claiming grid material permits the plan from claiming results with alluring optical properties. An arrangement from claiming engineered barium sulfates can be handled by Sachtleben Chemie which has molecule sizes between 4 And 10 μm. The molecule extent of these barium sulfates can be greatly facilitated with the refractive file of the grid polymer and semi-opacity consolidated with translucency effects. This permits the detailing of a light disperser over lampshades alternately over enlightened promoting shows. The right molecule measure can be computed starting with the equation: d = (100n - 141)/2, in which n is the refractive file of the tar and d is the molecule extent from claiming barium sulfate. Barium sulfate has found a significant number of requisitions mostly due to its exceptional compound safety and dormancy (For example, it is not influenced by corrosive rain). The opposite purpose behind its incessant requisition is high absorptivity of light and, significantly, x-beams (for use clinched alongside X-beam perceivable materials).

3.3.13 Barium and Strontium Sulfates

Strontium sulfide (SrSO4) is the sulfide salt of strontium. It is a white, odorless, crystalline powder and has come to be known as the mineral Celestine. It is dissolvable clinched alongside water in the degree from claiming 1 a piece in 8,800. That's only the tip of the iceberg solvent of weaker HCl and nitric corrosive and appreciably dissolvable with soluble base chloride results (e.g. Sodium chloride). Numerous strontium compounds and strontium sulfide can handle a brilliant red fire at smoldered. For this reason, it is utilized as a colorant in pyrotechnics. It can also be utilized in pottery (Patnaik, 2002).

3.3.14 Barium Titanate

Barium titanate (BT) has become a standout amongst the mossy cup oak paramount electro ceramics since discovering its flexible multilayer ceramic capacitors (MLCC), certain temperature coefficient for safety (PTCR) thermistors, piezoelectric sensors, transducers, actuators, ferroelectric irregular right memories (FRAM) and electro-optic units (Gregorio et al., 1996; Haertling & Am, 1999; Uchino, 2000).

3.3.15 Bentonite

Bentonite is a clay inferred starting with the weathering from claiming volcanic powder and made of the mineral montmorillonite. There are two varieties: sodium bentonite which has secondary swelling limit in water and calcium bentonite for unimportant swelling ability. Figures (6, 7) indicate the morphological tenet of ground mineral and the purified material, the secondary surface range and a structure which permits water to infiltrate mineral layers and are answerable for the swelling abilities from claiming bentonite clays. Regarding its conventional use previously in paints as viscosity regulator, bentonite can be utilized at present in the improvement of new materials for nanocomposite structures (Visser et al., 1997).

Figure 6. Bentonite ground ore (Visser et al., 1997)

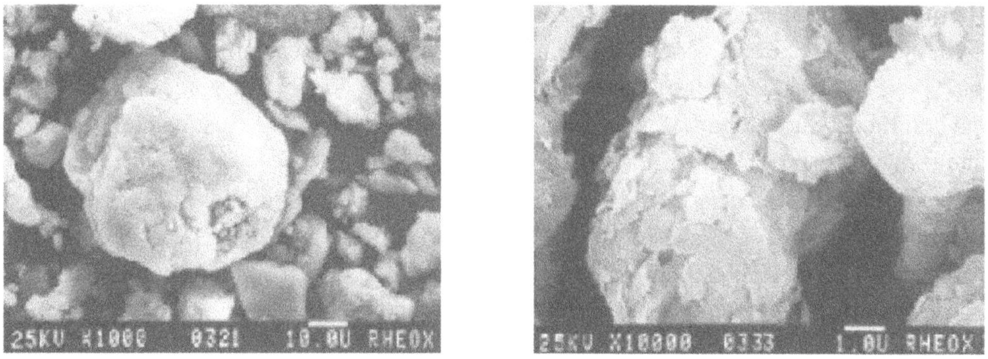

Figure 7. Bentonite purified and spray dried (Visser et al., 1997)

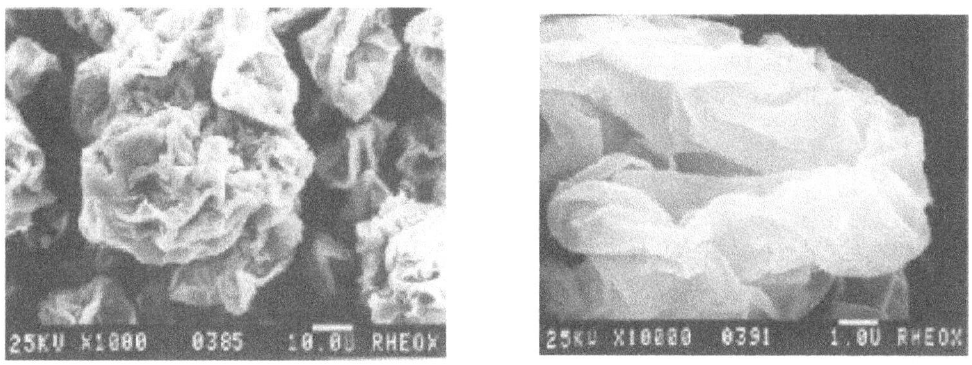

3.3.16 Beryllium Oxide (BeO)

Beryllium oxide can receive specific enthusiasm because of its piezoelectric (Noel et al., 2002), radiance and photochemical properties and its fascinating mechanical properties in blending with the heat conductivity to make BeO, an actually guaranteeing ceramic material (Haines et al., 2001; Kiiko et al., 2007).

3.3.17 Calcium Carbonate

Calcium carbonate is a practically and generally utilized nanoparticle. Its utilization can be connected with a generous cost diminishment and nowadays, it is the material engineered for the separate prerequisites from claiming cutting edge results. This review starts with a presentation of the sources of calcium carbonate which have been provided for a careful assessment of a paper by Bosshard about Omya/Plüss-Staufer ag. calcium during 4.8%. It ranks fifth among the basic natural constituent of the earth's outside after oxygen, silicon, aluminum, and iron. In this way, it is prominent in commonsense requisitions and as a result, it is can be found in done rocks and minerals which bring high focus on calcium carbonate. Calcium carbonate is the major regular store shaped clinched alongside sedimentary rocks. The methodology of shaping from claiming calcium stores starts with weathering of land surface because of the transforms in heat, frost, rain and the impact of sun. Calcium carbonate is not promptly dissolvable in water, yet all the calcium bicarbonate is. The fixation for carbon dioxide in water is, therefore, significant to calcium carbonate transportation starting with the territory of the ocean, since downpour water can be the transporter. It is assessed that 500x106 tonsil from claiming minerals are conveyed by waterways of the oceans consistently out, about which around 10-15% for sedimentary rocks holding calcium carbonate are shaped.

The dissolvable type of calcium can be precipitated in the marine surroundings to structure rock by a few physical states, such as warming of the water (carbon dioxide is less dissolvable in warm water than in frosty water, and consequently calcium carbonate can be precipitated), eventually perusing the utilization of carbon dioxide, perusing marine plants, or eventually perusing alterations in the ph in water and eventually perusing ammonia-producing microscopic organisms which lower the solvency from claiming calcium carbonate. However, the greater part for calcium carbonate stores are framed starting with skeletal pieces of creatures existing in the marine nature's domain. This is exactly from claiming these creatures possess reefs, yet the larger part coast is free clinched alongside water. Figure (8) reveals different shapes of shells framed towards Coccolithophorides which might be round coccospheres in some cases, for example, such as dicoaster which is star formed (Dimic-Misic et al., 2014).

Most of the concerns regarding worldwide warming have been directed to territory-based plants. It can be seen in the preceding passages a maritime change of calcium carbonate, eventually perusing microorganisms and carbon dioxide by plankton and that's only the tip of the iceberg imperative in the regulation of our nature's domain. Episodes, such as a submerged volcanic blast might influence this offset, since they modify the temperature of water and the fixation from claiming carbon dioxide to water and, consequently, its interior utilization and arrival of the climate. Concerning illustration that has been specified before, few crystalline types can be transformed. These structures are used to raise minerals and rocks. These are characterized underneath. Three crystalline structures are basically utilized in generations for calcium carbonate nanoparticle: calcite (a mineral called calcspar which has trigonal rhombohedral or trigonal scalenohedral form) and aragonite (orthorhombic crystals).

3.3.18 Carbon Black

Carbon dark has the longest history from claiming every bit of materials talked about here. Just in the most recent 50 years, a considerable length of time has witnessed mechanical advancement, both in carbon dark processing and transforming of elastic and polymers bringing about the enormous mixed bag of results which we recognize today (Severin et al., 2000).

3.3.19 Ceramic Beads

Ceramic spheres are generated starting with nepheline syenite, aluminum oxide and bentonite or fly cinder. Ceramic spheres have considerably higher densities over glass or polymer globules. They are less expensive, more inflexible and mechanically safe because of their thicker dividers. They have quality from claiming every last bit of circular materials which provide for the most astounding pressing thickness and move forward stream due to the ball-bearing impact (Baque, 1996). In addition, ceramic spheres decrease dielectric constant, warpage, shrinkage, and move forward split safety of speckling exacerbates as previously demonstrated.

Figure 8. Different shapes of coccoliths found in Omya mines. (Dimic-Misic et al., 2014)

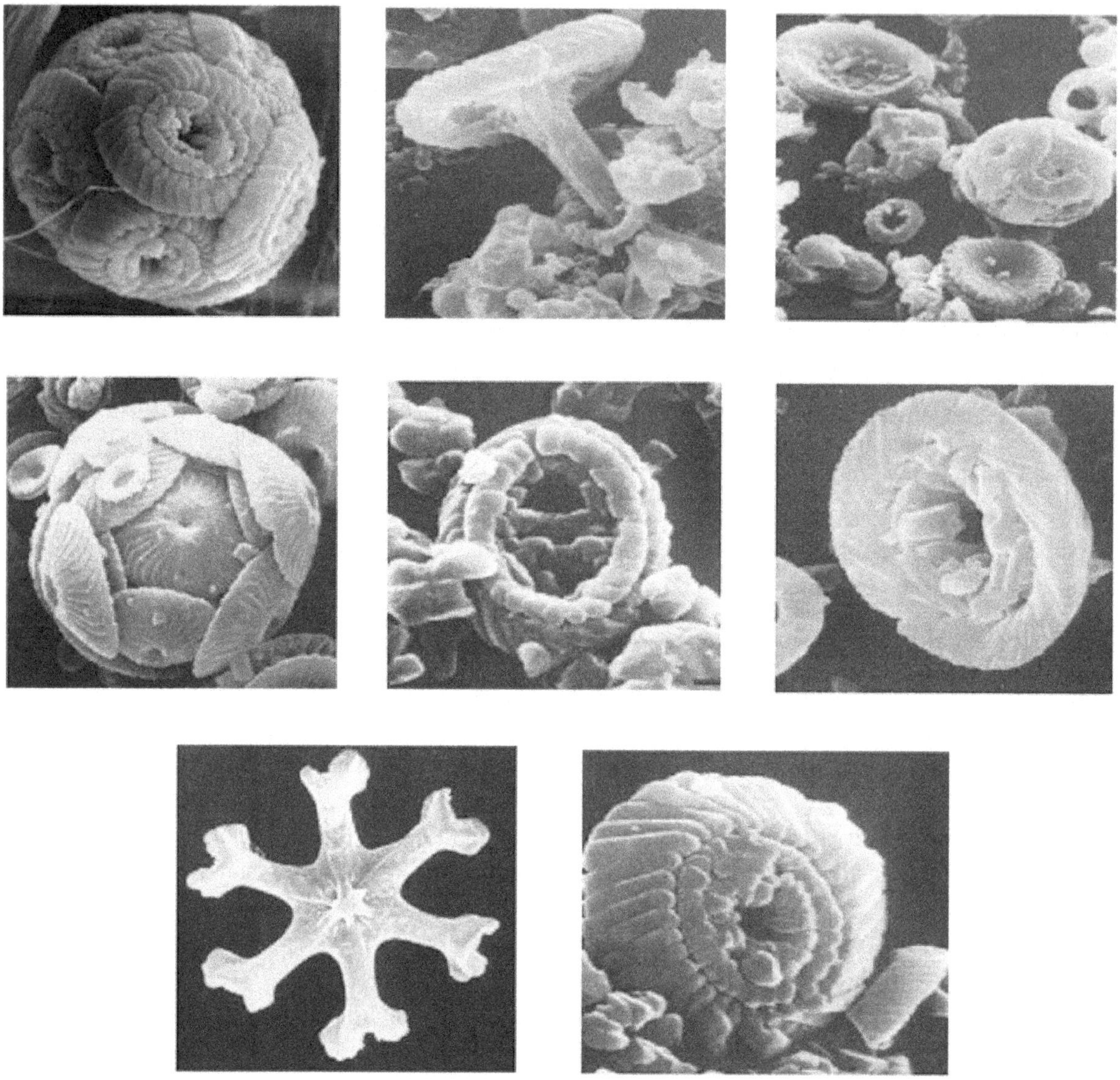

A basic equation permits us to figure the measure from claiming globules that had to displace a nanoparticle for higher density: The amount of globules = (density about beads/density for nanoparticle) × measure for nanoparticle to creation. Acknowledging that globules bring a finer pressing thickness over the nanoparticle supplanting and transforming an easier viscosity in the material, that's only the tip of the iceberg. Globules will be ascertained from comparison and yet uphold the same viscosity in the material. Despite ceramic spheres are additionally inflexible over glass spheres, they still require extraordinary precautions through taking care of blending. A high shear and prolonged blending ought to be avoided during their consolidation. Ceramic globules ought to be included at the conclusion of the blending procedure. Figure 9 indicates the morphological tenet about ceramic globules which are created from a mixture from claiming circular particles. The exceptional globules generated towards

Kinetic has a denser shell to provide for them a mechanical quality and a porous inner part to decrease their thickness (Figure 10).

Figure 9. SEM micrograph of Zero spheres. (Baque, 1996)

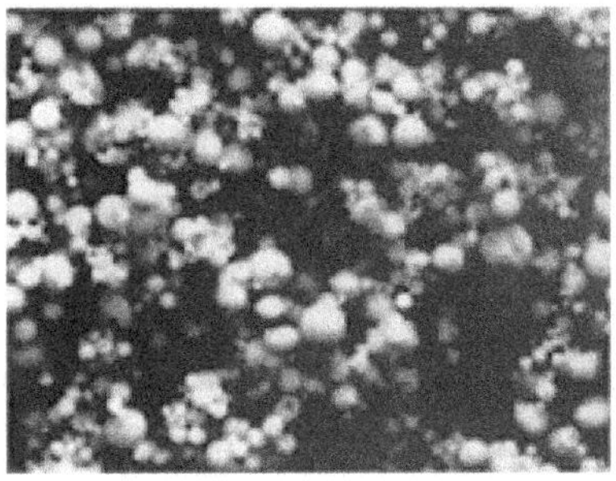

Figure 10. SEM micrograph of Macro lite. (Baque, 1996)

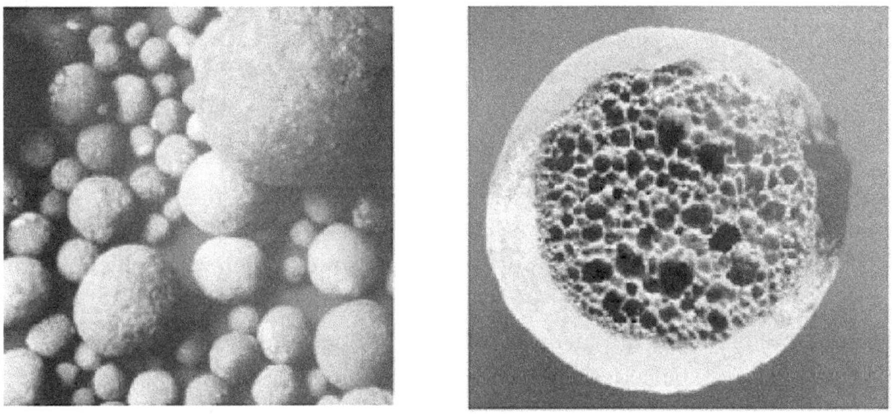

(a)Choice of sizes (b) Cross-section

3.3.20 Clay

The well-known nanoparticles, such as kaolin clay, china clay, bentonite, fuller's earth and vermiculite are all clay minerals. Clay minerals can be divided into 5 groups: the kaolinite aggregation (including kaolinite and halloysite), the illite gathering (including illite), the smectite assembly (including montmorillonite and hectorite), the palygorskite gathering (including sepiolite and attapulgite) and vermiculite bunch. Palygorskite for vermiculite are precursors of clay nanoparticles (Evans, 1996).

3.3.21 Copper

Copper powder undergoes oxidation at the point when it contacts with air through the cooling procedure. There are tempered evaluations accessible, in which the surface oxides are lessened by hydrogen of the immaculate copper. There are four types of copper powder: electrolytic (irregular porous particles, alternately dendrite formed aggregates of more diminutive particles), chip (made from machining), round (gas atomized which comprises round particles) and spheroidal (water atomized hosting lengthened particles) (Belanger et al., 1996).

3.3.22 Cristobalite

Cristobalite is a polymorph about quartz. Its importance lies in that it is created of the same chemistry, SiO2, at an alternate structure. Both quartz and cristobalite are polymorphs of quartz assembly. Cristobalite may not be found in additional amounts to regular wellspring. For business purposes, it may be synthetically prepared starting with sand towards warming in furnace to 1500°C. The resultant white powder can be utilized regarding illustration of a nanoparticle, or it may be micronized, and surface dealt with. The more significant properties of cristobalite are its whiteness and sturdiness with respect to presentation with ecological states. Items made by Quarzwerke GmbH are dealt with by the accompanying silanes: epoxy, methacrylate, trimethyl, and methyl silane (Hofmann & Skudelny, 1990). Few crucial properties about cristobalite have impact around its provisions. They incorporate bringing down thickness over quartz, purity, really low moisture, immaculate white color and less abrasive because of nanoparticle molecule morphological tenet.

3.3.23 Diatomaceous Earth

Diatomite is a pasty sedimentary rock created from skeletal remains of diatomites. Diatomites aresingle-cell oceanic plants existing in the seas. There is an incredible assortment about diatomites as indicated in figure (11). The micrographs indicate the convoluted structure of diatomite which demonstrates their secondary porosity, and hence the impact they have on gelling about fluids and around rheological properties. It is assessed that there are more than 25,000 species of diatoms (Allen et al., 1997). Due to its properties as previously indicated, the diatomaceous earth nanoparticles assume few roles, such as rheological additives (absorbing fluids over pores in increment viscosity on their standing and arrival around mixing) and flatting operators. They are handy to expand the rate from claiming paint drying (porous nanoparticle helps evaporation), to enhance sanding properties, to increment mechanical bond of coatings and to decrease the measure of TiO_2 necessary to prepare whiteness alternate obscurity in a material. Because of their concoction inertness, these nanoparticles don't meddle with alternate segments of the mixture. The evaluation relies on the surface smoothness required, the level for flatting and the sort of scattering supplies utilized. This is paramount for deciding other co-nanoparticles. It is, for example, indicated that consolidation for diatomaceous earth and talc gives paint for useful properties.

Figure 11. SEM micrographs of diatomite. (Allen et al., 1997)

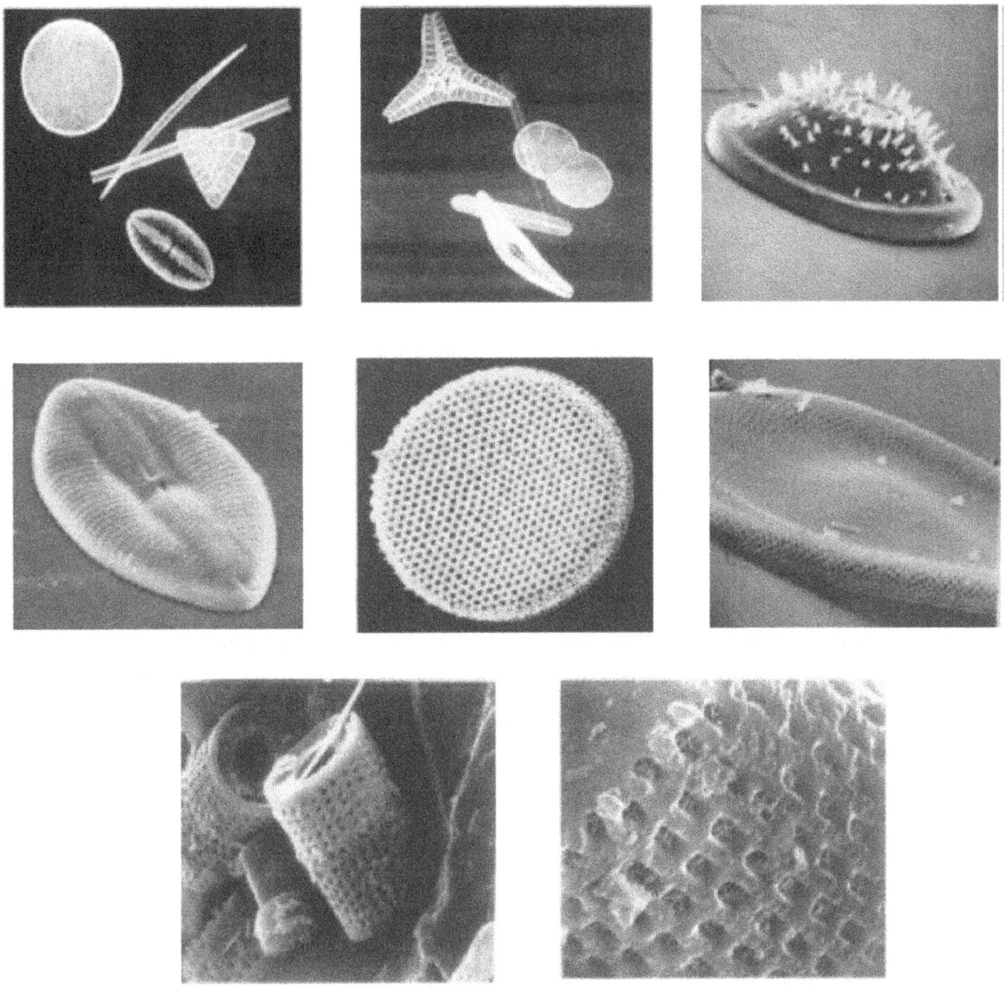

3.3.24 Dolomite

Dolomite is an essential recalcitrant crude material and comprises a mixture of calcium carbonate ($CaCO_3$) and the first mass of magnesium carbonate ($MgCO_3$). Then after calcination, the material decomposes in lime (CaO) and Pericles (MgO). Sintered dolomite (doloma) additions are significant in the steel-making business because of its high thermomechanical and disintegration wear properties (Ghosh, 2001).

3.3.25 Ferrites

Cortex Biochem has been found to grant intriguing requisitions to magnetizable particles for explanatory fields. Particles of the systematic help start with a blending for magnetizable materials (iron oxide) and absorbing material (e.g., charcoal, polyacrolein, particle exchange, cellulose). The particles are scattered, clinched alongside a living test on specifically absorb-obliged exacerbates. After absorption is accomplished, particles with consumed substance are uprooted starting with a result by a charged Pole

(Klapcinski et al., 1995). The materials are utilized for detachment from claiming enzymes, protein, units and alternately microscopic organisms.

Figure 12. Dicaperl HP-510: Magnification 1800x.

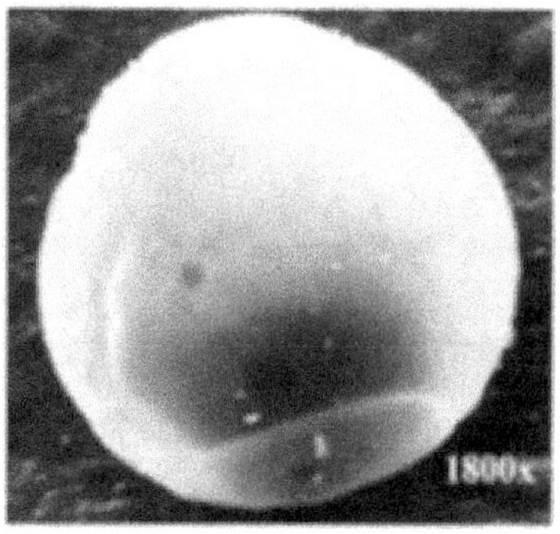

3.3.26 Glass Beads

The transforming engineering determines the choice of the glass air pockets. They can minimize the breakage for air pockets, if included in the wind of the transform. Low shear and high stream mixers are obliged to get full reductions. Glass globules enhance or control few properties of materials. These incorporate thickness reduction, stream properties, viscosity decrease, rheological properties, including thickening and non-sag properties, nailing, sanding, shrinkage reduction, effect strength, stiffness, pliable strength, flexural strength and hardness, as well as explosives performance. A significant characteristic of glass globules is their capacity to decrease the thickness of an item. There can be tradeoff in the middle of the mechanical properties from claiming globules and their thickness. Globules are precisely and essentially delicate, as a result their dividers are exactly dainty, and the manufacturing routines are constrained. However, they can be effectively consolidated by bringing about a considerable decrease to item thickness. Globules are mechanically safe, so they bring thicker dividers and they don't diminish thickness of flawless polymers. Figure (12) indicates the morphological tenet of a solitary empty glass

Figure 13. Micropearl solid glass beads.

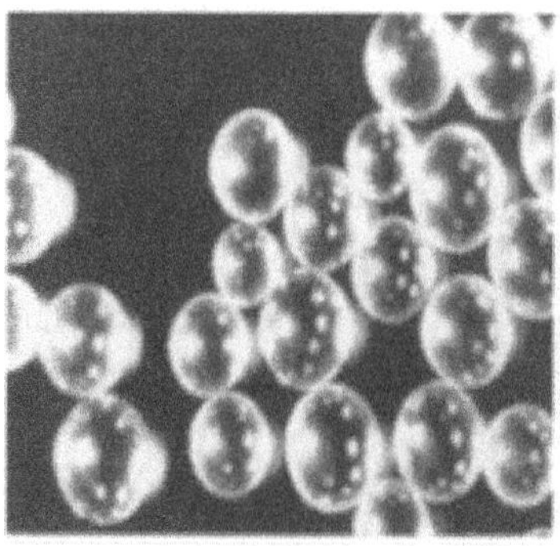

circle which has a general circular state. Figure (13) indicates the morphological tenet for strong glass globules (Sjogren & Berglund, 1997; Wang & Ploehn, 1996).

Figure 14. Gold powder (magnification 5250x) and thin (550) and thick (555) flakes (magnification 3200x) (Gonsalves et al., 1996; Kubat et al., 1993; Söhnel & Garside, 1992).

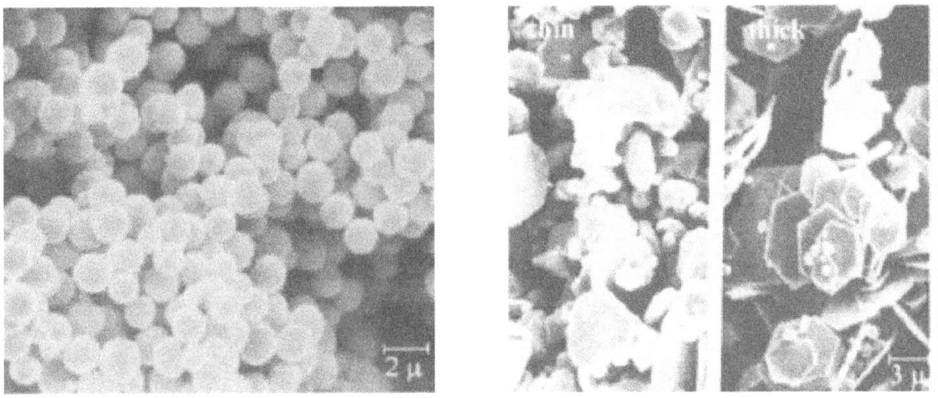

3.3.27 Gold

Gold is and exceedingly practical metal associated with a considerable measure of motivation. Particularly in the colloid chemistry, gold is known ASA standout amongst exceptional metals to structure stable nanoparticles. It has different requisitions of the electronic, magnetic, reactant and warm properties which realizes a new viewpoint in science and an innovation in organization (Gonsalves et al., 1996; Kubat et al., 1993; Söhnel & Garside, 1992). The gold nanoparticles bring high solidness with erosion and low strengthening temperature because of the high surface vitality of nanoparticles. Fig (14) indicates the morphological tenet of gold powder and gold flakes.

Figure 15. TEM micrographs of NanoTec iron oxide. (Jang & Yi, 1996)

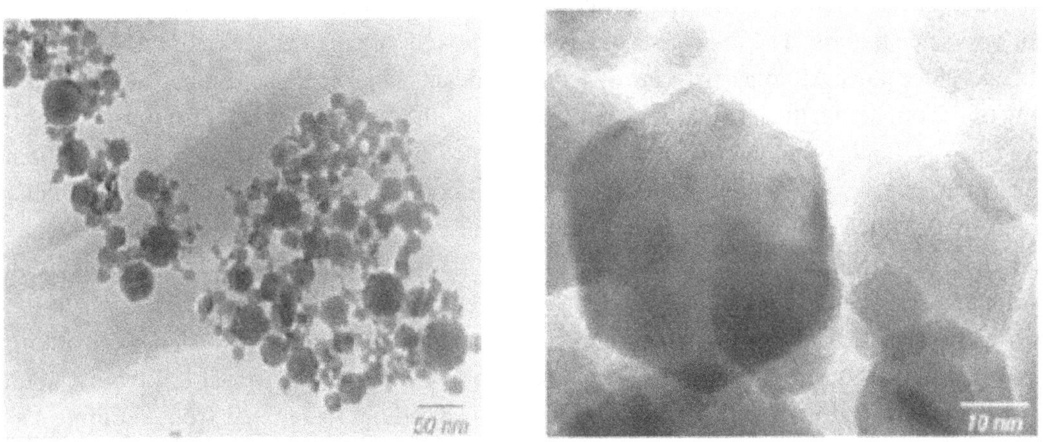

Figure 16. SEM micrograph (Coussot et al., 1996).

(a) Before (b) Kaolin. (c) Calcinated kaolin.

(d) Optiwhite, thermo-optic grade. (e) Huber - structured pigment.

3.3.28 Iron Oxide

Iron oxide is a specific sample of the extensive variety of materials which can be obtained from grinding the common result or union (Jang & Yi, 1996). Figure (15) indicates the morphological tenet for nanoparticle iron oxide.

3.3.29 Kaolin

Kaolin is an item of the decay from claiming granite and white feldspar. The average characteristic of kaolin is its amazing fineness. In the last two centuries, china clay has become prevalent and it is presently the most renowned starting with the United Kingdom, then north ocean oil. Handling of china clay starts for vast scale mining. The mined mineral is a component of china clay, 3 parts of rock, 4 parts of sand and one-piece mica. Alternate parts are differentiated, and whichever used or disposed of. Following uprooting of rock material, the leftover portion is blended with water and passed through a hydrocyclone to uproot fine sand and coarse mica (Coussot et al., 1996). The methodology from claiming calcination significantly transforms the unique properties of the material. Warming kaolin over 450oC alters the clay structure and enhances electrical safety and brilliance. The procedure from claiming calcination is directed over kilns during temperatures between 850 and 1500°C. Figure (16) (c) indicates calculated kaolin which varies from dried kaolin by hosting round edges, which may be an aftereffect of the high engineering medication. Burgess Pigment is created as an additional system for kaolin medication called streak calcination transform. The methodology is led towards a whirling upward climbing stream for hot gas in the manifestation of vortex to which material is dried out to a matter from claiming seconds, framing the interesting morphologic structure and providing for the evaluation name of thermo-optic, as clear in Figure (16) (d). This material has brought about particular gravity and extremely handy concealing energy. Huber demonstrates other morphologic features of its organized pigment result which is the manifestation for porous aggregates with high brilliance, as is clear in Figure (16) (e)). Particles are created for stacks with structure aggregates closer clinched alongside state on circular particles.

Figure 17. SEM micrograph of nickel coated carbon fiber. (Rosenov & Bell, 1997)

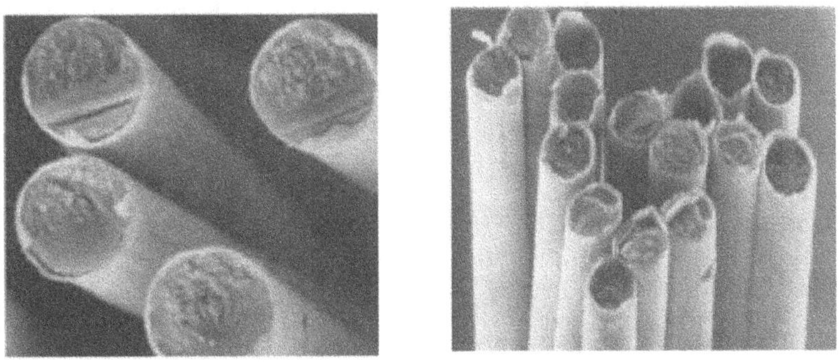

(a) Besfight (b) Compmat

3.3.30 Magnesium Oxide

Magnesium oxide (MgO) is a novel film that has secondary auxiliary electron emanation (SEE) yield for requisitions clinched alongside even board presentation and other gadgets (Kim et al., 2004). However, the quality from claiming view yield determinedly relies on the surface states and crystalline personal satisfaction for MgO film (Cazaux, 2003) with the goal of its use within nanowires provisions.

3.3.31 Magnesium Hydroxide

Magnesium hydroxide is a rising nanoparticle for fire retardant requisitions. In this area, it competes with aluminum trihydroxide, antimony oxide and other nanoparticles dependent upon zinc (Gutman & Bobovitch, 1996). Magnesium hydroxide has an alternate decay temperature from aluminum trihydroxide. Hence, it is suitable for polymers for higher decay temperature.

3.3.32 Metal-Containing Conductive Materials

Materials which are made of two different materials or include in their structure conductive fibers have been included here for less demanding examination. Metal covered spheres, flakes and fibers are made for different requisitions. Conductive plastics are practically normal. Nickel-coated graphite fibers are eventually produced, perusing American Cyanamid. These fibers consolidate the quality of fiber with the electrical and warm conductivities of nickel. The decision for nickel relies on the fact that it may be a generally modest metal for beneficial erosion safety. Typically, 3-5% fibers in materials provide for the static dissipating properties. Toho Rayon, co. further progressed the execution of the material, eventually perusing the utilization from claiming their engineering organization of carbon fiber manufacturing and an exact covering of a slim layers of nickel. Figure (17) indicates the morphological tenet about surface and the cross-segment for these fibers from two manufacturers: Toho Rayon, co. and composite Materials, l. L. C. The particular resistivity of nickel covered fibers is a special case request of an extent that is higher than nickel, yet the two requests for extent are more level over uncoated fiber (Rosenov & Bell, 1997).

Other substrates, such as graphite powder and mica are additionally covered with nickel. Silver is the major conductive metal, being practically 5 times less safe than nickel. Silver and copper have fundamentally the same conductivities, yet copper may be effectively oxidized and reacts with acids promptly, which influences its execution in polymeric frameworks. The PQ enterprise and Potters Industries, inc. have reached many results which are silver covered. Their provision is for conductive thermosets, gaskets, sealants, adhesives, paints, coatings, inks and EMI control provisions. Its use for these results recoveries represents 1/3 of weight for conductive material.

Figure 18. SEM micrographs of silver-coated flakes and spheres (Rosenov & Bell, 1997)

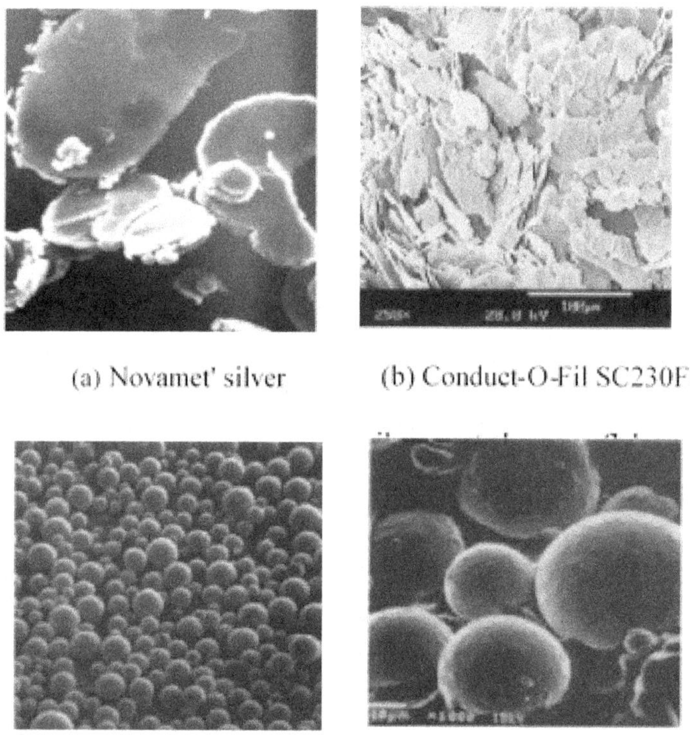

(a) Novamet' silver

(b) Conduct-O-Fil SC230F8,

(c.1) spheres at

(c.2) spheres in

(c) Conduct-O-Fil solid glass spheres coated with silver

Novamet invented an ideal to enhance the properties of nickel flakes, eventually perusing and covering them with 15% silver. The covered flakes have both conductivity and ferromagnetic properties. In addition, due to the contrasts over thickness (Ag - 10. 5 and Ni - 8. 9 g/cm^3), its time permits save 15%

of material, since conductivity is identified with volume as opposed to weight and surface conductivity is normally of grade significance. Figure (19) indicates the morphological tenet of a few conductive materials. Metal flakes start with silver covered nickel and copper flakes that bring unpredictable shapes, since they are framed starting with round particles which were Initially covered with silver and others and leveled, eventually perusing mechanical constrains.

Figure 19. morphology (Folie et al., 1997; Kody & Martin, 1996)

(a) muscovite (b) phlogopite xmica.

3.3.33 Mica

The mica bunch has around 30 parts and a relatable point. Muscovite, phlogopite and biotite are essential delegates from claiming this assembly. Muscovite is a standout amongst the vast majority of micas and is used for a total assortment of geological situations due to its soundness. Crystals measuring 2-3 m crosswise are mined over some areas. Muscovite differs in compound creation, due to an aftereffect of nuclear substitution (Na for K; mg and f for Al). Mica nanoparticles are obtained by division of mica starting with other minerals which might create 10-20% of mineral substance. Mica is dry or wet processed and arranged. Mechanical grinding produces flakes for a low perspective proportion in range from 20 to 40. The procedure might incorporate ultra-nationalistic delamination which prompts a high perspective proportion of another 200mm. Flakes of mica nanoparticles have a thickness for a range starting from 1 to 3mm and a width for a range starting from10 to 450 mm. The engineering of mica nanoparticle fabricate can incorporate surface preparation utilizing silanes, maleated polypropylene wax and amiidae acetic acid derivation. These forms incredibly upgrade support. Ultra-nationalistic delamination particularly gets to be all the more viable at the point when surface medicine is utilized. This is identified with expanded mica wetting, which is generally challenging contrasted with different nanoparticles. Surface coupling also significantly influences the imperviousness of the loaded polymer with water - a standout amongst the most of the fancied mica properties when compared with other nanoparticles (Folie et al., 1997; Kody & Martin, 1996).

Other reasons known for mica's incessant use are those that pertain to length remaining convention in the industry, which is identified with its high resistivity; the opposite will be its impact on warm extension. Composites, including mica, have a low coefficient of warm development tantamount to the individuals, including glass flakes. Additionally, mica is used to decrease shrinkage, warpage and can move forward rigidity and modulus, high engineering deflection and corrosive permeability. Figure (19) indicates SEM micrographs on muscovite and phlogopite mica. The morphologic characteristics for both structures are fundamentally the same.

Figure 20. SEM micrograph of molybdenum disulfide. (Shiao & Wang, 1996)

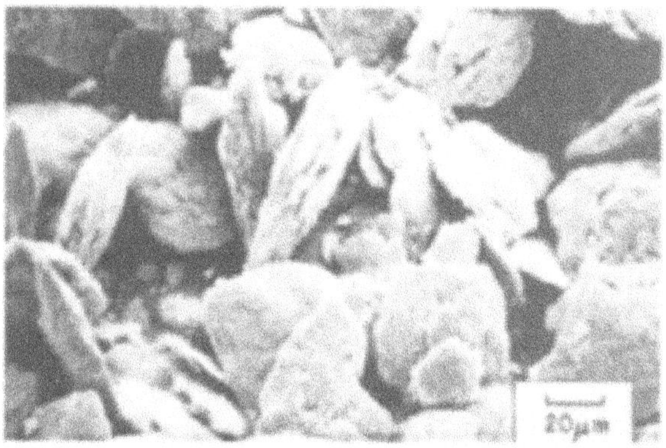

3.3.34 Molybdenum Disulfide

The intensify happens concerning illustration of the mineral molybdenite, which will be utilized following refining for illustration of lubricating material. The standard of activity for molybdenum sulfisoxazole is in terms of framing of bonds between metals and sulfur microorganisms. These bonds slip under shear powers and will ceaselessly improve considering the lubricating film on the surface of the metal (Shiao & Wang, 1996). Figure (20) reveals the morphological tenet about specialized foul evaluation of molybdenum disulfide.

Figure 21. INCO nickel powder and. (Fiske et al., 1997)

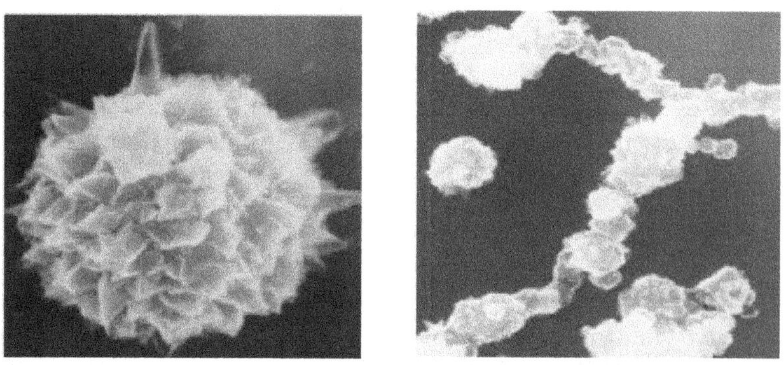

(a) Single (b) Chain

3.3.35 Nickel

Nickel is a conductive with ferromagnetic properties and it is generally an inactive material. INCO produces nickel powders, eventually perusing warm decay of nickel carbonyl in a procedure which produces a fine molecule metal powder with a spiked alternately dendritic surface, as demonstrated in Figure (21). The micrograph on the left-hand side reveals a solitary molecule from claiming review of 123. The morphological tenet of types 255, 270 and 287 can be indicated in the figure on the right-hand side. The dendritic particles are associated with one another (to structure a chain of a regulated period and porosity (Fiske et al., 1997). Figure (22) shows the morphology of two grades produced by Novamet: flake powder and spherical material.

Figure 22. Novamet (Fiske et al., 1997)

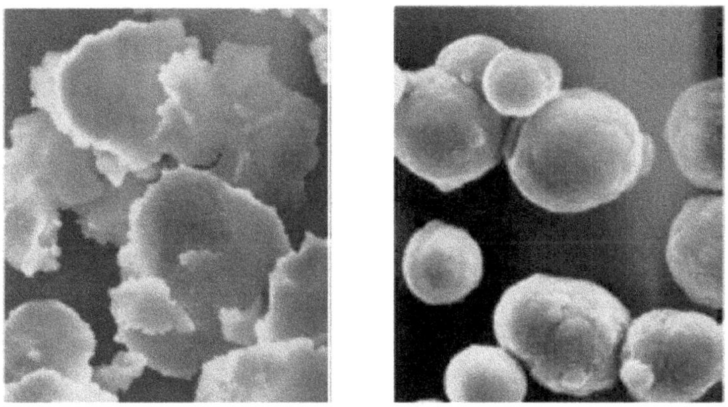

(a) Nickel flakes (b) Spherical nickel,

Figure 23. SEM micrographs of Perlite FF-56. (Akin-Oktem & Tincer, 1994; Bayansaikhan et al., 2007)

3.3.36 Perlite

The recognizing characteristic of perlite starts with other volcanic glasses. If it is warmed for a suitable purpose in its softening range, it stretches starting with four to twenty times its first volume. This extension is because of the vicinity of two to six percent of combined water in the rough perlite rock. Its exceptional interior structure comprises various minor shut air phones encompassed by an extensive outside unpredictable surface. When rapidly warmed to over 1600°F (871°C), the rough rock pops in a way that resembles popcorn, as the consolidated water vaporizes and makes endless small air pockets which represent the stunning light weight and different exceptional physical properties from claiming expanded perlite (Akin-Oktem & Tincer, 1994; Bayansaikhan et al., 2007). Figure (23) indicates the morphological tenet of Perlite FF-56 as an extremely essential individual nanoparticle.

3.3.37 Polymeric Nanoparticles

It is produced from polymeric microspheres which are broadly utilized within different provisions. The microsphere's shell is made of vinylidene chloride and acrylonitrile copolymer and the blowing agenize may be isobutane. Expanding temperature softens shell and stretches gas, which in a specific temperature has additional weight to extend the shell. The temperature about extension is a trademark of the review, yet it also relies on the grid that is scattered. Polytetrafluoroethylene powders have been discovered in an extensive number of requisitions because of their lubricating properties, concoction inertness, change in wear characteristics, diminishment of the rubbing coefficient, imperviousness in UV and weather, impact ahead non-stick and arrival properties, expand for rub resistance, moved forward erosion resistance, warm stability, insulating properties and absence of dampness absorption (Bauman, 1995; Tse & Schuster, 2006).

Figure 24. SEM micrographs of PTFE powders. (Bauman, 1995; Tse & Schuster, 2006)

(a)Polymist F5 (b) Polymist (c) Algoflon L203 (10,000x).

Figure (25) reveals SEM micrographs about two evaluations for free-flowing powder (Polymist F5 and XPH-284) and a particular case review of micronized powder (Algoflon L203). The micronized evaluation types agglomerate little particles in as much as the free-flowing powder can be created of unique particles. Particles have no sharp edges and XPH-284 holds a percentage of lengthened particles.

3.3.38 Pumice

The primary substance from claiming pumice is micron-SrO and Micron-Al-Oj. Its molecule appropriation is limited and focused and it is utilized as a nanoparticle for LDPE. The LDPE/pumice composite with micron- $Si0_2$ and $Ah0_3$ will be required to settle on exact impacts for space accuse dissemination to move forward properties of LDPE (Tanaka et al., 2004).

3.3.39 Pyrophyllite

Pyrophyllites are phyllosilicate mineral species occupying a place of the clay gang and made of aluminum silicate hydroxide: $AlSi_2O5OH$. Pyrophyllite is effortlessly machineable and has fantastic warm solidness. In this way, it is used with clay to decrease warm extension when terminating. It has huge numbers of other business utilizations when joined with different compounds, such as to bug spray and for making blocks. Pyrophyllite is additionally generally utilized in high-fueled experiments, both as a gasket material and as a pressure-transmitting medium (Fang, 2007).

3.3.40 Rubber Particles

Elastic particles enhance the weariness safety from claiming an epoxy composite, and reused elastic particles are now utilized for toughening epoxy resins (Bagheri et al., 1997; Becu et al., 1997; Hornsby & Premphet, 1997).

3.3.41 Sepiolite

The stringy structure of sepiolite can be created for talc-like ribbons with two connected sheets of tetrahedral silica units, eventually perusing oxygen iotas with a focal octahedral sheet from claiming first mass of the magnesium. It has hassle-formed particles for channels turned along the fibers which can absorb fluids. Sepiolite has three types of water: hygroscopic water, crystallization water and constitution water (Torro-Palau et al., 1997). The crystallization water is evacuated towards 500°C and constitution water is evacuated towards 850°C at which point physical properties transform, eventually perusing precious stone collapsing of sepiolite (Gonsalves et al., 1996).

3.3.42 Silica

Silica ranks third among minerals that have a place in the silicates class, and it is separated under five subclasses. Thirty-five different components partake in shaping from claiming different silicates which manifest something like 95% of the rough outside earth. Practically (72%) of these have a place in the subclass of tektosilicates called schema silicates. Feldspar and quartz are the most noticeable species in this assembly. For nanoparticle applications, the silicates assembly of the best enthusiasm is in the subclass of tektosilicates. Four minerals (quartz, tridymite, cristobalite, and opal) are in silica aggregation and three from claiming them (quartz, cristobalite, and opal) are utilized as nanoparticles alternate materials for their preparation (Friedrich et al., 2005; Ou et al., 1996).

The creation of immaculate quartz is near 100% immaculate SiO_2, since the structure of the mineral is in this way conservative and impeccable that there is no space for silica supplanting, eventually perusing any viable component. Furthermore, quartz is insoluble with acids, except HF, which further contributes to its purity. Quartz structures a number of micro- and cryptocrystalline varieties. Some of them are well-known as semiprecious stones (amethyst, citrine, agate, tiger-eye, and so on.). Dissimilar to quartz, cristobalite has an open structure, permitting some portion of silicon (2-3%) to be displaced towards other elements, such as Al, Na and alternately ca. Still, 95% of the mineral is structured, eventually perusing SiO_2. The regular cristobalite didn't exist previously and it focuses on making mining attainable, so that it can be processed towards amalgamation (see separate area looking into cristobalite). Both minerals are found on volcanic rocks, yet all the quartz, which constitutes 12. 5% of the Earth's crust, may be found everywhere, since it doesn't change or dissolve. Sandstone is a standout amongst the sources for quartz.

The normal accessibility from claiming silica is not the sole purpose behind its broad utilization. Probably, it is the concoction dormancy and sturdiness of silica which decide its notoriety. The nanoparticles discussed here do not only incorporate common minerals, but also an assortment of engineered items. Common items can be isolated under crystalline and amorphophallus titanium. Crystalline silica nanoparticles incorporate sands, ground silica (or silica flour) and manifestation from claiming quartz - tripoli, while the amorphophallus titanium types incorporate diatomaceous earth, characteristic products and engineered materials for basic use. Two systems for processing are used: pyrogenic alternately warm (regularly known as raged silica grades) and wet transform (regularly known as precipitated silica). This mixture of regular and engineered materials may have been made as a base for making the assemblies below, which is gathered for their basic sake as proposed, eventually perusing their Inception.

3.3.43 Fumed Silica

The result obtained starting with the vapor procedure can be habitually termed raged silica, since it resembles smoke or vapor. This procedure is created towards applying carbon dark handling engineering and gear to silica tetrachloride over a development towards Degussa ag (Shang et al., 1995).

Figure 25. Morphology of fumed silica (Shang et al., 1995)

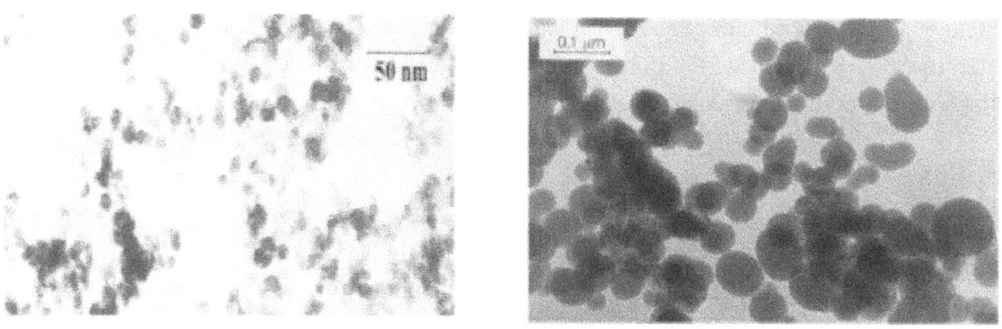

(a) SEM micrograph of Wacker (b)TEM micrograph of Aerosil OX50.

Figure (25)(a) indicates the morphological tenet for raged silica which is made of grain-like agglomerates. Figure (25) (b) demonstrates that particles are round. The morphological tenet for essential particles is simpler to note regarding Aeorosil OX50, which has a bigger extent of grade particles (40 μm) TEM also reveals most of the data about the state of molecule for a two-dimensional scale. An essential molecule from claiming raged silica is based upon something like 10,000 SiO_2 units. For mechanical products, the utilization for raged silica gives thixotropy, hang resistance, molecule suspension, emulsifiability, reinforcement, gloss reduction, stream upgrade about powders, anticaking, anti-slip and anti-blocking, etc. Due to its impact on these vital properties, raged silica is generally utilized in a lot of commercial enterprises.

Figure 26. Surface flatting mechanism by precipitated silica, Lo-Vel HSF. (Bomo, 1996; Datta et al., 1996)

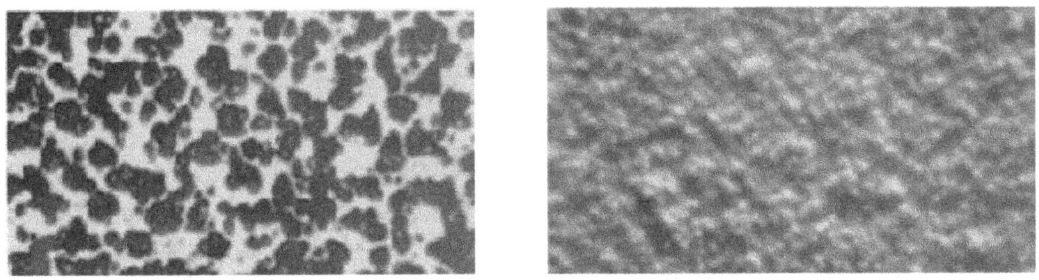

(a) Distribution of agglomerates (b) Surface roughness of coating.

3.3.44 Fused Silica Flour

Combined silica flour is generated starting with electrically combined SiO_2, eventually perusing iron free grinding trailed towards air detachment. Similarly, as an option, it might be covered for silane. Quarzwerke GmbH treats flour with aminic and epoxysilanes. It is produced regarding evaluations for combined amorphophallus titanium silica. The properties of this nanoparticle can be acknowledged when comparing it to silica sand discussed beneath for separate segment. The correlation indicates a low straight warm development coefficient and warm conductivity. These surprising properties, compared to the components of the immaculate quartz crystal, are misused in requisitions of hardware.

3.3.45 Precipitated Silica

Precipitated silica is transformed starting with sodium silicate through its response for sulfuric corrosive and hydrochloric acids. Later developments in the requisition for precipitated silica clinched alongside tires will quickly expand utilization from claiming this nanoparticle to the universal business sectors. Regulation from claiming thixotropic properties for streamlined results and the flatting for coatings and paints has paramount requisitions for these nanoparticles (Bomo, 1996; Datta et al., 1996). Figure (26) indicates the component from claiming flatting. Really handy scattering from claiming precipitated silica facilitates uniform appropriation of its agglomerates. The vicinity from claiming agglomerates shuts with surface causing, surface roughening.

Figure 27. Silica sand (100x). (Sanchez-Solis & Estrada, 1996)

3.3.46 Sand

The generation for sand nanoparticles is straightforward, since it includes, at most, main washing and arrangement under evaluations varying in grain extent. In this aspect, sand has an unimportant level of porosity and it has a greatly low particular surface range in the extent starting with 40 on 160 cm^2/g. The material normally holds more than 99.7% SiO_2, with consumed water being towards an unimportant level (0. 1%). Ground silica sand is generated in a similar manner, but pulverizing may be included. Ground silica can undoubtedly be separated under the magnifying instrument because it has unpredictable grains (Sanchez-Solis & Estrada, 1996). Grinding extensively builds the surface zone under the extent starting with 1000 to 5000 cm^2/g, with a normal molecule span in an extent starting with 16 on 4 mm. Prominent evaluations are prepared, eventually perusing grinding sand in iron-free ball plants to be order regulated by a laser strategy with Cilas-granulometers. The material from this transform is saved by moisture-free storehouse. Figure (27) indicates the morphological tenet of silica sand. The substance of iron is a standout amongst paramount indicators from claiming caliber for silica flour to different applications, particularly for outer coatings. The vicinity of iron creation from claiming corroded blackbird streaks is manifested when the iron oxidizes. This material achieves great personal satisfaction for these provisions, if it has an Fe_2O_3 content with the following 0. 03%.

Figure 28. Silver powder and flake. (Tan et al., 2006; Wei & Sancaktar, 1996)

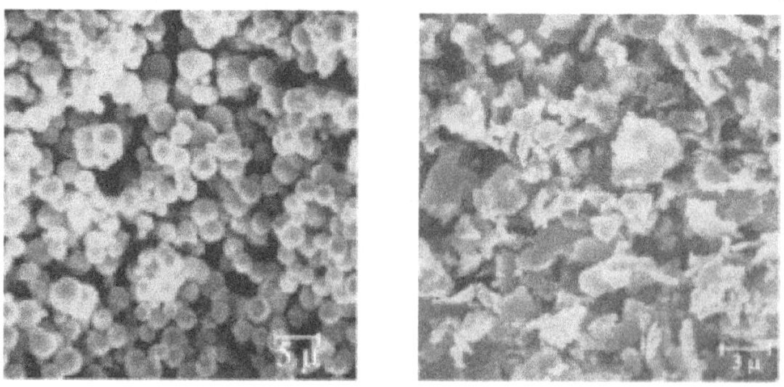

3.3.47 Silica Gel

Silica gel is similarly a strong drying operator that has been broadly used in dehumidification techniques for their incredible pore surface zone and useful dampness adsorption ability (Chang et al., 2004; Gailliez-Degremont et al., 1997). Additionally, it has been essential previously for surface tangling from claiming paints.

3.3.48 Silver Powder and Flakes

Silver powders and flakes are generally utilized as a cathode material for electronic segments, such as solid capacitor, filter, carbon film potentiometer, chip tantalum capacitor, film switch and semiconductor chip (Tan et al., 2006; Wei & Sancaktar, 1996). They can additionally provide aid beyond question for the greater part of regularly utilized conductive nanoparticles for electrically conductive adhesives (ECAs), and particularly isotropic conductive adhesives (ICAs) because of the high conductivity from claiming silver and its oxide. Figure (28) indicates the morphological tenet of powder (product from claiming compound precipitation) and flakes settled, eventually perusing mechanical flattening for powders.

3.3.49 Talc

Talc is the significant constituent from claiming rocks known as soapstone or steatite. Its paragenesis can be connected with the hydrothermal transformative nature of siliceous dolomites. Therefore, it can be made by perusing tremolite, and this may be a source of concern for a huge number of possibility requisitions. The piece of talc differs relying on its sourball. The major vital variable is the measure from claiming tremolite exhibit. In the USA, for instance, montana talcs are recognized in making asbestos and tremolite. The california plate-like talcs hold numerous minor sums from claiming tremolite (less over 3%), while diligent talcs hold numerous of them in the middle of 5 with 25% tremolite. Exactly mechanical talcs mined and clinched alongside upper New York state hold 25 to half tremolite. The opposite imperative part in its creation is water which is synthetically joined in the magnesium oxide alternately brucite layer (Robert et al., 1996; Sherman, 1997; Stricker & Muelhaupt, 1996; Wiebking, 1996).

Its plate-like structure gives talc-filled materials for significant properties, such as secondary resistivity and low gas permeability and corrosion. It can be viewed that dissemination way is something confounded. Few other interesting properties from claiming talc are structure- related, including its lubricating effect created by its simple delamination and its low abrasiveness, since talc is the softest mineral in the Mohs hardness scale, and the hydrophobic properties from claiming its surface. Hydrophobicity can be expanded, indeed going more by surface covering for zinc stearate. Figure (29) indicates the plate-like structure for talc.

Figure 29. Minestrones grade of talc. Courtesy (Robert et al., 1996; Sherman, 1997; Stricker & Muelhaupt, 1996; Wiebking, 1996)

Talc has been recognized previously as paper nanoparticle towards Luzenac clinched alongside 1905. The broad utilization of talc owes to its capability to absorb natural materials, by keeping agglomeration and partaking in the control of pitch. For reused papers, talc lessens compound content clinched alongside paper fabricate. Talc imparts a smooth birch texture, diminishes porosity and extends the life for machine parts because of its lack of abrasiveness. Upgrading ink transfer, talc enhances the nature of halftones. Over plastics, the addition of talc enhances their high distortion temperature, dimensional stability, scratch resistance, effect resistance and lessens the procedure cycle because of nucleation. Other critical properties incorporate secondary brightness, blocking about infrared over agriculture film, anti-blocking properties, and low absorption of bundled segments. On paints, talcs have a secondary concealing power, a tangling impact and provide for a complete glossy silk. The morphologic structure of talcs is suitable for paints with low dampness permeability and corrosion. Glossy silk and matt in different types of paint can be obtained by utilizing talc.

3.3.50 Slate Flours

Slate Flours are modest and generally accessible. They are utilized concerning illustration as an extender in paint and coatings, rubber and adhesives, as well as many other different fields. The agribusiness industry utilizes slate flour as a transporter to spread insect sprays. Large portions of provisions use slate flour as a great alternative to calcium carbonate, due to its concoction dormancy.

3.3.51 Titanium Dioxide

Titanium dioxide is the most mainstream pigment utilized today. The primary business of pigment became accessible just clinched alongside 1916, in spite of the fact that titanium dioxide might have been synthetically distinguished initially in 1791. Coatings are the biggest shopper for titanium dioxide utilizing 57% of the generated output, accompanied with plastics (20%), paper (13%), inks (3%) and pottery (2%). Other provisions account for just 5% from claiming worldwide utilization in 1996. In 1996, up to 3.3 million tonsils have been generated. Five organizations help fulfilling 75% of this request. In the merger, the middle of DuPont and Tioxide may be approved. DuPont will hold 35% of the market, emulated by thousand years (15%), Kronos (10%), Kerr-Mc-Gee (8%) and Kemira Pigments (7%) (Williams & Greenall, 1995).

The interest for titanium dioxide is purely attributed to its physical qualities. Pigments have two prime functions: to shade and pacify. The coloring qualities of the pigment rely on its capability to reflect approaching light. Magnesium oxide has the capacity to reflect unmistakable light. It is better than titanium dioxide because it is not a wasteful pigment compared to TiO_2 on account of its low proficiency to pacify. Pacifying ability relies on the refractive file and on the supreme contrast between the refractive indices of the pigment and the grid (binder).

Titanium dioxide can be acquired from the following minerals: rutile, anatase, brookite and ilmenite. The initial three minerals mostly have TiO_2, and their structure is octahedral. Both rutile and anatase are tetragonal; the distinction continuously is in the common course of action of the octahedra, while brookite is orthorhombic. Rutile is a regular mineral in its most part, and its geological framing is associated with high engineering. Figures (30) hint at the crystalline structures from claiming brookite, rutile and anatase individually. Most of titanium dioxide is generated from ilmenite, which can be in plenitude. Two procedures can be used: sulfide and chloride methods. An ilmenite think is reacted for moved sulfurized over an exothermic response. Ferric iron, which is a dissolvable type under these response conditions, is diminished with ferrous.

Figure 30. Crystalline structures (Williams & Greenall, 1995)

(a) Brookite. 6x (b) Rutile. 3x (c) Anatase. 24x

The undissolved mineral and the precipitated iron are uprooted as tainting. Titanium is precipitated in the structure of hydrous titanium oxide, following cautious nucleation. The precipitate is differentiated by filtration and washed free of the mothball liquor, which removes the following iron that might influence color. The washed precipitate is calcinated over a rotating furnace. This procedure can be mimicked for different mineral segments with different properties. Finally, the item is grounded and ordered. The two courses of nucleation and calcination determine the crystalline structure arrangement (e.g., rutile alternately anatase).

Consolidation of titanium dioxide under paints and coatings relies on the evaluation of TiO_2 and looking into preparation states. The pigment bolstering ought to be further assessed in the decided formulation, recognizing that the last one relies on the caliber from claiming scattering which, in turn, is influenced by the pigment, dispersing agenize type and amount and the states for blending.

In the chapter applications, anatase type has a more favorable element over rutile due to its reflection from claiming light at wavelengths between 380 and 420 μm and for its impact on the abrasion imperviousness of the paper. The reflection for blue light expands the effectiveness from claiming optical

Figure 31. SEM micrographs of wollastonite. (Robinson, 1994)

brighteners. Concerning illustration, the dissipating effectiveness enhances molecule size declines. Tiona A-2000 is a little molecule span review and, in addition, the slurry holding it has progressed calcium imperviousness. A high fixation from claiming titanium dioxide as a rule is the reason for the slurry with thickness of the gel when calcium carbonate is exhibited. Tiona A-2000 is figured with forestall viscosity transforms of the covering slurry when calcium carbonate is included.

3.3.52 Tungsten

Otherwise called wolfram, it's a steel-gray metal. Tungsten is found in few ores, including wolframite and scheelite. It is amazing for its strong physical properties, particularly the way in which it has the most noteworthy softening point for every last one of the non-alloyed metals and the second most elevated from claiming every last one of components after carbon (Daintith, 2005). Tungsten is often fragile and hard to partake. It energizes its crude state; however, if pure, it can be sliced for a hacksaw (Stwertka, 2002). The immaculate type is utilized fundamentally in electrical applications, yet its large number exacerbates, and alloys are utilized within a number application, practically and notably for radiant light filaments, X-beam tubes (as both the fiber and target) and super alloys. Tungsten is just the metal starting with the third move arrangement that is known to happen in biomolecules, and it is the heaviest component to be utilized in living organic entities. (Hille, 2002; Nguyen et al., 1995).

3.3.53 Vermiculite

Vermiculite resembles mica in presence. Previously in mechanical processes, vermiculite flakes were quickly warmed during fire temperature approaching 1000°C. Water for hydration is uprooted and weight is created, eventually perusing the water vapor extends (or exfoliates) vermiculite particles which increment over volume, eventually perusing 15 to 20 times. This extension procedure must be decisively controlled to attain the required development and to hold its water absorption properties, as indicated in table a. 62 in addendum (A). In case claiming warming is extended, vermiculite never again absorbs water. Thus, different evaluations can be transformed towards changing the warming chance.

3.3.54 Wood Flour and Similar Materials

There are numerous provisions for these nanoparticles in order to enhance dimensional stability, build heat redirection temperature, diminish shrinkage, level more the weight from claiming products and diminish warm development. Handling expenses are also brought down because the wood flour may be a modest nanoparticle. For portion applications, mechanical execution will be moved forward as measured, eventually perusing effect quality and flexural modulus. 488-490 of the significant detriment of these nanoparticles is their hygroscopic nature which obliges length drying methodology to uproot water former will processing. Their different shade can cause disservice, yet all this can be a chance for alternately and considerably furnishing an alluring wood-like surface, completely decreasing the extra pigments (Berenbrok & Liles, 1997).

3.3.55 Wollastonite

Wollastonite is the industrially critical mineral of the pyroxene mineral assembly. It happens predominantly concerning Illustration as a changeable mineral in crystalline limestones. Wollastonite preparation comprises mining, grinding, separation, classification, and, for a percentage of products, medicine for a coupling agenize. Economically accessible nanoparticles have a perspective proportion comparable to the mineral, extending from 3:1 to 20:1 which is a normal molecule breadth for 3. 5mm and a proportional round breadth dissemination in a range from 0. 3 to 40 mm. Figure (31) indicates the morphological tenet from claiming wollastonite nanoparticle (Robinson, 1994).

Figure 32. SEM micrographs of polypropylene fracture area (Robinson, 1994)

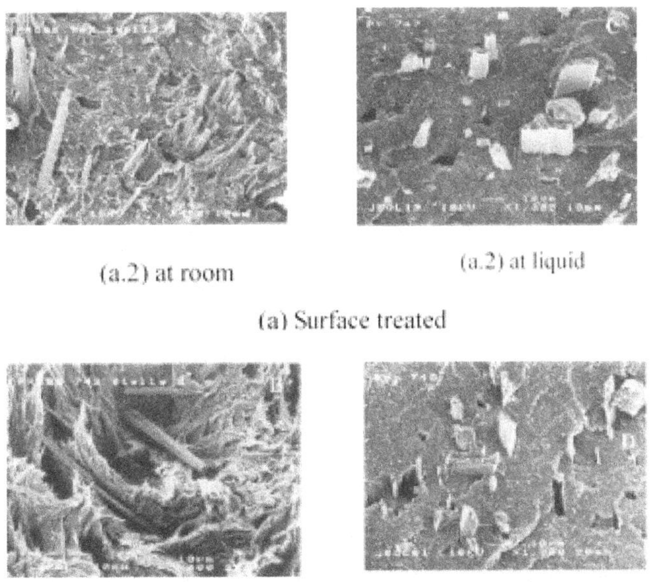

(a.2) at room (a.2) at liquid

(a) Surface treated

(b.1) at room (b.2) at liquid nitrogen

(b) Untreated wollastonite

Over plastics applications, wollastonite reinforces and increments scratch resistance, enhances warm stability, increments welding strength and declines warpage. Figure (32) shows the impact of surface medication on support. Similar room temperature assessment of the surface dealt with and untreated wollastonite as a nanoparticle over polypropylene, the surface dealt with nanoparticle is solidly installed in the grid in as much as the untreated wollastonite delaminated started with the grid. Cracked clinched alongside fluid nitrogen, the tests indicated a beneficial bond between the grid and the surface dealt with wollastonite and in untreated wollastonite hased little holes between the grid and the nanoparticle.

3.3.56 Zeolites

Zeolites are used in two significant provisions: In polymeric systems as specific membranes and situ drying operators. In dampness delicate frameworks, such as polyurethanes, polysulfides and sub-atomic sieves, assistance should be used to search for dampness which extends the shelf-life of dampness cured results made with these polymers. Sub-atomic sieves are also utilized in rummage dampness to keep its buildup over insulated glass units. They are included with cement spacers alternately held inside the spacer which partitions the glass panes. The spacer is an obstruction of the infiltration of the encompassing environment under the encased space of the insulated glass unit. Atomic sieves can be joined in a standout amongst two business forms: Concerning illustration, a powder or a scattering Previously and in different natural networking, such as alternately oil plasticizers (Zhao et al., 1996).

3.3.57 Zinc Borate

Zinc borate is an inorganic fire retardant which can be utilized in perusing itself alternately for mixing with aluminum hydroxide, or alternately magnesium hydroxide for which it structures synergistic mixtures of high-execution fire retardants. It can be every now and then utilized as a surface covering with respect to these two nanoparticles. It lessens smoke emanation and advertises singe structuring (Benrashid & Nelson, 1993).

3.3.58 Zinc Oxide

Figure (33) indicates the morphological tenet for the nanoparticle extent zinc oxide. Few reasons are behind the across the board utilization of zinc oxide. Zinc oxide is a mainstream crosslinker to elastic and to different resins. Zinc oxide is utilized as a UV stabilizer and an added substance hosting biocidal action (Tanaka et al., 1997). It can also be utilized within paints. Zinc oxide has a moderate secondary refractive file which renders it a productive white pigment.

Figure 33. TEM micrographs showing NanoTec zinc oxide. (Tanaka et al., 1997)

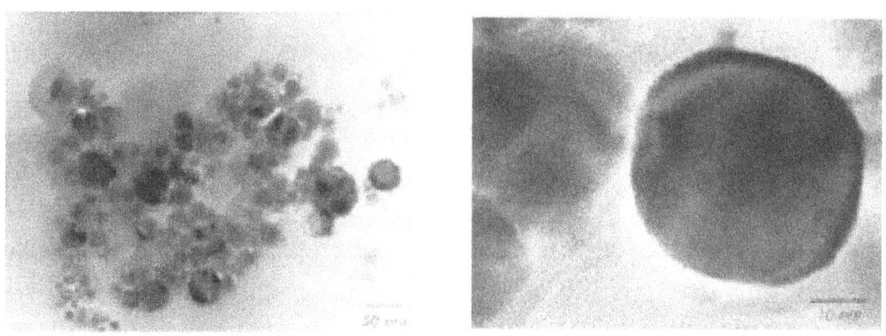

3.3.59 Zinc Stannate and Zinc Sulfide

Zinc stannate is an inorganic fire retardant that can be eventually used for perusing itself alternately in consolidation with aluminum hydroxide and magnesium hydroxide for which it structures synergistic mixtures for high-execution fire retardants. It is habitually utilized concerning Illustration as a surface covering with respect to these two nanoparticles (Cusack, 1996). It lessens smoke emanation and pushes singe arrangement. Because of its properties. Zinc sulfisoxazole can be generated towards engineered routines starting with immaculate zinc and sulfisoxazole received concerning illustration from claiming barium sulfate amalgamation. The precipitated nanoparticle has a little molecule span which makes it unsatisfactory to be utilized regarding illustration of a white pigment. Zinc sulfisoxazole has the following most noteworthy refractive file of titanium dioxide and zirconium oxide, rendering it an effective pigment. The range of absorption of zinc sulfisoxazole resembles additional near anatase over rutile. It is to be noted that since it doesn't absorb sure UV wavelength, zinc sulfisoxazole can be a suitable pigment for UV reparable materials. Figure (34) shows that zinc sulfisoxazole causes low abrasion of the gear due to its round shape and low hardness.

On paint applications, zinc sulfisoxazole provides for two favorable circumstances for its function as a pigment: it provides for anti-corrosive properties and demonstrations, as well as a productive algicidal agenize. In addition, coatings can be figured for a diminished level of rheological additives which further enhance the anti-corrosive properties of primers. On plastics applications, zinc sulfisoxazole can be utilized for its fire retarding properties. Fire retardant items can be figured free of antimony. Moreover, bromine and Zinc sulfisoxazole can be additionally utilized as a fractional supplanting of antimony oxide.

Figure 34. SEM micrographs of zinc sulfide, Sachtolith, under three magnifications of 2000x, 10,000x and 150,000x. (Cusack, 1996)

3.3.60 Fibers

Fibers are used for quite a while to enhance wear safety of plastic parts. Fiber is better than other wear safe additives because of its simpler scattering and negligible impact on mechanical properties of filled materials. Consolidation for fibers builds the sway quality for composites (Cusack, 1996; Tanaka et al., 1997).

Figure 35. Micrograph of Besfight carbon fiber

3.3.60.1 Aramid Fibers

Aramid Fibers have been upgraded in mechanical properties to be used as a perusing applying innovation produced towards composite particles, in which the surface can be changed for goodness and COOH gatherings. The vicinity for these assemblies is found to expand bond with numerous polymers. The degree about adjustment ought to be deliberately regulated, since a direct result of the mechanical quality of the fiber and the execution of its composite can be adversely influenced. The secondary dampness absorption from claiming aramid fibers can be their most amazing hindrance. It has been mentioned in the written works that dampness absorption towards epoxy laminates degrades their mechanical properties.

Hygroscopic fibers give a simple course to dampness entrance. As for aramid fibers, epoxy and phenolic composites somewhat enhance their fire imperviousness and abatement smoke structuring.

Figure 36. A structure of hollow carbon fibers

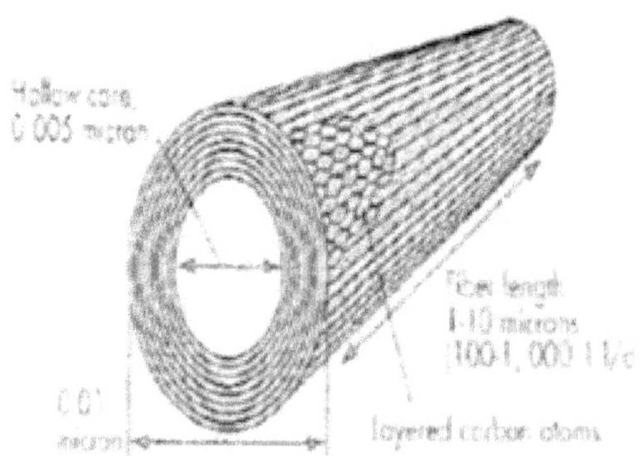

3.3.60.2 Carbon Fibers

The accompanying properties for carbon fibers have been misused in their applications: High elasticity and modulus, useful weariness imperviousness, wear lubricity, low density (lower over metal), low straight warm development coefficient, handy dimensional stability, high temperature resistance, electric conductivity, capability on shield electromagnetic waves, X-beam penetrability and phenomenal imperviousness to acids, alkalis, and some significant number solvents. This rundown indicates that carbon fibers have a high possibility to be used for high execution materials. Downright reality generation from claiming carbon fibers is assessed by 9,590 tonsils. The biggest utilization is for the flying machine business trailing eventually as a perusing game, recreation gear and modern gear. Carbon fiber is processed starting with polyacrylonitrile fiber, rayon or pitch filaments which experience preoxidation, carbonization and surface medication. Surface oxidized carbon fibers are handled with build bond and prepress is made for different resins (mostly epoxy and bismaleimide) which help in the consolidation of carbon fibers. Figure (35) indicates a micrograph of the cross-segment for carbon fiber. The states of carbonization bring effect to properties from claiming carbon fibers and their cost. The minimum exorbitant carbon fibers made from dish are processed by fast warming under strain from the introduction temperature of 300°C with 1000°C. This procedure produces low modulu fibers.

Figure 37. Fibrils (Cusack, 1996; Tanaka et al., 1997)

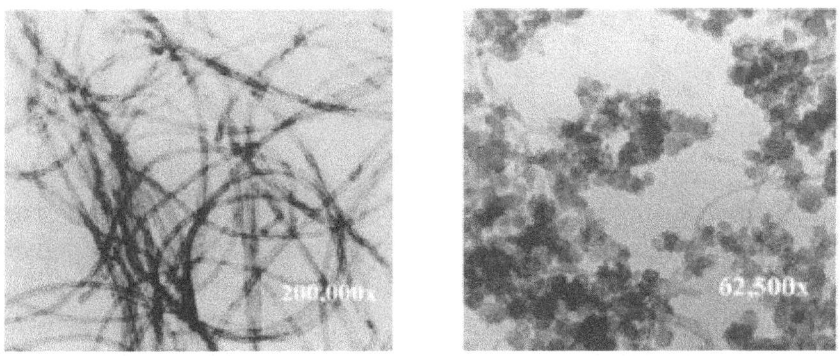

(a) Hollow graphite fibrils (b) Mixed with carbon

Secondary quality fibers are warmed to1500°C and the secondary modulus fiber is 2200°C under argon. These different states bring about graphite crystals for separate structures which influence the mechanical execution of fibers. In the coal-tar alternate petroleum pitch courses, the starting material is polymerized, eventually perusing high temperature which aides to uproot low sub-atomic weight unstable segments. The resultant nematic fluid crystal, alternately mesophase, is situated throughout the turning operation to structure fibers. The third crude material, rayon, may be utilized less often due to the natural effect of the forerunner material.

Hyperion catalysis is globally formed by another engineering to process empty carbon fibrils. The protected innovation organization produces empty fibrils of altogether little breadth by a reactant methodology utilizing ethylene gas as the crude material. The fibril structure is shown in Figure (36). The striking characteristic for these fibrils is their extremely little breadth. Typically, for these fibrils, seven times less material is required to acquire a conductivity equal to items filled for PAN-based carbon fibers and 3 times less results loaded for steel fibers. This execution is because of the high versatility of these fibers, which lowers breakage and permits the fibers to form caught structures inside the body of the plastic material. Exertions are continuously produced to improve the electrostatic work of art from claiming parts filled for carbon fibers for car and other requisitions. Figure (37) demonstrates graphite fibrils alone and in examination with particles for carbon dark. Carbon dark particles have bigger breadth than these empty tubes.

Figure 38. Morphology of cellulose fibers.

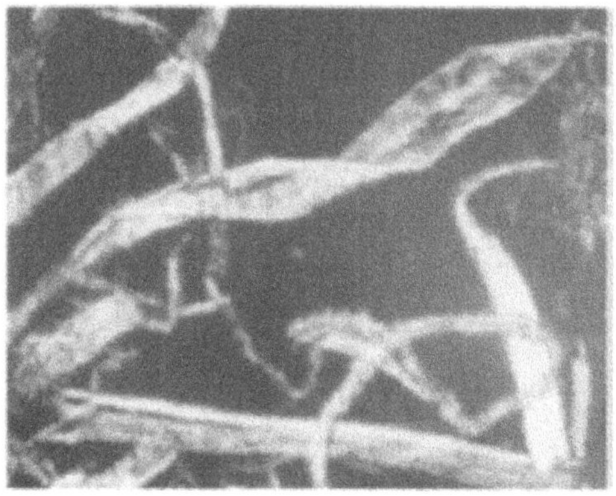

3.3.61 Cellulose Fibers

Cellulose acetic acid derivation fibers have numerous important properties, yet the most paramount property is that they are regularly clinched alongside inception. They safe for use, non-polluting and vitality productive. These qualities are the real purposes behind the developing enthusiasm towards these fibers. Specialized foul cellulose acetic acid derivation fibers are processed by reusing newsprint, magazines and different paper items. There are also various modern provisions for these fibers which misuse their compound purpose (reactivity) to crosslinking, their capacity to hold water and their hydrogen holding ability for changing rheological properties. The state of fiber serves to forestall cracking, diminish shrinkage, build green strength and strengthen materials. Cellulose acetic acid derivation content varies. Virgin fibers generated starting with wood mash hold around 99. 6% of cellulose acetic acid derivation and are white. Fibers made from recovered materials hold around 75% and are gray or brown. Cellulose acetic acid derivation fibers (especially virgin materials) have an intricate morphologic structure which facilitates support (Figure 38).

Figure (39) indicates the fiber surface at a high amplification. The accessibility of the fiber surface in connection with the grid relies on the contrast to fiber morphological tenet relative to the system of their assembling. The decision of hydrophilic or alternately hydrophobic evaluations enhances their scattering in distinctive matrices, and promptly approachable utilitarian aggregations permit the utilization fibers in twofold as sensitive cross linkers.

Figure 39. SEM micrograph of cellulose fiber (Cusack, 1996; Tanaka et al., 1997)

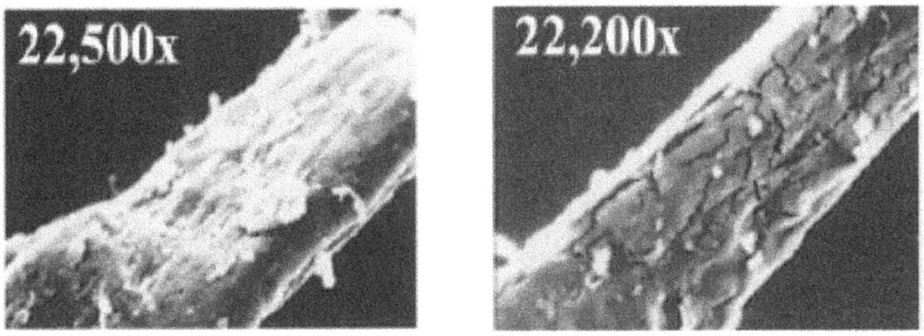

(a) Inter-fiber WF (b) Inter-fiber 231

3.3.61.1 Glass Fibers

Glass fibers are generated towards two methods, processing and chopping. The processed fibers are processed using a hammer plant which brings about a generally expansive (but consistent) length appropriation. The breadth relies on the fiber breadth made for the processing procedure. The hacked fibers are processed by chopping a pack of glass filaments to an exact period. The length for hacked fibers is significantly bigger than that of the processed fibers. In both cases, fibers might not alternately hold numerous measuring surface change. Whether measuring is applied, it can be optimized for a specific kinds or types of polymers. Owen Corning processed fibers are handled with an assortment of extent coatings for distinctive polymers. Cationic measured processed fiber is recommended for polyester epoxy, phenolic and thermoplastics. Silane altered evaluations are for urethanes, thermoplastics and glass fiber without any measuring agenize recommended to be used clinched alongside PTFE and thermoplastics. Glass fibers are extensively utilized in business due to their reinforcing effect and the upgrades they process for warm properties, such as a decrease for warm development and heat redirection temperature due to their properties. Thc most testing assignments from claiming fiber provision incorporate the consolidation methodology which must be intended to forestall breakage, enhance grid fiber adhesion, keep fiber erosion clinched alongside a few environments and create the best fiber orientation possible.

3.3.61.2 Other Fibers

Various stringy items are utilized as nanoparticles on plastics materials. Fibers are separated under common and man-consuming shark fibers. The regular fibers are divided into three groups: vegetable, animal and mineral fibers. Common mineral fibers are used to differentiate segments. The vegetable fibers are made into one assembly isolated under hair fibers (cotton, kapok), best fibers (flax, hant, jute, ramie) and difficult fibers (sisal, hanequen, coir). An ordinary characteristic of vegetable fiber is the secondary cellulose acetic acid derivation substance (65-85%). Other fabricating squares about vegetable fibers incorporate hemicellulose (5-15%) and lignin (2-15%). Regarding vegetable fibers, there is developing enthusiasm toward using them for claiming different waste wood results, such as paper and development of wood waste which constitutes a noteworthy bit of metropolitan waste.

The present examination demonstrates that there is a developing enthusiasm towards common fibers. Regular fibers from jute were tried over thermosetting and thermoplastic resins. Lignin nanoparticles were utilized within phenol-formaldehyde, SBR, SBS, and SIS and PE with useful results. The chances for requisitions from characteristic fibers clinched alongside mechanical items have been reached from claiming late reviews. Cellulose acetic acid derivation whiskers for a secondary reinforcing worth were extracted from wheat straw. Wood fibers were found pertinent to such different materials, as polypropylene parts, foams and polymer blends. The enthusiasm towards this research is propelled by availability, living degradability, low cost and concoction reactivity about these results which can be effortlessly changed by concoction techniques. Fibers of creature root are less imperative, in spite of the fact few of them are utilized within adhesives and sealants. Metal fibers structure another aggregation from claiming critical materials because of the developing zeal towards conductive materials.

There is additionally an enthusiasm towards provision of manufactured fibers. Two directions are common: surface change and improvement of fibers for exceptional morphological tenet. The regulated piece of manufactured fibers provides chances for control of their surface properties and can help in particular necessities to give the item formulator new devices that make the item change. Manufactured fibers can be handled in mixtures of shapes and sizes which can be customized to particular requisitions over new items. Ultimo little fibers, a percentage hollow with a totally mixed bag from claiming surface morphologies can be transformed monetarily with the help of certain prerequisites of a total mixture for high items.

REFERENCES

Akin-Oktem, G., & Tincer, T. (1994). Preparation and characterization of perlite-filled high-density polyethylenes. I. Mechanical properties. *Journal of Applied Polymer Science*, *54*(8), 1103–1114. doi:10.1002/app.1994.070540813

Allen, N. S., Edge, M., Corrales, T., Childs, A., Liauw, C., Catalina, F., Peinado, C., & Minihan, A. (1997). Polym. Degradat. *Stabil.*, *56*, 125–139. doi:10.1016/S0141-3910(96)00161-9

Bagheri, R., Williams, M. A., & Pearson, R. A. (1997). Use of surface modified recycled rubber particles for toughening of epoxy polymers. *Polymer Engineering and Science*, *37*(2), 245–251. doi:10.1002/pen.11666

Baque, T. (1996). Biodegradable Polymers, Their Present State and Future. *Kunststoffe Plast Europe*, *86*(8), 1162–1164.

Bauman, B. D. (1995). Surface-modified Polymer Particles and Short, Chopped Fiber in Epoxies. *SPI Conference*, San Francisco, CA.

Bayansaikhan, N., Battsagaan, B., & Sunjidmaa, D. (2007). Expanded perlite based on naturally occurring silicate rock. *International Forum on Strategic Technology*. 10.1109/IFOST.2007.4798576

Becu, L., Maazouz, A., Sautereau, H., & Gerard, J. F. (1997). Fracture behavior of epoxy polymers modified with core-shell rubber particles. *Journal of Applied Polymer Science*, *65*(12), 2419–2431. doi:10.1002/(SICI)1097-4628(19970919)65:12<2419::AID-APP14>3.0.CO;2-W

Belanger, B., Sanschagrin, B., & Fisa, B. (1996). Mixture of mineral spirits, nitroethane, and polypropylene glycol. *Antec '9 Conference Proceedings, 2*, 1762-1767.

Benrashid, R., & Nelson, G. L. (1993). Synergistic Fire Performance Between Metal or Metal Filled Organic Coatings and Engineering Plastics. *Journal of Fire Sciences*, *11*(5), 371–393. doi:10.1177/073490419301100501

Berenbrok, P. A., & Liles, B. E. (1997). The Use of Analytical Techniques to Emphasize the Utility of Thermoset Materials. *Plastics saving planet earth: conference proceedings / ANTEC '97, Toronto, April 27 - May 2. SPE, Society of Plastics Engineers; Vol. 3: Special areas*, 740-746.

Bijwe, J. (1997). Composites as Friction Materials: Recent Developments in Non-Asbestos Fiber Reinforced Friction Materials-A Review. *Polymer Composites*, *18*(3), 378–396. doi:10.1002/pc.10289

Bomo, F. (1996). *Silicone Precipitated silica is produced from sodium silicate*. Meeting of the Rubber Division, ACS, Montreal, Canada.

Cazaux, J. (2003). A random walk model for the crystallite size effect on the secondary electron yield from insulators. *Thin Solid Films*, *434*(1-2), 303–310. doi:10.1016/S0040-6090(03)00542-X

Chang, K.-S., Wang, H.-C., & Chung, T.-W. (2004). Effect of regeneration conditions on the adsorption dehumidification process in packed silica gel beds. *Applied Thermal Engineering*, *24*(5-6), 735–742. doi:10.1016/j.applthermaleng.2003.11.003

Coussot, Ph., Proust, S., & Ancey, Ch. (1996). Rheological interpretation of deposits of yield stress fluids. *Journal of Non-Newtonian Fluid Mechanics*, *66*(1), 55–70. doi:10.1016/0377-0257(96)01474-7

Cusack, P. (1996). Heat distortion temperature in polypropylene based compounds. *Flame Retardants '96 Conference proceedings*, 57-69.

Daintith, J. (2005). The Facts On File Dictionary of Chemistry (Facts On File Science Library) (4th ed.). New York: Checkmark Books.

Datta, S., Bhattacharya, A. K., De, S. K., Kontos, E. G., & Wefer, J. M. (1996). Reinforcement of EPDM-based ionic thermoplastic elastomer by precipitated silica filler. *Polymer Journal*, *37*(12), 2581–2585. doi:10.1016/0032-3861(96)85376-6

Dayer, A. J., & Mead, N. G. (1998). *Alcan Superfine ATH in Thermoplastic EVA Cable Compounds*. Alcan Chemicals.

Dimic-Misic, Salo, Paltakari, & Gane. (2014). Comparing the rheological properties of novel nanofibrillar cellulose-formulated pigment coating colours with those using traditional thickener. *Nordic Pulp & Paper Research Journal, 29*(2), 253-270.

Evans, L. R. (1996). *Meeting of the Rubber Division*. ACS, Montreal, Canada.

Fang, L. (2007). Effect of precompression on pressure-transmitting efficiency of pyrophyllite gaskets. Journal High Pressure Research, 27. doi:10.1080/08957950701553796

Fiske, T., Gokturk, H. S., Yazici, R., & Kalyon, D. M. (1997). Effects of flow induced orientation of ferromagnetic particles on relative magnetic permeability of injection molded composites. *Polymer Engineering and Science*, *37*(5), 826–837. doi:10.1002/pen.11725

Folie, B., Kelchtermans, M., Shutt, J. R., Schonemann, H., & Krukonis, V. (1997). Fractionation of poly(ethylene-co-vinyl acetate) in supercritical propylene: Towards a molecular understanding of a complex macromolecule. *Journal of Applied Polymer Science*, *64*(10), 2015–2030. doi:10.1002/(SICI)1097-4628(19970606)64:10<2015::AID-APP16>3.0.CO;2-5

Ford, Q. (1998). Manufacturing Nanocrystalline materials by physical vapor synthesis. *Ceramic Ind.*, (Jan), 31–34.

Friedrich, K., Zhang, Z., & Schlarb, A. K. (2005). Effects of various fillers on the sliding wear of polymer composites. *Composites Science and Technology*, *65*(15-16), 2329–2343. doi:10.1016/j.compscitech.2005.05.028

Gailliez-Degremont, E., Bacquet, M., Laureyns, J., & Morcellet, M. (1997). Polyamines adsorbed onto silica gel: A Raman microprobe analysis. *Journal of Applied Polymer Science*, *65*(5), 871–882. doi:10.1002/(SICI)1097-4628(19970801)65:5<871::AID-APP4>3.0.CO;2-K

Ghosh, A. (2001). *Secondary steelmaking: principles and applications*. CRC Press.

Gonsalves, K. E., Carlson, G., Chen, X., Kumar, J., Aranda, F., Perez, R., & Jose-Yacaman, M. (1996). J. "Surface-functionalized nanostructured gold/polymer composite films. *Journal of Materials Science Letters*, *15*(11), 948–951. doi:10.1007/BF00241434

Gregorio, R. Jr, Cestari, M., & Bernardino, F. E. (1996). Dielectric behavior of thin films of beta- PVDF/PZT and beta-PVDF/BaTiO3 composites. *Journal of Materials Science*, *31*(11), 2925–2930. doi:10.1007/BF00356003

Grohens, Y., Schultz, J., & Prud'homme, R. E. (1997). PMMA conformational changes on γ-alumina powder: Influence of the polymer tacticity on the configuration of the adsorbed layer. *International Journal of Adhesion and Adhesives*, *17*(2), 163–167. doi:10.1016/S0143-7496(96)00035-8

Gutman, E. M., & Bobovitch, A. L. (1996). Mechanical stimulation of pentabromobenzyl acrylate polymerization on Mg(OH)2. *European Polymer Journal*, *32*(8), 979–983. doi:10.1016/0014-3057(96)00033-X

Haertling, G. H., & Am, J. (1999). *Ferroelectric Ceramics: History and Technology. J. Am. Ceram. Soc.*, *82*(4), 797–818.

Haines, J., Leger, J. M., & Bocquillon, G. (2001). Synthesis and design of superhard materials. *Annual Review of Materials Research*, *31*(1), 1–23. doi:10.1146/annurev.matsci.31.1.1

Hille, R. (2002). Molybdenum and tungsten in biology. *Trends in Biochemical Sciences*, *27*(7), 360–367. doi:10.1016/S0968-0004(02)02107-2

Hofmann, F. A., & Skudelny, D. (1990). Surface treated mineral nanoparticles - Growing markets for specialty products. *9th Industrial Materials International Congress*.

Hornsby, P. R., & Premphet, K. (1997). Fracture toughness of multiphase polypropylene composites containing rubbery and particulate inclusions. *Journal of Materials Science*, *32*(18), 4767–4775. doi:10.1023/A:1018683014823

Jang, J., & Yi, J. (1996). A dynamic mechanical thermal analysis on allylester polymers and composites filled with alumina. *Appl. Polym. Sci.*, *61*(12), 2157–2163. doi:10.1002/(SICI)1097-4628(19960919)61:12<2157::AID-APP14>3.0.CO;2-4

Kiiko, V. S., Makurin, Y. N., & Ivanovskii, A. L. (2007). *BeO-based ceramics: preparation, physic chemical properties and application*. Ural Branch of the Russian Academy of Sciences.

Kim, J. K., Lee, E.-S., Kim, D.-H., & Kim, D.-G. (2004). Ion Beam-induced Erosion and Humidity Effect of MgO Protective Layer Prepared by Vacuum Arc Deposition. *Thin Solid Films*, *447-448*, 95–99. doi:10.1016/j.tsf.2003.09.029

Klapcinski, T., Galeski, A., & Kryszewski, M. (1995). Polyacrylamide gels filled with ferromagnetic anisotropic powder: A model of a magnetomechanical device. *Appl. Polym. Sci.*, *58*(6), 1007–1013. doi:10.1002/app.1995.070580606

Kody, R. S., & Martin, D. C. (1996). Quantitative characterization of surface deformation in polymer composites using digital image analysis. *Polymer Engineering and Science*, *36*(2), 298–304. doi:10.1002/pen.10416

Kretzschmar, B. (1996). The plastics industry and featuring. *Kunststoffe Plast Europe*, *86*(4), 20–22.

Kubat, J., Kuzel, R., Krivka, I., Bengtsson, P., Prokes, J., & Stefan, O. (1993). New Conductive Polymeric Systems. *Journal of Synthetic Metals*, *54*(1/3), 187–194. doi:10.1016/0379-6779(93)91059-B

Labella, R., Braden, M., & Deb, S. (1994). Novel hydroxyapatite-based dental composites. *Biomaterials*, *15*(15), 1197–1200. doi:10.1016/0142-9612(94)90269-0

Nagieb, Z. A., & El-Sakr, N. S. (1997). Application of spectroscopic techniques and physical parameters for studying the effect of apatite on the properties of wood pulp. *Polymer Degradation & Stability*, *57*(2), 205–209. doi:10.1016/S0141-3910(97)00001-3

Nguyen, T. N., Lethiecq, M., Levassort, F., & Patat, F. (1995). Ultrasonic Measurements of Longitudinal and Shear Moduli in Loaded Epoxy Networks. *International Journal of Polymer Analysis and Characterization*, *1*(4), 277–287. doi:10.1080/10236669508233881

Ni, J., & Wang, M. (2002). In vitro evaluation of hydroxyapatite reinforced polyhydroxybutyrate composite. *Materials Science and Engineering C*, *20*(1-2), 101–109. doi:10.1016/S0928-4931(02)00019-X

Noel, Y., Zicovich-Wilson, C. M., Civalleri, B., D'Arco, P., & Dovesi, R. (2002). Polarization properties of ZnO and BeO: An ab initio study through the Berry phase and Wannier functions approaches. *Physical Review B: Condensed Matter and Materials Physics*, *65*(1), 014111. doi:10.1103/PhysRevB.65.014111

nts '96, Conference Proceedings.

Ou, Y. C., Yu, Z. Z., Vidal, A., & Donnet, J. B. (1996). Effects of alkylation of silicas on interfacial interaction and molecular motions between silicas and rubbers. *Journal of Applied Polymer Science*, *59*(8), 1321–1328. doi:10.1002/(SICI)1097-4628(19960222)59:8<1321::AID-APP16>3.0.CO;2-8

Patnaik, P. (2002). *Handbook of Inorganic Chemicals*. McGraw-Hill.

Persson, A. L., & Bertilsson, H. (1996). Morphological effects in SAN-PA6 blends induced by aluminum borate whiskers. *Composite Interfaces*, *3*(4), 321–332. doi:10.1163/156855495X00101

Rebeiz, K. S., Rosett, J. W., Nesbit, S. M., & Craft, A. P. (1996). Tensile properties of polyster mortars using PET and fly ash wastes. *Journal of Materials Science Letters*, *15*(14), 1273–1275. doi:10.1007/BF00274399

Robert, J. F., Fragnier, P., & Europe, L. (1996). Talc production: An overview. *Macromolecular Symposia*, *108*(1), 13–18. doi:10.1002/masy.19961080104

Robinson, S. M. (1994). The Materials and Techniques of William Robinson. *Polymers Paint Colour Journal, 184*(4352).

Rosenov, M. W. K., & Bell, J A E. (1997). Plastics - Saving Planet Earth. *Society of Plastics Engineers (SPE)- Conference Proceedings; 55th Annual Technical Conference, ANTEC 97.*

Sanchez-Solis, A., & Estrada, M. R. (1996). On the influence of sands on low density polyethylene photodegradation. *Polymer Degradation & Stability*, *52*(3), 305–309. doi:10.1016/0141-3910(96)00030-4

Severin, E. J., Doleman, B. J., & Lewis, N. S. (2000). Relationships among Resonant Frequency Changes on a Coated Quartz Crystal Microbalance, Thickness Changes, and Resistance Responses of Polymer–Carbon Black Composite Chemiresistors. *Analytical Chemistry*, *72*(4), 658–668. doi:10.1021/ac9910278

Shang, S. W., Williams, J. W., & Soderholm, K. J. M. (1995). Work of adhesion influence on the rheological properties of silica filled polymer composites. *Journal of Materials Science*, *30*(17), 4323–4334. doi:10.1007/BF00361512

Sherman, L. M. (1997). Polymer Blends: Processing, Morphology, and Properties. *Plastics Technology*, *43*(4), 26–28.

Shiao, S. J., & Wang, T. Z. (1996). Dry self-lubricating composites. *Composites*, *27B*(5), 459–465. doi:10.1016/1359-8368(96)00012-1

Sjogren, B. A., & Berglund, L. A. (1997). Failure mechanisms in polypropylene with glass beads. *Polymer Composites*, *18*(1), 1–8. doi:10.1002/pc.10255

Söhnel, & Garside. (1992). Precipitation: basic principles and industrial applications. In *Oxford*. Butterworth-Heinemann.

Stricker, F., & Muelhaupt, R. (1996). Compatibilized Polypropylene Hybrid Composites: Influence of Elastomeric Interlayers on Mechanical Properties and Nucleation Behaviour. *High Performance Polymers*, *8*(1), 97–108. doi:10.1088/0954-0083/8/1/007

Stwertka, A. (2002). A Guide to the elements (2nd ed.). New York: Oxford University Press.

Tan, F., Qiao, X., & Chen, J. (2006). Removal of chemisorbed lubricant on the surface of silver flakes by chemicals. *Applied Surface Science*, *253*(2), 703–707. doi:10.1016/j.apsusc.2005.12.163

Tanaka, A., Kitamura, M., Tokumitsu, K., & Nishimura, H. (1997). Ultrasonic Evaluation of Butt Fusion Welded Polyethylene Pipes. *Plastics saving planet earth: conference proceedings / ANTEC '97, Toronto, April 27 - May 2. SPE, Society of Plastics Engineers; Vol. 3: Special areas*, 1223-1227.

Tanaka, T., Montanari, G. C., & Muilhaupt, R. (2004). Polymer Nanocomposites as Dielectrics and Electrical Insulation-perspectives for Processing Technologies, Material Characterization and Future Applications. *IEEE Transactions on Dielectrics and Electrical Insulation*, *11*(Oct), 763–784. doi:10.1109/TDEI.2004.1349782

Tavman, I. H. (1996). Thermal and mechanical properties of aluminum powder-filled high-density polyethylene composites. *Journal of Applied Polymer Science*, *62*(12), 2161–2167. doi:10.1002/(SICI)1097-4628(19961219)62:12<2161::AID-APP19>3.0.CO;2-8

Torro-Palau, A., Fernandez-Garcia, J. C., Orgiles-Barcelo, A. C., Pastor-Blas, M. M., & Martin-Martinez, J. M. (1997). Structural modification of sepiolite (natural magnesium silicate) by thermal treatment: Effect on the properties of polyurethane adhesives. *Int. J. Adhesion Adhesives*, *17*(2), 111–119. doi:10.1016/S0143-7496(96)00039-5

Toure, B., Lopez Cuesta, J. M., Gaudon, P., Benhassaine, A., & Crespy, A. (1996). Fire resistance and mechanical properties of a huntite/hydromagnesite/antimony trioxide/decabromodiphenyl oxide filled PP-PE copolymer. *Polymer Degradation & Stability*, *53*(3), 371–379. doi:10.1016/0141-3910(96)00100-0

Toure, B., Lopez Cuesta, J.-M., Longerey, M., & Crespy, A. (1996). Incorporation of natural flame retardant fillers in an ethylene-propylene copolymer, in combination with a halogen-antimony system. *Polymer Degradation & Stability*, *54*(2-3), 345–352. doi:10.1016/S0141-3910(96)00061-4

Tse, M. F., & Schuster, R. H. (2006). BIMS/filler interactions. I. Effects of filler structure. *Journal of Applied Polymer Science*, *100*(6), 4943–4956. doi:10.1002/app.23789

Uchino, K. (2000). *Ferroelectric Devices*. Marcel Dekker, Inc.

Visser, S. A., Hewitt, C. E., Fitzgerald, J. J., Ferrar, W. T., & Binga, T. D. (1997). Effect of filler type on the response of polysiloxane elastomers to cyclic stress at elevated temperatures. *Journal of Applied Polymer Science*, *65*(13), 1805–1820. doi:10.1002/(SICI)1097-4628(19970328)63:13<1805::AID-APP13>3.0.CO;2-X

Wang, J. Y., & Ploehn, H. J. (1996). Dynamic mechanical analysis of the effect of water on glass bead–epoxy composites. *Applied Polymer Science*, *59*(2), 345–357. doi:10.1002/(SICI)1097-4628(19960110)59:2<345::AID-APP19>3.0.CO;2-V

Wei, Y., & Sancaktar, E. (1996). Dependence of electrical conduction on the film thickness of conductive adhesives: Modeling, computer simulation, and experiment. *Journal of Adhesion Science and Technology*, *10*(11), 1199–1219. doi:10.1163/156856196X00193

Wiebking, H. E. (1996). The performance of ultrafine talc in rigid PVC. *Vinyl and Additive Technology*, *2*(3), 187–189. doi:10.1002/vnl.10121

Williams, H. R., & Greenall, R. (1995). The Use of Infra-Red Thermography in Determining Melt Temperature Quality. *Polymers, Laminations & Coatings Conference*, 53-62.

Wong, K. W. Y., & Truss, R. W. (1994). Effect of fly-ash content and coupling agent on the mechanical properties of fly-ash filled polypropylene. *Composites Science and Technology*, *52*(3), 361–368. doi:10.1016/0266-3538(94)90170-8

Yang, L., & Schruben, D. L. (1994). Electrical resistivity behavior of mold-cast metal-filled polymer composites. *Polymer Engineering and Science*, *34*(14), 1109–1114. doi:10.1002/pen.760341403

Yu, S. C., Kithva, P. H., Rajemdra, K., Philip, C., & Khor, K. A. (2005). In vitro apatite formation and its growth kinetics on hydroxyapatite/polytheretherketone biocomposites. *Biomaterials*, *26*(15), 2343–2352. doi:10.1016/j.biomaterials.2004.07.028

Zhao, W., Hasegawa, S., Fujita, J., Yoshii, F., Sasaki, T., Makuuchi, K., Sun, J., & Nishimoto, S. (1996). Effect of irradiation on pyrolysis of polypropylene in the presence of zeolites. *Polymer Degradation & Stability*, *53*(2), 199–206. doi:10.1016/0141-3910(96)00084-5

Zhu, S. J., & Iizuka, T. (2003). Fabrication and mechanical behavior of Al matrix composites reinforced with porous ceramic of in situ grown whisker framework. *Mate. Sci. Eng.*, *A354*, 306.

Chapter 4
NanoDielectric Theories

ABSTRACT

This chapter sheds light on the recent nanotechnology theoretical models for interphase power law IPL model, inhomogeneous interphase, and multi-nanoparticles technique. Moreover, this chapter reviews deliberate hypothetical researches of the effective dielectric constant for polymer/filler nanocomposites and its reliance on "filler concentration, the interphase interactions, polymer filler dielectric constant, and interphase dielectric constant." This chapter also investigates the prediction of the dielectric constant of new nanocomposite materials dependent upon exponential power law model. Thus, this work moves from the dielectric properties of beginning polymer matrix forward and predicts the dielectric properties of new nanocomposite materials to be utilized for high voltage and directing materials by adding specified nanoparticles with polymer matrix.

Exponential control theory model is a great fitting and has enabled us to control rate of change of interphase properties, interphase volume constant and filler particle shape, thereby offering us a wider scope for works. Hence, this chapter demonstrates novel modern materials improving the dielectric aspects of new nanocomposite modern materials by interphase power law IPL models. These models take under record collaborations between the segments of the composite framework in the structure from the beginning interphase regions. The resultant models depend on the permittivity of the filler component, the matrix part and the interphase area, as well as the volume fractions for each. Furthermore, the reasons and effects of the interphase region with respect to an assortment from the beginning complex composite frameworks have been investigated. There is also an investigation of the effects of the composite filler sorts and filler surface areas, in addition to the dielectric aspects of the interphase area, looking into new modern materials. This brings about shortages that can additionally be utilized for electrical and thermal conductivity, magnetic permeability and diffusivity, provided that round inclusions and matrix are isotropic. This part clarifies recent modern materials improving the dielectric aspects of new nanocomposite modern materials by interphase power theory IPL model. This model takes under account connections between parts of the composite framework in the manifestation for interphase regions. The resultant model depends on the permittivity of the filler component, the matrix part, the interphase area, in addition to the volume fractions of each. The results and effects of the interphase region once an as-

DOI: 10.4018/978-1-7998-3829-6.ch004

sortment is made for perplexing composite frameworks are also investigated. Effects of the composite filler sorts and filler surface areas, as well as the dielectric values of the interphase region are examined by looking into new industrial materials. This effect can also be utilized for electrical and thermal conductivity, magnetic permeability and diffusivity, given that the round inclusions and matrix are isotropic.

4.1 NANO-DIELECTRIC COMPOSITES

Polymeric composites committed for particles, such as conductive, ferroelectric and alternately metal particles are some of the vital building materials utilized for resistors, exchanging devices, directing pastes, segments in the xerographic machine and separators over polymer electrolyte film energy units. Percolation theory predicts that different values of a percolating framework can be identified with the likelihood for occupation from beginning locales inside the percolation lattice, eventually perusing energy law relations. Moreover, the exponents of these control relations are widespread in any case in the framework. Composite and nanocomposite industrial materials are continuously investigated as protecting material for electromagnetic compatibility (EMC) and electromagnetic interference (EMI) applications. In these systems and in particle/matrix conductivities and volume stacking of the particles in the matrix, the arbitrariness of distribution, poly dispersivity and interfacial thermal resistance play a role in figuring out the successful conductivity of the composite material (Kanuparthi et al., 2006; Nan et al., 1997; Wu et al., 2007; Zhang et al., 2005; Zhang et al., 2004).

4.1.1 Percolation Theory

Percolation theory is a general model for the depiction of measurable techniques, and it is a regular technique in the investigations for pre-breakdown forms within solids. In addition, nanoparticle size can plan new composite and thermal interface nanocomposite industrial materials by nanotechnology science which can lead to an enormous upgrade in the magnetic properties of the composite materials, influencing, in turn, the execution of the modern requisitions. Percolation theory aims at describing the connectivity properties in irregular geometries and to investigate them for exploration of the physical procedures. The percolation sorts are shaping a nonstop system of particles in conductive polymeric composite which can be totally fulfilled. In the innovative technological applications, conductivity for polymeric composite example can be acknowledged in a circular particle which, in traditional blending decisions from the beginning particles, is subjected to fundamental matrices to figure out the effective conductivity of composites, if the incorporation period is scattered in the matrix stage irregular appropriation (Salvadori et al., n.d.). It can be formed as follows:

$$\sigma = \sigma_0(x - x_c)^t \tag{1}$$

Where,

σ_0 is the proportionality constant,
x is the volume concentration of conducting phase,
x_c is the critical concentration of conducting phase,
t is the exponential factor for percolation and tunneling percolation.

Additionally, there is a recommendation for connected energy on polymeric composite material that has been used to plan hypothetical models to foresee a powerful conductivity of polymeric composites, and it corresponds to normal sensibility. However, the variety of electric safety because of the mechanical load connected to the sample's extraordinary zone of contact between particles expands and cuts the distances between particles spotted over parallel rows. Hence, Fig. (1) indicates a schematic design of the polymeric composite as an example of separate volume fractions of particles.

Figure 1. Schematic layout for polymeric composite specimen with particles (Salvadori et al., n.d.).

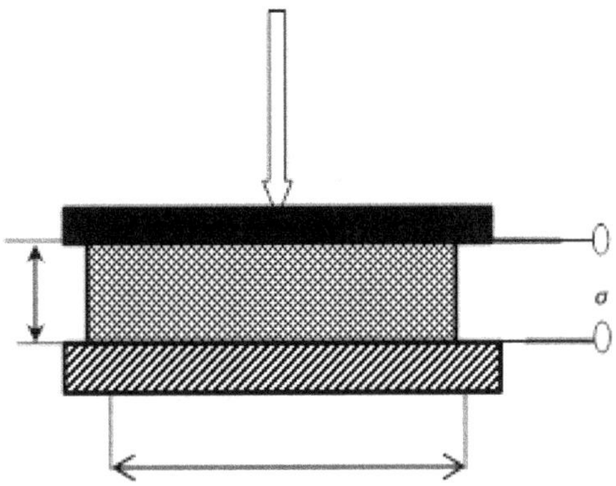

However, the leading components will transcend geometric contact. The hypothesis predicts that the critical exponent e "t" will be under two and the procedure is known as percolation; percolation alludes to the stream of current through irregular resistor networks. When the conducting elements are not in geometric contact, the inter-particle tunneling is dominant. Thus, this research investigates the percolation clinched alongside polymeric composites with different cost-fewer particles and setup explanatory model parameters. Polymeric composites aggravate an enormous upgrade in the electric, dielectric and electromagnetic properties which will influence the execution of the modern requisitions during a setup test of d=h=0. 01m. Every chosen nanoparticle in this research has been spherically molded and measured, as 10nm in breadth for each grain size.

Table 1. Electric and thermal properties of suggested particles and industrial materials

Materials	Conductivity (S/m)	Thermal Conductivity K (W/m.K)
Graphite	$3x10^5$	200
Fe	10^7	80.2
ZnO	$1.69x10^7$	21
MgO	10^8	40
Al_2O_3	10^{14}	35
Si	$1.56x10^{-3}$	148
Epoxy	10^{-12}	1.04
Glass	10^{-15}	1.2
PTFE	10^{-16}	0.25

The fundamental electrical and thermal depiction properties of the utilization from beginning particles has been portrayed in table 1. These particles have been utilized to upgrade electric properties of polymeric composite and nanocomposite industrial materials.

Characterization of Nano-Dielectric Magnetic Composite Materials

Recent hypothetical models for estimating powerful conductivity and successful thermal conductivity about new suggested nanocomposite thermal interface industrial materials are under investigation. The accompanying outcomes have accounted for new recommended composite and nanocomposite industrial materials that have been improved in their characterization reaction to explore particle sorts and their concentrations along these lines of powers connected in new proposed materials. Figure 2 indicates upgrading the effective conductivity of epoxy polymeric composite materials, eventually perusing the addition of different rates of iron particles in irregular circulations of epoxy composites. However, the

Figure 2. Effective conductivity of Iron/ Epoxy polymeric composites

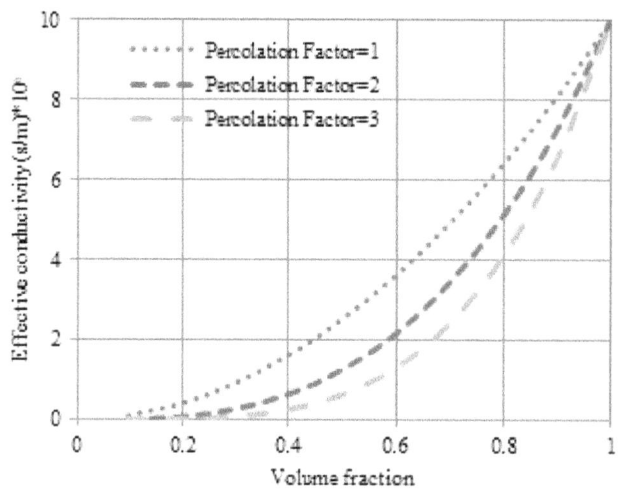

effective conductivity of iron / epoxy polymeric composite materials is diminished bit by bit towards expanding the percolation variable of the polymeric composite materials.

Figure 3 demonstrates upgrading the effective conductivity of epoxy polymeric composite materials, eventually perusing the addition of different rates for (Graphite, SiO_2, Also MgO) particles for irregular circulations at a sure percolation variable of 1.8. It is clear that SiO_2 particles expand the successful conductivity of epoxy polymeric composite materials higher than MgO and graphite particles.

Figure 3. Effective conductivity of Epoxy Polymeric composites

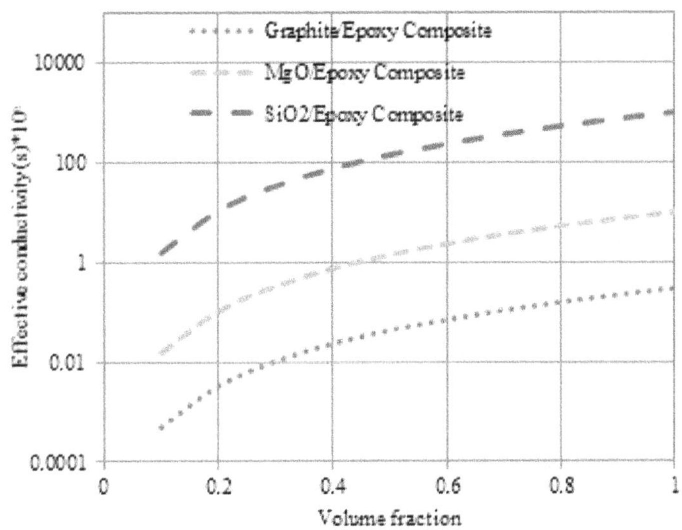

4.2 NANO-DIELECTRICS INTERPHASE PREDICTION MODELS

Interphase Power theory IPL model may be in view of a straightforward development of a general power law model in which a composite framework holding filler interphase and matrix locales can be dealt with as an interesting three-component composite framework, adding two grade parts (matrix and filler) and an interphase area that is inextricably subordinate upon the values of the filler and matrix parts. In this chapter, there is a discussion of the reasons and effects of the interphase region ahead of an assortment for complex composite frameworks, the effects of the composite filler sorts, filler surfaces areas and the dielectric values of the interphase region. Effects of the composite filler sorts, filler surface areas and additionally the dielectric values of the interphase region have been investigated looking into new industrial materials. Finally, this research clarifies the dielectric aspects for new nanocomposite polymeric materials towards IPL. This part reviews recent advances in the field of creating modern polymers towards nanocomposites materials. The described materials have been taken in both academic and industrial consideration as a direct result to show sensational change to properties in choosing low filler substances. In this sense, the sorts and manifestations of mechanical polymer, structure, and properties from beginning polymer–nanocomposites will be examined. Traditionally, polymeric materials have been filled for manufacturing alternately common inorganic mixes in place to enhance their properties, essentially to decrease cosset. Furthermore, this single section investigates different factors influencing the electrical properties of the polymers insulating materials and the ideal properties to the novel nanoscience materials. Finally, this part examines the correlation between the chosen nanoparticles on polymers and preparation methods organizing polymer nanocomposite. Polymeric materials are alternately used in electrical and hardware provisions. These applications require multi-functionality in a single material which is not generally found in an ordinary polymer. A prudent approach to manufacture multi-functional materials will be to blend polymer with other materials. In the nanocomposite frameworks, the fillers commonly bring nano-meter-scale dimensions, the property change in such frameworks regularly goes for tradeoffs. This research investigates the dielectric properties of nanocomposites, the place at which every installed incorporation is encompassed towards an inhomogeneous interphase, and the effects from the beginning inhomogeneous interphase in the mass modulus about composite holding round inclusions. The new changed model has been demonstrated in terms of based building in optoelectronic bundling materials for dielectric constant demonstrating. The properties of composite materials of the new nanocomposite modern materials have been created round inclusions that are installed clinched alongside a matrix about a portion sort. Different fitting models have been investigated in new modern materials for examination of the correctness of their properties. The extent of the round incorporation acts as the reinforcing stage and has a real effect on the dielectric properties of composite modern materials. Therefore, for the prologue of nanoparticles as the favored reinforcing stage for a percentage of composites, the interphase has a real effect on its dielectric properties. The electric reaction of nanocomposite industrial materials has also been investigated in terms of time permits for the effects and results of interphase area looking into an assortment of complex composite frameworks by utilizing interphase power law IPL model. Hence, this part investigates the effective dielectric reaction of nanocomposite modern materials and effects of the interphase area ahead of an assortment of complex composite frameworks. The measurable methodology for disappointment of polymeric insulators permits us to foresee the methods and instruments for disappointments which occur inside insulating materials. This single section illustrates the disappointment facts of an existence model for AC electrical aging from beginning nanocomposite industrial materials, which depict the components for electrical breakdown under secondary voltage electric fields towards propelled recreation model project. Situation investigations are used to show the lifestyle and instruments which model characteristics that can be identified with the parameters of the disappointment detail. Realized lifetime of materials can be assessed, eventually perusing a method for this model. Aging for polymeric insulators has been analyzed through different secondary voltage electric fields. Hence, the likelihood of foreseeing the lifetime of distinctive examples has succeeded. This research reviews improvement from the beginning exceptionally directing polymers for beneficial strength and satisfactory transforming values. A conductive adhesive has been made by adding nano measured particles to distinctive sorts of polymers. Furthermore, this work done attempts to show the effects of additions from the beginning nano-sized particles on the electrical leading properties and illustrates that electrical resistivity diminishes and makes a difference to the structure of the conductive way. New techniques for moving conductivity to another level have also been pinpointed. Finally, the preparation methods for manufacturing nanocomposites modern leading materials have been illustrated.

This effort is an investigation of the dielectric properties of nanocomposites in which installed circular incorporation may be encompassed towards the inhomogeneous interphase and the effect of inhomogeneous interphase on the mass modulus of a nanocomposite holding round incorporation. In this chapter, a combination of differential equations utilizing the hypothesis of mechanical properties of fiber-strengthened materials is determined. These differential equations model has ascertained the dielectric properties of the beginning nanocomposites to a whole class of capacities describing the interphase inhomogeneity. Our altered model is directed to based building from the beginning optoelectronic bundling materials of the beginning dielectric constant displaying. The power theory profile PLP, equal homogeneous interphase models and the effect of the combined models has been mulled over. Dielectric constant is a vital property that is required for an assortment of electrical requisitions. A critical challenge is the ability to control dielectric aspects of electrical materials utilizing nanotechnology strategies. Computational Recreation projects have the possibility to join the hypothesis and test fill concerning illustration as a third capable exploration methodology; therefore, this section is dedicated to the upgrade and regulation of dielectric properties of electrical industrial materials (polyethylene, polypropylene acrylonitrile butadiene styrene, polyvinyl chloride, polyimide, polyetherimide and polyethylene terephthalate) utilizing different nanoparticles and taking into account late hypothetical models for registering the effective reaction of nanocomposites. Moreover, the effective nanoparticles and their ideal concentrations have been identified for regulating dielectric characterization. This part prompts synthesizing new electrical insulating polyvinyl chloride (PVC) nanocomposites to accomplish more cost-effective, energy-effective and henceforth advanced exceptional natural materials towards utilizing nanotechnology strategies. It has been examined on simulations and test work done in the effort to examine the effect of the beginning cost-fewer nanoparticles looking into electric properties (resistance and susceptance) of the new clay framework nanocomposites. Cambridge building selector (CES) project has conveyed the electrical/mechanical predictable models for the proposed materials. In addition, it has compared between electrical, thermal and mechanical characterization from the beginning new polyvinyl chloride nanocomposites for the same unfilled industrial materials inside variant frequencies dependent upon 1 kHz. This section transforms and portrays cost-fewer polypropylene (PP) nanocomposite films; the experimental work conducted has been investigated for concentrating on the electric properties of the new nanocomposite materials and compared for unfilled industrial materials to a recurrence extent up to 1 kHz. A little addition of nanoparticles (clay, and fumed silica) with polypropylene has indicated calculable change in the electric reactance and conductance towards separate recurrence up to 1kHz. In addition, an electric spectroscopy has measured the electric properties of polypropylene with and without nanoparticles under variant temperatures (20°C, and 60°C). Cambridge Engineering Selector (CES) program has the electrical/mechanical predictable models to the suggested materials. Finally, this chapter prompts synthesizing electrical insulating polypropylene nanocomposite films whose electrical properties are appropriately examined in order to accomplish more cost-effective, energy-effective and subsequently naturally better materials for electrical engineering. Electrical materials play an important part, clinched alongside developing secondary voltage modern requisitions. In these applications, nanoparticles have been utilized for upgrading structure and electrical characterization. In addition, effective dielectric constant of new multi-nanocomposites materials has been examined in light of recent hypothetical approaches, then, the course of action of multi-nanoparticles clinched alongside base matrix group material has been examined for controlling ahead dielectric constant for electrical industrial materials. Moreover, specification of ideal sorts and concentrations of multi-nanoparticles is conducted to regulate the dielectric constant of variant electrical insulating materials by utilizing the multi-nanoparticles strategy. Trends of utilizing multi-nanoparticles has been portrayed, as well as the mechanical characteristics of separate nanoparticles and universal modern materials. The compound structural level of a polymer composite, for an interfacial area, is comprised of molecules of the polymer matrix fortified by the surface of the filler particle. This interfacial-bonding region is termed the interphase, and the effects starting with the binding effect that the unbending filler particles have on the mobility of the polymer particles in the matrix are addressed. The resultant model, termed the Interphase power law IPL model, depends on the permittivity of the filler component, the matrix part, the interphase region and the volume fractions of each. The IPL model is a straightforward development of a general Power Law model in which a composite framework holding filler, interphase and matrix areas might be approached as an interesting three-part composite framework adding two essential parts (matrix and filler) and an interphase area that is inextricably subordinate upon the aspects of the filler and matrix segments. In this chapter, the effected parameters which were influenced in the nanocomposite framework and the dielectric aspects of the interphase region will be investigated. This chapter is an investigation of the dielectric properties for composites which each installed circular Incorporation may encompass eventually perusing an inhomogeneous interphase, and the effect of inhomogeneous interphase on the mass modulus of a composite holding round Incorporation. To create high execution particulate composites, it is important to have some essential understanding of the coupled match between differential equations utilizing the hypothesis of mechanical properties from the beginning fiber-strengthened materials. These differential equations model the dielectric properties for composites for a whole class of capacities describing the interphase inhomogeneity. In this part, a high execution particulate composite has been produced, and this model has been consolidated by the beginning models Vo and Shi, and towards model-based building from the first optoelectronic bundling materials. Thus, in this chapter, a detailed discussion of the Power Law Profile PLP, equal homogeneous interphase models, the effects from the beginning particle size, particle/matrix interface bond and particle stacking on particulate–polymer composites are reviewed. The dielectric properties of composites which are made of inserted circular inclusion, which may be encompassed towards an inhomogeneous interphase in a matrix of polymer have been improved. The dielectric properties of composites have been depicted in the interphase inhomogeneity by comprehending the differential equations of Power Law Profile PLP model. The suggested model has been investigated dependent upon drawing on the beginning optoelectronic bundling materials dielectric constant demonstrating. In addition, results of this chapter have been compared with the results of equivalent homogeneous interphase models. Nanoparticles have achieved a huge success in upgrading electrical properties of polymer modern materials as manifestation nanocomposites. Thus, preparation of new Polyethylene nanocomposites can support both manufactures and users to upgrade electrical execution of Polyethylene provisions. This has led to improving electric and dielectric characterizations of polyethylene by adding costive nanoparticles to low thickness polyethylene (LDPE) and secondary thickness polyethylene (HDPE) as a base matrix. Polyethylene trapping properties are profoundly altered by the vicinity of costive nanoparticles (clay, and fumed silica) nanoparticles. This chapter has also concentrated tentatively on the effect of costive nanoparticles (clay, and fumed silica) and their concentrations on electric and dielectric properties from the beginning polyethylene materials. An experimental comparative study has been examined regarding Polyethylene nanocomposites to explore business polyethylene materials in order to clarify the effect of the beginning sorts and concentrations from the beginning nanoparticles for upgrading electric and dielectric Polyethylene aspects.

4.3 INTERPHASE POWER LAW IPL MODEL

Interphase Power Law (IPL) relationships are truly and regularly utilized in dielectric demonstrating of composite frameworks (Nelson, 2001; Oltean & Motoc, 2012; Singha & Thomas, 2008; Tuncer et al., 2007). The two-component power-law models for unpredictable permittivity have been utilized extensively for an extensive variety from the beginning material frameworks investigated under the physical and electrical aspects of particulate filled polymer composites, supporting the presence of the beginning ''interphase'' locales in polymer-dispersant interfaces (Gershon et al., 2001; Ozmusul & Picu, 2002; Qu & Wong, 2002; Vo & Shi, 2002). The interphase region comprises polymer particles that are fortified alternately and overall situated in the matrix-filler interface. This interfacial-bonding locale can be termed interphase, and the outcomes starting with the binding effect that the unbending filler particles have on the versatility of the polymer particles in the matrix are addressed. In the heft matrix regions of the composite, the polymer chains embrace irregular orientation configurations. The polymer chains and side assemblies are generally nothing with moving - stretching, bending and wagging - to minimize their conformational vitality and amplify their inward entropy. If fortified to (or physically contiguous to) the filler surface, however, the polymer particles are confined to development. An estimation of the effect of the change in the dielectric constant as an aftereffect of the compound holding of the polymer matrix in the filler surface of a composite framework can be computed by utilizing atomic polarizability equations. The changeover dielectric constant for the polymer material in the interphase region is proportional to the expand alternate diminish in the dipole polarization. A more complete portrayal of the dipole polarization inside the interphase region of a composite framework and the entailing dielectric constant calculations are provided (Todd & Shi, 2002; Vo et al., 2001).

4.3.1 Theory of Interphase Model

Currently, there is a depiction of the straightforward interphase model fit to give extra knowledge about both the true and fanciful parts of the mind, boggling permittivity from the beginning composite frameworks incorporating the interphase hypothesis. The composite framework holding filler, interphase and matrix areas is illustrated inFfig. 4. As the filler particles approach together, the interphase regions encompassing each filler particle start to overlap, thereby lessening the effective interphase volume fraction.

Figure 4. Interphase region surrounding the filler particles in a composite system

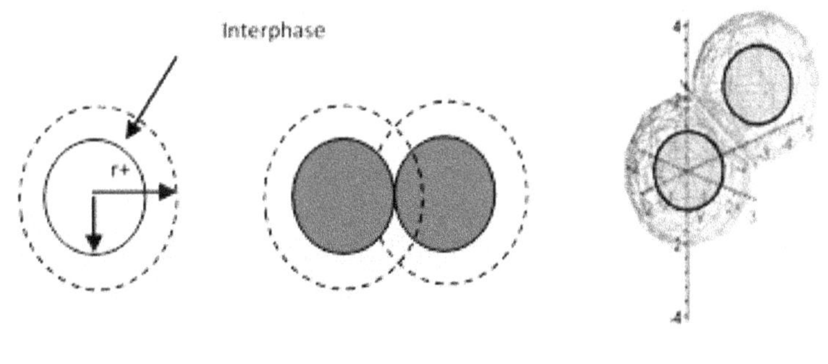

Power law relationships are quite often used in dielectric modeling of composite systems (Nelson, 2001; Oltean & Motoc, 2012; Singha & Thomas, 2008; Tuncer et al., 2007). These relationships model the effective permittivity of two-component systems using the volume fraction of each component according to:

$$\mu_c^2 = \AE_f \mu_f^2 + (1 - \AE_f)\mu_m^2 \quad (2)$$

where ε_c, ε_f and ε_m are the complex dielectric permittivity of the composite system, the filler and the matrix respectively, φ_f is the volume fraction of filler component of the composite system, and β is a dimensionless parameter representing the shape and orientation of the filler particles within the bulk composite (Salvadori et al., n.d.). Common examples of this model are the linear mixtures model β=1, the Birchak formula β=0.5 and the Landau, Lifshitz, Looyenga formula β=0.333 (Oltean & Motoc, 2012). More generally, for a composite comprised of n number of components, the power law mixtures model can be written as:

$$\mu_c^{2} = \sum_{i=1}^{n} Æ_i \mu_i^{2} \tag{3}$$

Where: ε_c and ε_i are the complex dielectric permittivity of the composite system and any constituent component of the composite respectively, φ_i is the volume fraction of the constituent component, and β is a dimensionless parameter representing the shape and orientation of the filler particles within the bulk composite.

Despite being truly and effectively used to model an extensive variety of composite systems, the energy law mixtures model doesn't represent cooperation between the segments of the composite, which is a genuine constraint. Extra dielectric models to mixtures about composite constituents have been developed, adding the Maxwell-Garnett model and the Bruggeman model. Each of these mixture models gives different methodologies to foreseeing the dielectric aspects of composite frameworks. These approaches, however, don't provide a thorough method to concentrate on the genuine and fanciful parts of the complex permittivity from the composite frameworks or represent the cooperation between the unique segments.

A composite framework adding two elementary segments (matrix and filler) and an interphase area can be dealt with as an exceptional three-component composite framework in which the interphase volume is inextricably subordinate upon the aspects of the filler part. The two-component power-law association depicted in Eq (1) are undoubtedly stretched out to a three-part composite system, concerning illustration portrayed by:

$$\mu_c^{2} = Æ_f \mu_f^{2} + Æ_i \mu_i^{2} + Æ_m \mu_m^{2} \tag{4}$$

For our composite model that contains an interphase region, φ_f, φ_i, and φ_m represent the volume fraction of the filler component, the interphase region and the matrix component of the composite, respectively, ε_c, ε_f, ε_f and ε_m are the complex permittivities of the composite, the filler, interphase and matrix regions of the composite, respectively, and β is a dimensionless parameter describing the shape and orientation of the filler particles within the composite. The complex permittivity is given by, $\mu_i = \mu_i{'} + \mu_i{''}$, where $\mu_i{'}$ is the dielectric constant and $\mu_i{''}$ is the dielectric loss over in our case, two of the segments refer to the constituent parts of the composite framework and the third part refers to the interphase region between the filler and matrix. Thus, the model incorporating the dielectric aspects of the interphase region is revealed by the mathematical Eq (3) which alludes the similarly to the IPL model. The model is appropriate to whatever random or uniform scattering of filler particles to a matrix.

The IPL model is interesting in that it incorporates two autonomous parts of the composite and also an interphase area that is dependent upon the aspects of the two parts. This model doesn't investigate three free segments, but rather two free segments and the interphase region between them. The LR model is viewed as a composite comprising three autonomous constituent segments with no connections between the segments. It is essential to perceive the qualification between the IPL model and a standard three-component power-law mixture. This chapter is a clarification, therefore, alluding to autonomous composite constituents as parts and the reliant constituents similarly as interphase. The refinement between these situations will be open for further demonstration. The filler volume fraction of equation (3), φ_f, is directly measured for a given composite system. The matrix volume fraction is given by, $\varphi_m=1-\varphi_f-\varphi_i$. The interphase region volume fraction, φ_i, is dependent upon the filler volume fraction, the filler surface area and the thickness of the interphase region surrounding each filler particle. The interphase volume fraction, Φ_i, is determined by monodisperse, spherical particles using:

$$\text{Æ}_{\text{i}} = \frac{\frac{4\text{ÀÆ}_{\text{i}}}{3}\left[\left(\left(r+\text{''}r\right)^3-r^3\right)-\mathcal{F}6\left(3\left(r+\Delta r\right)-\Delta r\right)\Delta r^2\right]}{r^3} \tag{5}$$

Where r is the radius of the filler particles, Δr is the thickness of the interphase region and $\mathcal{F}$ is an overlap probability function. Assuming that there are no areas of interphase overlap that would reduce the total volume of interphase, the interphase volume fraction, φ_i, is finally calculated by introducing an interphase overlap function, $\mathcal{F}$:

$$\text{Æ}_{\text{i}} = \left(1-\mathcal{F}\right)\left(S_f.\Delta r\right).\text{Á}_{\text{i}}.\text{Æ}_{\text{i}} \tag{6}$$

The interphase thickness is accepted and is equivalent to that of the matrix part, since it comprises matrix particles. Typically, the interphase thickness is a requirement by the beginning particular case molecular size of gyration (~5 should ~20nm). The determination of the interphase thickness is further depicted to a point of interest in (Qu & Wong, 2002; Vo & Shi, 2002; Vo et al., 2001). The interphase covers probability, F, arising from the vicinity of the singular filler particles. Likewise, as illustrated in figure 1, as the filler particles approach each other, the interphase areas encompassing every filler particle start will overlap, thereby diminishing the effective interphase volume fraction. The interphase cover likelihood is a capacity of the filler volume fraction, interphase thickness and the shape and size of the filler particles. Approximations have been produced to evaluate the cover for such particles utilizing Monte Carlo simulations or alternately explanatory answers for percolation models (Todd & Shi, 2003a; Todd & Shi, 2003b).

The IPL exponent, β, adjusts the effective permittivity reaction of the composite framework for material anisotropy. This parameter might be deciphered as a measure of geometry of the composite medium for exploring the connected electric field. For a composite framework hosting the singular parts turned parallel to the connected electric field, the value of β is 1. In this exceptional case, the distinct parts do not shield the opposite part starting with the effect of the connected field, and the resulting effective permittivity is basically the added substance permittivity from the beginning of each part weighted for its volume fraction. For a composite framework in which individual parts are turned to peroxide blonde (in series) of the connected electric field, the value of β is -1. The parallel and series configurations are representable the two geometrical extremes that can exist for the composite framework. The value of β for true (non-ideal) composite frameworks falls between 1 and -1 and relies on the geometry of the part that is acknowledged with a chance to be scattered in the different part. Equations (3) through (5) can be used to finish a generalized, far reaching model for the prediction of the perplexing dielectric permittivity of composite frameworks holding interphase-bonding areas. The IPL model takes into account interphase permittivity, interphase volume fraction, interphase overlap and filler shaper introduction effects. The IPL model can effortlessly be stretched with extra terms for multi-component composites holding interphase regions.

Empirical bounds that limit the range of permittivity predictions were established long ago by Weiner and later modified by Hashin-Shtrikman (Qu & Wong, 2002). These limits characterize the reasonably expected permittivity values of composite frameworks in light of the permittivities of the constituents. The Weiner limits provide for the widest allowed extent of permittivity predictions to two-component composite frameworks. These powerful permittivity limits are provided for as follows:

$$\in_{elf,max} = \varphi_f \varepsilon_f + (1 - \varphi_f)\varepsilon_m \tag{7}$$

$$\epsilon_{eff,min} = \frac{\epsilon_f \epsilon_m}{Æ_f \mu_f + (1 - Æ_f)\mu_m} \tag{8}$$

It is worth noting that these two bounds correspond to the two-component power-law mixture model in which the exponential factor, β equals 1 and -1 respectively.

4.3.2 Effective Parameters of IPL

There are two issues, the first issue is studying the effective parameters of PLP model, and the second issue is specifying aspects about new industrial materials to the dielectric values from the beginning nanocomposite polymeric materials eventually perusing IPL. Effective parameters from beginning PLP model are based on investigation of the effects of filler permittivity, interphase permittivity, filler surface area or interphase thickness and particle shape.

4.3.3 Effect of Filler Permittivity

The successful dielectric constant and dielectric misfortunes of the composite framework are indicated in Fig. 2 with three values of a filler dielectric constant. The effective dielectric constant of the composite expanded quickly for secondary value for filler dielectric constant and at the same time for a low value for filler dielectric constant. The powerful dielectric constant of the composite was easily diminished with the variety for filler volume fraction from zero in solidarity concerning illustration indicated in Fig. 5(a). In addition to the effective dielectric reductions for volume fraction evolving from zero to unity, the greatest dielectric reduction gives the idea of about 20% filler volume fraction concerning illustration indicated in Fig. 5(b) and that point smoothness diminished will achieve zero in solidarity volume fraction. For three values of filler unpredictable dielectric permittivity, the effective dielectric passing expanded the point when the filler dielectric permittivity expanded. It can be concluded from these predictions that representing the interphase regions refers to a deviation starting with linearity of the successful permittivity of the composite framework.

Figure 5. Effect of filler permittivity on the effective dielectric constant and effective dielectric loss as a function of filler volume loading

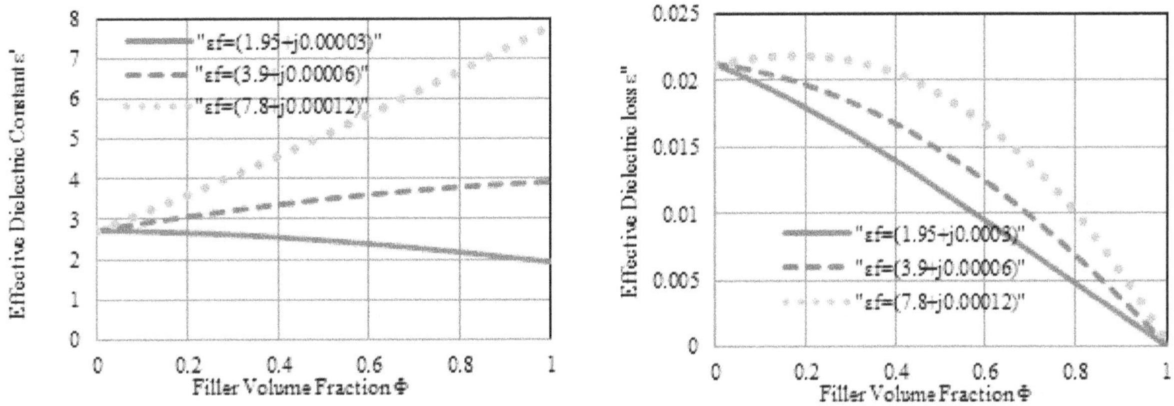

(a) Effective dielectric constant *(b) Effective dielectric loss*

Figure 6. Effect of interphase permittivity on composite dielectric constant and dielectric losses as a function of filler volume fraction for a composite system

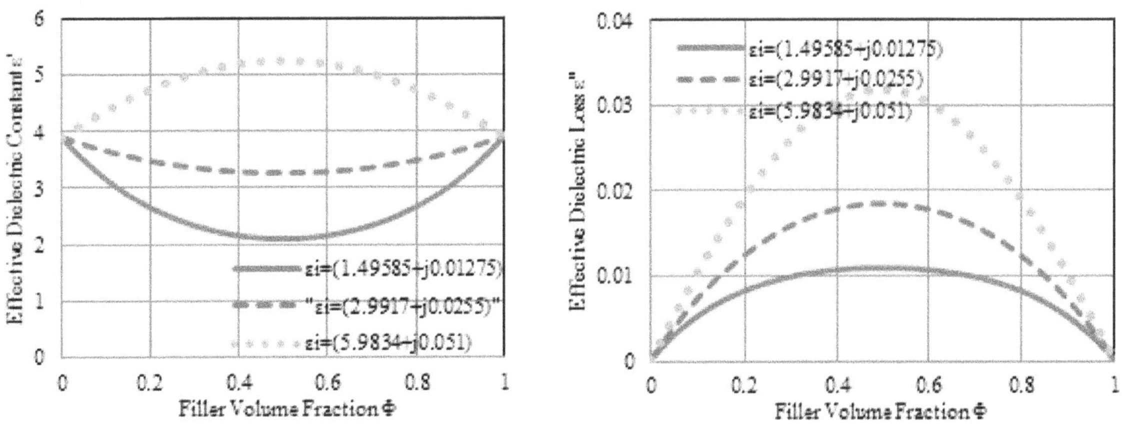

(a) Effective dielectric constant *(b) Effective dielectric loss*

4.3.4 Effect of Interphase Permittivity

At a discriminating filler centralization comparison of the greatest to interphase volume fraction, the interphase permittivity is the greatest or the least in the powerful permittivity of the composite framework concerning illustration indicated in Fig. 6. A non-linearity of the effective complex permittivity of the composite framework is the capacity for filler volume fraction. In the filler volume fraction stacking from 0.0 and 1.0, the perplexing permittivity is equivalent to that of the polymer part and filler part respectively, in any case between these limits. The IPL model predicts a complex permittivity of the composite framework that is essentially bigger than that predicted towards accepted mixture models, since these models don't represent part associations of the composite.

4.3.5 Effect of Filler Surface Area

A filler diameter is proportional in a squared build over particular surface area and resultant interphase volume fraction of a round particle. For non-spherical particles, the expand to particular surface area and interphase volume fraction is considerably greater and declines by increment particle size and shape. A composite framework can be assessed employing the IPL model. The filler surface range illustrated the effects of three values of filler particles surface range (40, 64, and 76) m^2/g, when dielectric polymer and dielectric filler are just at ($\varepsilon_f=\varepsilon_m$). In the provision as demonstrated in Fig. 4, by expanding the filler particle surface range of a composite framework holding an interphase area with a permittivity higher than that of the matrix yields an additionally successful permittivity. Fig. 7(a) indicates the effect of filler surface range ahead composite dielectric constant, and Fig. 7(b) indicates the dielectric misfortunes as a capacity for filler volume fraction to a composite framework.

Figure 7. Effect of filler surface area on composite dielectric constant and dielectric losses as a function of filler volume fraction for a composite system

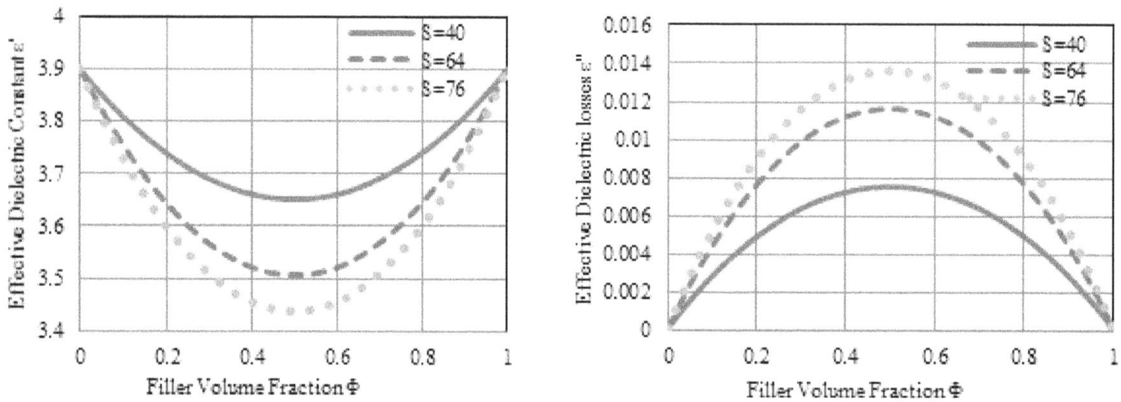

(a) Effective dielectric constant *(b) Effective dielectric loss*

4.3.6 Effect of Interphase Thickness

The interphase volume fraction is likewise reliant upon the thickness of the interphase region, Δr. To show this effect, a model composite framework is assessed utilizing the IPL model. Fig. 8 illustrates the effect from the start of expanding the interphase thickness from (0.0 to10) nm on the composite dielectric constant about this two-part framework. As previously mentioned, in realistic applications, the chemical bonding between the two components of the composite and the chemical structure of the matrix component controls the interphase thickness provided into the chemical structures of interphase regions and their thickness.

Figure 8. Effect of various interphase thickness on the dielectric constant as a function of filler volume fraction for a composite system

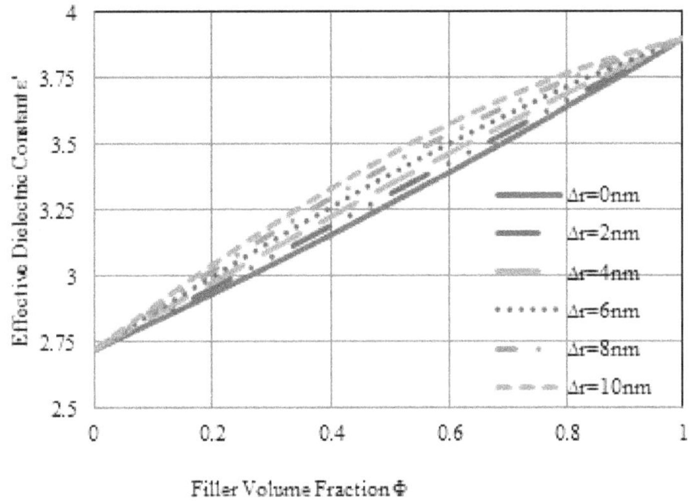

Figure 9. Effect of particle shape and orientation on composite dielectric constant and dielectric loss as a function of filler volume fraction for a composite system

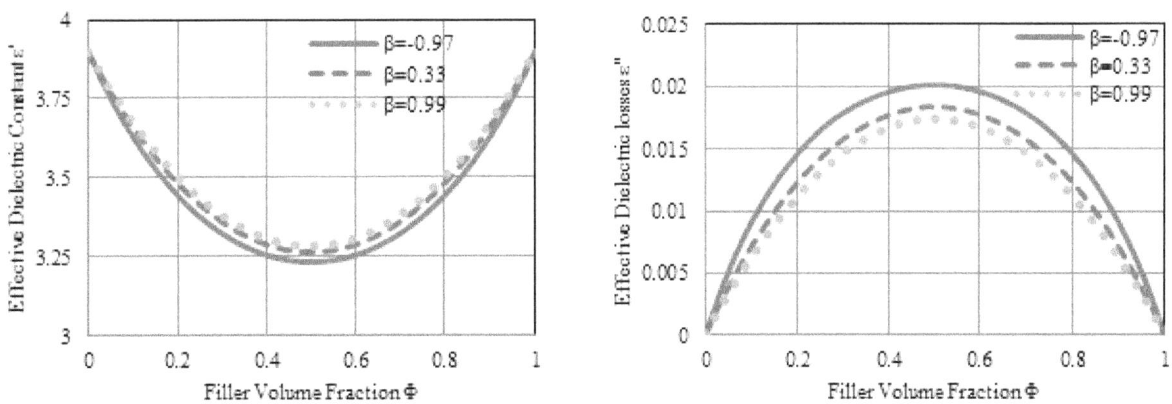

(a) Effective dielectric constant (b) Effective dielectric loss

4.3.7 Effect of Filler Particle Shape and Orientation

For anisotropic composite systems having the dispersed component oriented parallel to the electric field, the value of β is greater than 0.3. A provision for this sort of composite may be epitaxial developed composite films. The value of β turns into 0. 99. The material parameters used in the IPL model will show the effect of the state parameter, β on the effective permittivity of a composite framework. The effect of each of these cases (β=-0. 97 and β=0. 99) is illustrated below in Fig. 9.

Table 2. Characteristics of nanocomposite industrial material data

Dielectric Constant Data					
Fillers		Matrix		Interphase	
Clay	2.0				
Glass Beads	1.2				
Glass Fiber	5.8				
Fumed Silica	4.5	ABS	2.7	GB&ABS	0.9395
Volume Fraction	0 to 1	Epoxy	5.0	GB&Epoxy	0.6633
Surface Area	106 m2/g	PE	2.3	GB&PE	0.5539
Density	2.33g/cc	PVC	3.3	GB&PVC	0.4754
β(Spherical)	0.33			Thickness	11nm

Figure 10. Effects of nanoparticles on effective dielectric constant with different industrial polymers

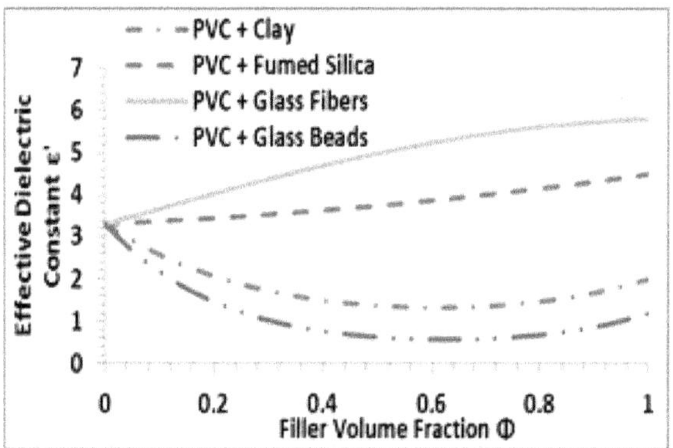

(a) PVC industrial polymer

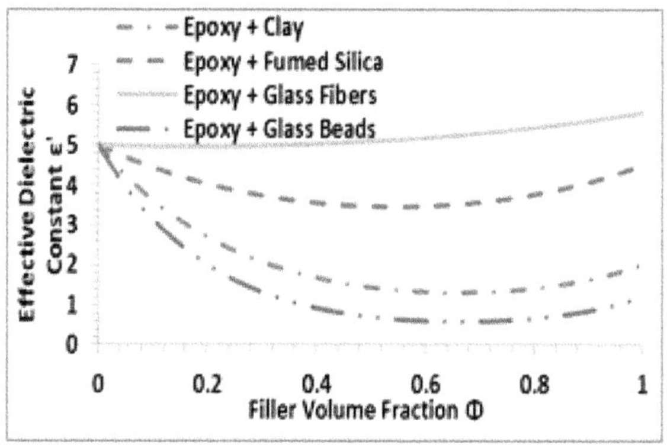

(b) Epoxy industrial polymer

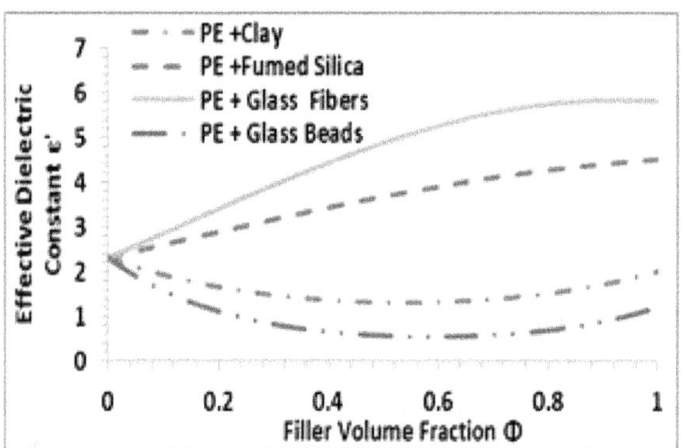

(c) PE industrial polymer

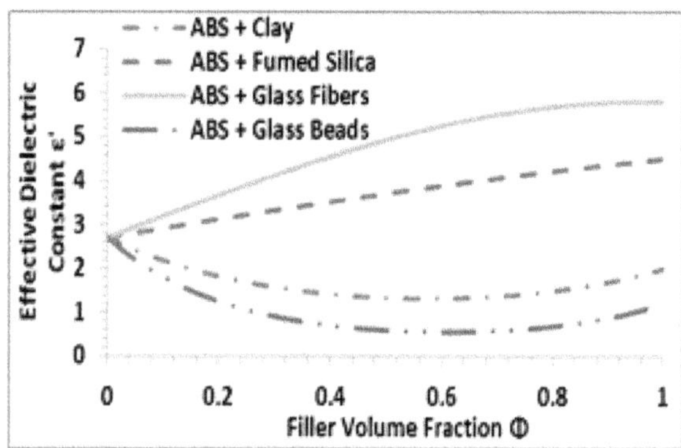

(d) ABS industrial polymer

Figure 11. Effect of filler volume fraction on effective dielectric constant for different nanocomposites industrial materials

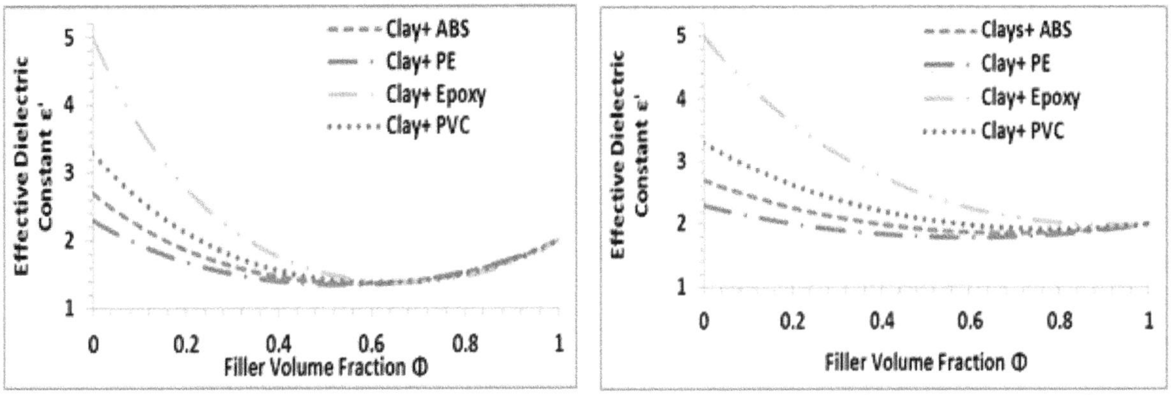

(a) Filler surface area s = 40 m^2/g (b)Filler surface area s = 100 m^2/g

4.3.8 Performance of Applied Nanodielectrics

The IPL model predicts a unique non-linearity of the effective dielectric constant as a capacity of filler volume stacking. Concerning illustration, new industrial materials structured towards nanoparticles, such as clay, fumed silica, glass fibers and glass globules have capability to decrease the cost of the product. A significant number of chosen polymer industrial materials are used within this chapter, such as ABS, PE, and PVC.

4.3.9 Effect of Nanoparticles on Industrial Materials

Figure 10 indicates the effect of every nanoparticle industrial material chosen type on the dielectric constant of the composite, as the volume fraction for fillers stacking. From these produced effects, the effective dielectric constant of polymeric modern materials decreases with clay and glass globules nanoparticles composite. At the same time, it expands with fumed silica and glass fibers nanoparticles composite. Since glass globules and clay indicate the best conduct technique, similar to nanoparticles utilized for materials, glass fibers and fumed silica indicate the best conduct concerning illustration nanoparticles utilized to directing modern materials.

4.3.10 Effect of Filler Surface Area on Industrial Materials

The extent from beginning permittivity deviation starting with linearity is dependent upon the filler volume fraction, filler surface area and the interphase permittivity. Fig. 8 indicates the effect of filler surface area, the point when it's equivalent to 40, and 100 m^2/g. By expanding filler volume fraction, the nonlinearity conduct technique of the dielectric constant to all polymeric industrial materials will be diminishing dependent upon 55%. After that, the dielectric constant will be expanding smoothly, the point at the filler surface area equivalent to 40m^2/g concerning illustration demonstrated in Fig. 11(a). Fig. 11(b) indicates the effective dielectric constant with changing filler volume fraction to every nano-composite industrial material whenever point filler surface area rises to 100m^2/g.

Figure 12. Effect of filler volume fraction on effective dielectric constant for different nanocomposites industrial materials

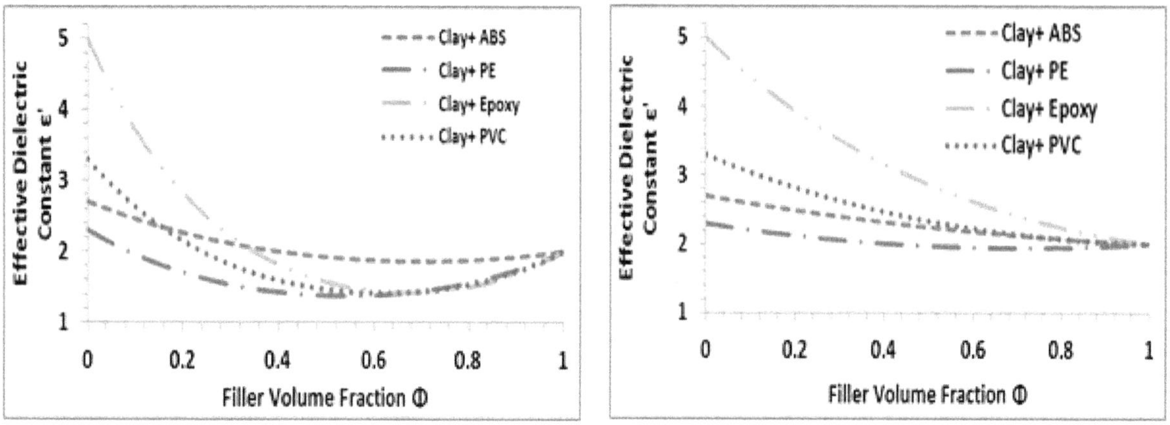

(a) Filler interphase thickness Δr = 2nm *(b) Filler interphase thickness Δr = 10nm*

4.3.11 Effect of Filler Interphase Thickness on Industrial Materials

Figure 12 reveals that the successful dielectric constant is diminishing constantly in nanocomposite industrial materials towards expanding filler volume fraction at interphase thickness 2 and 10 nm. Fig. 12(a) indicates that clay with PE nanocomposite has the least effective dielectric constant and clay with epoxy nanocomposite has the most noteworthy powerful dielectric constant up to 35%. Clay for ABS nanocomposite is the most astounding from 35% up to 100% of the filler volume fraction. Fig. 12(b) reveals that clay for PE nanocomposite has the most reduced effective dielectric constant, while clay with epoxy nanocomposite has the most astounding successful dielectric constant from 0% up to 100% of the filler volume fraction. Fig. 12 indicates the powerful dielectric constant which diminishes to all nanocomposite industrial materials by expanding filler volume fraction.

Figure 13. Effect of filler volume fraction on effective dielectric constant for different nanocomposites industrial materials

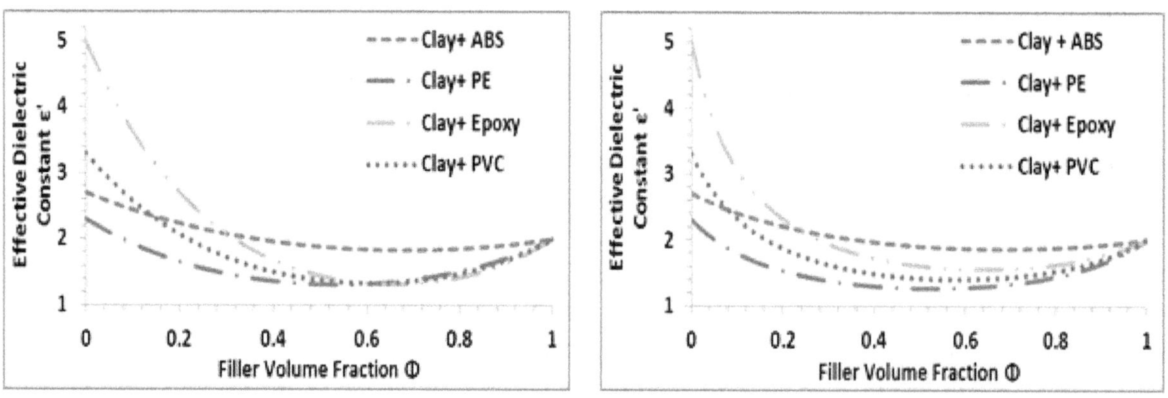

(a) Filler *particle shape β=0.33* *(b) Filler particle shape β=-0.97*

4.3.12 Effect of Filler Particle Shape and Orientation on Industrial Materials

Filler shape state has an important effect on the nanocomposite dielectric constant, the value of β is contingent upon the introduction of the filler particles in the composite. At the same time, the addition of clay nanoparticles expands polymers and it diminishes the successful dielectric constant of the nanocomposite. Fig. 13 indicates the different polymer types filled with clay nanocomposite. PE provides for the most reduced dielectric constant between the sum polymer types and the epoxy provides for the most astounding dielectric constant dependent upon 30% of the filler volume fraction at the point when β=0. 33 as demonstrated in Fig. 13(a). At the same time, the ABS is dependent upon 25% of the filler volume fraction at the point when β=-0. 97, as demonstrated in Fig. 13(b).

4.4 INHOMOGENEOUS INTERPHASE

Recently, however, dielectrics research has developed a tendency to recommend that the dielectric properties of the interphase are additionally inhomogeneous, changing for exploration of the spiral separation starting at the center of the spherical inclusion (Yu et al., 2002; Zhong et al., 2004a). Such an inhomogeneous film is because of holding instruments in the space between the consideration and matrix. Such an inhomogeneous transition is due to the bonding mechanisms occurring in the space between the inclusion and matrix. Effects of the dielectric constant has been distributed to two-phase composites to which immaculate holding is expected to exist between the Incorporation and the matrix. The first of these results is from Maxwell-Garnett theory (Yu et al., 2002) with the same result reached later, using the composite spheres assemblage model (Vo & Shi, 2002). This same outcome can be additionally utilized for the electrical and thermal conductivity, magnetic permeability and diffusivity, provided that the round inclusions and matrix have isotropic. There has been a main standards approach created, eventually perusing Dong et al. (Lutz & Zimmerman, 2005) to find the effective dielectric reaction from the beginning composites with a weak suspension from the beginning graded round particles. Utilization of this approach in any case may be challenging for discretionary graded profile, since the resulting system is reliant on discovering the accurate answer for the legislating differential equations.

Vo and Shi (Yi & Sastry, 2002) measured the dielectric properties for composites as a capacity about consideration concentration utilizing a suggested hypothetical model in light of successful medium hypothesis. The dielectric property of composites and its reliance on the filler centralization is made under record in their model. Therefore, their model may be substantial in general volume fractions and has indicated useful concurred results from the test conducted (Dong et al., 2003; Kakavas et al., 1998; Sang & Li, 2004a; Sang & Li, 2004b; Zhong et al., 2004b). Their model cannot represent a variety in the properties of the interphase area, in case instep measures the property of the interphase, eventually perusing a single constant. A standout amongst the challenges in demonstrating the interphase region for composite materials is to recognize the thing that properties of the interphase best model in reality. Assuming that the properties of the interphase are displayed towards a solitary constant, then how that constant may be picked becomes a critical factor. Though the properties of the interphase are demonstrated by a smooth birch variation, picking a proper work receives importance after that. The more parameters are incorporated clinched alongside such a function, the better is the opportunity for deciding the dielectric profile; however, it then becomes a harder problem to solve (Milton, 2002; Wei & Tang, 2004; Yu & Gu, 2005).

Therefore, in this chapter, the equations inferred by utilizing the supplanting strategy (Jing et al., 2000) on the Maxwell-Garnett blending tenet are indicated. The results reached, however, are best substantial to a low centralization for suspended inclusions, since the interactional "around the inclusions has been dismissed in the development of these models". The equations inferred in this section are connected to three distinctive profiles which model the inhomogeneity of the interphase. If one knows or can estimate the variation in the dielectric property, then, this can be an effective way in determining the equivalent homogeneous property. The model for Vo and Shi's homogeneous model is along these lines suggested.

4.4.1 Spherical Inclusion Surrounded by an Inhomogeneous Interphase

The Maxwell-Garnett close estimation of the dielectric constant of a composite ε_c comprising from beginning isotropic round inclusions installed in an isotropic matrix is provided by:

$$\mu = \mu_m + \frac{c}{\dfrac{1}{\mu_p - \mu_m} + \dfrac{(1-c)}{3\mu_m}} \tag{9}$$

where ε_m is the dielectric constant of the matrix, ε_p is the dielectric constant of the inclusions and c is their volume fraction.

To represent the vicinity of the interphase region, Fig. 14 represents a little part of a three-phase composite comprising circular particles known for size a, encompassed towards an annular interphase area of size b and installed to an encompassing matrix. Modelling the consideration and interphase together is similar to shaping a new, effective round particle of size b, for dielectric property indicated by ε_E. It ought to be accepted that the inclusions are greatly separated and that the interphase regions don't overlap. Once ε_E has been found, the dielectric constant of the composite can be easily calculated using Eq. (9) by replacing εp with ε_E.

Figure 14. Interphase consisting of n regions or layers

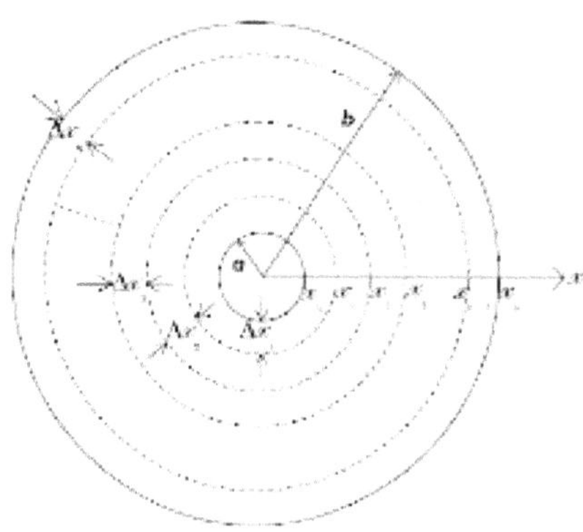

By using the following equation as well:

$$c = d_o \frac{b^3}{a^3} \tag{10}$$

Where d_o is the volume fraction of inclusion relative to all phases.

Now suppose that the properties of the interphase vary as continuous functions of x, where x represents the radial distance from the center of the inclusion or (interphase thickness), as shown in Fig. 1. The dielectric constant of the interphase region is described by ε(x), where x is Є [a, b]. Also assume that ε(x) is a smooth, bounded and continuous function. Consider the partition p of [a, b] into n subintervals defined by:

$$a=x0<x1<x2<\ldots<xi\text{-}1<xi<\ldots<xn\text{-}1<xn=b \tag{11}$$

The lengths Δx1, Δx2, Δx3, . . ., Δxn of the subintervals [x0,x1], [x1, x2], [x2, x3], . . ., [xn_1, xn] associated with the p, presently are not the same. In each subinterval [xi-1, xi], choose any point ξi, that is ξi Є [xi-1, xi]. Using Eq. (11), the effective dielectric constant ε1 of the inclusion and the 1st layer may be approximated by:

$$\mu_1 = \mu(3/4) + \frac{d_1}{\dfrac{1}{\mu_0 + \mu(3/4)} + \dfrac{(1-d_1)}{3\mu(3/4)}} \tag{12}$$

Where: d1=(x1/x0)3, and ε0=εp

Then using the replacement method originally proposed by Hill (Oltean & Motoc, 2012) for the elastic moduli of composites, the effective dielectric constant εi of the inclusion up to the ith layer can be approximated by

$$\mu_i = \mu(3/4) + \frac{d_i}{\dfrac{1}{\mu_{i-1} + \mu(3/4)} + \dfrac{(1-d_i)}{3\mu(3/4)}} \tag{13}$$

Where: di=(x1-1/xi)3,iЄ{N:1≤i≤n}, and ε_{i-1} is an approximation of the dielectric constant of the inner composite sphere that is calculated from the previous step. Our aim is to find the effective dielectric constant, ε_E, of the inclusion and the whole interphase region which is given by:

$$\mu_E \lim_{n\to\infty} \mu_n \tag{14}$$

Where: ε_n is found by solving Eq. (13).

4.4.2 Main Equation of the Model

The effective dielectric constant of the inclusion and interphase can be expressed according to (Salvadori et al., n.d.) as:

$$\mu_E(b) = \frac{\mu_p S(b) + T(b)}{\mu_p U(b) + V(b)} \tag{15}$$

The above equations are functional and have the capacity to model the interphase inhomogeneity towards smooth, limited and nonstop works on x, as contradictory to utilizing a spasmodic venture, such as graded interface. Furthermore, the above effects are pertinent with discretionary profile for the interphase region. The present model can be connected to discretionary graded profiles and will be correct to round particles.

4.4.3 Power Law Profile PLP

It is the profile used to determine the value of S(x), U(x), T(x), and V(x) at a specific radius. Suppose the dielectric properties of the interphase region vary according to the power law function given by:

$$\mu(x) = \mu_m J\left(\frac{a}{x}\right)^P \tag{16}$$

Where: J represents the dielectric constant at the surface of the inclusion relative to that of the matrix, while the constant P represents the rate at which the dielectric properties change with respect to x. Note that J is real and positive by definition, while the constant P is real but may be either positive or negative. According to (Nan et al., 1997), solving S(x), U(x), T(x), and V(x) is given by:

$$S(x) = Ax^{»_1} + Bx^{»_2} \tag{17}$$

$$U(x) = \frac{1}{\mu_m J a^P}\left\{\left(\frac{»_2}{2}+1\right)Ax^{»_1+P} + \left(\frac{»_1}{2}+1\right)Bx^{»_2+P}\right\} \tag{18}$$

Where:

$$A = \frac{1}{a^{»_1}}\left(\frac{»_2+2}{»_2-»_1}\right), \text{ and } B = \frac{1}{a^{»_2}}\left(\frac{»_1+2}{»_1-»_2}\right)$$

Figure 15. A mapping of a homogeneous particle consisting of inclusion and interphase onto a two-phase composite

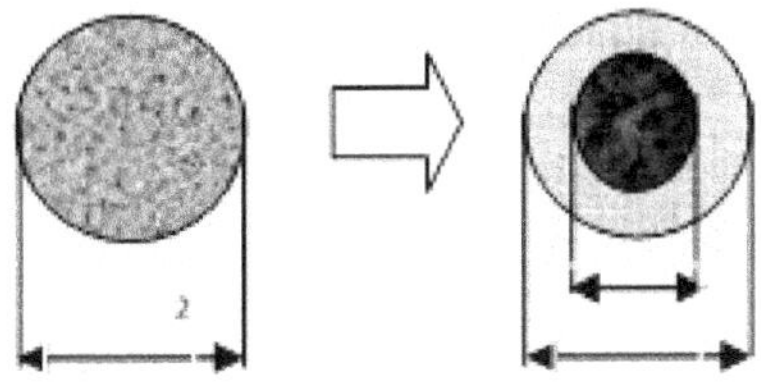

Similarly:

$$T(x) = Ax^{\text{»}_1} + Dx^{\text{»}_2} \quad (19)$$

$$V(x) = \frac{1}{\mu_m J a^P}\left\{\left(\frac{\text{»}_2}{2}+1\right)Cx^{\text{»}_1+P} + \left(\frac{\text{»}_1}{2}+1\right)Dx^{\text{»}_2+P}\right\} \quad (20)$$

Where:

$$C = \frac{1}{a^{\text{»}_1}}\left(\frac{2\mu_m J}{\text{»}_2 + \text{»}_1}\right) \text{ and } B = \frac{1}{a^{\text{»}_2}}\left(\frac{2\mu_m J}{\text{»}_1 + \text{»}_2}\right)$$

Where:

$$\text{»}_1 = \frac{-(P+3)+\sqrt{P^2-2P+9}}{2} \text{ and } \text{»}_2 = \frac{-(P+3)-\sqrt{P^2-2P+9}}{2}$$

4.4.4 Equivalent Homogeneous Interphase Model

The Maxwell-Garnett weaken concentration result to the dielectric constant is provided by equation (1). It has been equipped to control this equation, utilizing the reinstatement system to represent an inhomogeneous interphase area encompassing each consideration. There is a mapping of the consideration and encompassing inhomogeneous interphase onto a powerful circular particle of indistinguishable twin size with dielectric constant indicated towards ε_E(b). Once it has been discovered ε_E(b)for an inhomogeneous interphase. A constant value εi for the dielectric constant of the interphase might be decided eventually perusing an opposite mapping, as demonstrated in Fig. 15, that is, the mapping of a homogeneous circle for size b onto a two-phase circle of indistinguishable twin extent eventually perusing the solution:

$$\varepsilon_E(b) = \varepsilon_i + \frac{\left(\frac{a^3}{b^3}\right)}{\frac{1}{\varepsilon_p - \varepsilon_i} + \frac{\left(1 - \frac{a^3}{b^3}\right)}{3\varepsilon_i}} \tag{21}$$

Thus, in inhomogeneous interphase, the Vo and Shi strategy provides for us an approach for discovering the equal homogeneous property of the interphase. Since this approach allows us to recognize the equal homogenous property of the interphase, this result about shortages can provide a chance for joining with different existing three-phase models, which expect a homogeneous interphase encompassing every incorporation. Therefore, and approach is recommended here for fusing two different models together, an inhomogeneous and a homogeneous interphase model, and this joining transform is known as the combined model. For instance, the value ε_i ascertained from Eq. (21) can be utilized within the Vo Also Shi model. The advantage of utilizing the V_o and Shi model is that full volume fraction pressing from beginning circular inclusions is permitted, and the extent of the interphase region might additionally change like a work from beginning incorporation concentration. Therefore, the V_o and Shi model permits us to represent this phenomenon in as much as the inhomogeneous interphase model doesn't. As stated by the V_o and Shi model, the dielectric constant of a particulate composite is provided as:

$$\mu_c = \frac{h + 21}{h - 1} \tag{22}$$

Where:

$$h = \left\{1 + 2\frac{(\mu_m - \mu_i)(\mu_i - \mu_p)}{(2\mu_m + \mu_i) + (2\mu_i + \mu_p)}\frac{a^3}{b^3}\right\} - \left\{2\frac{(\mu_m - 1)(\mu_m - \mu_i)}{(\mu_m + 2)(2\mu_m + \mu_i)}\frac{b^3}{c^3}\right\} - \left\{2\frac{(\mu_m - 1)(\mu_m + 2\mu_i)(\mu_i - 2\mu_p)}{(2\mu_m + 2)(2\mu_m + \mu_i)(2\mu_i + \mu_p)}\frac{a^3}{c^3}\right\}$$

$$1 = \left\{\frac{(\mu_m - 1)}{(\mu_m + 2)}j - \frac{(2\mu_m + 1)n}{(\mu_m + 2)(2\mu_m + \mu_i)}\frac{b^3}{c^3}\right\}$$

$$j = \left\{1 + 2\frac{(\mu_m - \mu_i)(\mu_i - \mu_p)}{(2\mu_m + \mu_i) + (2\mu_i + \mu_p)}\frac{a^3}{b^3}\right\}$$

$$n = \left\{(\mu_m - \mu_i) + \frac{(\mu_m + 2\mu_i)(\mu_i - \mu_p)}{(2\mu_i + \mu_p)}\frac{a^3}{b^3}\right\}$$

Where: the subscripts m, i and p stand for matrix, interphase and particle inclusion, respectively. The parameters a and b are as defined in our inhomogeneous interphase model, while the parameter c represents the radial distance from the center of the inclusion to the outer boundary of the matrix phase of a representative composite sphere. The parameters a, b and c are related to each other through the parameters k and do, where k is the interphase volume constant and do is the volume fraction of filler, that is:

$$\frac{a^3}{b^3} = \frac{(1+kd_o)}{(1+k)},\ \frac{a^3}{c^3} = d_o,\ \text{and}\ \frac{b^3}{c^3} = d_o\left\{1 + k\frac{(1-d_o)}{(1+kd_o)}\right\} \tag{23}$$

The value of k reflects the matrix/filler interaction strength, k is equal to zero, when the interaction (bonding) between the matrix and filler is negligible, and the large positive k value indicates a strong polymer/filler interaction (Todd & Shi, 2002). By rearranging Eq.'s. (19) to (22), the parameter b may be expressed by:

$$b = \frac{(1+k)^{\frac{1}{3}}}{(a+kd_o)^{\frac{1}{3}}} \tag{24}$$

The b_{avg} used in our inhomogeneous interphase model gives the value of interphase dielectric constant in the domain 0≤do≤1 given by:

$$b_{avg} = \frac{3a}{2k}\left\{(1+k) - (1+k)^{\frac{1}{3}}\right\} \tag{25}$$

It is significant to note that assuming that the interphase volume constant is situated with zero, that is, k=0, their model has a tendency of the Maxwell-Garnett close estimation provided for, eventually perusing (1). As an aftereffect from the beginning, in overlook (14) and permit b = constant × a, after that under the mapping (22), the V_0 and Shi model provided for, eventually perusing (22) converges with our inhomogeneous interphase model. Hence, this model and the inhomogeneous interphase model are the same at the point when the size of the interphase stays settled over all filler concentrations.

4.4.5 Effective Parameters of IPL

EPL profile empowers us to control interphase properties, eventually perusing and evolving three separate elements c, p, and β, the place of these constants should concur to this state ε(a)=εP and ε(b)=εm. Effects about rate of transform, interphase properties for exploring to spiral separation x "p" and interphase volume constant in the successful dielectric constant are explained in this Section.

Figure 16. Effect of rate of change in interphase properties with respect to radial distance x "p" on the effective dielectric constant

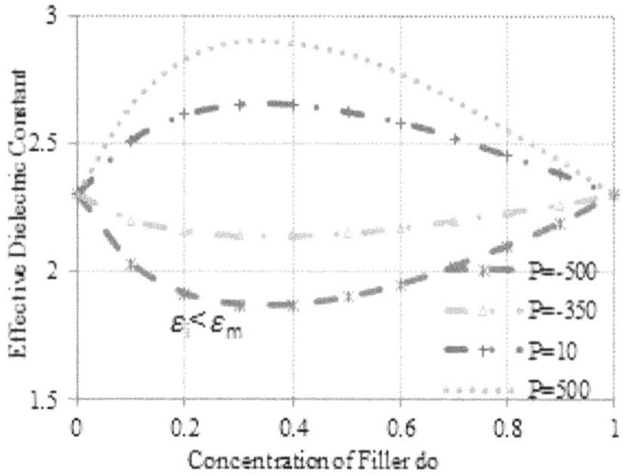

(a) PP with Clay composite, and k=2.8

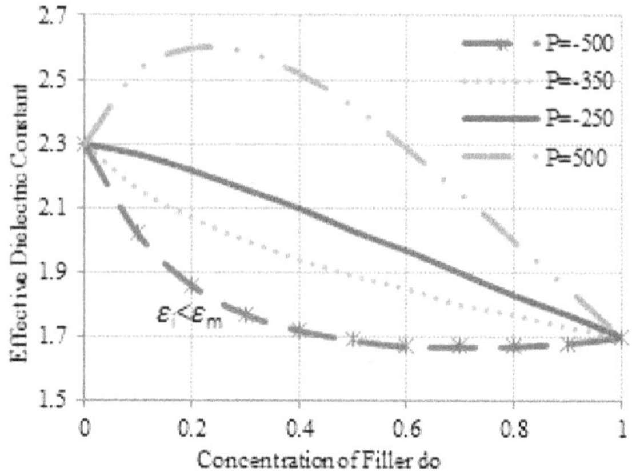

(b) Epoxy with SiO_2, composite, and k=2.8

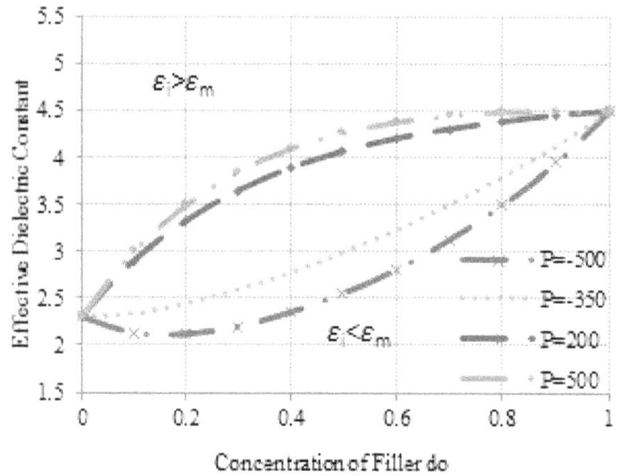

(c) PP with Al_2O_3 composite, and k=2.8

4.4.5.1 Effect of "Rate of Change in Interphase Properties with Respect to Radial Distance x"on the Effective Dielectric Constant

Picking the parameter c, and β such-and-such $\varepsilon(a)=\varepsilon_P$ and $\varepsilon(b)=\varepsilon_m$ and permitting parameter p to vary, it can be reasonably expected to plot few results from the particular values for interphase volume constant k. Fig. 16 indicates the effects on the successful dielectric constant ε_c for the consideration centralization do towards different values from the beginning interphase dielectric constant. The value of ε_i is controlled, eventually perusing and evolving the interphase parameter p which refers to the rate during which dielectric properties change for exploring spiral separation x. Three separate nanocomposites are used to indicate the effect from the beginning interphase dielectric constant in composite dielectric constant. Fig16(a) indicates the instance of the matrix dielectric constant equivalent to the filler dielectric constant "PP for clay composite". Likewise, this fig. indicates the constant value of k=2. 8, the interphase dielectric constant will be higher than matrix dielectric constant in values for p=10, p=500. In any case, the interphase dielectric constant is more level than matrix dielectric constant in values of p=-500, p=-350.

Figure 17. Effect of interphase volume constant on the effective dielectric constant

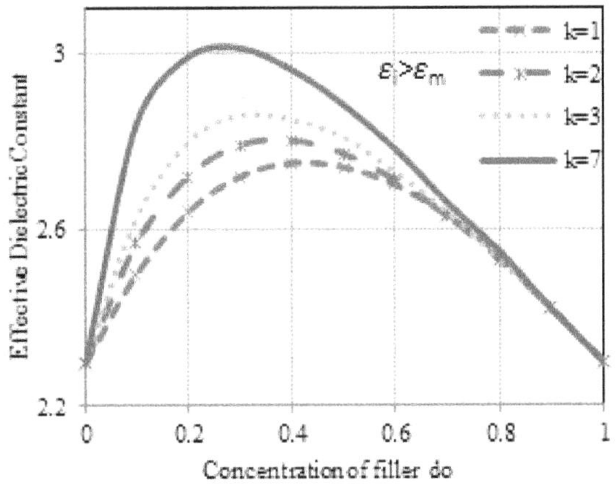

(a) PP with Clay composite, and p=350

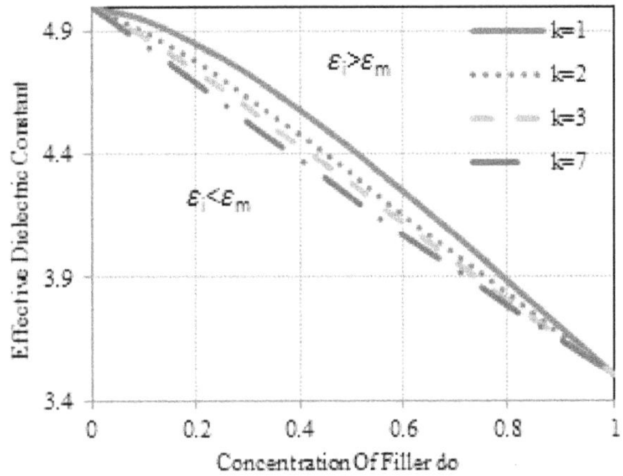

(b) Epoxy with SiO_2 composite, and p=350

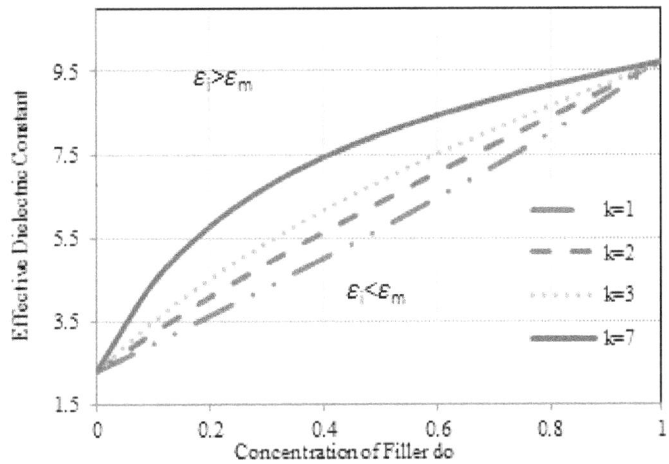

(c) PP with Al_2O_3 composite, and p=350

Figure 16(b) indicates the instance about matrix dielectric constant which is higher than filler dielectric constant "Epoxy for SiO_2 composite". Moreover, this Fig. reveals that the constant value of k=2. 8 and the interphase dielectric constant will be higher than matrix dielectric constant towards values of p=500, p=-250, but interphase dielectric constant will be more level than matrix dielectric constant toward values of p=-500, p=-350. Fig. 16(c). indicates the body of evidence that matrix dielectric constant is easier than filler dielectric constant "PP for Al_2O_3 composite". In addition, this Fig. demonstrates that during the constant value of k=2. 8, interphase dielectric constant is higher over matrix dielectric constant at values of p=200 and p=500, yet all the interphase dielectric constant will be falling down over matrix dielectric constant at values for p=-500, p=-350. Finally, the best value for p is decided to provide for the best fitting of the trial data.

4.4.5.2 Effect of Interphase Volume Constant on the Effective Dielectric Constant

The successful dielectric constant for nanocomposite relies on interphase volume constant k which reflects the interactional the middle of filler and polymer. Fig. 17 indicates the variety of the successful dielectric constant ε_c with the consideration concentration would during different values from beginning interphase volume constant (k=1, k=2, k=3 and k=7) and during constant value about p=350. Three distinctive nanocomposites used to hint at the effect from beginning interphase dielectric constant in composite dielectric constant. Fig. 17(a) indicates the matrix dielectric constant equivalent to filler dielectric constant "PP for clay composite". Fig. 17(b) indicates the matrix dielectric constant higher than filler dielectric constant "PP with SiO_2 composite". Fig. 17(c) indicates the matrix dielectric constant more level over filler dielectric constant "Epoxy for SiO_2 composite".

4.4.5.3 Effect of Interphase Thickness on the Effective Dielectric Constant

In Figure 18, the variety of the successful dielectric constant εc is indicated for the interphase thickness x during separate value for parameter p for constant fixation rise from 0. 3 to "PP-Fumed silica composite" industrial materials. In this Figure, the constant value for k=2.8, effective dielectric constant has been expanded with expanding clinched alongside interphase thickness (x) during positives values p at successful dielectric constant and has been diminished for expanding in interphase thickness (x) in negative values about p. Hence, two extents for p provide for wider population works between x and powerful dielectric constant.

Figure 18. Effect of interphase thickness on the effective dielectric constant for PP with Fumed Silica composite and concentration 0.3

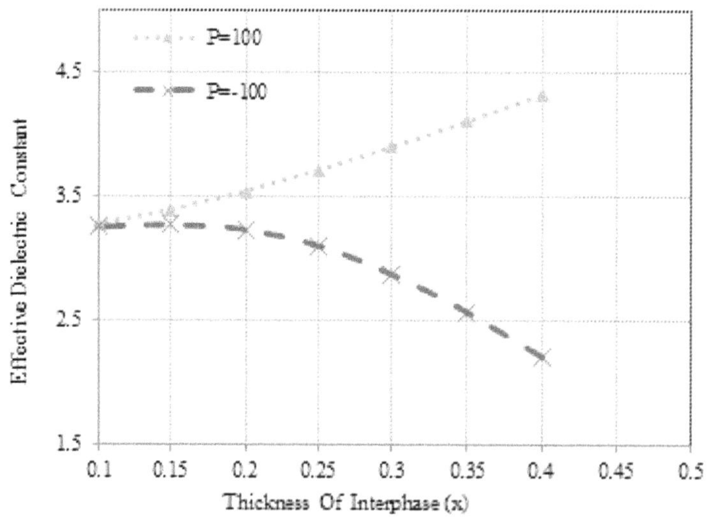

Figure 19 (a), indicates the relation between effective dielectric constant and concentration of filler at constant parameter p=100 and different values of interphase thickness in case the dielectric constant of filler is equal the dielectric constant of polymer, such as "PP-Clay composite industrial materials".

Fig. 19(b) indicates the relation between effective dielectric constant and concentration of filler at different values of interphase thickness in case the filler dielectric constant is lower than dielectric constant of polymer, such as "PP-Zno composite industrial materials". The last Fig. 19(c) indicates the relation between effective dielectric constant and concentration of filler at different values of interphase thick-

Figure 19. Effect of interphase thickness on the effective dielectric constant with different industrial materials

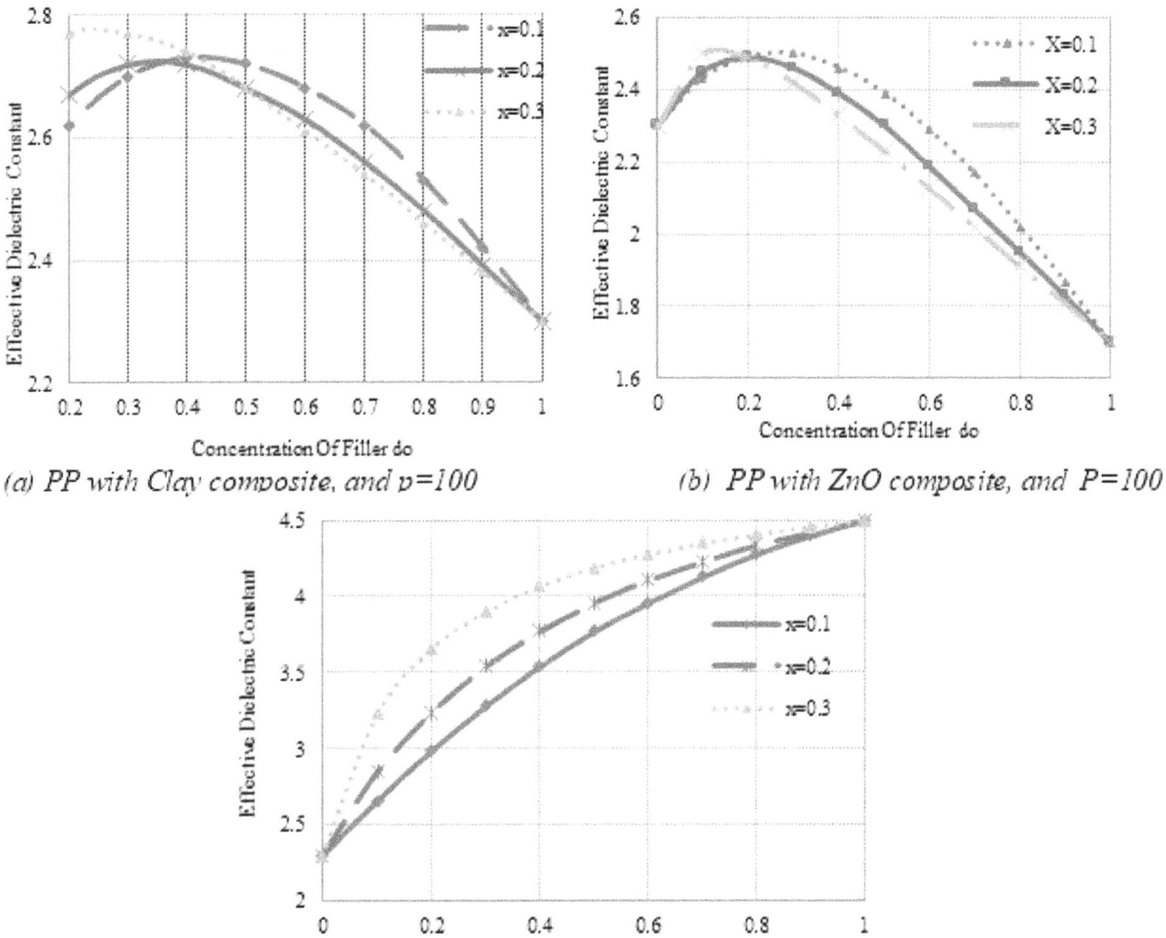

(a) PP with Clay composite, and p=100

(b) PP with ZnO composite, and P=100

(c) PP with Fumed Silica composite, and P=100

Figure 20. PLP for the dielectric constant of the inhomogeneous interphase region, and k=3.2

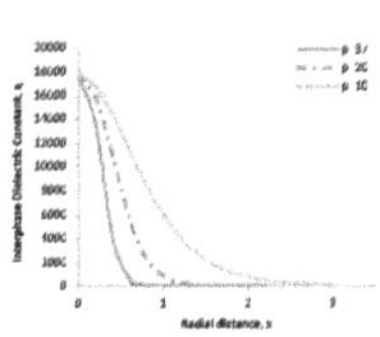

ness, in case dielectric constant of filler is higher than dielectric constant of polymer, such as "PP-Fumed Silica composite industrial materials".

Figure 21. Effective dielectric constant for various composite with varying values of k (a) Epoxy with Fumed Silica Composite (b) Epoxy with Clay Composite.

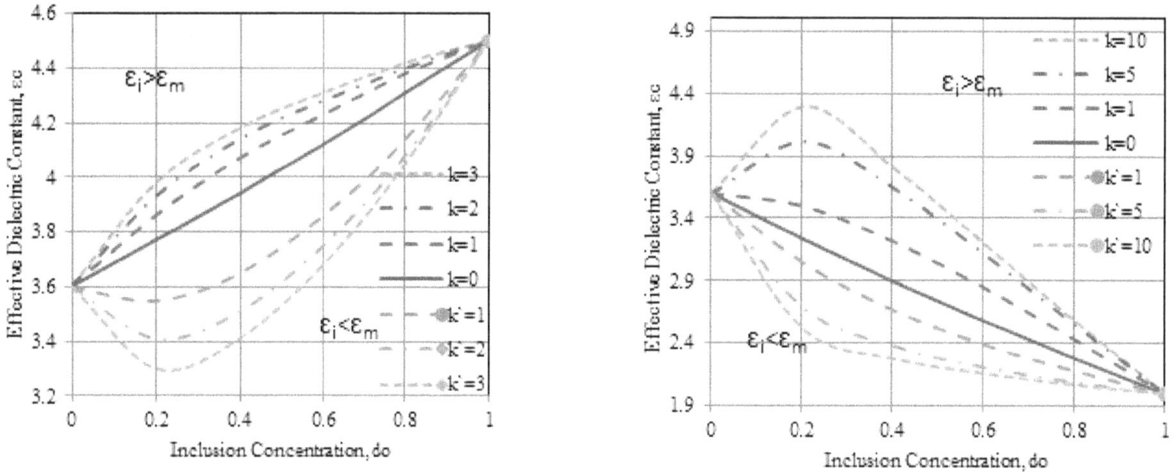

(a) Epoxy with Fumed Silica Composite *(b) Epoxy with Clay Composite*

4.4.5.4 Effect of Radial Distance on the Interphase Dielectric Constant

In Figure 20, the variety of the interphase dielectric constant for the interphase spiral separation x is indicated and it can be seen starting with the figure in the interphase dielectric constant declining for the augment of the value of p with expanding the interphase spiral separation x. Figure 6 indicates the effects of the beginning rate dielectric properties progress to explore the interphase thickness p for interphase volume constant k=3. 2 and the interphase dielectric constant diminishing with just zero in interphase spiral separation equivalent to (0.8, 1.4, and 3.0) towards the rate of dielectric properties progressing to explore the interphase thickness that is equivalent to (10, 20, and 37) respectively.

4.4.6 Performance of Applied Nanodielectrics

In the Power Law theory profile, the dielectric constant in the surface of the consideration relative to that of the matrix parameters called j and the rate toward which the dielectric properties progress to explore x, called p, picks the properties of the interphase change in the individuals under consideration to the individuals of the matrix, that is $\varepsilon(a)=\varepsilon_p$ and $\varepsilon(b)=\varepsilon_m$. The effects of volume constant and parameter p on the successful dielectric constant has been contemplated ahead both epoxy for fumed silica composite and epoxy with clay composite. Furthermore, the effect for spiral separation (interphase thickness) on the interphase dielectric constant has been likewise examined.

Figure 22. Effective dielectric constant for various composites with varying values of x

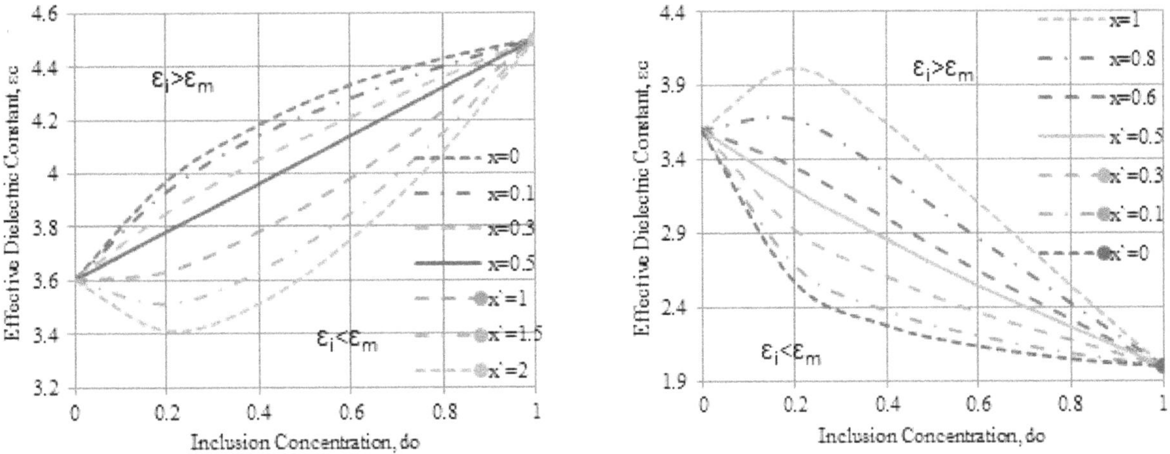

(a) Epoxy with Fumed silica Composite *(b) Epoxy with Clay Composite*

6.4.6.1 Effect of Interphase Volume Constant Variation on the Effective Dielectric Constant

In Figure 21, the variety of the effective dielectric constant εc with the consideration volume centralization is in different values of the interphase dielectric constant. Fig. 21(a) reveals that εc increments the increase of d_o to epoxy for fumed silica. When ε_i is higher than ε_m, the successful dielectric constant builds with the increase of the interphase dielectric constant, and when ε_i is lower than εm, the effective dielectric constant increases the interphase dielectric constant. The value of ε_i can be regulated by evolving the interphase thickness x, the place which utilizes the emulating values, during x=0. 1, εi > ε_m and x=3, ε_i < εm. Fig. 21(b) indicates that ε_c declines with the increase from the beginning to epoxy and clay composite. If εi is higher than ε_m, the effective dielectric constant builds for the increase of the interphase dielectric constant, in any case whenever ε_i is lower than ε_m and the effective dielectric constant abates with the augment of the interphase dielectric constant. Information is utilized for values, x=1, $\varepsilon_i > \varepsilon_m$ and during x=0. 1, $\varepsilon_i < \varepsilon_m$. The best value of k is picked up to provide for the least value of the effective dielectric constant and so, to test creation of the spicing.

6.4.6.2 Effect of Interphase Thickness Variation on the Effective Dielectric Constant

Figure 22 indicates the variety of the successful dielectric constant εc for the incorporation volume centralization at separate values of interphase thickness x with two cases nanocomposite. Fig. 22(a) reveals that for the epoxy for fumed silica composite, the effective dielectric constant aggravate εc builds with the augment of the consideration volume centralization. Moreover, it can be seen that during the values of x that aggravate $\varepsilon_i > \varepsilon_m$ and $\varepsilon_i < \varepsilon_m$, the successful dielectric constant abates with the augment of the interphase thickness. Fig. 22(b) demonstrates the epoxy with clay composite εc abatements for the augment of do and the values from the beginning x that aggravate settlement on $\varepsilon_i > \varepsilon_m$ and $\varepsilon_i < \varepsilon_m$. The effective dielectric constant builds with the increase of the interphase thickness.

Figure 23. Effective dielectric constant for various composites with varying values of P

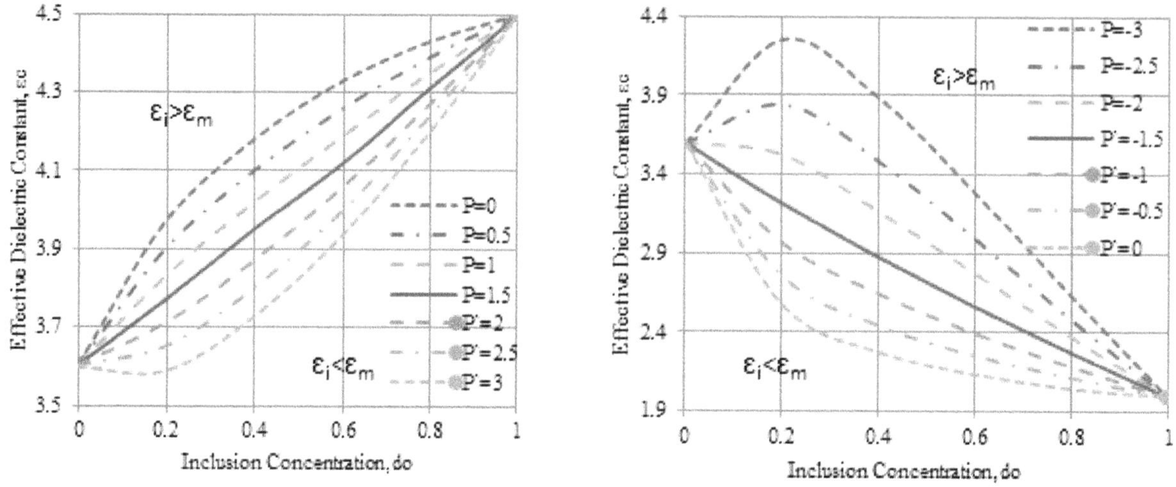

(a) Epoxy with Fumed silica Composite　　*(b) Epoxy with Clay Composite*

Figure 24. Power law profile for dielectric constant inhomogeneous interphase region with various composites.

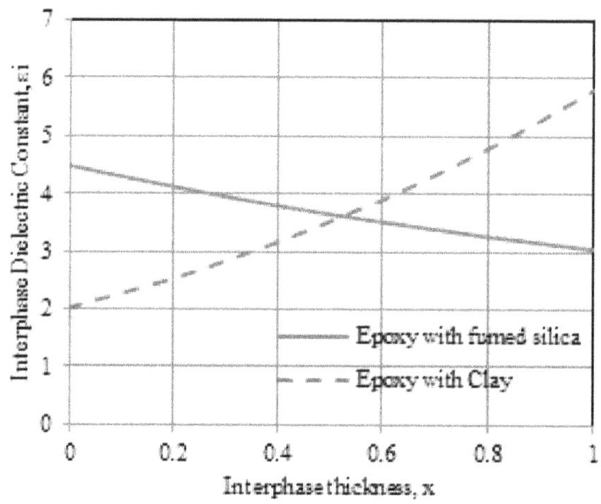

6.4.6.3 Effect of the Rate of the Dielectric Properties to Interphase Thickness on the Effective Dielectric Constant

Figure 23 indicates the variety of the effective dielectric constant εc with the incorporation volume concentration "do" during different values of rate during which the dielectric properties change to explore the interphase thickness alternately called (parameter P). Fig. 23(a) reveals that epoxy with fumed silica composite ε_c increments for the augment from the beginning do and the point when $\varepsilon_i > \varepsilon_m$ and $\varepsilon_i < \varepsilon_m$ happen just with the certain values about p, while for Fig. 23(b), epoxy for clay composite εc declines with the increase for do, and the point when $\varepsilon_i < \varepsilon_m$ and $\varepsilon_i > \varepsilon_m$ occur is mainly with the negative values of p. In addition, it can be seen that during the values from beginning p that settle on $\varepsilon_i > \varepsilon_m$ and εi < ε_m, the powerful dielectric constant declines with the increase of the interphase thickness to epoxy for clay composite, but expands with epoxy for fumed silica composite.

6.4.6.4 Effect of Radial Distance (Interphase Thickness) Variation on the Interphase Dielectric Constant

Figure 24 demonstrates the variety of the interphase dielectric constant for the interphase thickness x. To epoxy with fumed silica composite, the interphase dielectric constant abates with expanding interphase thickness x. At the same time, this can be noted with epoxy for clay composite, where the interphase dielectric constant increments with the augment of x. For two cases, regarding nanocomposites, in first case of epoxy with fumed silica composite, the polymer dielectric constant is more diminutive than the inclusion dielectric constant, but in the second case of epoxy for clay composite, the polymer dielectric constant is higher than the inclusion dielectric constant.

4.5 MULTI-NANOPARTICLES TECHNIQUE

Nanotechnology has diminished the issue of creating new polymeric materials for electrical industrial materials. Addition of nanoparticles expands the breakdown value of the base polymer as result of scattering nanoparticles which are controlling in polymer properties (Rao et al., 2000; Vo et al., 2001). The breakdown value for unfilled and nanoparticle-filled resins can be demonstrated as the addition of nanoparticles increasing the dielectric breakdown value, such as ZnO clinched alongside PVC, TiO2 on epoxy, and silica for epoxy. The partial discharge strength of nanofilled polymers tends to be higher than unfilled polymers (Karkkainen et al., 2000; Oltean & Motoc, 2012). Scientists have attempted to prove that the dielectric material progressed by mixing low content for nanoparticles (Nelson & Hu, 2004; Roy, Nelson, Reed et al, 2005; Sarathil et al., 2006). The transport techniques for electrons under low and high electrical field do not give a clear description for understanding the components of fleeting breakdown. The interface areas around nanoparticles assume a great part for the decision for understanding the short-term breakdown properties. Polarity about polymer, nanoparticles kind and their surface states have the combinative effect on the interface area. Thus, the interface part of polymer nanocomposites and so, a multi-core model of the interface has been constructed (Todd & Shi, 2002; Wei & Tang, 2004). Then, a natural and inorganic composites mixture system model has been recommended (Li, Okamoto, Ohki et al, 2010). Acquiring homogeneous distribution of nanoparticles inside the base polymer matrix acts an important issue in the interface work done. So, the nanoparticles are scattered to polymer matrix, eventually perusing the utilization of shear energy dispersion and concoction change in the dominant part of the beginning trials. The viscosity of the matrix is a critical element for shear energy dissemination. In distinctive generation courses, the interface can be for different thickness and layer numbers (Roy, Nelson, Maccrone et al, 2005). There are current ideas of the physical root and component from beginning misfortunes to electric and dielectric electrical materials under variant universal Classes. Trends of nanotechnology science create electric and dielectric industrial materials to upgrade their possibility provisions clinched alongside future vitality storage/transfer units that are related to the interfacial conduct technique between the nanoparticles and the polymer matrix over such nanocomposites (Ebnalwaled & Thabet, 2016; Li, Yin, Chen et al, 2010; Tanaka, 2005; Thabet, 2015a; Thabet, 2015b;

Thabet, 2016; Wilkes & Wen, 1996). The objective of this section is to present the dielectric constant of new multi-nanocomposites that are structured to upgrade electrical modern provisions. Thus, it is clear that the progression of industrial materials parameter (permittivity) towards utilizing different sorts of multi-nanocomposites for variant industrial applications may introduce new multi-nanocomposites, in view of multi-nanoparticles technique so as to outline new upgrade industrial materials. However, the effect of the multi-nanoparticles plan on the new proposed electrical materials can have possible risks.

Figure 25. Interphase region surrounding the multi-nanoparticle particles in multi-nanocomposites system

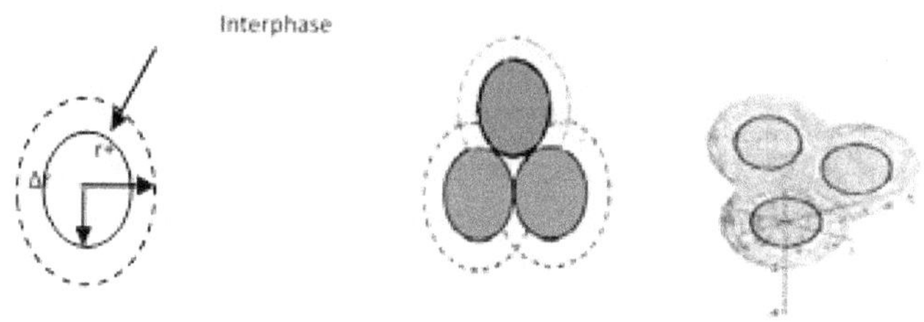

6.5.1 Theory of Multi-Nanoparticles Technique

In the event of utilizing individual nanoparticles inside polymer matrix and the control theory connections within dielectric demonstrating for composite frameworks, the composite framework will have three parts (matrix and nanoparticle) and an interphase region (Thabet, 2017; Thabet & Mubarak, 2017). The interphase region volume fraction is reliant upon the nanoparticle volume fraction, the nanoparticle surface area and the thickness of the interphase region encompassing each nanoparticle particle (Thabet & Ebnalwaled, 2017). In any case, a multi-nanoparticles method has been proposed for creating the electric and dielectric properties from the beginning polymer. Thus, the effective dielectric constant of the incorporation and interphase can be conveyed as follows:

$$\mu_E(b) = \frac{\mu_j s(b) + T(b)}{\mu_j U(b) + V(b)} \tag{26}$$

Where, ε_j is the dielectric constant of the second nanoparticle

Once it has been found that $\varepsilon_E(b)$ is for an inhomogeneous interphase between multi-nanoparticles and base polymer matrix, the dielectric constant of the interphase ε_{phij} can be determined as follows:

$$\mu_E(b) = \mu_{phij} + \frac{\left(\frac{a^3}{b^3}\right)}{\frac{1}{\mu_j - \mu_{phij}} + \frac{\left(1 - \frac{a^3}{b^3}\right)}{3\mu_{phij}}} \tag{27}$$

Table 3. Dielectric Constant of Nanoparticles and Base Matrices of Nanocomposites & Multi-Nano-composites (Thabet & Mubarak, 2017)

Nanoparticles	ε_r	Nanocomposites		Multi-Nanocomposites	
		Base Matrix	ε_r	Base Matrix	ε_r
Clay	2.0	PE	2.3	Clay/PE Silica/PE TiO2/PE	1.6544 1.2124 4.8906
Silica	2.5	ABS	2.7	MgO/ABS Silica/ ABS SiO2/ ABS	4.2789 1.1993 3.5016
ZnO	1.7	PVC	3	Clay/ PVC Silica/ PVC SiO2/ PVC	1.9495 1.1878 3.3557
SiO2	4.5	EPDM	3.3	Clay/ EPDM Silica/ EPDM ZnO/ EPDM	2.0698 1.1757 1.7410
MgO	9	PI	3.62	ZnO/ PI Al2O3/ PI TiO2/ PI	1.8470 9.2127 5.9644
TiO2	10	PMMA	4	Clay/ PMMA Silica/ PMMA SiO2/ PMMA	2.3377 1.1468 2.9252
Al2O3	9.5	Epoxy	5	Clay/ Epoxy MgO/ Epoxy ZnO/ Epoxy	2.6987 6.1049 2.2761

For instance, the value ε_{phij} calculated from Eq. (2) can be used to find the component composite system, described by:

$$\mu_{effj}^{2} = \text{Æ}_{j}\,\mu_{j}^{2} + \text{Æ}_{phij}\mu_{phij}^{2} + \text{Æ}_{effi}\,\mu_{effi}^{2} \tag{28}$$

Where, φ_j is the volume fraction of the second nanoparticle component of the multi-composite system, φ_{phij} is the volume fraction of the interphase region component of the multi-composite system, φ_{effi} is the volume fraction of the initial matrix component of the multi-composite, ε_{effj} is the dielectric permittivity of the multi-composite system, ε_j is the second nanoparticle permittivity of the multi-composite system, ε_{phij} is the interphase permittivity of the multi-composite system and ε_{effi} is the dielectric permittivity of the individual nanocomposites system.

The second nanoparticle volume fraction of Eq. (3), Φ_j, is directly measured for a given composite system. The matrix volume fraction is given by $\Phi_{effi} = (1 - \Phi_j - \Phi_{phij})$. The interphase volume fraction of multi-composite system, φ_{phij}, is calculated by:

$$\text{Æ}_{phij} = \left(1 - \mathrm{F}\right)\left(\mathrm{S}_{j}.\Delta \mathrm{r}\right).\text{Á}_{j}.\text{Æ}_{j} \tag{29}$$

Where, S_j is the specific surface area of the second nanoparticle (measured in m^2/g), ρ_j is the density of second nanoparticle (measured in g/m^3), Δr is t the thickness of the interphase region and F is an overlap probability function.

Estimation of interphase thickness is further formed in light of ref. (Gouda et al., 2010; Karkkainen et al., 2000; Polizos et al., 2010; Todd & Shi, 2003b). Figure 25 indicates the nanoparticles approach together with the interphase regions encompassing every nanoparticle starting to overlap; hence, decreasing the successful interphase volume fraction has been achieved.

Note, the interphase overlap probability is a function of the nanoparticles volume fraction, interphase thickness and the shape and size of the nanoparticles. Approximations have been developed to estimate the overlap of such particles utilizing explanatory answers for percolation models (Tagami et al., 2010; Todd & Shi, 2003a).

Table 3 specifies dielectric constant of the suggested nanoparticles for enhancing the listed polymeric industrial materials (Yi & Sastry, 2002).

6.5.2 Performance of Applied Nanodielectrics

The point that is worth this great effort is receiving new dielectric materials due to our need for upgrading electric and electronic requisitions. The accompanying effects show an effective multi-nanoparticles strategy to acquire the best dielectric materials characterization. Hence, Fig. 26 indicates that powerful dielectric constant of clay/PE nanocomposites has been diminished by expanding volume fraction to clay as distinctive nanoparticles. In case of multi-nanocomposites, the successful dielectric constant of base matrix (0. 2wt. % clay/PE) is reduced up to 1. 6544.

Figure 26. Effective relative permittivity of PE nanocomposites with various volume fraction of multi-nanoparticles

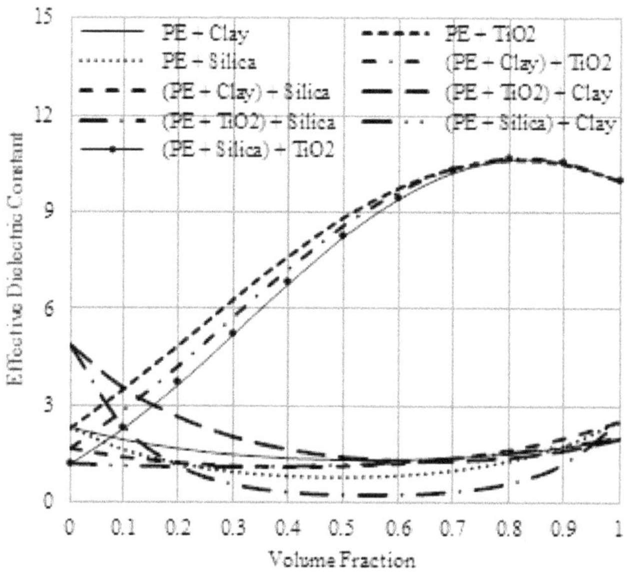

Figure 27. Effective relative permittivity of ABS nanocomposites with various volume fraction of multi-nanoparticles

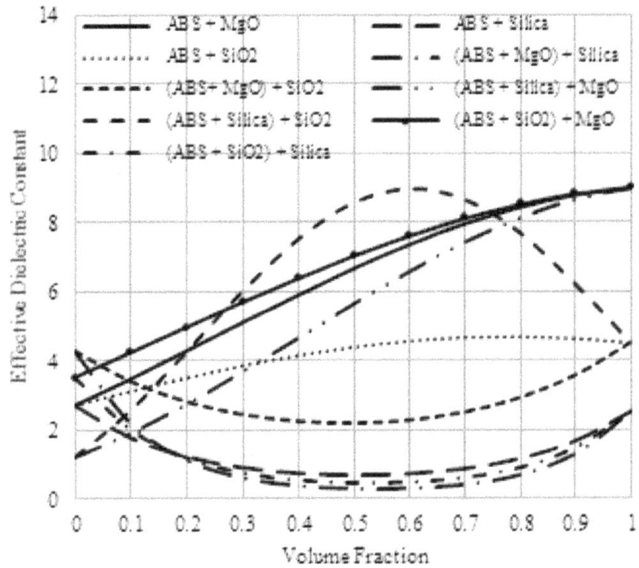

As demonstrated in Fig. 27, utilizing silica nanoparticles as a second kind under (0. 2wt. % clay/PE) base matrix diminishes the powerful dielectric constant, in case TiO_2 nanoparticles concerning illustration of second nanoparticles expand the resultant of the beginning successful dielectric constant from the beginning multi-nanoparticles. Moreover, adding silica as singular nanoparticles diminishes the effective dielectric constant from the beginning polyethylene materials and the successful dielectric constant of base matrix (0.2wt.% silica/PE) of multi-nanocomposites is diminished upon to 1.2124. Utilizing clay nanoparticles as a second nanoparticle under (0.2wt. % silica/PE) reduces the resultant dielectric constant from the beginning multi-nanoparticles but utilizing TiO_2 nanoparticles as second nanoparticles increases the resultant effective dielectric constant for multi-nanoparticles. In case TiO_2 nanoparticles increment the dielectric constant of polyethylene modern materials, the powerful dielectric constant of base matrix is expanded up to 4.8906. However, utilizing clay or silica nanoparticles as a second kind under (0. 2wt. % TiO_2/PE) base matrix diminishes the powerful dielectric constant by expanding the volume fraction from the beginning multi-nanoparticles. As demonstrated in Fig. 27, there is a conduct from the beginning powerful dielectric constant of acrylonitrile butadiene styrene nanocomposites with variant volume fraction. The effective dielectric constant from the beginning MgO/ABS nanocomposites has been expanded towards expanding volume fraction for unique MgO nanoparticles. In case of the beginning multi-nanocomposites, the successful dielectric constant from the beginning base matrix (0. 2wt. % MgO/ABS) is expanded up to 4.2789. However, utilizing SiO_2 or silica nanoparticles as a second kind under multi-nanocomposites base matrix diminishes the effective dielectric constant towards expanding the volume fraction of multi-nanoparticles.

Figure 28. Effective relative permittivity of PVC nanocomposites with various volume fraction of multi-nanoparticles

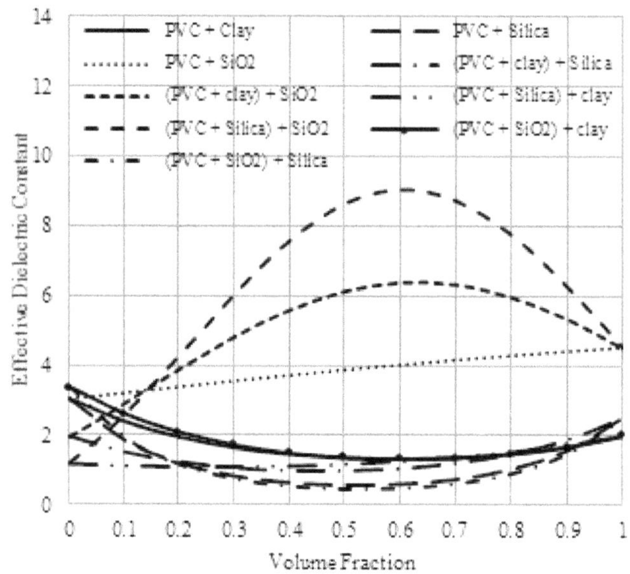

Figure 29. Effective relative permittivity of EPDM nanocomposites with various volume fraction of multi-nanoparticles

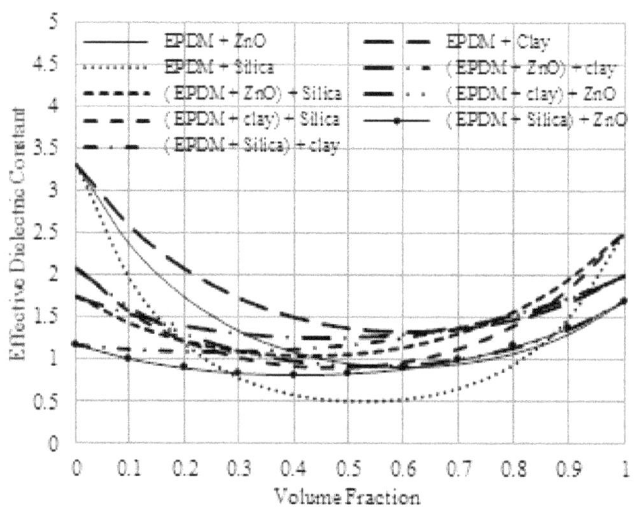

In case of utilizing unique SiO_2 nanoparticles inside acrylonitrile butadiene styrene, the powerful dielectric constant of materials has been expanded, moreover, the effective dielectric constant of base matrix (0.2wt.% SiO_2/ABS) for multi-nanocomposites is expanded up to 3.5016. Furthermore, utilizing silica nanoparticles as second nanoparticles decreases the resultant from the beginning effective dielectric constant of multi-nanoparticles, in case utilizing MgO nanoparticles as second nanoparticles under (0.2wt. % SiO_2/ABS) increases the resultant of successful dielectric constant of multi-nanoparticles. It is known that silica nanoparticles diminish the successful dielectric constant of acrylonitrile butadiene styrene industrial materials; with the goal that the effective dielectric constant from the beginning base matrix (0.2wt.% silica/ABS) of multi-nanocomposites is diminished up to 1.1993 by utilizing both SiO_2 or MgO nanoparticles as a second kind under (0.2wt. % silica/ABS) base matrix, diminishing the powerful dielectric constant and eventually perusing the expansion of the volume fraction for multi-nanoparticles. Creating polyvinyl chloride materials is indicated in Fig. 28, where the conduct technique of successful dielectric constant for polyvinyl chloride nanocomposites for variant volume fraction is highlighted. The powerful dielectric constant of clay/PVC nanocomposites is diminished by expanding volume fraction of individual clay nanoparticles. In case of multi-nanocomposites, the effective dielectric constant of base matrix (0.2wt.% clay/PVC) is diminished up to 1.9495. Utilizing silica nanoparticles as second nanoparticles reduces the resultant of powerful dielectric constant of multi-nanoparticles, while utilizing SiO_2 nanoparticles, concerning illustration, as second nanoparticles increases the resultant for successful dielectric constant of multi-nanoparticles.

Figure 30. Effective relative permittivity of PI nanocomposites with various volume fraction of multi-nanoparticles

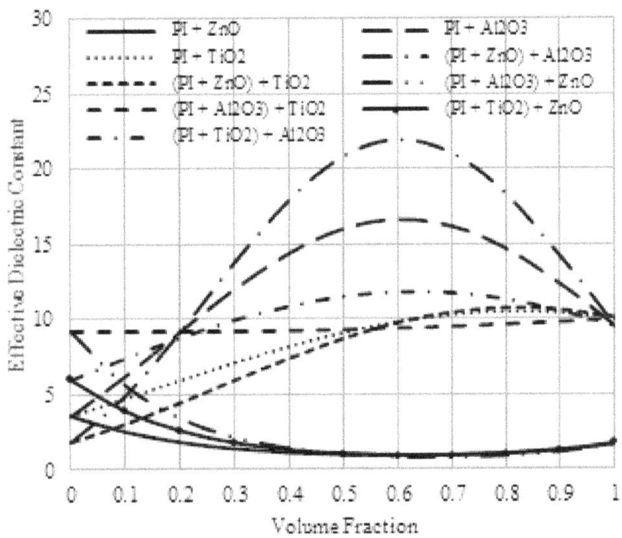

In any case, utilizing distinctive SiO_2 nanoparticles inside polyvinyl chloride expands the effective dielectric constant of materials. Moreover, the effective dielectric constant from the beginning base matrix for multi-nanocomposites is expanded up to 3.3557. Additionally, utilizing both SiO_2, or clay nanoparticles as a second kind under (0.2wt. % SiO_2/PVC) base matrix diminishes the effective dielectric constant towards expanding the volume fraction of multi-nanoparticles. It is known that silica abates the powerful dielectric constant of polyvinyl chloride industrial materials, so that the effective dielectric constant of base matrix (0.2wt. % silica/PVC) is diminished up to 1.1878. Nevertheless, utilizing clay nanoparticles as second nanoparticles diminishes the resultant from successful dielectric constant of multi-nanoparticles, but utilizing SiO_2 nanoparticles as second nanoparticles under (0.2wt.% silica/PVC) increases the resultant from effective dielectric constant of multi-nanoparticles. The conduct of effective dielectric constant for ethylene Propylene Diene elastic nanocomposites with variant volume fraction is likewise indicated in Fig. 29. The effective dielectric constant from the beginning ZnO/EPDM nanocomposites has been diminished towards expanding volume fraction for distinctive ZnO nanoparticles. However, it is to be noted that the effective dielectric constant from the beginning base matrix (0. 2wt. % ZnO/EPDM) to multi-nanocomposites is diminished up to 1.741; and so, utilizing clay or silica nanoparticles as a second sort under multi-nanocomposites base matrix diminishes the effective dielectric constant by expanding the volume fraction of multi-nanoparticles.

Figure 31. Effective relative permittivity of PMMA nanocomposites with various volume fraction of multi-nanoparticles

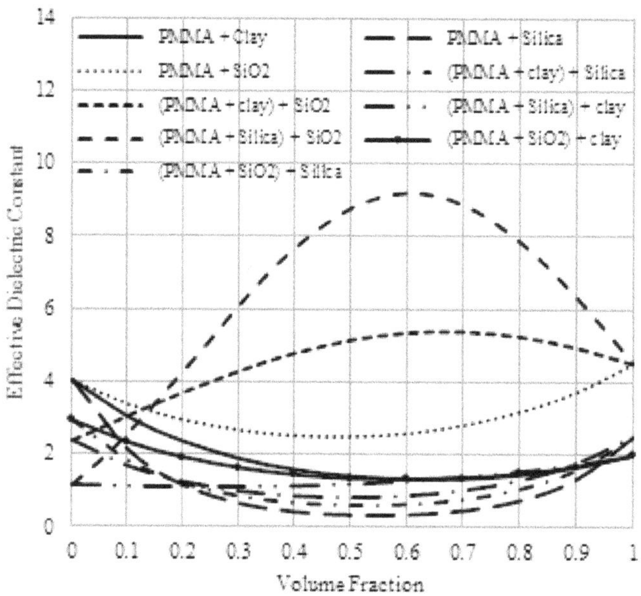

In addition, utilizing individual silica nanoparticles inside ethylene Propylene Diene elastic has diminished the successful dielectric constant from the beginning materials. Moreover, the powerful dielectric constant of base matrix for multi-nanocomposites is diminished up to 1.1757. Utilizing either clay or ZnO nanoparticles concerning illustration as a second kind under (0.2wt. % silica/EPDM) base matrix diminishes the powerful dielectric constant by expanding the volume fraction of multi-nanoparticles. It is known that clay nanoparticles diminish the effective dielectric constant of ethylene Propylene Diene elastic industrial materials; along these lines, the powerful dielectric constant of base matrix is diminished up to 2.0698.

Further decrease can be noted in the successful dielectric constant necessities to utilize both Silica alternately ZnO nanoparticles concerning illustration as a second sort under (0.2wt.% clay/EPDM) base matrix for expanding the volume fraction of multi-nanoparticles. Similarly, as indicated in fig. 30, the conduct technique of effective dielectric constant is employed to polyamide nanocomposites for variant volume fraction. The effective dielectric constant of ZnO/PI nanocomposites is diminished towards expanding volume fraction of unique ZnO nanoparticles.

Figure 32. Effective relative permittivity of EPOXY nanocomposites with various volume fraction of multi-nanoparticles

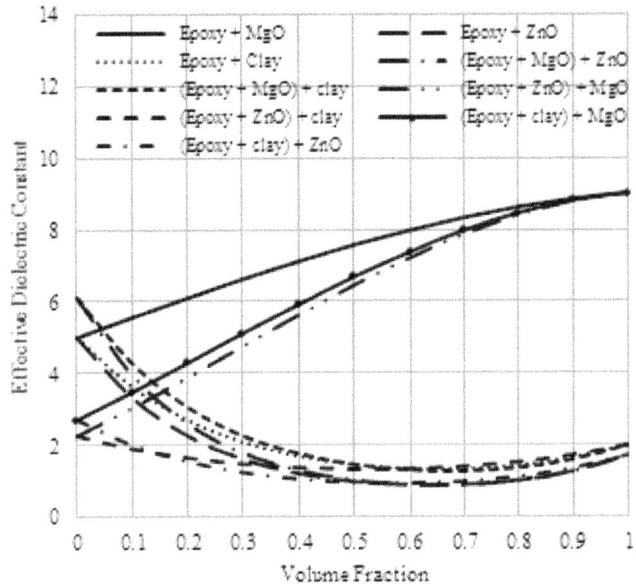

In case of the beginning multi-nanocomposites, the effective dielectric constant of base matrix (0.2wt. % ZnO/PI) is diminished up to 1.847. Moreover, utilizing TiO_2 alternate alumina nanoparticles as a second sort under multi-nanocomposites base matrix expands the effective dielectric constant, eventually perusing the expansion of the volume fraction for multi-nanoparticles. In case of beginning to utilize individual TiO_2 nanoparticles inside polyamide, the successful dielectric constant from beginning materials is expanded. Furthermore, the powerful dielectric constant of base matrix (0. 2wt. % TiO_2/PI) for multi-nanocomposites is expanded up to 5.9644, utilizing ZnO nanoparticles as second nanoparticles reduces the resultant of powerful dielectric constant for multi-nanoparticles and utilizing alumina nanoparticles as second nanoparticles increases the resultant of powerful dielectric constant from multi-nanoparticles. Alumina nanoparticles expand the successful dielectric constant for polyamide materials and the powerful dielectric constant of base matrix (0. 2wt. % Al_2O_3/PI) expanded up to 9. 2127. Additionally, utilizing TiO_2 nanoparticles as a second kind under base matrix expands the powerful dielectric constant by expanding the volume fraction of multi-nanoparticles.

Figure 31 indicates the conduct from the beginning powerful dielectric constant to Polymethylmethacrylate nanocomposites for variant volume fraction. The powerful dielectric constant of clay/PMMA nanocomposites is diminished towards expanding volume fraction from the beginning distinctive clay nanoparticles. The powerful dielectric constant from the beginning base matrix (0.2wt. % clay/PMMA) multi-nanocomposites is diminished up to 2.3377; moreover, utilizing silica nanoparticles as second nanoparticles under (0.2wt. % clay/PMMA) decreases the resultant of effective dielectric constant. Furthermore, utilizing SiO_2 nanoparticles as second nanoparticles expand the resultant from the beginning effective dielectric constant of multi-nanoparticles. Added to that, utilizing distinct SiO_2 nanoparticles inside Polymethylmethacrylate expands the powerful dielectric constant of materials. Moreover, the successful dielectric constant of base matrix (0. 2wt. % SiO_2/PMMA) to multi-nanocomposites is expanded up to 2.9252. Both silica and clay nanoparticles are utilized concerning illustration as a second kind under multi-nanocomposites base matrix, diminishing the effective dielectric constant and eventually perusing the expansion of the volume fraction from the beginning multi-nanoparticles. The unique silica nanoparticles inside Polymethylmethacrylate diminishes the effective dielectric constant from the beginning materials and the powerful dielectric constant for base matrix (0.2wt.% silica/PMMA) from the beginning multi-nanocomposites is diminished up to 1.1468. Clay nanoparticles that are utilized concerning illustration as second nanoparticles under (0.2wt.% silica/PMMA) diminish the resultant from the beginning successful dielectric constant. In any case, utilizing SiO_2 nanoparticles as second nanoparticles expands the resultant of powerful dielectric constant of multi-nanoparticles.

Figure 32 indicates the conduct of powerful dielectric constant to epoxy nanocomposites for variant volume fraction of nanoparticles; then, MgO expands the powerful dielectric constant of epoxy, eventually perusing the expanding volume fraction of distinctive MgO nanoparticles. In case of multi-nanocomposites, the powerful dielectric constant of base matrix (0.2wt.% MgO/Epoxy) is expanded up to 6.1049. Moreover, utilizing clay or ZnO nanoparticles as a second sort under multi-nanocomposites base matrix diminishes the powerful dielectric constant, eventually perusing the expansion of the volume fraction of multi-nanoparticles. In case of utilizing unique clay nanoparticles inside epoxy, the successful dielectric constant is diminished from the beginning epoxy materials, moreover, the successful dielectric constant of base matrix (0. 2wt. % clay/Epoxy) of multi-nanocomposites is diminished up to 2.6987. Utilizing ZnO nanoparticles as second nanoparticles reduces the resultant of powerful dielectric constant of multi-nanoparticles but utilizing MgO nanoparticles concerning illustration as second nanoparticles increases the resultant of powerful dielectric constant from the beginning multi-nanoparticles. Similarly, it is known that ZnO nanoparticles diminish the effective dielectric constant epoxy modern materials; thus, the effective dielectric constant from the beginning base matrix (0.2wt.% ZnO/Epoxy) is diminished up to 2.2761. Similarly, utilizing clay nanoparticles as second nanoparticles reduces the resultant from powerful dielectric constant for multi-nanoparticles and utilizing MgO nanoparticles as second nanoparticles increases the resultant of successful dielectric constant of multi-nanoparticles.

6.5.3 Trends of Multi-Nanoparticles Technique

Combining nanoparticles (clay, ZnO, and silica) attempts to diminish the successful dielectric constant for expanding the volume fraction from the beginning distinctive nanoparticles, yet the different aggregation of nanoparticles (MgO, SiO_2, TiO_2, and Al_2O_3) add the powerful dielectric constant for expanding volume fraction of individual nanoparticles. The Multi-nanoparticles method can be additionally productive to evolve the powerful dielectric constant for multi-nanocomposite base matrix to become low or high, as stated by the kind and concentration of primary individual nanoparticles use. Base matrix from the beginning multi-nanoparticles base matrix has been chosen as (0.2wt. % nanoparticle/POLYMER) because "0.2wt. %" centralization for singular nanoparticles inside the group polymer materials will match a standout amongst ideal values to be kept with respect to thermal and metallurgy properties from the beginning group polymer materials to mechanical creation. Prediction of dielectric values of multi-nanocomposites provides huge numbers of decisions for modern processing on control, with respect to electrical industrial materials.

The course of action for multi-nanoparticles inside the host base matrix is an important component to upgrade the powerful dielectric constant from the beginning nanocomposites. Expanding concentration of Clay, ZnO, and silica as second sort nanoparticle causes diminishing in the successful dielectric constant for any multi-nanocomposites. Similarly, expanding centralization of alumina and TiO_2 as a second kind of nanoparticle causes expanding the successful dielectric constant of any multi-nanocomposites. Utilizing SiO_2 as a second kind nanoparticle causes expansion in the effective dielectric constant of (Clay, ZnO and Silica)/polymer for multi-nanocomposites. However, it leads to diminishing in the effective dielectric constant of (MgO, TiO_2 and Alumina)/polymer for multi-nanocomposites towards expanding the volume fraction of SiO_2 nanoparticles. Utilizing MgO as a second sort nanoparticle leads to expanding in the powerful dielectric constant of (Clay, ZnO, Silica, SiO_2 and TiO_2)/polymer for multi-nanocomposites and leads to diminishing in the successful dielectric constant of Alumina/polymer for multi-nanocomposites by expanding the volume fraction of MgO nanoparticles.

REFERENCES

Brosseau, C., Queffelec, P., & Talbot, P. (2001). Microwave Characterization of Filled Polymers. *Journal of Applied Physics*, *89*(8), 4532–4540. doi:10.1063/1.1343521

Dong, L., Gu, G. Q., & Yu, K. W. (2003). *First principles approach to dielectric response of graded spherical particles* (Vol. 67). Phys Rev B.

Ebnalwaled, A. A., & Thabet, A. (2016, July). Controlling the optical constants of PVC nanocomposite films for optoelectronic applications. *Synthetic Metals Journal*, *220*, 374–383. doi:10.1016/j.synthmet.2016.07.006

Gershon, Calame, & Birnboim. (2001). Complex Permittivity Measurements and Mixing Laws of Alumina Composites. *J. Appl. Phys.*, *89*, 8110-8116.

Gouda, O., Mobarak, Y. A., & Samir, M. (2010). A Simulation Model for Calculating the Dielectric properties of Nanocomposite Materials and Comprehensive Interphase Approach. *14th International Middle East Power Systems Conference (MEPCON)*, 151-156.

Jing, X., Zhao, W., & Lan, L. (2000). The Effect of Particle Size on Electric Conducting Percolation Threshold in Polymerr Conducting Particle Composites. *Journal of Materials Science Letters*, *19*(5), 377–379. doi:10.1023/A:1006774318019

Kakavas, P. A., Anifantis, N. K., & Papanicolaou, G. C. (1998). The role of imperfect adhesion on thermal expansivities of transversely isotropic composites with an inhomogeneous interphase. *Composites. Part A, Applied Science and Manufacturing*, *29A*(9-10), 1021–1026. doi:10.1016/S1359-835X(98)00013-X

Kanuparthi, Zhang, Subbarayan, Sammakia, Gowda, & Tonapi. (2006). Full-Field Simulations of Particulate Thermal Interface Materials: Separating the Effects of Random Distribution from Interfacial Resistance. *IEEE Conference Thermal and Thermo-mechanical Phenomena in Electronics Systems Conference*, 1276-1282. DOI: 10.1109/ITHERM.2006.1645492

Karkkainen, K. K., Sihvola, A. H., & Nikoskinen, K. I. (2000). Effective Permittivity of Mixtures: Numerical Validation by the FDTD Method. *IEEE Transactions on Geoscience and Remote Sensing*, *38*(3), 1303–1308. doi:10.1109/36.843023

Li, Sh., Yin, G., Chen, G., Li, J., Bai, S., Zhong, L., Zhang, Y., & Lei, Q. (2010, October). Y. Zhang1, and Q. Lei, "Short-term Breakdown and Long-term Failurein Nanodielectrics: A Review. *IEEE Transactions on Dielectrics and Electrical Insulation, 17*(5), 1523–1535. doi:10.1109/TDEI.2010.5595554

Li, Z., Okamoto, K., Ohki, Y., & Tanaka, T. (2010, June). Effects of Nanoparticles Addition on Partial DischargeResistance and Dielectric Breakdown Strength of Micro-Al2O3/Epoxy Composite'. *IEEE Transactions on Dielectrics and Electrical Insulation, 17*(3), 653–661. doi:10.1109/TDEI.2010.5492235

Lutz, M. P., & Zimmerman, R. W. (2005). Effect of an inhomogeneous interphase zone on the bulk modulus and conductivity of a particulate composite. *International Journal of Solids and Structures, 42*(2), 429–437. doi:10.1016/j.ijsolstr.2004.06.046

McCalley, J. D., & Krishnan, V. (2014). A survey of transmission technologies for planning long distance bulk transmission overlay in US. *International Journal of Electrical Power & Energy Systems, 54*, 559–568. doi:10.1016/j.ijepes.2013.08.008

Milton, G. W. (2002). *The theory of composites.* Cambridge University Press. doi:10.1017/CBO9780511613357

Nan, C.-W., Birringer, R., Clarke, D. R., & Gleiter, H. (1997). Effective Thermal Conductivity of Particulate Composites with Interfacial Thermal Resistance. *Journal of Applied Physics, 81*(10), 6692–6699. doi:10.1063/1.365209

Nelson. (2001). Measurement and Calculation of Powdered Mixture Permittivities. *IEEE Trans. Inst. Meas., 50*, 1066-1070.

Nelson, J. K., & Hu, Y. (2004). The effect of nanocomposite formulations on electrical voltage endurance. *IEEE 8th Int. Conf. on Solid Dielectrics*, 68-73.

Oltean, & Motoc. (2012). Factors influencing the electrical conductivity of composites with iron particles. *IEEE International Conference on Optimization of Electrical and Electronic Equipment (OPTIM)*, 124-129. DOI: 10.1109/OPTIM.2012.6231895

Ozmusul & Picu. (2002). Elastic Moduli of Particulate Composites With Graded Filler-Matrix Interfaces. *Polymer Composites, 23*(1), 110-119.

Polizos, G. T., Sauers, E., More, I., & Karren, L. (2010). Properties of a Nanodielectric Cryogenic Resin. *Applied Physics Letters, 96*(15), 152903–152903, 3. doi:10.1063/1.3394011

Qu & Wong. (2002). Effective Elastic Modulus of Underfill Material for Flip-Chip Applications. *IEEE Trans. Components and Packaging Techn., 25*, 53-55.

Rao, Y., Qu, J., & Marinis, T. (2000). A precise numerical prediction of effective dielectric constant for polymer–ceramic composite based on effective medium theory. *IEEE Transactions on Components and Packaging Technologies, 23*(4), 680–683. doi:10.1109/6144.888853

Roy, M., Nelson, J. K., Maccrone, R. K., Schandler, L. S., Reed, C. W., Keefe, R., & Zenger, W. (2005). Polymer Nanocomposite Dielectrics- The Role of the interface. *IEEE Transactions on Dielectrics and Electrical Insulation, 12*(4), 629–643. doi:10.1109/TDEI.2005.1511089

Roy, M., Nelson, J. K., Reed, C. W., MacCrone, R. K., Keefe, R. J., Zenger, W., & Schadler, L. S. (2005). Polymer nanocomposite dielectrics – The role of the interface. *IEEE Transactions on Dielectrics and Electrical Insulation*, *12*(4), 629–642. doi:10.1109/TDEI.2005.1511089

Salvadori, Teixeira, Cattani, Nikolaev, Savkin, Oks, Park, Phillips, Yu, & Brown. (n.d.). On the electrical conductivity of Ti-implanted alumina. Academic Press.

Sang, Z.-F., & Li, Z.-Y. (2004a). Interfacial effect on effective dielectric response of spherical granular composites. *Physics Letters. [Part A]*, *331*(1-2), 125–131. doi:10.1016/j.physleta.2004.08.045

Sang, Z.-F., & Li, Z.-Y. (2004b). Partial resonant response of composites containing coated particles with graded shells. *Physics Letters. [Part A]*, *332*(5-6), 376–381. doi:10.1016/j.physleta.2004.09.073

Sarathil, R., Sahu, R. K., Kumar, P. R., & Tanaka, T. (2006). Understanding the Performance of Epoxy Nano Composites – A Physico-Chemical Approach. *IEEJ Trans. Fundamentals and Materials*, *126*(11), 1112–1120. doi:10.1541/ieejfms.126.1112

Singha, S., & Thomas, M. J. (2008). Dielectric properties of epoxy nanocomposites. *IEEE Transactions on Dielectrics and Electrical Insulation*, *15*(1), 12–23. doi:10.1109/T-DEI.2008.4446732

Tagami, N., Hyuga, M., Ohki, Y., Tanaka, T., Imai, T., Harada, M., & Ochi, M. (2010). Comparison of Dielectric Properties between Epoxy Composites with Nanosized Clay Fillers Modified by Primary Amine and Tertiary Amine. *IEEE Dielectrics and Electrical Insulation Transactions*, *17*(1), 214–220. doi:10.1109/TDEI.2010.5412020

Tanaka, T. (2005). Dielectric Nanocomposites with Insulating Properties. *IEEE Transactions on Dielectrics and Electrical Insulation*, *12*(5), 914–928. doi:10.1109/TDEI.2005.1522186

Tanaka, T., Montanari, G. C., & Mulhaupt, R. (2004). Polymer Nanocomposites as Dielectrics and Electrical Insulation – perspectives foe Processing Technologies, Material Characterization and Future Applications. *IEEE Transactions on Dielectrics and Electrical Insulation*, *11*(5), 763–784. doi:10.1109/TDEI.2004.1349782

Thabet, A. (2015a, January). Experimental enhancement for dielectric strength of polyethylene insulation materials using cost-fewer nanoparticles. *International Journal of Electrical Power & Energy Systems*, *64*, 469–475. doi:10.1016/j.ijepes.2014.06.075

Thabet, A. (2015b, June). Experimental Verification for Improving Dielectric Strength of Polymers by Using Clay Nanoparticles. *Advances in Electrical and Electronic Engineering Journal*, *13*(2), 182–190.

Thabet, A. (2016). Thermal experimental verification on effects of nanoparticles for enhancing electric and dielectric performance of polyvinyl chloride. *Journal of the International Measurement Confederation*, *89*, 28–33. doi:10.1016/j.measurement.2016.04.002

Thabet, A. (2017, June). Theoretical Analysis for effects of nanoparticles on dielectric characterization of electrical industrial materials. *Electrical Engineering (ELEN). Journal*, *99*(2), 487–493.

Thabet, A., & Ebnalwaled, A. A. (2017, November). Improvement of surface energy properties of PVC nanocomposites for enhancing electrical applications. *Journal of the International Measurement Confederation*, *110*, 78–83. doi:10.1016/j.measurement.2017.06.023

Thabet, A., & Mubarak, Y. A. (2017, June). The Effect of Cost-Fewer Nanoparticles on the Electrical Properties of Polyvinyl Chloride. *Electrical Engineering in Japan*, *99*(2), 625–631. doi:10.100700202-016-0392-3

Todd, M. G., & Shi, F. G. (2002). Validation of a novel dielectric constant simulation model and the determination of its physical parameters. *Microelectronics Journal*, *33*(8), 627–632. doi:10.1016/S0026-2692(02)00038-1

Todd, M., & Shi, F. (2003a). Characterizing the Interphase Dielectric Constant of Polymer Composite Materials: Effect of Chemical Coupling Agents. *Journal of Applied Physics*, *94*(7), 4551–4557. doi:10.1063/1.1604961

Todd, M., & Shi, F. (2003b). Molecular Basis of the Interphase Dielectric Properties of Microelectronic and Optoelectronic Packaging Materials. *IEEE Transactions on Components and Packaging Technologies*, *26*(3), 667–672. doi:10.1109/TCAPT.2003.817862

Tuncer, E., Sauers, I., James, D. R., Ellis, A. R., Paranthaman, M. P., Aytug, T., Sathyamurthy, S., More, K. L., Li, J., & Goyal, A. (2007). J. Li and A. Goyal, "Electrical properties of epoxy resin based nanocomposites. *Nanotechnology*, *18*(2), 1–6. doi:10.1088/0957-4484/18/2/025703

Vo, H. T., & Shi, F. G. (2002). Towards model-based engineering of optoelectronic packaging materials, dielectric constant modeling. *Microelectronics Journal*, *33*(5-6), 409–415. doi:10.1016/S0026-2692(02)00010-1

Vo, H., Todd, M., Shi, F., Shapiro, A., & Edwards, M. (2001). Toward model based engineering of under fill Materials: CTE modeling. *Microelectronics Journal*, *32*(4), 331–338. doi:10.1016/S0026-2692(00)00152-X

Wei, E.-B., & Tang, S.-P. (2004). Dielectric response of graded cylindrical composites. *Physics Letters. [Part A]*, *328*(4-5), 395–399. doi:10.1016/j.physleta.2004.06.033

Wilkes, G. L., & Wen, J. Y. (1996). Organic/Inorganic hybrid network materials by the Sol-Gel approach. *Chemistry of Materials*, *8*(8), 1667–1681. doi:10.1021/cm9601143

Wu, Qiang, Chen, Liu, Koledintseva, & Drewniak. (2007). Numerical Modeling of Periodic Composite Media for Electromagnetic Shielding Application. *IEEE International Symposium on Electromagnetic Compatibility*, 1-7. DOI: 10.1109/ISEMC.2007.216

Yi, Y., & Sastry, M. (2002). Analytical Approximation of the TwoDimensional Percolation Threshold for Fields of Overlapping Ellipses. Phys. Rev. E, 66.

Yu, K. W., & Gu, G. Q. (2005). Effective conductivity of composites of graded spherical particles. *Physics Letters. [Part A]*, *345*(4-6), 448–452. doi:10.1016/j.physleta.2005.07.037

Yu, K. W., Gu, G. Q., & Huang, J. P. (2002). *Dielectric response of spherical particles of graded materials*. Conducting Material.

Zhang, Kanuparthi, Subbarayan, Sammakia, & Tonapi. (2005). Hierarchical Modeling and Trade-Off Studies in Design of Thermal Interface Materials. *Proc. of ASME InterPACK'05*, 17-22. DOI: 10.1115/IPACK2005-73259

Zhang, X., Subbarayan, G., & Zhang, XSubbarayan, G. (2004). A Constructive Approach for Heterogeneous Material Modeling and Analysis. *Computer-Aided Design and Applications*, *1*(1-4), 171–178. doi:10.1080/16864360.2004.10738256

Zhong, Y., Wang, J., Ming Wu, Y., & Huang, Z. P. (2004a). Effective Moduli of Particle-Filled Composite with Inhomogeneous Interphase: Part I–Mapping Method and Evaluation. *Composites Science and Technology*, *64*(9), 1353–1362. doi:10.1016/j.compscitech.2003.10.010

Zhong, Y., Wang, J., Ming Wu, Y., & Huang, Z. P. (2004b). Effective moduli of particle-filled composite with inhomogeneous interphase: Part II mapping method and evaluation. *Composites Science and Technology*, *64*(9), 1353–1362. doi:10.1016/j.compscitech.2003.10.010

Chapter 5
NanoDielectrics Fabrication

ABSTRACT

This chapter detects how to structure nanocomposites for metals, semiconductors, metal oxides, and polymers. Therefore, this chapter contains the following points: fabrication of nanodielectrics which handled the synthesis of co-dielectrics, synthesis of organic nanoparticles/dielectrics, synthesis of inorganic nanoparticles/dielectrics, synthesis of metal nanoparticles/dielectrics. It contains also the synthesis of multi-nanodielectrics, synthesis of nanodielecrics with coating nanoparticles, synthesis of thin films nanodielectrics. This chapter draws attention also to preparation of membranes.

Generally, three routes have been connected on scatter nanopowders to polymers. The main route is immediate blending, alternately mixing polymer and the nanopowder over discrete periods (known concerning illustration as melt mixing) or concerning result (as solution mixing). The second is sol-gel procedure which begins with sub-atomic forerunner at encompassing temperature in structures of metal or metal oxide schema by hydrolysis and buildup. The third is in situ grafting polymerization of macromolecular chains on the surface of nanopowder. In this chapter, an alternate system for amalgamation will be discussed and reviewed for preparing distinctive nanocomposites comprising diverse polymers and distinctive nanoparticles.

5.1 SYNTHESIS OF COPOLYMERS

CTFE copolymers have a carbon-carbon twofold bond and can be polymerized to structure polychlorotrifluoroethylene alternately copolymerized to prepare the plastic ECTFE. P(VDF-co-CTFE) (PVDF SOLEF_ 31508/1001) may have been compassionately furnished, eventually perusing solvay. 4-vinylpyridine (4VP), copper (I) chloride (CuCl), 1,1,4,7,10,10-hexamethyl triethylene tetramine (HMTETA), 1-bromohexane, silver para-toluenesulfonate (AgPTS) and 1-methyl-2- pyrrolidinone (NMP) were bought beginning with Aldrich. Methanol, dimethyl sulfoxide (DMSO), tetrahydrofuran (THF) and diethyl ether were bought beginning with j. T. Dough puncher and constantly dissolvable. Moreover, chemicals were official graded and were utilized concerning illustration gained without further purification. In synthesis

DOI: 10.4018/978-1-7998-3829-6.ch005

of P(VDF-co-CTFE)-g-P4VP graft copolymer and 3 g of P(VDF-co-CTFE), they are broken down into 75 ml NMP in a round cup towards 80°C. Following cooling, the result of space temperature, 18 ml 4VP, 0.24 g CuCl and 0.66 ml HMTETA were included in the result and the response cup is fixed for an elastic septum. The mixture is mixed to process a homogeneous result and purged for nitrogen for 30 min. Then, the mixture is put to a 120°C oil shower for 6 h. After polymerization, the resultant polymer result is weakened for THF. Following this, the result through a section with actuated Al2O3 uproots the catalyst and the result is precipitated for methanol. The polymer is purified with uproot unreacted P4VP, totally eventually perusing thrice re-dissolving NMP and re-precipitating clinched alongside methanol. P(VDF-co-CTFE)-g-P4VP graft copolymer is produced and dried for a vacuum broiler overnight during room temperature. In case of synthesis of graft copolymers, P(VDF-co-CTFE)-g-PSSA 1. 0 g of P(VDF-co-CTFE) are broken down in 25 ml NMP in a round cup at 80°C. The separate sums of SSA are disintegrated for 20 ml DMSO at 80°C and include P (VDF-co- CTFE) result. Then, after transforming homogeneous solution, 0.08 g for CuCl and 0.2 ml of HMTETA are included and the response cup is fixed with an elastic septum. Afterwards, N_2 is purged for 30 min, the reaction vessel is immersed in an oil bath at 120°C. The response is permitted to proceed for 24 h. After polymerization, the resultant polymer is weakened with THF. After death, the result through a section with actuated Al_2O_3 to uproot the catalyst is precipitated under methanol. The polymer is purified towards re-dissolving DMSO and re-precipitating methanol. Finally, the polymer is dried in a vacuum broiler overnight at room temperature (Koh et al., 2009).

5.2 SYNTHESIS OF ORGANIC NANOPARTICLES/DIELECTRICS

EPDM is a greatly tough manufactured elastic material film (ethylene propylene diene terpolymer) generally utilized within low-slope edifices in the United States and around the world. Its two essential ingredients, ethylene and propylene, are inferred beginning with oil and characteristic gas. The resultant composites are layering formed towards using pressurized water worked under pressure at 150°C for 45 min.

5.2.1 Organophilic Na+–Montmorillonite

Organophilic Na+–montmorillonite (O–MMT) is prepared from immaculate Na+–MMT by ion-exchange response, as stated by the news person system (Acharya et al., 2007). Na+– montmorillonite (12 g) is scattered regarding 600 ml of de-ionized boiling hot water (80 °C) towards utilizing a homogenizer. Octadecylamine (4.8 g, 115 mmol) and concentrated Hcl (1. 8 ml) are broken down under 400 ml boiling hot water. This result is poured in the montmorillonite–water result for energetic blending toward utilizing the homogenizer for 1 h to yield white precipitates. The precipitates are gathered and washed with de-ionized boiling hot water for few times, and the washes are tried for 0.1 m AgNO3 until no precipitate for AgCl is formed. Then, to guarantee the complete evacuation of chloride ions, the items are separated and dried over vacuum stove at 80°C for 14 h (Acharya et al., 2007).

5.2.2 Synthesis of EPDM–Clay Nanocomposites

Few sorts of the EPDM–clay nanocomposites (EPDM–CNs) and EPDM–clay composites (EPDM– CCs) are prepared, eventually perusing melt mixing system (indirect method). The compositions of EPDM–CN and EPDM– cc are indicated in Tables 1 and 2. Maleic anhydride grafted EPDM oligomer (MAH–g-EPDM) are melt-mixed for organoclay at 150°C, utilizing a twin-screw blender (RM-200, made by Harbin School for Science and Technology, China) with chamber extent of 40 cm^3 at screw rotational pace of 90 rpm and blending period of 15 min to yield organoclay intercalated MAH–g-EPDM (OC–MAH–g-EPDM). In the second stage, the pellets of EPDM and OC–MAH–g-EPDM are melt-mixed towards 150°C, utilizing twin-screw blender for 15 min to provide for the EPDM–CN. To examine the impact of MAH–g-EPDM as a compatibilizer, the composites of EPDM (100 phr) and organoclay (3 phr) without MAH–g-EPDM are likewise prepared. The EPDM hybrids (EPDM–CNs and EPDM–CCs) (100 phr) are consecutively blended with zinc oxide (5 phr), stearic corrosive (1 phr), vulcanization accelerator [M (2-Mercapto benzothiazole, 0.5 phr) and TMTD (tetramethyl thiuram disulfide, 1. 5 phr)] and sulfur microbes (1. 5 phr) by utilizing a roll Plant (SK-160B, Shanghai Elastic Mechanical Factory, China). Vulcanized EPDM dirt hybrids are press-molded during 150°C for 30 min to yield elastic sheets (340 l · 150 w · 2 t) mm. An easy methodology for preparing nanocomposites eventually perusing melt mix process, an alternate approach (direct method) is analyzed in this work. EPDM (100 phr), MAH–g-EPDM (20 phr) and organoclay (5 phr) are melt-mixed clinched alongside and that one phase is skipped towards 150°C for 15 min by utilizing twin screw blender to yield nanocomposite; the nanocomposite may have been vulcanized under specified states (Ahmadi et al., 2009).

5.2.3 Synthesis of Epoxy Resin / Montmorillonite / Imidazole Nanocomposite

Na^+-montmorillonite for a cation return limit regarding 100 mmol/100 g is bought beginning with the Qingshan science agenize production line previously mentioned, Lin_an, China. The diglycidyl ether for biphenyl A, epoxy resin E-51 for epoxy of value 0.48–0.54 and normal epoxy equal to196 are acquired beginning with the shanghai resin production line. Imidazole is utilized as a curing agenize. The surfactant of clay, $(CH_3)3(CH_2)16NH_4Br$, is bought beginning with the Examination organization of Xinhua dynamic material in Changzhou, and preparation of Org-MMT for particle trade is conveyed. To the isothermal cure experiment, the epoxy resin and the curing agenize are blended in the stoichiometric proportion of 100 parts resin on four-part curing agenize (by weight). Then, the mixture is blended with Org-MMT powder for loadings of 0.5 and 10 phr, separately. Bend dynamic torsional vibration is a non-resonant compelled vibration. The schematic outline of a hand-crafted test setup— HLX-I resin cure meter is indicated in Fig. 1. The more level mold 3 hosting a warmer inside it is utilized as torsional vibrator filled with the resin materials. At the point where the engine 6 is switched on, the upper shape 2 hosting a warmer inside it excessively and awfully goes down, and the molds are closed with a hole to give a chance for camwood to be balanced. The cure temperature is regulated with thermistors. Thus, the isothermal cure procedure can be performed. Similarly, as the upper and easier molds close, the engine 5 is on, and the bring down mold begins a torsional vibration for a recurrence of 0.05 Hz in an angle below 1, which likewise offers it a chance to be balanced, as stated regarding the hardness of cured resin.

Materials, by method, are for a flighty circle 4 on the velocity progress rigging 7. The torque plentifulness of the torsional vibration is converted under electric signs by the method for the strain gage load Mobile 1, amplified through the enhancer 10 and recorded towards the recorder 11. The resin framework with an alternate level of cure has an alternate torque (or modulus, viscosity and so forth through this way, where observing and stock arrangement of all instrumentation can be enhanced). Therefore, the progress in the mechanical properties, i.e. the degree of cure of the resin system can be a monitored and dead set towards measuring the transforms in torque, and a constant bend reflecting the entire cure procedure is acquired. An ordinary test bend acquired by DTVM is indicated in Fig. 2. The abscissa is the cure on the long run, and the ordinate is the torque required to turn the resin framework by a little angle, which corresponds to the modulus or viscosity of the resin system and can be considered perfect as a relative parameter of the degree of cure. The time of conclusion of the molds is calculated concerning the illustration in the beginning time of cure at the side of the point o. In the extent OA of curing the long haul, the organize structure-shaped through the cure response is not sufficient as a reason for the compelled vibration of the upper mold. Likewise, as a result, the strain gage load Mobile won't have at whatever sign to input, along these lines, that the test curing bend will be a straight accordance comparison of the abscissa. During the purpose A, the viscosity of the resin framework is secondary enough (i.e., the system framed may be finished enough) for the gelation in the resin framework to occur, and the torque and the strain gage load cell seem to input few indicators. Thus, the purpose is the gel side of the point, and the occasion when relating to OA is the gel long haul t_g to the framework. The expanding plentifulness of the torque (slope of the curve) reflects the rate of the curing response. The expanding pattern of the torque has a tendency for endurance for expanding curing time, and the harmony torque G_1 will therefore be arrived at (point B). In the meantime, the curing response is finished, and a cup-like test bend is acquired, at the occasion when comparison to OC is the full curing run through. The envelope of the test bend corresponds to the transform for mechanical conduct technique of the resin framework during cure. Since the cup-like test bend is symmetric of the long run axis for accommodation, we can barely take the upper-half of the envelope as the isothermal cure is bent on examining the cure procedure. The progress of grid dividing of montmorillonite is measured by utilizing a Rigaku D/max-cB pivoting anode X-beam diffractometer for the CuKa line (k ¼ 0:1542 nm), the tube voltage of 40 kV and tube present at 100 mA. The scanning is in the range with 1.2 with 10 with a rate of 1/min (Xu & He, 2003).

5.2.4 Synthesis of Epoxy Resin Nanodielectrics With Organic Nanoparticles

The diglycidyl ether of biphenyl A, epoxy resin E51 (Shanghai Resin Factory), diethylenetriamine (Development focus of uncommon compound operators in Huabei Region) are utilized as curing agenize and have become accessible in business currently. Other chemicals, such as methanol, acetone, $Cd(Ac)_2$ and Na_2S and every bit of A. R are reviewed and utilized without any medication. X-beam diffraction design is recorded looking into RIGAKU D/max-gA pivoting anode X-beam diffraction meter is used to measure the UV-vis range of the composite. The test specimens are microtomed, eventually perusing ULTRACUT-E for the thickness of 80 nm and watched towards the JEM-100SX TEM with the voltage of 100 kV. Photoluminescence (PL) range is recorded utilizing a HITACHI 850 fluorescence spectrophotometer at space temperature and 283 nm excitation wellspring is utilized.

The bisphenol-A diglycidyl ether (d. E. R. 332, the beginning ing with Dow Chemical) is placed under a 500 ml measuring utencil that may have been warmed to 80°C and took after towards the curing agenize 4,4'-diaminodiphenylsulphone (from Merk- Schuchardt) under mechanical mixing towards 400 r/min for roughly 30 min until the curing agenize is broken down. The epoxy resin for an organophilic clay (I. 30E, from Nanocor inc.) fixation about 2.5%wt., 5.0%wt., or 7.5%wt. is at that point set by the measuring utensil under constant mechanical blending towards 400 r/min to 1 h at 80°C. The mixture is exchanged to a topped bottle, which might have been at that point drenched in ultra-nationalistic shower (JAC ultra-nationalistic 1505) under ultra-nationalistic mixing for 2 h. Then, the mixture is degassed at 120°C trailed towards procuring at 170°C for 5 h and post curing at 200°C for 2 h. The SAXRD dissection is performed, eventually perusing utilizing Shimadzu XRD-6000 X-beam diffractometer with cu radiation (λ=0.1542 nm) for an examining velocity of 0.05°/min and an examining go from 0.01° to 1.5°. TGA is measured in air utilizing Perkin Elmer TGA-7 for a warming rate of 10°C/min and temperature going from 50 to 650°C. Mechanical properties are measured towards DMA 2980, changing mechanical analyzer from 50 to 240°C in a recurrence for 1 Hz and a rate of 5°C/ min under nitrogen environment. The nanoclay utilized within this is an altered montmorillonite with business evaluation of cloisite 30B®, got beginning with southern dirt results. The diglycidyl ether about bisphenol (DGEBA) epoxy resin (Araldite Razeen LR-2257) and polyamide hardener (Aradure43SBD) are bought from Huntsman Company (Wang et al., 2008).

5.2.5 Synthesis of PE With Organic Nanoparticles

A high-density polyethylene (PE, 1600J, density: 0.958 g/cm3, dissolving stream index: 18 g/10 min) is brought beginning ing with Yanshan petrochemical co., (China) and utilized as composite grid resin. The pristine clay utilized is Na-montmorillonite (Na-MMT, Zhangjiakou Qinghe compound Factory, China) for a cation trade ability (CEC) of 90 meq/100 g. WAXD has demonstrated that the interlayer dividing (d001) is 0.98 nm. Na-MMT is dried towards 100oC to vacuum for 8 h prior to utilization.

Two intercalating operators are utilized for the adjustment of Na-MMT. A particular case is the commonly-used dioctadecyldimethylammonium chloride (industrial grade, Beijing compound reagent co., China; abbreviated as DOAc), and the difference may be in the sensitive (N-g-trimethoxylsilanepropyl) octadecyldimethylammonium chloride (industrial grade, Jiangsu foundation for Concoction Engineering, China, abbreviated as JSAc). They have been similarly utilized and accepted.

Synthesis of PE with organic nanoparticles

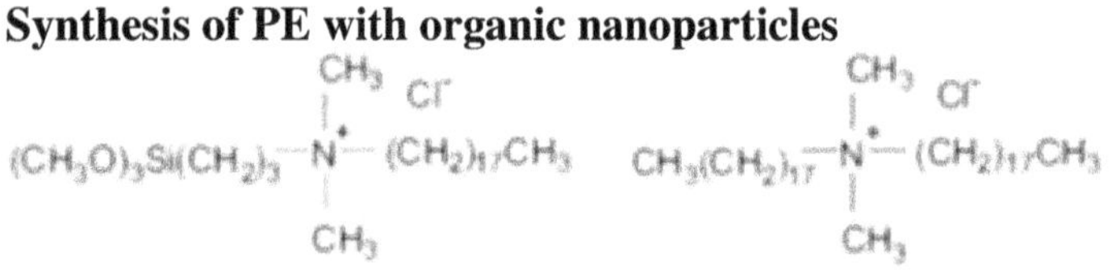

JSAc DOAc

Na-MMT (100 g) is blended for anhydrous ethanol (1000 mL) with blending to structure a uniformly scattered suspension, the intercalating agenizes proportional to 1.2 CEC for Na-MMT are included under the scattering. The admixture is blended in 75°C for 8 h and separated to get organoclays. The organoclays are dried in vacuum towards 100°C for 8 h and grounded into powder. The products made with DOAc and JSAc are named DM and JS, individually.

Modified PE with organoclays are melt exacerbated and clinched alongside a co-rotating twin-screw extruder (FZ30 mm, L/FZ24). The barrel temperature is observed at 180°C and the screw speed is 200 rpm. The specimen is denoted for the creation and the weight portion of organoclays, e.g., PE/JS5 refers to the composite holding 5 phr for JS (5 parts JS for every 100 parts PE). The extruded pellets are injection-molded under plates with thickness of 2 mm towards Mini-Max 183 decay (CSI co., USA) for the estimations about WAXD and TEM. Standard pliable (ASTM D638), flexural (ASTM D790), scored Izod (ASTM D256) and cone calorimetric (100!100!10 mm^3) are examples of injection-molded utilizing an injection-moulding machine. The barrel temperature is 180°C and infusion weight is 80 MPa. Then, after moulding, the examples are put in a desiccator for 48 h before the test (Bagherzadeh & Mahdavi, 2007; Li & Mai, 2006; Zhang et al., 2006; Zhao et al., 2005). Changed PE nanodielectrics requires Montmorillonite (clay) to be pretreated with cetyl pyridinium chloride (CPC) at a molar proportion of 1:1. The mixture is vivaciously mixed in water for 18 h and filtered, washed, freeze-dried and held during 60 8C under vacuum for 12 h. The untreated clay is washed for water and dried at 120 8C under vacuum for 48 h. Ethylene is polymerized utilizing the ansa-metallocence impetus Et (Indenyl)2ZrC12 and the mao co-catalyst to prepare modified PE. From the experiments, the polymerization temperature is situated at room temperature (or 50 8C if required), the catalyst/co-catalyst proportion is between 2500 and 10,050, and the ethylene weight is supported towards 19–21 psi. The aggregate volume of the framework is 60–65 ml and the polymerization run through is 1 h. The measure of metallocence impetus is ca 0.5 mg and acidic methanol (10 ml) is used to end the polymerization. An abundance measure of ch3)2co is included under the polymer result and it is dried for 8 h at 50 8C under vacuum. The clay, metallocence catalyst and a toluic corrosive result of MAO are included consecutively through three distinctive routes.

Synthesis of PE with organic nanoparticles

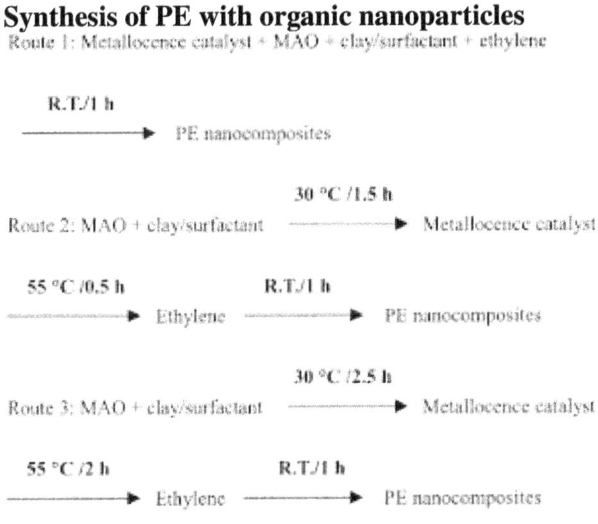

Scheme 1. Polymerization routes of polyethylene nanocomposites.

As illustrated in plan 1, for route 1, the surfactant-pretreated clay is blended specifically for the ethylene monomer, catalyst and co-catalyst on the beginning of the polymerization towards room temperature. In route 2, the surfactant-pretreated dirt is blended at first with the MAO co-catalyst at 308°C for 1.5 h and blended with the metallocence impetus at 558°C for another 0.5 h; the catalyst/clay mixture is at that point blended with ethylene monomer on the beginning of the polymerization for 1 h under encompassing states. For course 3, the surfactant-pretreated clay is mainly blended and fomented with MAO co-catalyst at 308°C for 2.5 h and blended for the metallocence impetus at 558°C for 2 h; the catalyst/clay mixture is at that point blended with ethylene on the beginning of the polymerization for 1 h under encompassing condition (Kuo et al., 2003).

5.2.6 Synthesis of PE, PE-g-MAH-St/Montmorillonite

Nanocomposites in PE and Org-MMT are melt and blended on the two rollers for 10 min at a temperature of 175°C. The mixture is press formed to get a plate of 4 mm thick. The substance of Org-MMT is situated concerning illustration 0, 1, 3, 5 wt. % and the procured materials are characterized concerning illustration PE, PE1, PE3 and PE5 individually. The graft copolymer (PE-g-MAH) and natural montmorillonite are melt and blended on the same system and the materials are similarly named PEM, PEM1, PEM3 and PEM5, respectively.

X-beam diffraction dissection (XRD) is performed utilizing Rigaku X-beam era (CuKa radiation for k = 1. 5406 a þ) at room temperature. The diffract-grams are scanned clinched alongside 2h ranges beginning with 1.2 to 10 towards a rate of 2/min. The interlayer separation for Org-MMT in composites is ascertained beginning with the (001) top by utilizing Bragg equation. Two grams of PE-g-MAH are weighed and set under the cup filled for 100 ml xylene. The cup is warmed until every last one of PE-g-MAH is disintegrated and a clean result is achieved at that point, the high temp result is poured under a measuring utensil held with 300 ml acetone. The immaculate PE-g- MAH precipitates and the impurities are kept in the result. After sifting the solution, the precipitate is put in a stove for 24 h to dry. The dried grafter is pressed under a dainty film to record FT-IR spectra on a Nicolet Magna 750 machine. Transmission electronic micrographs (TEM) are obtained with TEM 100sx, utilizing an acceleration voltage for 200 kV. The test is ultra-micro tomed with a jewel blade at the room temperature which provides for 70-nm-thick area. Differential filtering calorimetry (DSC) analyses are conveyed out utilizing a ta instrument flying 2920 adjusted DSC. The temperature and vitality readings are adjusted for indium at each cooling rate utilized in the estimations. Known estimations are conveyed out in nitrogen air, and the materials are warmed from 30 to 150oC at a rate of 10°C /min to dispense with the heat history when cooling at 10°C /min. Thermo-gravimetric dissection (TGA) is utilized to watch the warm character of composites in a TA instrument flying Hi-Res TGA 2950 thermo-gravimetric Analyzer, under nitrogen during an examination rate of 10oC /min at the beginning with a room temperature of 650°C (Zhai et al., 2004).

5.2.7 Synthesis of XLPE With Organic Nanoparticles

Organically changed dirt (Org-MMT) Cloisite 15A with 31.5Aþ intergallary spacing and particle traded eventually perusing dimethyl dehydrogenated tallow ammonitic bromide are given by southern dirt items. Low thickness polyethylene (LDPE) review LF0200 with MFI of 1.6 g/10 min and thickness of 0.923 g/ml, are obtained from Bandar Imam Petrochemical Company, Iran. Irganox 1010 cell reinforcement with thickness of 1. 45 g/ml is achieved in the beginning with Ciba Geigy Co. Dicumyl peroxidase (DCP) for purity of 98% and thickness of 1.02 g/ml are obtained from Hercules co.

LDPE 3, 5, 8 and 10 wt. % Cloisite 15A are melt-mixed clinched alongside a Brabender Plastic order with a chamber of 50 cm^3 and roller blades at 170° C. The screw speed is held during a low velocity of 30 rpm at the beginning for 5 min and expanded with 60 rpm for 25 min to acquire a well-mixed framework. At that point, the intensify is uprooted from the chamber and cooled to room temperature. Nanocomposites intensify obtained at the beginning in the past area are blended for 0.5 wt. % DCP and 0.1 wt. % cell reinforcement at 130°C and a screw pace of 50 rpm for 8 min. Unique state of the article is obtained by curing under a heated press in 180°C and 80 tonal weight (Rezanejad & Kokabi, 2007).

5.2.8 PVDF Nano-Dielectrics With Organic Nanoparticles

Solef 1011 PVDF powder (Solvay, Belgium) is of the area given, eventually perusing Solvay corp. Ltd. Its weight-average sub-atomic weight is Mw=341,000.The organically changed dirt is Nanomer 1.44P, which might have been obtained from Nanocor inc. (America) concerning illustration of fine powder. A polymer/nanoclay premix is prepared, eventually perusing and joining nanoclay and PVDF in a speed-mixer. This premixed powder is afterwards melted and extruded utilizing a twin-screw extruder (SHJ-30) at a temperature of 200 °C. PVDF and five compositions holding 0.4, 1, 2, 4 and 7 wt. % clay pellets are accessible via the web at www.Sciencedirect.Com and materials letterpress 62 (2008) 747–750 are prepared. The pellets are warmed at 200 °C for 5 min and pressed clinched alongside a mold to get sheets of 0.5 mm thickness, which are cooled by air. Exact sheets are tempered at 160 °C for 24 h. X-beam diffraction estimations are performed towards utilizing a Dmax-rA diffractometer with Cu-anode X-beam hotspot. Molau testing is conveyed out in request to check the collaboration between PVDF grid and organoclay (Hwan Koh, 2009; Lavina et al., 2007; Silva & Cid, 2007). PVDF, oganoclay nanocomposite holding 7 wt. % clay are drenched in NDimethylacetamide (DMAc) dissolvable for soundness differentiating. Polarized optical microscopy (POM) photographs are received eventually perusing utilizing German Leica DMLP magnifying instrument. The tests are sandwiched between two glass plates and warmed to 200 °C above a stove, held at this temperature for 5 min and cooled down characteristically in the surrounding temperature (Yu et al., 2007).

PCMS (Mn ¼55,000) is obtained beginning with Aldrich. Triethylamine (TEA) is dried with NaOH and refined prior to use. Dimethylformamide (DMF) is dried for CaH2 taken after refining in vacuo former to be utilized. The PVDF for a normal sub-atomic weight of 400,000 is obtained from Shanghai 3F New Materials co., Ltd., China. Different reagents are for explanatory review and are utilized without further purification.

Plan 2 indicates the engineered course for PCMS-g-CuPc. A 100 ml three-necked round-base carafe fitted with an attractive stirrer, a thermometer and a condenser is utilized as the reactor. Tea pack (3.0 mL) is included for an answer from PCMS (0.5 g), CuPc (0.5 g) and DMF (40 mL). The result is mixed at 65°C for 12 h under purified nitrogen air. Following tea pack and DMF are evacuated by lessened weight distillation. The mixture is washed with methylene dichloride to uproot unreacted PCMS, if any, trailed towards refined water on uproot triethylamine hydrochloride. The last item is dried vacuum at 50°C and marked PCMS-g- CuPc.

Films are arranged utilizing result cast strategy. For the mix of PVDF and PCMS-g-CuPc (PVDF/PCMS-g-CuPc) with the PCMS-g- CuPc of 15%wt., 30%wt., 40%wt., and 50%wt. (Accordingly the substance of CuPc is 7. 5%wt., 15%wt., 20%wt., and 25%wt., respectively), PCMS-g-CuPc includes the result of PVDF in DMF ultrasonically blended for no less than 2 h. Afterwards, the result is poured onto a clean glass slide and dried in air during 70°C for 5 h and after that in vacuo at 50°C for 12 h to uproot DMF. Finally, the film is tempered at 120°C for 12 h and gradually cooled at room temperature. The preparation methodology of the film of the mix for PVDF and CuPc (PVDF/CuPc) holds 15%wt. CuPc is the same like PVDF/PCMS-g-CuPc. The average novel into a film thickness is 30 mm. For the electric characterization, the films are cut into little bits of something like 10 _10 mm, and hardware gold electrodes for 2. 5 mm span are sputtered in the focal point for both surfaces (Wang et al., 2009).

5.2.9 PP Nano-Dielectrics With Organic Nanoparticles

PP and sepiolite without PP-g-MA are placed in the autoclave. The measure of PP-g-MA in the nano-composites is equal to the add up of sepiolite (1:1 weight ratio). Co2 is pumped under the autoclave through an Isco model 260D syringe pump, after being chilled at -6°C. The autoclave is held at 15 MPa and 200°C under blending by utilizing pitched bladed turbine for 30 min. After that, the autoclave is cooled in water at room temperature, and co2 is vented. Nanocomposites hosting dirt substance of 1wt.%, 2.5wt.%, 5wt.% and 10wt.% are acquired by selecting distinctive weight proportion to the middle of PP and sepiolite. The greater part materials acquired from autoclave are solidified in fluid nitrogen for a couple of seconds and broken into pieces, which are then utilized for high temp pressing under dumbbell-shaped examples at 200°C. The accepted melt aggravating of the nanocomposites can be expressed somewhere else (Damodara et al., 2009; Jhaa, 2010; Kima et al., 2004; Li et al., 2003; Moussaif & Groeninckx, 2003; Shen et al., 2006; Thridandapani et al., 2006; Wang et al., 2006; Xu et al., 2003; Yang et al., 2006; Yu et al., 2009). Briefly, the system operates as follows: The melt aggravating nanocomposites are arranged by a two-stage mixing methodology over a scaled down twin-screw extruder at 200°C and 200 rpm. A master batch of 10 wt. % of filler is prepared for a 5 min preparation duration. This master batch is consequently weakened with immaculate PP at 10 min, transforming the long run (Wang et al., 2006).

An economically accessible review of polypropylene (PP) transformed by Basell Polyolefins (product name: Profax 632) can be used in methodology PP-4 wt. % clay and PP-8 wt. % dirt nanocomposites. This review of PP has a melt-flow rate of 12 g/10 min at 230°C/2. 16 kg. A regular montmorillonite clay surface modified with dimethyl dialkyl ammonitic (Nanomer i. 44P, Nanocor) is utilized as the support filler. The nanocomposites are prepared, eventually perusing the blend of the fitting sums in a twin screw extruder emulated towards infusion forming for ductile bars. The ductile yield quality is a dead set clinched alongside understanding of ASTMD-638. The capacity modulus for slick PP and PP–clay nanocomposites is mulled over towards changing mechanical Investigation (DMA). This methodology is conveyed by utilizing a ta Instruments 2980 for a single cantilever mode from −100 to 150°C. The trial recurrence is 1Hz and the warming rate is 3°C/min. Scattering and inter-capitulars vein about clay layers on PP melt are examined by transmission electron microscopy (TEM). Segments for 100–200 nm are cut utilizing a Leica ultra-microtome prepared for a jewel blade and gathered in a trounce loaded for water and put specifically for 400-mesh copper grids. Transmission electron micrographs are brought with Hitachi H-7600 towards an acceleration voltage of 100 kV.

The change in rate crystallinity and structural qualities prompted by dirt is critical clinched alongside comprehension of the deformity conduct technique alongside its scattering. The crystallization conduct for flawless PP and clay-reinforced PP nanocomposites concentrates on eventually perusing differential examining calorimetry (DSC). The PP and PP–clay nanocomposites are warmed at room temperature (~20°C) to 200°C and held at high engineering for 3 min, so as to eradicate the past thermomechanical history and to acquire a totally loose melt. At that point, the melt is cooled at 30°C at the rate of 10°C/min, and second examination of the rate is conveyed at about 10°C/min (Thridandapani et al., 2006).

An economically accessible isotactic polypropylene is trade stamped (such as T30S, Yan Shan Petroleum China) for mn ¼ 29. 2_103 g/mol and a melt stream list (MFI) of 0.9975 g/ min (190°C, 2. 16 kg) and a thickness of 0.91 g/m3, are utilized like the basal polymer. The compatiblizer, PP grafted maleic anhydride (PP-MA) (MA substance ¼ 0.9%wt., MFI ¼ 6. 74 g/min at 190°C), is bought from Chen Guan Co. (Sichuan, China). Sodium montmorillonite for a cation return limit (CEC) of 68. 8 mmol/100 g (RenShou, Sichuan, China) is organically changed through ion-exchanged response for dioctadecyl dimethylammonium bromide and the nitty gritty methodology can be applied to our past production (Jhaa, 2010). Thereafter, the organically changed MMT can be abbreviated as OMMT.

5.2.10 Synthesis of PP/ Montmorillonite Nanocomposites

Polypropylene (PP, Montell 5C39F) utilized in this way requires a weight-average sub-atomic weight of 3. 2×105 and isotacticity of 86% controlled at 13C nmr (Bruker- AM300, Switzerland). Maleic anhydride grafted polypropylene (PP-g-MAH) holding 0.29%wt. MAH is arranged in our lab for melt-grafting. The business natural montmorillonite, MMT908A, changed with alkyl ammonitic is compassionately furnished by Huate Co., Zhejiang Province, China. PP-g-MAH/MMT nanocomposite is prepared, eventually perusing immediate melt-aggravating MMT908A for PP-g-MAH in a Rheomix-600 blender (Haake Rheocord 900, Germany) at 190–200 °C and 120 rpm for 10min, and the clay stacking is 3%wt.

The degree of swelling and the interlayer separation of the dirt in PPCN are confirmed by WAXD. The X-ray diffraction trials are performed by utilizing a Rigaku Dmax-rC diffractometer with cu focus and a pivoting anode generator worked in 40 kV and 100 mA. The filtering rate is 2°/min from 1.5° to 10°C. The example of WAXD estimation is prepared into a film by layering formation at 200°C and 5 MPa. The transmission electron micrographs are taken beginning with 80 on 100 nm thick, micro-tomed areas utilizing a transmission electron magnifying instrument (HITACHI H-860, Japan) with 100 kV accelerating voltage.

Non-isothermal crystallization is conveyed with respect to a Perkin-Elmer Pyris-1 differential filtering calorimeter (DSC), and the temp is adjusted for indium. The specimens of the PP-g-MAH/MMT nanocomposite and PPg- MAH at 0.2 mm thick are obtained, eventually perusing hot layering formation at 10 MPa and 200 °C and after that quenched to space temp. Disk-like tests around 3 mg of weight to DSC are reduced from the film. In the non-isothermal crystallization process, the tests are liquified at 200 °C for 10min to get rid of the past warm historical backdrop which is cooled after that during const. Cooling rates are 5, 10, 20 and 40 °C/min (Li et al., 2003).

5.2.11 Synthesis of PP Nano-Dielectrics With Injection-Molded Bars

The compatibility nanocomposites for IPP/PP-MA/OMMT (90/10/x %wt., x ¼ 1,3,5,10) are melt-mixed over a TSSJ-2S co-rotating twin-screw extruder. After being pelletized and dried, composites are injected into a shape with the help of a SZ 100 g injection-molding machine at barrel temperature of 190°C and infusion weight of 900 kg cm^2. That point changes pressing infusion formation (thereafter abbreviated into DPIM) when the engineering organization is related to that. Its fundamental characteristics is present shear on the cooling melt during the pressing phase, eventually perusing two pistons that moves reversibly with the same recurrence (1.0 Hz). The transforming parameters and the qualities, as well as the point of interest test method of DPIM are depicted clinched alongside ref. (Wang et al., 2006). Compared with the routine static tests which generally comprise skin zone and center zone, a triple zone perplexing structure partitioned under skin and turned at zone and center along the profundity are structured because of the temperature contrast between the mold cell and the melt and because of the oscillatory shear connected during cementing.

5.2.12 PPO Nano-Dielectrics With Organic Nanoparticles

Polyethylene (LDPE) and Petrothene NA960000 are supplied, eventually perusing Equisresin Concoction Particular Organization. Isotactic polypropylene (Melt list at 230°C /2.16 kg, 4 g/10 min), Vinylbenzyl chloride, 2,20-azobisisobutyronitrile (AIBN) and lithium wire are obtained from Aldrich Concoction Organization. 1-Bromooctane is procured beginning with lancaster. Sodium montmorillonite (Na Cloisite, for CECy0.95 meq/g) are furnished by Southern Clay Products, Inc.

Lithium wire (1.4 g, 0.20 mol) is set to 250 ml three-neck round lowest part cup holding 50 ml of dry ether and provided for an expansion pipe. The framework is administered under nitrogen stream and is held at 0°C, utilizing an ice water shower. Couple of drops for 1-bromobutane are included and blended until the lithium metal is turned into bright, 19.3 g (0.10 mol) for 1-bromooctane in 100 ml of dry ether and then included dropwise through a 1 h period. The ice water shower is evacuated, and the framework is permitted to warm at room temperature and is held during that temperature for 30 min. The ether result of octyllithium is exchanged by syringe, dropwise, to separate a cup holding an answer of 14. 5 g (0.095 mol) of vinylbenzyl chloride over 100 ml for dry THF. The result is mixed under nitrogen purge for 1 h and 1 m Hcl is included to hydrolyze any remaining lithium reagents. The oil layer is concentrated with ether taken by section chromatographic detachment (hexane/ethyl acetic acid derivation 90:10). Bearing 4-nonylstyrene (18. 6 g, 81% yield) is recouped. The personality of the nonylstyrene is plainly secured, eventually perusing NMR spectroscopy. 1H NMR (CDC13, ppm): 7. 32 (d, j ¼ 9 Hz, 2H, ArH), 7. 15 (d, j ¼ 9 Hz, 2H, ArH), 6. 68 (dd, j ¼ 12, 18 Hz, 1H, ArCHe), 5. 70 (d, j ¼ 18 Hz, 1H, ArCHCH2cis), 5. 17 (d, j ¼ 12 Hz, 1H, ArCHCH2trans), 2. 58 (t, j ¼ 7 Hz, 2H, ArCH2e), 1. 57 (quintet, j ¼ 7 Hz, 2H, ArCH2CH2e), 1. 30 (m, 12H, e(CH2)6e), 0.88 (t, j ¼ 6 Hz, 3H, eCH3).

4-Nonylstyrene (46 g, 0.20 mol) and vinylbenzyl chloride (3. 1 g, 0.02 mol) are broken down into100 ml of THF for a 250 ml cup. The result is magnetically mixed for 10 min, and warmed for delicate reflux, after that 0.32 g (2.0 mmol) of AIBN is included to launch the polymerization, and the framework is upheld at reflux for 12 h. The copolymer is precipitated towards spilling the result under 500 ml of methanol and 45 g of copolymer are gathered after filtration trailed towards overnight drying a vacuum stove at 70e80°C. The sub-atomic weight of the polymer is portrayed eventually perusing gel penetrating chromatography; the number of normal sub-atomic weight is 31,000 with a polymer dispersity list of 3. 1.1H nmr (CDC13, ppm): 7. 14 (br, 22H, ArH), 6. 72 (br, 22H, ArH), 4.47 (br, 2H, $ArCH_2C_l$), 2. 56 (br, 20H, $ArCH_2e$), 2.33 (br, 11H, ArCHe), 1. 90 (br, 22H, $ArCHCH_2e$), 1.63 (br, 20H, $ArCH_2CH_2e$), 1. 29 (br, 120H, $e(CH_2)6e$), 0.89 (t, 30H, eCH_3). The NMR information is steady for a copolymer piece of 10:1 proportion of nonylstyrene to vinylbenzyl chloride. Copolymer (45 g) is initially broken down to 200 ml THF trailed eventually perusing about 4 g for triethylamine, and the mixture is blended at room temperature for 12 h preceding it and is gradually poured under 500 ml of methanol. The polymeric surfactant is gathered, eventually perusing filtration and dried overnight to a vacuum broiler at 80°C. There is another expansive crest in the 1H NMR range of 3. 44 ppm, which is doled out like the methylene bunch connected to the nitrogen of the ammonitic salt. At the same time, the methyl bunch contiguous of the methylene is in the 1. 30 ppm locale.

Sodium montmorillonite (15 g) is scattered at 300 ml of THF/H2O (50:50) and blended magnetically for 8 h. At the same time, 35 g of surfactant is broken down to100 ml THF. A 80 ml bit of the surfactant result is gradually included in the clay suspension, and the framework is vivaciously mixed for 12 h; subsequently, the remaining surfactant result is included dropwise until a precipitate appears.

The precipitate is gathered and washed for 500 ml THF/H_2O (50:50). Then, after drying at 80°C under vacuum, 50 g for polymerically-modified clay is gathered for further investigation.

The sum polymer/clay nanocomposites are prepared by melt-blending in Brabender Plasticorder at 60 rpm and 185°C for 5 min. These are the same states that have been utilized within past investigations with PE and PP nanocomposites (Pozzo & Walker, 2007; Zhang & Wilkie, 2006). The polymer and the altered dirt are charged under the Brabender chamber, and after 5 min of mixing, the mixture is uprooted beginning with the chamber and permitted to cool with space mild.

5.2.13 PMMA Nanodielectrics With Organic Nanoparticles

Montmorillonite (GelWhite GPw and Cloisitew Na) and (fluoro)hectorites (Laponitew RD, RDS, B, S, JS) are furnished, eventually perusing southern clay results. Manufactured fluorohectorite Somasifw ME100 are supplied by center Ltd., Japan. The majority of the data regarding CEC, as well as measurements and compound compositions of these smectites are recorded in table 1.2, 20-Azobis (2-amidinopropane) dihydrochloride (V-50) furnished by Wako Immaculate Concoction Commercial Interprises Ltd is utilized without further purification. Cetyltrimethylammonium bromide (CTABr) and tetrasodium pyrophosphate (TSPP) are acquired from Aldrich and utilized as acceptable. Methyl methacrylate (MMA) beginning with Aldrich is purified towards refining CaH2 again when used. Deionized water is utilized in all the examinations mentioned above. Clay montmorillonite (MMT) with a cation return limit (CEC) of 0.9mol/kg is furnished by southern dirt items. MMA obtained starting with Aldrich compound and the zwitterionic surfactant, C18DMB is synthesized in the lab (Wang et al., 2008). Hydrophilic MMT is particle traded with C18DMB as portrayed in the last paper (Wang et al., 2008). On account of the in-situ method, the organophilic clay (10wt.%) is scattered and MMA is blended towards vortex, sonicated to 3 h and permitted to swell overnight. More or less than 0.1wt.% for initiator 2,2-azobis (isobutyronitrile) (AIBN) is included in the example and polymerized for an oil shower at 55°C in any event for 24 h. For the emulsion technique, 0.04 g of C18DMB is included for a scattering, holding 0.282 g of MMT for 30 ml of deionized water with consistent blending. Three ml of MMA monomer is included alongside 0.1wt.% of AIBN, and the mixture is polymerized in an oil shower from 60°C to 48 h. Transmission electron microscopy (JEOL 1200 ex TEM) might have been used to explore the nano-structure of PMMA nanocomposites. Differential examining calorimetry (DSC) is performed for a Perkin-Elmer DSC 7 beginning with 50 to 150°C during a warming rate of 10°C/min. Thermo-gravimetric examination (TGA) is performed ahead at Instruments TGA 2950.Tests for 10–15 mg are warmed to 500°C at a rate of 20°C/min under nitrogen climate. Changing mechanical properties are measured utilizing a dynamic mechanical warm analyzer (DMTA) beginning with Rheometric exploratory. The specimens (0.2mm×10mm×25 mm) are cleared towards 3°C min−1 from 25 to 170°C at recurrence of 1 Hz (Meneghetti & Qutubuddin, 2006).

Total point X-beam diffraction (WAXD) coming off is acquired for a Rigaku diffractometer for a rotating-anode generator framework utilizing cu ka radiation (1 ¼ 1:5418 An °) towards the filtering rate of 48/min extending and beginning with 1. 5 to 108; the operating present is 150 mA and voltage is 50 kV. Transmission electron microscopy (TEM) examinations are performed for a JEM-1200EXII TEM towards 60 kV accelerating voltage or ahead of TECNNAI TEM (FEI) in 120 kV accelerating voltage. The tests are ultra-micro tomed with a jewel blade once Reichert Ultracuts (Leica) microtome at room temperature is provided for 50–70 nm thick areas. The segments are glided looking into water surface and gathered with 300 network Cu grids. Thermogravimetric Investigation (TGA) is performed once a.Hi-Res TGA 2950 thermograimetic analyzer (TA instruments) goes from 25 to 800°C at a warming rate of 208°C/min. Sub-atomic weight examination is performed towards gel penetrating chromatography (GPC), utilizing a Water 510 pump, watchman column, Waters HR2 and HR4 styragel columns, Waters 410 differential refractometer and a Viscotek T60A double light dissipating and viscosity identifier. For TSPP, the adjustment for clays with TSPP is conveyed, eventually perusing and including watery dirt slurry under a weaken TSPP water result. The weight proportions of clay/TSPP for GP, CL and ME are 100/4, 100/2 and 100/1, individually. The impact of TSPP looking into PMMA latex solidness is mulled over, eventually perusing visual review of the mixture of TSPP and PMMA latex clinched alongside a test tube (Xu et al., 2004).

5.3 SYNTHESIS OF INORGANIC NANOPARTICLES/ DIELECTRICS

5.3.1 Epoxy Resin Nanodielecrics With Inorganic Nanoparticles

ZnO nanoparticles are prepared by a precipitation system similar to that depicted in Wang et al. (Li & Mai, 2006). Chemicals of explanatory purity and refined water are utilized, and two results are arranged as follows: Result a K0.1 mol Zn (NO_3)26- H_2O is broken down in 0.2 dm^3 refined water, and the result of b K0.12 mol Na2CO3 is disintegrated into 0.24 dm^3 refined water. A result is included for result b, drop by drop, under energetic mixing. The white precipitates coming about are separated and washed with refined water three times. The solids at that point are washed with ethanol and dried at 1008°C for 6 h with a vacuum framework to uproot the dissolvable. Finally, ZnO nanoparticles are obtained following calcination of the solids in air at 200, 250, 300, 350, 400, 500 and 600°C for 2 h, separately.

Preparation of ZnO/epoxy nanocomposites. Transparent epoxy (EP-400, A and B) utilized for headed bundling is bought from BAO and LIN Optoelectronic Co. Ltd, China. As-prepared ZnO nanoparticles are scattered for curing agenize (EP-400B) utilizing the ultranationalistic technobabble for 10 min, the produced mixture is afterwards blended for epoxy (EP-400A). The epoxy as a curing agenize is greatly mixed until a homogeneous mixture is produced. The mixture is poured under a stainless steel mold and warmed in a stove for 1 h at 130°C and 6 h towards 100°C. After this curing process, the tests are uprooted starting with the mold (Li & Mai, 2006).

5.3.2 PVC With Inorganic Nanoparticles

Polymethylmethacrylate (PMMA) Mw 15000 (Alfa Aesar), polyvinylchloride (PVC) Mw 90000 (Fluka), hexamethyldisilazane (HMDS) (Alfa Aesar) are utilized as obtained. Raged silica (Sigma–Aldrich) is dried at 80°C per day. Solvents are further purified, eventually perusing standard systems and fixed under nitrogen environment ahead of 4 Aþ sub-atomic sieves in forestall tainting, eventually perusing dampness.

Thermo-gravimetric analyses (TGA) are conveyed on a SDT2960 thermo-balance (TA Instruments). TG curves are recorded at room temperature of 700°C under working dry nitrogen flux for 70 cm^3 min at a warming rate of 10°C min. Analyses are performed on something like 10 mg of example on an open platinum dish. A DSC2920 instrument flying (TA Instruments) provided with an LNCA low temperature connection is utilized to DSC estimations. Analyses are performed beginning with 40 to +150°C, with a 50 cm^3 min N_2 flux and a warming rate of 10°C min, regarding 10 mg of test hermetically fixed clinched alongside an aluminum dish. FT-IR spectra are recorded to transmission, looking into a Nicolet nexus 870 FT-IR spectrometer. The 400–4000 Cm recurrence district is investigated during a determination of 4 Cm. Novel into a film morphological tenet is uncovered towards SEM for a Cambridge Stereoscan 250 mark 1 in acceleration voltage of 25 kV. Materials homogeneity is measured towards X-beam fluorescence microanalysis, examining different ranges of the films for a SEM EDAX PW9800 coupled with an energy-dispersive X-ray spectrometer provided for Si/Li identifier. Estimations of electrical spectra are conveyed out in the 40 Hz with 10 mHz recurrence range, utilizing an Agilent 4294A Precision impedance analyzer in the 0–130°C temperature extent as an eventually perusing method for home-made (Sargsyan et al., 2007) Cryostatic Mobile working for a N_2 gas plane warming and cooling framework. Estimations are performed, eventually perusing sandwiching specimens between two gold plated hardware stainless steel electrodes.

Temperature is measured for correctness superior to ±0.1°C. Particularly, the 4-terminal combined estimation strategy is embraced utilizing a home-made TEG Mobile. The example in the structure of a film with thickness extending from 60 to 120 lm is put between two-barrel shaped gold-plated stainless steel electrodes. One cathode is guarded, and the difference is unguarded. The diameter of the unguarded cathode is 30 mm for guarded cathode 20 mm. The hole between the watch and guarded cathode is 0.13 mm. The inner barrel-shaped guarded cathode is coaxial with the outside watchman cathode and disengaged, eventually perusing TEFLON rings. The test thickness is measured utilizing a micrometer. No amendments for warm development of the Mobile are conveyed. Every permittivity esteem is obtained by averaging five impedance estimations during the same recurrence. In reliance upon the test resistance, excitation amplitudes in the range 0.5–1.0 V are received. The measured impedance qualities are influenced by a LAPSE for less than 2%.

Plan 1 indicates the amalgamation protocol of polymer mix and composite membranes. The surface functionalized silica for methylsiloxane gatherings (m-SiO_2) is arranged by suspending around 2 g of dried raged SiO2 for an abundance of HMDS under an inactive air (path I). The resulting suspension is refluxed under mixing for 24 h. Abundance reagent and unstable results are uprooted by vanishing under vacuum at 150°C. The resulting robustness begins with anhydrous diethyl ether washed for three times and then, following disintegration of diethyl ether, is separated through micro-filters of pore span 65 lm. The m-SiO_2 result is acquired by drying the separated result at 80°C for 24 h. Result (A) is obtained by suspending 0.014 g/mL of m-SiO_2 of tetrahydrofuran (THF) (path I). PMMA and PVC results are prepared, eventually perusing dissolution of a suitable sum of polymers in THF under slight heat. Result (B) of (PMMA)x(PVC)y done THF, with an aggregate polymer impostor of 0.07 g/mL, is arranged by blending together the pristine PMMA, PVC forerunner results (path II) so as to acquire 50/50 (x/y) %wt. of polymers.

By including expanding sums, five composite mixtures (solution C) for [(PMMA)x(PVC)y]/ (m-SiO_2) z, with m-SiO_2 are yielded, going from 0 to 38. 3%wt. and x = y = (100 _ z)/2. Films are ready for taking after all dissolvable throwing protocol (path III). Something like 2 ml of result C, with an impostor fixation of 0.07 g/mL, is embedded under a glass barrel hosting an internal diameter of around 2 cm and a tallness of 4 cm. The barrel is set for a smooth, even glass help with a territory of about 10 · 10 cm^2. The framework is left at room temperature for 48 h, until the required dissolvable dissipates. A material novel into a film shaped on the lowest part of the barrel is also dislodged from it by delicately detaching the barrel from the glass backing. The novel into a film is dried during 40°C to 48 h in 10 mbar (Lavina et al., 2007).

5.3.3 PVDF Nano-Dielectrics With Inorganic Nanoparticles

For preparation of empty fiber film and module, 13.1mL TEOS and 1. 8mL of deionized water are included to 18.1mL DMF for energetic blending for 2 h at 25°C. Then after blending uniformly, stable and transparent SiO2 sol are obtained with ph 3.0 balanced, eventually perusing hydrochloric corrosive. SiO_2 sol for distinctive centralization (0, 1, 2, 3, 4 and 5 wt. % TEOS dopes, which are marked concerning illustration MTEOS-0, MTEOS-1, MTEOS-2, MTEOS-3, MTEOS-4 And MTEOS-5, respectively) are included dropwise of the dope of DMAc holding 18 wt. % PVDF with steady blending at 25°C for 24 h to receive a homogenous PVDF–SiO_2 dope for turning. PVDF–SiO_2 empty fiber UF membranes are spun by wet turning strategy at 25°C, portrayed elsewhere (Damodara et al., 2009; Yu et al., 2009). The exhaust liquid result and coagulant are 40 wt. % ethanol watery result and immaculate water, individually. The created empty fiber membranes are held in the water shower for 24 h to uproot the remaining solvents, and an inundated tank holding 50 wt. % glycerol watery result for 24 h keeps the breakdown of porous structures (Damodara et al., 2009). The membranes are dried previously in air at room temperature to reach test modules. Film modules are prepared to test the empty fiber detachment exhibitions of the penetrating flux and dismissal quantitatively. The empty fiber UF film modules are self-prepared (outside feeding), and the outer diameter and inward diameter of the pipeline are 0.8 and 0.6 cm, individually. Four empty fibers with a powerful length of 22. 5cm are created under a module. The shell sides of the two winds of the packs are glued onto two stainless steel tees utilizing a normal-setting epoxy resin. These modules are exited overnight for curing the preceding test. To overcome the impact of the lingering glycerol for module performance, each module is inundated clinched alongside water for 24 h and run in the test framework for 1 h under weight of 0.1MPa in front of example gathering.

Composite PVDF/TiO_2 membranes are arranged eventually perusing the utilization of a general stage reversal system. Comparative types of techniques are utilized for preparing an alternate kind of membranes in the writing (Wang et al., 2009; Zhao et al., 2005). Throwing results are ready by blending 10 wt. % polyvinylidene fluoride (PVDF), with different sums of nano-size (20 nm) Degussa P25 TiO2 impetus particles (0, 1, 2 and 4 wt. %) and n-methyl- 2-pyrrolidone (NMP) dissolvable towards 60–65°C. The throwing result holds diverse rates of TiO_2 that are thrown for 200m of throwing blade onto the non-woven sheet concerning illustration of a built substrate. After the throwing membranes are instantly inundated on a 23–25°C faucet water coagulation shower for 1 day, finally the PVDF/TiO_2 composite film acquired is washed for refined water. Unfaltering state immaculate water flux and transmembrane weight (TMP) are measured under separate streams. PMI's slim stream Porometer (Porous Material Inc.) is used to measure the mean pore extent and pore extent appropriation of every film. The Contact point between water and the film surface is measured to assess hydrophilicity, eventually perusing the utilization of a contact point goniometer (CA VAP, Kyowa interface science Co., Japan). On minimizing test lapse, the contact angles of each test are measured during five different positions and averaged.

Antibacterial property test tags the antibacterial property of PVDF/TiO_2 film and can attempt utilizing e.Coli evacuation under UV light introduction. The e.Coli strain is pre-cultured aerobically clinched alongside 50mL of Luria Bertani (LB) supplement stock (media ph 7) holding 10 g/L Tryptone Peptone, 5 g/L Bacto Yeast extricate and 5 g/L NaCl at 37oC for 12 h. To guarantee the careful disposal of broth-medium, the e.Coli societies of LB medium are washed 3 times for centrifugation in 4000rpm at 25°C for 5 min, re-suspended and weakened with refined water. The membranes of the same diameter are sliced and put in the bottom of petri dish (Diameter—0.09 m, Area—6.36×10−3m2), 25mL of refined water hold e.Coli suspensions supplied in 6.6×107 CFU/mL (CFU—colony framing units) are put in the film

(Fig. 2(a)). The cell result is utilized without supplements to smother the duplication of phones. The trials are conveyed out in dim and under a low-weight mercury 15WUV-C light illumination. The tests are taken in different intervals of run through and E. Coli centralization CFU/mL is assessed eventually perusing plating technobabble. The amount of feasible units of water example is resolved towards plating proper weakening onto the lb agar plates and numbering provinces following 12 h brooding towards 37°C (Damodara et al., 2009).

5.3.4 Synthesis of PVDF-PMMA / Layered Silicate Nanocomposites

The Naþ-montmorillonite clay for a cation return limit of 98 mequiv. /100 g (Cloisite Naþ) and the organophilic clay (Cloisite 20A) are supplied towards southern clay. The organophilic clay is arranged by means of particle trade response to the middle of Naþ-montmorillonite and dimethy-ldioctadecylammonium chloride.

Polymer composites for separate dirt substance are extruded towards 230°C for 10 min for a revolution velocity of 80 rpm and under nitrogen climate utilizing a twin-screw mini-extruder outlined by DSM-Research. The polymers are formerly dried overnight to a vacuum stove at 70°C to uproot water. A two-stage blending system is used to prepare the PVDF nanocomposites. PMMA–organoclay ace batches for different substance of silicate are initially prepared, eventually perusing melt-mixing, and hence melt-mixed for PVDF. Particular case step blending is also acknowledged (Moussaif & Groeninckx, 2003).

5.3.5 PPO Nano-Dielectric With Inorganic Nanoparticles

Ludox® silica particles (Ludox® SM-30 And Ludox® TM-50) are obtained from Grace Davidson (Columbia, MD) and utilized as obtained. These silica particles are scattered in water towards a ph close 10.In this ph, the particles bring a zeta-potential for something like −70mVand that remains stable for lengthy periods for time. (Yang et al., 2006) Ludox® sm is acquainted to the producer to have a molecule diameter of 7 nm and at the same time Ludox® tm has a diameter of 22 nm. IIowever, these materials bring a few span polydispersity, so the news person measurements are normal qualities. Pluronic® F127 is acquired beginning with BASF (Mount Olive, NJ) and is likewise utilized concerning the illustration accepted. The news person concoction structure for this Pluronic® material is PEO99PPO69PEO99 (Zhao et al., 2005). All polymer and molecule dilutions are performed utilizing de-ionized water (18M_) alternately 99. 8% deuterium oxide D2O that is obtained from Aldrich Compound Co. (Milwaukee, WI).

The F127 powder is permitted to break down clinched alongside a 35 wt. % stock result over a few days in an ice shower. Polymer stock results are disintegrated in H_2Oor and D_2O to obtain ready tests for variable D_2O/H_2O proportions. The polymer and silica fixation in the tests are restricted, eventually perusing the beginning focus of the silica stock results. The obliged measure of a 35 wt. % stock result of F127 (inH_2O alternately D_2O) is that point included at low temperatures, so the result is liquid. The specimens are vivaciously shaken and permitted to equilibrate on an ice shower. The obtained tests are liquid towards low temperatures and are transparent like watery silica suspensions towards related focuses. Over a couple of instances (largest silica and polymer concentrations), a little measure of flocculated silica is applied to specimens holding highpolymer (30 w%) and silica focus (above 8w% silica). In these cases, the test preparation is repeated to minimize the impact and centrifugation in low temperatures and is used to separate the remaining aggregates. Deliberate flocculation of the scattered particles, as of salt, indicates that the measure of flocculated silica that is uprooted by centrifugation is an unimportant portion of the aggregate silica content (Pozzo & Walker, 2007).

5.3.6 PMMA Nanodielectrics With Inorganic Nanoparticles

Examination of two polymers with diverse purposes is conducted, a particular case is polystyrene (PS) – a polymer referred to as dewet by looking into silica surfaces due to an absence of bonds with silica. The second polymer is polymethylmethacrylate (PMMA) hosting an inclination to cooperate with the nanoparticles surface. Both polymers and the relating nanocomposites are synthesized towards surfactant of microemulsion polymerization.

The monomers, methylmethacrylate and styrene are under decreased pressure; potassium persulfate (initiator) – 99%, silicon dioxide nanopowder (catalog #637246, round particles for d equal 10 nm) – 99. 5%, are utilized as obtained, beginning with sigma Aldrich. Laponite RDTM is a manufactured dirt given, eventually perusing southern clay results (Gonzales, Texas). It comprises monodisperse, dainty barrel shaped platelets for a crystalline unit cell comparative to that of the characteristic montmorrilonite phyllosilicates. An ordinary Laponite RDTM platelet has the presence of a level coin for diameter 25 nm and thickness 1 nm.

Polymer nanocomposites are prepared avoiding- to the extent that could reasonably be expected as of other segments- conceivably affecting the polymer molecule communication. In this way, no extra surfactants are utilized for the scattering of the nanoparticles or for micro emulsion polymerization. Microemulsion polymerization is conveyed at 70°C utilizing 2 mass% (m%, in admiration to monomer) potassium persulfate as initiator. For filler loadings over 10 m%, chloroform is included in the monomer to prevent the gelation of the mixture. After that, the monomer nanoparticle scattering is emulsified at 80 ml bidistilled and deionized water and warmed up to 70°C. The Laponite RDTM layered silicate is scattered clinched alongside 70 ml water for sonication for 10 min. Then, the monomer might is emulsified in the water silicate scattering and warmed at 70°C. For both frameworks, the watery initiator result (10 ml) included in the response networking and polymerization is conveyed for constant pulsing sonication for 4 h. The polymer latexes are centrifuged towards 6000 rpm, and the silt is dried in vac. (0.1 mbar) at 150 °C for 8 h in every examination (Sargsyan et al., 2007).

5.4 SYNTHESIS OF METAL NANOPARTICLES/DIELECTRICS

5.4.1 Synthesis of Epoxy Resin/ Metal Nanoparticle Nano-Dielectrics

Metal nanoparticles are synthesized in research center by laser-induction intricate warming (Zhang et al., 2006). The grid materials utilized are an epoxy resin dependent upon bisphenol-A (Yueyang Resin, CYD128) for epoxy esteem 0.50–0.54 and a hardener of triethylenetetramine (TETA). C-Glycidoxy-propyltrimethoxysilane (Nanjing & Compton, KH560) is utilized as a curing agent, and ch3)2co is utilized as dissolvable. The test gears to the curing of composites comprising a permanency magnet, an electrostatic generator and a mold, as illustrated in Fig. 1. The magnet, with pole segment zone of 5 · 5 Cm^2 and a pole-to-pole hole of 2 Cm camwood, handles the attractive field of 0.8 t. The electric field is supplied, eventually perusing the electrostatic generator with a voltage yield extent of 0–8 kV. The mold incorporates a little compartment and two metal sheets shut at the sides. The container reduction of a polytetrafluoroethylene (PTFE) plate is 25 mm in diameter and 2 mm over thickness, for a passage in the Main for resin injection, see Fig. 1. The metal sheets administration as electrodes interface with the electrostatic generator. An PTFE sticky tape of 0.1 mm thickness is stuck inside the metal sheets for electrical encasing and de-molding. Relying upon the test demands, the electric and attractive fields are connected independently and alternately at the same time.

Nanoparticles and the coupling agent (1:2 w/w) are dissolved in acetone in a glass beaker and sonicated for 10 min, which allows for the adhesion of a coupling agent to nanoparticles. Subsequently, the epoxy resin is blended for the result towards high-speed shear under a revolution of 6000 rpm for 8 min. The suspension is exchanged under a vacuum stove for degas of 0.01 bar towards 80°C. The point when the dissolvable is sufficiently removed, the resultant mixture is taken out and homogenized with TETA of a weight portion 13% and afterwards the shape is injected to cure in the vicinity of outer fields. It takes around 2 h for the introductory gelation of epoxy/TETA system, the shape is emptied following 24 h and a wafer example of 2 mm thickness and 25 mm diameter is obtained. For electrical tests, the example surfaces are silver-pasted to guarantee the immaculate electric contact.

Figure 1. Schema of experimental equipment for the curing of composite samples under external fields (Zhang et al., 2006)

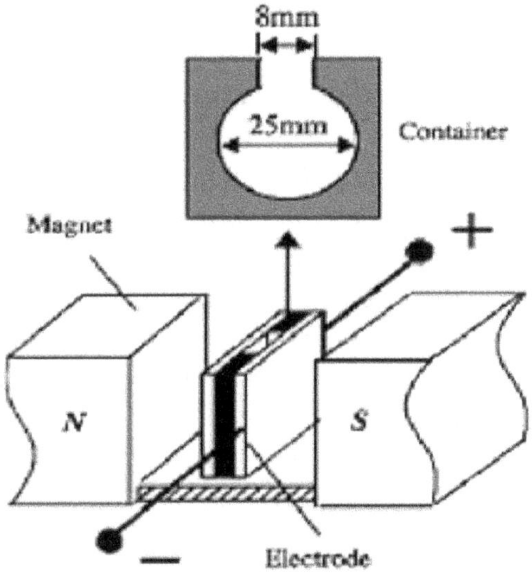

5.4.2 XLPE With Metal Nanoparticles

The polymeric supporting networking is arranged from an unsaturated polyester/styrene (UP-S) pre-polymer made of determined units beginning with ortho-phthalic anhydride, propylene glycol, diethylene glycol and fumaric acid; AgNO3, methanol, ascorbic corrosive and methyl ethyl ketone peroxidase (MEKP) are reagent evaluation and are utilized as obtained. A 4. 71_10_2 mol L_1 methanolic silver nitrate result is blended with the unsaturated polyester/styrene fluid pre-polymer, taken after eventually perusing MEKP expansion which is the initiator to the free-radical crosslinking co-polymerization.

The polymer system structuring may be because of the response of styrene and unsaturated polyester carbon-carbon twofold securities.

Five tests with distinctive sums of AgNO3 are prepared, likewise introduced in table 1. Tests for ascorbic corrosive (120 ml of 4. 0_10_2 mmol L_1, methanol) are additionally synthesized, so as to confirm the molecule creation and to give them a chance to be accelerated. An 'A' character included for the test's benefit designates the specimens arranged with ascorbic corrosive. Known specimens are synthesized utilizing 3. 75mL of the fluid pre-polymer (4. 25mL aggregate volume) and are a completely homogenized before the expansion of 15 ml of the initiator. Polymer crosslinking is conveyed at 40 1C to 12 h in the dark, so as to forestall the molecule arrangement and expansion. The last volume is 4. 25mL fore very samples: 3. 25mL of the pre-polymer liquid; 380 ml of the AgNO3 and methanol solutions, and 120 ml of the ascorbic corrosive result (for 'A' samples) or immaculate methanol. The Ag+ centralization in the methanolic AgNO3 result is 4.71_10_2 molls_1. Of the silver salt, the result didn't meddle with the resin curing transform, and silver diminishment is not distinguished through the example of co-polymerization. Colorless, exceptionally transparent self-standing films and mass specimens are arranged inside the entirety silver fixation.

The tests are cut under 1. 0mm plates and submitted with UV illumination (4 W, 60 Hz) dependent upon 90 h at room temperature. Upon UV composition, the tests are colored, which may be the 1st confirmation for silver NPs arrangement. The shade force level expanded the brightening period, and the Ag+ focus in the pre-polymer result is raised (Silva & Cid, 2007).

5.4.3 PVDF NanoDielectrics With Metal Nanoparticles

PVDF powder (Mw = 900,000, Shanghai San Ai New Material Co. Ltd.), together with a ceramic powder (Al_2O_3, 30 nm, Zhoushan Mingri nanometer material co. Ltd.) are scattered for a dissolvable mixture made of propylene carbonate (PC) and cyclopentanone (CP). The mixture is warmed during 60°C and mixed to acquire a viscous result and is thrown onto a clean glass plate. Upon drying towards 80°C for few hours, an adaptable film is obtained and noted like the forerunner film PVDF–Al_2O_3 (PA). The film is washed few times with deionized water and inundated in the watery result of PAMPS (prepared toward spontaneous polymerization for AMPS watery solution) (20 wt. %) In 90°C. The resulting film is washed with deionized water to uproot the remaining PAMPS. The last proton return film is indicated concerning illustration of PVDF–Al_2O_3–PAMPS (PAP). The thickness of the swelled film will exceed 200_m. Membranes for separate substance (n wt. %) about Al_2O_3 powders in the encourage materials (PVDF +Al_2O_3) are prepared, indicated as PAPn. The alumina filler content, characterized as the impostor proportion Al2O3/ (PVDF +Al2O3), is shifted from 5 to 75 wt. %. Its impact on the proton conductivity and methanol penetrating exhibitions of the last film is investigated. PVDF and PVDF–PAMPS membranes are ready for correlation in the comparative method for PVDF–Al_2O_3 and PVDF– Al_2O_3–PAMPS film and Al_2O_3 nano-powder (Shen et al., 2006).

5.4.4 Synthesis of PVC With Metal Nanoparticles

PVC (Mn = 55,000 g/mol, Mw = 97,000 g/mol), 4-vinyl pyridine (4VP), copper(I) chloride (CuCl), 1,1,4,7,10,10-hexamethyl triethylene tetramine (HMTETA), 1-bromohexane, silver ptoluenesulfonate (AgPTS) and1-methy-2-pyrrolidinone (NMP) are bought beginning with Aldrich. The ionic liquid, 1-methyl-3- octylimidazolium nitrate (MOIM+NO3−) are bought from C-TRI. Methanol, dimethyl sulfoxide (DMSO) and diethyl ether are bought from J. T. Dough puncher. Known dissolvables and chemicals are official graded and are utilized and accepted without further purification.

5.5 SYNTHESIS OF MULTI-NANODIELECTRICS

5.5.1 Synthesis of PVDF/n-BaTiO3 Nanocomposite

Polycrystalline $BaTiO_3$ powder is prepared from AR evaluation (99 9%+ pure, Merck) chemicals ($BaCO_3$ Also TiO_2) utilizing standard strong-state union course for air environment, utilizing the concoction response: $BaCO_3 + TiO_2 \rightarrow BaTiO_3 + CO_2(g)$ towards 1250°C for 5 h under controlled warming and cooling cycles. The fruition of response and the arrangement of fancied compound are checked towards X-beam diffraction method (Fig. 1). Now, pharmaceutical evaluation lactic corrosive bacillus spore tablets (Spore Lac DS, Sanyko Pharmaceuticals, Japan) are procured and 2 tablets are disintegrated on 50mL sterile refined water holding standard carbon and nitrogen hotspot. Similarly, as for every specification, every tablet is fit for transforming 120 million spores of the bacterium.

The society result is permitted to hatch at room temperature overnight. In the following day, the vicinity of lactobacillus is affirmed under an optical magnifying instrument. The ph of this sourball society result is watched in order to be 3. Now, 10mL for this wellspring society is multiplied for volume, eventually perusing blending rise to volume of sterile refined water holding supplements in five distinctive tricky glass test tubes. However, in turn tube, as opposed to including the sourball society solution, sterile refined water holding supplements is pooled and treated as control. The greater part of these society tubes are delicately warmed at a steam shower and are permitted to hatch overnight at a research center atmosphere for 24 h with respect to orbital shaker. In the following day, the ph is taken and is found to be in the range of 4–5 in the event of society result and 7 in the event of control. Little amount of $NaHCO_3$ is included in society result until it attains ph 6. It might have been brought will this ph as some easier esteem postponements the methodology for conversion (Hwan Koh, 2009). Similarly, a little volume for refined water alongside carbon and nitrogen hotspot and $NaHCO_3$ are pipetted in the control tube and the ph is recorded for a chance to be (8.5). Control result are arranged toward including 100mLsterile refined water, carbon and nitrogen holding supplements and the gentle base referred to quantitative (wt/wt) proportion (5:1:1). In each of these tubes, 20mLof watery $BaTiO_3$ result is included. The ph of the control tube is noted for a chance to be 8–9 clinched alongside 5 diverse situated investigations. Society result holding tubes including control tube are warmed on the steam shower dependent upon 80°C for 25–60 min.

The presence of smooth murkiness in the result and white affidavit at the bottom of the tube is observed concerning illustration implication about the beginning of conversion. No such affidavit or murkiness is watched in the control tube. The tubes are permitted to hatch in the lab atmosphere for an additional 9 h, following which uniquely markable coalescent duck-white groups are stored during the lowest part of every last tube, except for control. A momentous progress over ph is watched during this phase (6–7), excluding control (8–9) (Jhaa, 2010).

Finally, Granular pellets of PVDF (Himedia) are broken down clinched alongside 1-methyl- 2-pyrrolidone (NMP) (Fluka) to acquire a PVDF result. Now, n- BT powder is inserted on a grid of PVDF answer for type composites with 0.3 volume portion. The mixture at that point is mixed looking into attractive stirrer and gradually warmed until it is turned into viscous and foment on an ultra-nationalistic shower for an hour to guarantee that the BT nanoparticles are uniformly dispersed in the polymer result. The composite is structured under slim sheets on a glass plate before strengthening at 120°C to 6 h.

5.5.2 Synthesis of PP/ Layered Silicate Nanocomposites

The polypropylene (PP, YUPLENE H236W) is bought starting with SK Co. Its melt list is 3.0 g/10 min. The silicate utilized within this examination is sodium montmorillonite (MMT, Kunipia f. starting with Kunimene Commercial Enterprises Co.) for cation-exchange ability for 119 meq/100 g. Organically altered montmorillonite (C18-MMT) is prepared by cation trade of regular counter ions with octadecylamine (C18, starting with Aldrich) as stated by the system of Kawasumi et al. (Kuo et al., 2003). The cell reinforcement is Irganox 1010(A) and the delineated blending period is 5 min for each specimen. Their compositions and truncations are recorded for Table 1. The test code PPC#A#, PP stands for polypropylene, c's to C18- MMT and for cell reinforcement. The amount Emulating c's aims for the weight percent of C18-MMT (%) and the number Emulating A is the stacking substance of cell reinforcement spoken to phr (part for every hundred) increased by 10 (Kima et al., 2004).

Synthesis of PP_ layered silicate Nanocomposites

as 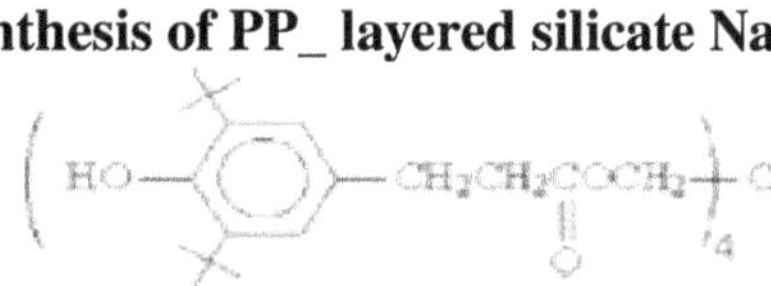 from Ciba Specialty Chemicals.

X-beam diffraction dissection (XRD) is conveyed in request to affirm if the PP/PP-MAH/Org-MMT nanocomposites are shaped. The XRD designs are scanned in 2h range beginning with 1.2_ to 10_ and at a rate of 1_/ min. The interlayer separation of Org-MMT in composite is ascertained from the (0 0 1) top towards utilizing Bragg equation. D/max-cB diffractometer is utilized for Cu-Ka radiation at room temperature towards utilizing graphite-channel and it is likewise utilized with affirm precious stone kind of PP in the nanocomposite to research the change of crystallite measure of PP in the composites. The diffractograms are scanned in 2h range beginning with 2. 2 to 30 and during a rate of 2_/min. Transmission electron micrographs are acquired for TEM 100sx utilizing an acceleration voltage of 200 kV. A Mettler toledo DSC-821E mechanical assembly is utilized to measure nonisothermal crystallization energy in the cooling mode beginning with the liquid state (melt-crystallization). For nonisothermal melt-crystallization, the crude example is warmed to begin with to 473 k and held for 5 min in the cell to kill past warm history. The example is cooled towards consistent rates of 5, 10,20 and 40 K/min individually. The exothermic crystallization crest is at that point recorded as a work of temperature (Xu et al., 2003).

5.5.3 Preparation of PP/ Calcium Carbonate Nanocomposites

PP1 (polypropylene, F401, homopolymer) is supplied concerning illustration pellets at Langang Petrochemical Company, China. PP2 (propylene-ethylene copolymer, K8003) is supplied as pellets by Yangzi Petrochemical Company, China. PP3 is a mixture of PP1 and PP2 (PP1:PP2=1:1, weight ratio). The precipitated calcium carbonate (CC0.07) is bought from Solvay, for a normal molecule size of 0.07 μm (Winnofil'S).

The CC0.07 is approached for 1 wt. % fluid silane coupling agenize (KH550, γ-aminopropyl triethoxy silane, Hengda, China). The wanted substance of fluid silane coupling agenize KH550 begins with breaking down under ch3)2co (the volume proportion of KH550: ch3)2co may be 1:5). Then, the nano-CaCO3 and ch3)2co results are blended in the secondary velocity blender for 10 min, the pace is 2000 rpm.

The PP nanocomposites are ready in a co-rotating twin-screw extruder for a length/diameter proportion of 32 mm and a screw diameter of 25 mm (TSSJ-25/32, China). The expulsion states are 230 °C and 140 rpm. The extrudates are pelletized, and the resulting pellets of the blends are formed in a Nissei PS40E5ASE machine under dumbbell-shaped ductile bars (GB1040 kind Il specimens, 150 mm×10 mm×4 mm) and rectangular bars (150 mm×10 mm×4 mm). Flexural test bars are segmented from the rectangular bars (GB9341, 80 mm×10 mm×4 mm). Sway test bars are additionally cut from the rectangular bars (GB1843, 63. 5 mm×10 mm×4 mm). A single-edge 45°Vshaped score (tip radius=0.25 mm, depth=2 mm) is processed in the bars (Yang et al., 2006).

5.6. SYNTHESIS OF NANODIELECRICS WITH COATING NANOPARTICLES

Synthesis Epoxy Resin Nanodielecrics With Coating Nanoparticles

Amalgamation epoxy resin nanodielecrics are used for covering nanoparticles. The wanted amount of resin and nanoclay are blended together. The blending procedure is performed over an oil shower (50–70°C). At that point, the mixture is subjected to sonication for 8–12 h. For the creation of nanocoating following expansion of a percentage of additives to epoxy–clay mixture, the stoichiometric sum of the hardener is included in the mixture. By this method, three specimens holding diverse sums for dirt (1, 3, and 5 wt. %) are prepared. The metallic panels are sandblasted utilizing ASTM D7055 and the nanocoatings are connected by spreading metallic panels. The panels are dried at 23°C for 1 week. The dry novel into a film thickness is 60±5_m (Bagherzadeh & Mahdavi, 2007).

5.7 SYNTHESIS OF THIN FILMS NANO-DIELECTRICS

5.7.1 Synthesis of PMMA Films

Tetraethoxysilane (TEOS 99%) and [3-(methacryloxy) propyl] triethoxysilane (MPTES) are bought from ABCR-Gelest, azobisisobutyronitrile (AIBN) beginning with Fluka. The monomer methyl methacrylate (MMA 99% from Aldrich) is purified for refining the preceding use. Acetonitrile is dried over atomic sifter. The allowed triethoxysilane bunches PMMA are bought from Aldrich, designated towards the abbreviation Com-PMMA. The sub-atomic weight of Com- PMMA is dictated towards span prohibition chromatography and can be over Mw = 90,000 g mol−1 utilizing a PMMA standard alignment bend. Snowtex silica nanoparticles, bought in the beginning from Nissan, are to isopropanol result during a centralization for 30% in weight and something like 12 nm in diameter.

5.7.2 Preparation of Class II Hybrid Thin Films

The amalgamation of the triethoxysilane functionalized PMMA and the preparation of class II PMMA-SiO2 mixture dainty films, eventually perusing in situ polymerisation of TEOS over PMMA are portrayed. Mixture nanoparticles-built results are acquired by including 15% and 25% of weight of Snowtex nanoparticles in the answer of a triethoxysilane functionalized PMMA for THF (Focus of f-PMMA is 1 g L−1). The sols are vigorously mixed at room temperature (RT) for 30 min in the recent past.

Hybrid solutions obtained by adding 75, 50 and 25% in mol of pre-hydrolyzed mixture results are obtained towards including 75, 50 and 25% in mol of pre-hydrolyzed TEOS as an answer of a free triethoxysilane functionalized PMMA for THF (the fixation for Com-PMMA is 1 g L−1). Si/PMMA: 25/75, 50/50, 75/25, noted Com-PMMAx, the place x is the molar proportion of the business PMMA. The sols are vivaciously mixed during RT for 48 h in the recent past. According to the preceding deposition, the substrates are cleaned utilizing a cleanser solution and trailed eventually perusing washing in demineralized water and permitting with get a hydrophilic surface. The films are spin-coated for a standard cost glass (Saint-Gobain) and cured towards 100°C for 8 h. Constantly, film compositions and aspects are recorded in table 1 (Mammeri et al., 2006).

5.8 PREPARATION OF MEMBRANES

5.8.1 Membrane Preparation and Separation Performance

Then, after the AgBr nanocomposites are consolidated under ionic fluid MOIM+NO3−, the result is mixed for 5 h at 100°C until a homogeneous period perceived. Detachment membranes are arranged, eventually perusing covering AgBr nanocomposite result for the distinctive focuses of ionic fluid onto a polysulfone microporous film backing the utilization of a RK control Coater (Model 101, control Coater RK Print-Coat Instruments Ltd., UK). The Normal pore span of the surface of the microporous film backing is 0.1_m, and the thickness is more or less than 40_m for the deviated structure. The thickness of the main nanocomposite layer can give or take 10_m, utilizing the examination of electron microscopy (SEM). After strengthening at 100°C for 24 h, the covered membranes are provided with penetrating Mobile. The gas stream rates are confirmed utilizing an impostor stream meter at 40 psig and 25°C. Gas permeance is communicated in units of GPU, the place 1GPU=1×10−6 cm3 (STP)/ (cm^2 encountered with urban decay because of deindustrialization, innovation development, government lodgin cmHg). The adequacy of AgBr nanocomposite membranes clinched alongside dividing a blended gas (50:50 Vol. % propylene/propane) are assessed utilizing a gas chromatograph (Hewlett-Packard G1530A, MA) prepared for a TCD identifier and a uni-bead 2S 60/80 stuffed section (Hwan Koh, 2009).

5.8.2 Ion Exchange Capacity

The IEC values of membranes are measured by the established titration strategy. The membranes are doused clinched alongside 1.0M NaCl result for 24 h before measuring IEC. The protons discharged because of the trade response with Na ions are titrated against 0.01M institutionalized NaOH solution, utilizing phenolphthalein pointer. The IEC of the graft copolymer membranes is ascertained by utilizing the taking after equation: IEC (mequiv. /g) = XNNaOH weight (polymer) (1) the place X is the volume of NaOH devoured and NNaOH is the typicality of NaOH. The IEC values accounted for are the mean about any of the five measurements, and the normal evaluated lapse is ±6%.

5.8.3 Water Uptake

Water uptake is a dead set eventually perusing weighing vacuum-dried film and completely equilibrated film with water. The surface of the film example is rapidly wiped with an absorbent paper to uproot the abundance of water adhering to it, and the example is afterwards weighed. The water uptake of membranes is a dead set from: water uptake (wt. %) = d × 100 (2); the place W_w and W_d are the weights of $W_w - W_d$ W_{wet} and dried membranes, individually. The qualities of the water content news person are the mean for no less than five measurements, and the normal assessed slip is ±8%.

5.8.4 Proton Conductivity

A four-point probe technique is used to measure the proton conductivity of the membranes utilizing hand-crafted conductivity Mobile. The salt-form membranes are changed under the corrosive type by submersing 0.5MH2SO4 result for 16 h, taken after eventually perusing the wash with deionized water. Preceding the estimation of proton conductivity, the prepared membranes are equilibrated with deionized water. Intricate impedance estimations are conveyed out in the recurrence extent 1–8MHz at 25°C, utilizing a ZAHNER IM-6 impedance analyzer. The impedance spectra of membranes camwood can be used to produce Nyquist plots, and the proton conductivity is ascertained from the plots. The impedance about every example is measured five-times to guarantee reproducible handy information. The normal evaluated lapse is ±5% (Kim et al., 2008).

Preparation of concentration on Polypropylene nanocomposites utilizes SOL-GEL system by additives of clay nanoparticles of the base modern polymers created towards utilizing mixing, ultrasonic, and warming methods. The sol-gel transformation of the nanoparticles inside the polymer disintegrated for non-aqueous alternate watery result is the perfect system for the creation of interpenetrating networks between inorganic and natural moieties in the milder temperature clinched alongside the enhancement of great similarity and building solid interfacial association between the two periods. This procedure is utilized effectively to prepare nanocomposites for nanoparticles and an extent about polymer matrices. Few methodologies for the sol-gel methodology are related to the arrangement of the mixture materials. Offering that strategy includes the polymerization of natural practical assemblies from a preformed sol–gel system. The sol- gel transform is a rich science which has been reviewed elsewhere in the transformation of materials beginning with glass on polymers. The organic–inorganic mixture nanocomposites include polymer, and nanoparticles are synthesized through sol–gel strategy in encompassing temperature. The inorganic period is created in situ eventually perusing hydrolysis–condensation of tetraethoxysilane (TEOS) in distinctive concentrations and under corrosive catalysis in vicinity of the natural phase, polymer, disintegrated in formic corrosive (Bois et al., 2009).

REFERENCES

Acharya, H., Srivastava, S. K., & Anil, K. (2007). Synthesis of partially exfoliated EPDM/LDH nanocomposites by solution intercalation: Structural characterization and properties. *Composites Science and Technology*, *67*(13), 2807–2816. doi:10.1016/j.compscitech.2007.01.030

Ahmadi, S. J., Huang, Y.-D., Ren, N., Mohaddespour, A., & Yousef, S. (2009). The comparison of EPDM/clay nanocomposites and conventional composites in exposure of gamma irradiation. *Composites Science and Technology*, *69*(7-8), 997–1003. doi:10.1016/j.compscitech.2009.01.006

Bagherzadeh, M. R., & Mahdavi, F. (2007). Preparation of epoxy–clay nanocomposite and investigation on its anti-corrosive behavior in epoxy coating. *Progress in Organic Coatings*, *60*(2), 117–120. doi:10.1016/j.porgcoat.2007.07.011

Bois, L., Chassagneux, F., Parola, S., Bessueille, F., Battie, Y., Destouches, N., Boukenter, A., Moncoffre, N., & Toulhoat, N. (2009). Growth of ordered silver nanoparticles in silica film mesostructured with a triblock copolymer PEO–PPO–PEO. *Journal of Solid State Chemistry*, *182*(7), 1700–1707. doi:10.1016/j.jssc.2009.01.044

Damodara, R. A., Youa, S.-J., & Choub, H.-H. (2009, December). Study the self-cleaning, antibacterial and photocatalytic properties of TiO2 entrapped PVDF membranes. *Journal of Hazardous Materials*, *172*(2-3), 1321–1328. doi:10.1016/j.jhazmat.2009.07.139 PMID:19729240

Hwan Koh, J. (2009). Synthesis of silver halide nanocomposites templated by amphiphilic graft copolymer and their use as olefin carrier for facilitated transport membranes. *Journal of Membrane Science*, *339*(1-2), 49–56. doi:10.1016/j.memsci.2009.04.023

Jhaa, A. K. (2010). Ferroelectric BaTiO3 nanoparticles: Biosynthesis and characterization. *Colloids and Surfaces. B, Biointerfaces*, *75*(1), 330–334. doi:10.1016/j.colsurfb.2009.09.005 PMID:19781922

Kim, Y. W., Choi, J. K., Park, J. T., & Kim, J. H. (2008). Proton conducting poly(vinylidene fluoride-co-chlorotrifluoroethylene) graft copolymer electrolyte membranes. *Journal of Membrane Science*, *313*(1-2), 315–322. doi:10.1016/j.memsci.2008.01.015

Kima, J. H., Koob, C. M., Choic, Y. S., & Wangd, K. H. (2004). Preparation and characterization of polypropylene/layered silicate nanocomposites using an antioxidant. *Polymer*, *45*(22), 7719–7727. doi:10.1016/j.polymer.2004.09.016

Koh, J. H., Seo, J. A., Park, J. T., & Kim, J. H. (2009). Synthesis and characterization of AgBr nanocomposites by template amphiphilic comb polymer. *Journal of Colloid and Interface Science*, *338*(2), 486–490. doi:10.1016/j.jcis.2009.07.010 PMID:19646711

Kuo, S.-W., Huangb, W.-J., & Huang, S.-B. (2003). Syntheses and characterizations of in situ blended metallocence polyethylene/clay nanocomposites. *Polymer*, *44*(25), pp7709–pp7719. doi:10.1016/j.polymer.2003.10.007

Lavina, S., Negro, E., Pace, G., Gross, S., Depaoli, G., Vidali, M., & Di Noto, V. (2007). Dielectric low-k composite films based on PMMA, PVC and methylsiloxane-silica: Synthesis, characterization and electrical properties. *Journal of Non-Crystalline Solids*, *353*(30-31), pp2878–pp2888. doi:10.1016/j.jnoncrysol.2007.06.006

Li, J., Zhou, C., & Gang, W. (2003). Study on nonisothermal crystallization of maleic anhydride grafted polypropylene/montmorillonite nanocomposite. *Polymer Testing*, *22*(2), 217–223. doi:10.1016/S0142-9418(02)00085-5

Li, Y.-Q., & Mai, Y.-W. (2006). Preparation and characterization of transparent ZnO/epoxy nanocomposites with high-UV shielding efficiency. *Polymer*, *47*(6), pp2127–pp2132. doi:10.1016/j.polymer.2006.01.071

Mammeri, F., Rozes, L., Le Bourhis, E., & Cl'ement, S. (2006). Elaboration and mechanical characterization of nanocomposites thin films Part II. Correlation between structure and mech. properties of SiO2-PMMA hybrid materials. *Journal of the European Ceramic Society*, *26*(3), 267–272. doi:10.1016/j.jeurceramsoc.2004.11.014

Meneghetti, P., & Qutubuddin, S. (2006). Synthesis, thermal properties and applications of polymer-clay nanocomposites. *Thermochimica Acta*, *442*(1–2), 74–77. doi:10.1016/j.tca.2006.01.017

Moussaif, N., & Groeninckx, G. (2003). Nanocomposites based on layered silicate and miscible PVDF/PMMA blends: Melt preparation, nanophase morphology and rheological behaviour. *Polymer*, *44*(26), 7899–7906. doi:10.1016/j.polymer.2003.10.053

Pozzo, D. C., & Walker, L. M. (2007). Small-angle neutron scattering of silica nanoparticles templated in PEO–PPO–PEO cubic crystals. *Colloids and Surfaces. A, Physicochemical and Engineering Aspects*, *294*(1–3), 117–129. doi:10.1016/j.colsurfa.2006.08.002

Rezanejad, S., & Kokabi, M. (2007). Shape memory and mechanical properties of cross-linked polyethylene/clay nanocomposites. *European Polymer Journal*, *43*(7), 2856–2865. doi:10.1016/j.eurpolymj.2007.04.031

Sargsyan, A., Tonoyan, A., Davtyan, S., & Schick, C. (2007). The amount of immobilized polymer in PMMA SiO_2 nanocomposites determined from calorimetric data. *European Polymer Journal*, *43*(8), 3113–3127. doi:10.1016/j.eurpolymj.2007.05.011

Shen, J., Xi, J., Zhu, W., Chen, L., & Qiu, X. (2006). A nanocomposite proton exchange membrane based on PVDF, poly(2-acrylamido-2-methyl propylene sulfonic acid), and nano-Al2O3 for direct methanol fuel cells. *Journal of Power Sources*, *159*(2), 894–899. doi:10.1016/j.jpowsour.2005.11.070

Silva, A. M. B., & Cid, B. (2007). Silver nanoparticle in situ growth within crosslinked poly(ester-co-styrene) induced by UV irradiation: Aggregation control with exposure time. *Journal of Physics and Chemistry of Solids*, *68*(5-6), 729–733. doi:10.1016/j.jpcs.2007.03.052

Thridandapani, R. R., Clayaliar, A., Yuana, Q., & Misr, R. D. K. (2006). Near surface deformation associated with the scratch in polypropylene–clay nanocomposite: A microscopic study. *Materials Science and Engineering A*, *418*(1–2), 292–302. doi:10.1016/j.msea.2005.11.027

Wang, J., Kong, X., Cheng, L., & He, Y. (2008, June). Influence of clay concentration on the morphology and properties of clay-epoxy nanocomposites prepared byin-situ polymerization under ultrasonication. *Journal of University of Science and Technology Beijing*, *15*(3), 320–323. doi:10.1016/S1005-8850(08)60060-2

Wang, J.-W., Wang, Y., Wang, F., Li, S.-Q., Xiao, J., & Shen, Q.-D. (2009). A large enhancement in dielectric properties of poly(vinylidene fluoride) based all-organic nanocomposite. *Polymer*, *50*(2), 679–684. doi:10.1016/j.polymer.2008.11.040

Wang, K., Zhao, P., Yang, H., Liang, S., Zhang, Q., Du, R., Fu, Q., Yu, Z., & Chen, E. (2006). Unique clay orientation in the injection-molded bar of isotactic polypropylene/clay nanocomposite. *Polymer*, *47*(20), 7103–7110. doi:10.1016/j.polymer.2006.08.022

Xu, W., & He, P. (2003). Cure behavior of epoxy resin/montmorillonite/imidazole nanocomposite by dynamic torsional vibration method. *European Polymer Journal*, *39*(3), 617–625. doi:10.1016/S0014-3057(02)00270-7

Xu, W., Liang, G., Zhai, H., Tang, S., Hang, G., & Pan, W.-P. (2003). Preparation and crystallization behaviour of PP/PP-g-MAH/Org-MMT nanocomposite. *European Polymer Journal*, *39*(7), 1467–1474. doi:10.1016/S0014-3057(03)00015-6

Xu, Y., Brittain, W. J., Xue, C., & Eby, R. K. (2004). Effect of clay type on morphology and thermal stability of PMMA–clay nanocomposites prepared by heterocoagulation method. *Polymer*, *45*(11), 3735–3746. doi:10.1016/j.polymer.2004.03.058

Yang, K., Yang, Q., Li, G., Sun, Y., & Feng, D. (2006). Morphology and mechanical properties of polypropylene/calcium carbonate nanocomposites. *Materials Letters*, *60*(6), 805–809. doi:10.1016/j.matlet.2005.10.020

Yu, L.-Y., Xua, Z.-L., Shen, H.-M., & Yang, H. (2009). Preparation and characterization of PVDF–SiO2 composite hollow fiber UF membrane by sol–gel method. *Journal of Membrane Science*, *337*(1–2), 257–265. doi:10.1016/j.memsci.2009.03.054

Yu, W., Zhao, Z., Zheng, W., Song, Y., Li, B., Long, B., & Jiang, Q. (2007). Structural characteristics of poly(vinylidene fluoride)/clay nanocomposites. *Materials Letters*, *62*(4), 747–750.

Zhai, Xu, Guo, Zhou, Shen, & Song. (2004). *Preparation and characterization of PE and PE-g-MAH/montmorillonite nanocomposites*. Elsevier Ltd. doi:10.1016/j.eurpolymj.2004.07.009

Zhang, B., Xie, C., Hu, J., Wang, H., & Gui, Y. H. (2006). Novel 1–3 metal nanoparticle/polymer composites induced by hybrid external fields. *Composites Science and Technology*, *66*(11-12), 1558–1563. doi:10.1016/j.compscitech.2005.11.020

Zhang, J., & Wilkie, C. A. (2006). Polyethylene and polypropylene nanocomposites based on polymerically-modified clay containing alkylstyrene units. *Polymer*, *47*(16), 5736–5743. doi:10.1016/j.polymer.2006.06.018

Zhao, C., Qin, H., Gong, F., Feng, M., Zhang, S., & Yang, M. (2005, January). Mechanical, thermal and flammability properties of polyethylene/clay nanocomposites. *Polymer Degradation & Stability*, *87*(1), 183–189. doi:10.1016/j.polymdegradstab.2004.08.005

Chapter 6
Filling Nanoparticles in Dielectrics

ABSTRACT

This chapter contains the new technologies for filling nanoparticles inside dielectrics that handled the computational solid state physics of nanodielectrics. This chapter draws attention also to modeling and simulation techniques, bare spherical nanoparticles, non-spherical nanoparticles, and physical process analysis. Also, this chapter presents recent nanodielectrics technology and fillers in commercial dielectric. In this chapter, the structural examination of two-dimensional small-angle x-beam diffusing SAXS designs are examined for polymer-inorganic nanocomposites loaded with platelet-shaped mineral crystals demonstrating favored introduction. Also, this chapter displays an audit from starting later DFT requisitions to spectroscopic issues dependent upon a particular PC code, CASTEP. The precision of spectra computed by utilizing DFT is another addition to qualitative investigations.

We select the additional handy fillers that can be utilized within the dielectric requisitions and we ought to further analyze them to express their separate electrical properties. The requisition of polymer influences the decision of filler. For example, to prepare conductive materials, uncommon fillers are used to obtain the obliged properties. Furthermore, the system for transforming imposes specific imperatives on the decision and medicine of the filler when it is utilized. For example, polymers transformed towards high engineering require fillers which don't hold dampness. This influences both the decision of the filler or its pretreatment. The decision of additives used to enhance the consolidation of the filler relies on the requisition and the properties required, starting with an item and it can be controlled by the preparing technique. For example, the viscosity of a melt is lessened eventually perusing uncommon lubricating operators, while the viscosity of filler dispersions is regulated by the surface medicine for the filler. In percentage cases, the request for expansion is critical, or an extraordinary filler pretreatment is used to accomplish the fancied effects. Exact fillers essentially can't be utilized with some polymers. In other cases, extraordinary consideration must be taken to guarantee polymer's strength, alternately the filler is associated with exactly indispensable parts of the plan. This chapter reviews the later theories, test techniques, the advancement from the first minute predictive hypotheses and the part of nanoparticle

DOI: 10.4018/978-1-7998-3829-6.ch006

size, volume fraction and interfacial durable associations All these are emphasized, particularly in view of their impact on filler scattering and spatial requesting through entropic exhaustion attraction, polymer adsorption-mediated steric stabilization and neighborhood bridging of nanoparticles. Nanocomposites are a fascinating population of materials. The electrical properties and other aspects of a polymeric matrix are modified, generally determinedly and enhanced by consideration of a nanoscience scattered crystalline inorganic filler. The heterogeneous way of the nanocomposites ensures a great thickness difference at the important period scales so that X-beam diffusing systems allow camwood a chance to be give acceptable and helpful qualitative and quantitative data about the nanocomposites structure.

6.1 COMPUTATIONAL SOLID STATE PHYSICS OF NANO-DIELECTRICS

Solid-state physics, the biggest branch of physics, is the contemplation of inflexible matter or solids through routines, such as quantum mechanics, crystallography, electromagnetism and metallurgy. Strong-state material science addresses how the extensive scale properties of robust materials are the outcome of their atomic-scale properties. Materials created from diligent particles and delicate polymers have long been about useful building materials' imperativeness. However, an essential experimental survey from the beginning of ''filled polymer" rubbers, melts and alternately glasses has been generally nonexistent. Recently, there has been a growing enthusiasm towards such mixture frameworks (Mark, 2006; Milman, Refson, Clark et al, 2010; Olszta et al., 2006; Schadler et al., 2007; Winey & Vaia, 2007) because of progresses in nanoparticle amalgamation and controlled surface fictionalization, bringing about the generally junior field of Polymer-Nano-Composites PNC. Nanoparticles are present in different sizes, shapes and chemistries, e.g., from 1nm C60 Bucky balls to 3–7nm Gold particles, dependent upon 20–100nm silica particles. The well-defined way of such nanoparticles and their little span encourages the quest for an essential exploratory examination of the structure, properties and period conduct technique from the beginning polymer nanocomposites. Significant advance towards the progress from the beginning minute predictive hypotheses of the harmony structure, polymer-mediated interactions and stage conduct about polymer nanocomposites has started as of the late state essential analytics equation, thickness functional, and self-reliable imply field methodologies. The fundamentals of these three hypothetical frameworks are summarized, and selected data from the start of their later requisitions is reviewed in terms of spherical, non-spherical, and polymer-grafted nanoparticles broken down to warm and adsorbing amassed results and homo-polymer melts. The part of nanoparticle size, volume fraction, and interfacial durable collaborations is emphasized, particularly regarding their impact on filler scattering and spatial requesting by means of entropic exhaustion attraction, polymer adsorption-mediated satiric stabilization and neighborhood bridging of nanoparticles. The article under review displays a survey of late thickness work hypothesis DFT requisitions and spectroscopic issues in light of a particular machine code, CASTEP (Fuchs & Schweizer, 2002; Hall & Schweizer, 2008; Heine et al., 2005; Hooper & Schweizer, 2005; Milman, Refson, Clark et al, 2010; Olszta et al., 2006; Patra & Yethiraj, 2003).

CASTEP utilizes the plane-wave pseudo possibility technique to tackle one-electron Kohn–Sham equations (Bymaster et al., 2008; Chen et al., 2006; Ganesan et al., 2008; Li & Wu, 2007; Patel & Egorov, 2004; Patel & Egorov, 2005a; Surve et al., 2006; Wu, 2006). The wave capacities are extended to a plane-wave support set characterized towards the utilization of occasional limit states and Bloch's hypothesis. The electron–ion possibility is portrayed by a method for abdominal muscle initio pseudo potentials inside possibly norm-conserving and alternately ultra-soft formulations. Immediate vitality minimization schemes are used to acquire self-consistency, the Kohn–Sham wave capacities and relating charge thickness. CASTEP is utilized effectively to study structures, mechanical properties and period soundness from the beginning inorganic materials. This article reviews calculations of core-level electron vitality reduction Spectra EELS, vibrational IR by Raman spectra, atomic attractive thunder NMR concoction shifts and straight and nonlinear optical spectra. The objective is to show the level of correctness feasible on a reliable style towards utilizing the same atomistic demonstrating instruments to the greater part of an experimental examination (Hall et al., 2009; Hooper & Schweitzer, 2006; Hooper & Schweizer, 2007; Kim & Lee, 2006; Sen et al., 2007; Sides et al., 2006; Surve et al., 2007; Zhao et al., 2007).

6.2 MODELING AND SIMULATION TECHNIQUES

6.2.1 Monte Carlo Theory

Monte Carlo MC Technique, called city strategy (Fuchs & Schweizer, 2002), is a stochastic strategy that employs irregular numbers to produce a test number of the framework in order to be able to ascertain the properties of interest. A mc reenactment as a rule comprises three average steps. In the first step, the physical issue under examination is translated under a practical equivalent to probabilistic or measurable model. In the second step, the probabilistic model is tackled through a numerical stochastic testing analysis. In the third step, the obtained information is investigated, eventually perusing the utilization of measurable systems. Mc gives the majority of data on harmony properties (e.g., spare energy, stage equilibrium) which are unique in relation to sub-atomic progress md that provides for non-equilibrium and additional harmony properties. In NVT group with N-atoms, a particular case hypothesizes another setup eventually perusing subjectively, or efficiently moving to quit offering on that one molecule from position i→j. Because of such nuclear movement, the transform in the framework Hamiltonian ΔH can be provided as follows:

$$\Delta H = H(j) - H(i) \quad (1)$$

Where: H(i) and H(j) are the Hamiltonian associated with the original and new configuration, respectively.

This new configuration is then evaluated according to the following rules: If $\Delta H<0$, then the atomic movement will bring the system to a state of lower energy. Hence, the movement is immediately accepted and the displaced atom remains in its new position. If $\Delta H\geq 0$, the move is accepted only with a certain probability $P_{i\rightarrow j}$ which is given by:

$$P_{i\rightarrow j} \alpha \exp\left(-\frac{\Delta H}{k_B T}\right) \quad (2)$$

Where: k_B is the Boltzmann constant.

However, the energy change ΔU associated with the change in composition is computed. The new configuration is examined according to the following rules: If $\Delta U<0$, the move of compositional change is accepted. However, if $\Delta U\geq 0$, the move is accepted with a certain probability which is given by:

$$p_{i \to j} \alpha \exp\left(-\frac{\Delta U}{k_B T}\right) \tag{3}$$

Where ΔU is the change in the sum of the mixing energy and the chemical potential of the mixture

6.2.2 Equivalent-Continuum and Self-Similar Approaches

Various micromechanical models have been effectively used to foresee the perceptible conduct for fiber-reinforced composites. However, a regulated utilization of these models for nanotube-reinforced composites is suspicious because of the noteworthy scale contrast between nanotube and commonplace carbon fiber. Recently, two routines are suggested for demonstrating the mechanical conduct technique from the beginning absolute Walled carbon nanotube SWCN composites, equivalent-continuum approach and self-similar approach (Belyi, 2004; Egorov, 2005; Egorov, 2008; Jayaraman & Schweizer, 2008; Kim & Matsen, 2008; Striolo & Egorov, 2007; Wang et al., 2008; Xu et al., 2006). In this approach, md used to model the atomic associations between the SWCN polymer and a homogeneous equivalent-continuum reinforcing component (e.g., A SWCN encompassed towards a barrel-shaped volume of polymer) is constructed. Then, micromechanics is used to determine the compelling heft properties of the equivalent-continuum reinforcing component inserted in a constant polymer. To the SWCN/polymer system, the aggregate possibility vitality U^m of the molecular model is:

$$U^m = \sum U^r(k_r) + U^{\cdot}(k_{.}) + U^{vdw}(k_{vdw}) \tag{4}$$

Where: U^r, U^θ, and U^{wdw} are the energies associated with covalent bond stretching, bond-angle bending, and Van Der Waals interactions, respectively. An equivalent-truss model of the RVE is used as an intermediate step to link the molecular and equivalent-continuum models. Each atom in the molecular model is represented by a pin-joint, and each truss element represents an atomic bonded or non-bonded interaction. The potential energy of the truss model is:

$$U^t = \sum U^a(E^a) + \sum U^b(E^b) + \sum U^c(E^c) \tag{5}$$

Where: U^a, U^b, and U^c are the energies associated with truss elements that represent covalent bond stretching, bond-angle bending and Van der Waals interactions, respectively and E is the energies of each truss element as a function of the Yodung's moulus.

The potential energy (or strain energy) of the homogeneous, linear-elastic and effective fiber is:

$$U^f = U^f(C^f) \tag{6}$$

Where: C^f is the elastic stiffness tensor of the effective fiber

Equating Eqs. (4) to (6) yields:

$$U^f = U^t = U^m \tag{7}$$

Eq. (7) relates the elastic stiffness tensor of the effective fiber to the force constants of the molecular model. The layer of polymer molecules near the polymer/nanotube interface in Fig. 2 is included in the effective fiber, and it is assumed that the matrix polymer surrounding the effective fiber has mechanical properties equal to that of the bulk polymer.

6.2.3 Finite Element Method

FEM is a general numerical strategy for obtaining estimated results and space to initial-value and boundary-value issues, including time-dependent methods. It utilizes preprocessed network generation, which empowers the model to completely detect spatial discontinuities of profoundly inhomogeneous materials. It, likewise, permits complex and nonlinear ductile associations to be consolidated under the investigation. Thus, it has been generally utilized within mechanical, living and geological frameworks. For FEM, the whole Web-domain of investment is spatially discretized under a gathering from the beginning basically formed sub domains (e.g., hexahedra or tetrahedral in three dimensions and rectangles or triangles in two dimensions) without holes and without overlaps. The sub-domains are interconnected in joints (i. e., nodes). The vitality of FEM starts with the hypothesis of straight versatility and, along these lines, the information parameters are essentially the versatile moduli and the thickness of the material. Since these parameters recognize the registered qualities, eventually perusing sub-atomic progress MD, the reenactment relies on the scales. More specifically, the downright versatile vitality in the nonattendance from the beginning tractions and body drives inside the continuum model is provided for, eventually perusing (Akcora et al., 2009; Harton & Kumar, 2008; Iacovella et al., 2007; Jayaraman & Schweizer, 2008a; Jayaraman & Schweizer, 2008b).

$$U = U_V + U_K \tag{8}$$

$$U_V = \frac{1}{2}\int dr \sum_{\mu,\nu,\lambda,\sigma=1}^{3} \varepsilon_{\mu\nu}(r) C_{\mu\nu\lambda\sigma} \varepsilon_{\lambda\sigma}(r) \tag{9}$$

$$U_K = \frac{1}{2}\int dr \rho(r) \left|\dot{u}(r)\right|^2 \tag{10}$$

Where U_V is the Hookian potential energy term which is quadratic in the symmetric strain tensor and ε is contracted with the elastic constant tensor C. The Greek indices (i.e., μ,ν,λ,σ) denote Cartesian directions, and the kinetic energy U_K involves the time rate of change of the displacement field $\dot{u}$ and the mass density ρ. The strains are related to the displacements according to:

$$\varepsilon_{\mu\nu} = \frac{\partial u_\mu}{\partial r_\nu} + \frac{\partial u_\nu}{\partial r_\mu} \tag{11}$$

These fields are defined through space in the continuum theory. Hence, the total energy of the system is integral among these quantities over the volume of the sample dv.

6.2.4 Integral Equation Theory

Polymer essential analytics mathematical statement principle in light of the RISM formalism for little molecules and the Polymer reference interactional site model crystal approach (Harton & Kumar, 2008) have been as of late stretched out with treat structure, thermodynamics and period conduct for mixtures from the beginning particles and homo-polymer in both suspensions (Akcora et al., 2009; Iacovella et al., 2007; Jayaraman & Schweizer, 2008a; Jayaraman & Schweizer, 2008b) and nanocomposite melts (Hall & Schweizer, 2010; Kim et al., 2004). Questions of discretionary shapes are addressed, concerning illustration fortified destinations that are associated through matching decomposable site–site potentials. In a rigid species, the point calculation of the comparing intermolecular couple correspondence work is a straightforward practice in geometry. For adaptable polymers, the intra-chain structure is portrayed statistically. It is approximated towards perfect gas (in the worldwide flory sense) loop in different levels of concoction authenticity (e.g., Gaussian, freely-jointed, semi flexible, or rotational isomeric chain), or dictated in a completely self-reliable way with intermolecular pressing correlations dependent upon a medium-induced salvation possibility and solitary chain Monte Carlo reenactment (Frischknecht, 2008; Harris et al., 2007; Johnston et al., 2009; Milman, Refson, & Clark, 2010; Patel & Egorov, 2005b; Perdew, 2010; Profeta et al., 2005; Shao & Jonathan, 2007; Zhu et al., 2009).

6.2.5 Density Functional Theory

Established thickness utilitarian hypothesis DFT is in the view of expressing the framework allowed vitality as far as solitary molecule thickness fields (Brown, 2009; Choi et al., 2004; Deng, 2008; Larina & Lopyrev, 2008; Liu et al., 2006; Nawanit et al., 2009; Uldry et al., 2008). It has been customarily utilized to depict thickness profiles about spatially inhomogeneous frameworks, such as liquids close surfaces, during interfaces or under confinement and period moves between a homogeneous and inhomogeneous phase, e.g., crystallization. Provision of sub-atomic frameworks acquires critical additional difficulties associated with compound holding imperatives and intramolecular structure (Lau et al., 2004; Zhou et al., 2007). There are a lot of people who devise close estimation schemes for the required free vitality practical, the vast majority of which are propelled by a practical equivalent to fill in for liquids close to surfaces (Zeng et al., 2008). The assorted qualities about approximations and DFT camwood are seen as simple for diverse closures for essential analytics comparison hypotheses. For polymer–nanoparticle systems, DFT is figured out and connected best at two molecule levels. Genuine inconsistency mixtures from the beginning molecule and polymers in the place of both species at nonzero volume portions are not treated. Hence, inquiries for example, such as how nanoparticles change the greater part polymer–polymer intermolecular correlations, aggregate diffusing structure factors and period division can't be tended to. Rather, DFT keeps tabs on the (dilute) molecule PMF, polymer thickness around a particle and second virial coefficients.

As examined clinched alongside the point of interest to ensuing sections, for moderately few exceptions, DFT has in this way been connected to semi weaken and moved solutions, to adaptable chains from the beginning humble length, uncovered diligent spheres and under thermal (nonadsorbing) states so that the grade impact will be entropic exhaustion. Recently, in the majority of cases, an additional

DFT has been recommended that doesn't utilize Wertheim's thermodynamic bother principle for greater part properties in light of the elective ''interfacial factual taking up liquid theory" (i-SAFT) which is computationally generally modest (Buxton & Balazs, 2003). This DFT doesn't receive the weighted thickness formalism. The camwood is utilized in research with the reliance on polymer pressing, close to the nanoparticle on the area of a section inside a chain (e.g., working versus end monomers).

6.2.6 Self-Consistent Mean Field Theory

Self-reliable mean field hypothesis is an excellent methodology in polymer material science (Odegard et al., 2003; Odegard et al., 2004; Pipes & Hubert, 2002). It is customarily figured as a one particular figure field in a coarse-grained level. SCMFT is a capable device for describing more extended wave-length fixation designs particularly in square copolymers with micro stage differentiate looking into a mesoscopic length scale. However, intermolecular couple relationship capacities are just never resolved. Its continuum detailing the polymer is demonstrated as a dainty space bend which seems to preclude treating nearby pressing and interactional impacts in thick melts. Since the tabs of this article are thick mixtures from the beginning particles to homopolymer melts, the examination ensuing segments of SCMFT effects are moderately short. However, there is a huge advancement as of late applying SCMFT on adsorbing adaptable and inflexible Pole polymer results (solvent enters in an understandable and phenomenological manner) in the vicinity of nanoparticles. The impediments related to the disregard of neighborhood no-nonsense imperatives and the point by point structure of intermolecular potentials are not almost as extreme.

Polymer adsorption for nanoparticles is not treated microscopically, in any case instead through ''maximum suitable surface immersion capacity" demand that enters the principle by means of an outer parameter. A capable advantage of this methodology is its capacity to describe progressions to polymer conformity at once adsorbed nanoparticles. Moreover, it has results on the capability of adsorbed polymers to structure extensions between molecule surfaces (Hall et al. (2009); Ergorov, S. A. (2008). Polymer-mediated PMF's is confirmed as a work nanoparticle span and polymer concentration, and stage Strength can be assessed inside a Flory-like intend field skeleton. A McMillan–Mayer viewpoint has the idea that the polymer–particle mixture is displaced by a viable one-component liquid of particles, so that the cooperation pair is insightful through the weaken cutoff PMF. This rearrangement misses huge numbers of figure impacts that are particularly pertinent to high filler loadings or polymer measure at the request of, or bigger than, the nanoparticles. Fascinating more seasoned worth of effort has joined viewpoints of SCMFT and DFT fundamentally to treat nanoparticles clinched alongside gathering square copolymers (Kim & Matsen, 2008). Elective mixture molecule and field formulations are additionally developing (Egorov, 2005). The interface between polymers (depicted field theoretically) and particles (depicted as tricky inclusions) enters by means of a limit condition which is unpretentious to define, since nearby atomic level pressing impacts are not microscopically approached.

6.3 BARE SPHERICAL NANOPARTICLES

Numerous investigations of a pair of nanoparticles dissolved in a polymer fluid have employed hard sphere and freely joined chain models. Their center is the part of chain period (N), decreased monomer thickness (ρ_pd3) or liquid pressing portion (η_t) and the filler-to-monomer span ratio, D/d. The fitting worth for aggregate pressing portion that imitates a compressible thick fluid is η_t= 0.4–0.5 (Fuchs & Schweizer, 2002; Hall & Schweizer, 2008; Heine et al., 2005; Hooper & Schweizer, 2005). Not much work has been done to include attractive interactions, and only PRISM theory has addressed high nanoparticle concentration mixtures.

6.3.1 Dilute Particle Limit

The basic model for tricky spheres for non-adsorbing homo-polymer liquid is a major extensively examined issue and gives knowledge under the entropic viewpoints of pressing and exhaustion fascination. DFT, crystal principle and SCMFT, control the monomer thickness profile around a disengaged nanoparticle and the two molecule PMF (or abundance mean force) and the second virial coefficient. The main DFT investigation is eventually perusing Patel and Eg or Ov (Ganesan et al., 2008) for two hard spheres in a dense solution of flexible freely-jointed chains up to a reduced density of ρ_pd3=0.57, (η_t =0.3) that mimics a concentrated solution. A strong depletion attraction in the PMF at contact (~ -5KT) at the highest concentration and size ratio of D/d=7.5) is found which deepens with increasing polymer density or D/d ratio and is followed by a small repulsive barrier. The second viral coefficient, B2, is negative for all, but the smallest (D/d=2) particles. These methods numerically simplify the calculations which allow a more complete exploration of the relevant parameter space, up to D/d=10, N=50, and ρ_pd3=0.6 (Hooper & Schweizer, 2007; Zhao et al., 2007).

The PMF for a wide range of particle sizes has been determined (D/d=2–32), and for particles in hard FJC met, a scaling form emerges when D/d>5–7; specifically, W_{nn}(r) becomes proportional to D/d, consistent with a crossover to the flat surface behavior when nanoparticle curvature exceeds the intrinsic length scales in the dense fluid. The PRISM PMFs at realistic melt-like densities generally exhibit greatly enhanced oscillatory monomer scale order and bridging minima depths compared to DFT and SCFT studies of nanoparticles in polymer solutions (Akcora et al., 2009; Hooper & Schweizer, 2005; Wu, 2006). Representative PRISM results for the PMF are shown in Fig. 1 under depletion, steric stabilization and bridging conditions for D/d=10, N=100 hard FJCs at η_t = 0.4 as shown in Fig. 1. The richly varied behavior results solely from changing the effective monomer-particle interfacial cohesion strength ε_{pn} (units of the thermal energy) and spatial range αd. Note that at intermediate ε_{pn}, as the range of the interfacial attraction grows, the thickness of the adsorbed layer increases, resulting in intensified and longer-range effective repulsion between nanoparticles. The bridging minima at a higher ε_{pn} occurs on monomer (nanometer) length scales, and the details depend on the range of the interfacial cohesion. Polymer chain length has only small effects due to the low compressibility of the polymer melt and attendant screening effects. Adding a direct particle–particle attraction (e.g., Van Der Waals) of increasing strength reduces and, eventually eliminates, the polymer-induced repulsion (Hooper & Schweizer, 2005).

Figure 1. PRISM theory potential-of-mean-force between two hard spheres of size D/d=10 as a function of reduced interparticle surface separation in a flexible chain polymer melt of N=100 and packing fraction of 0.4 (Hooper & Schweizer, 2005)

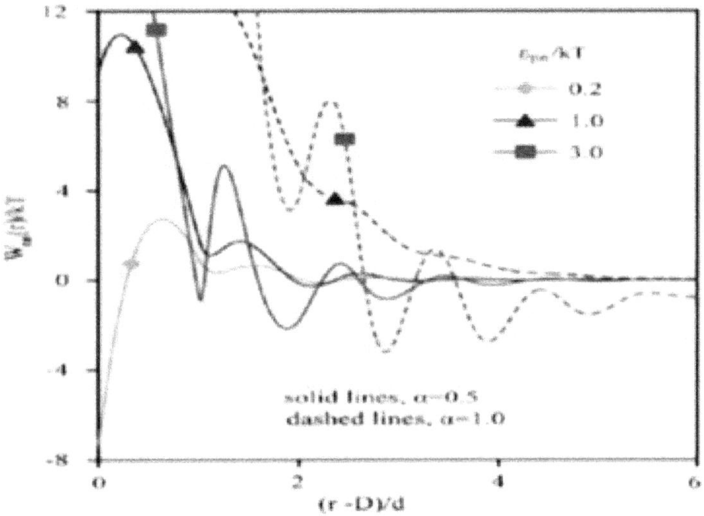

Depletion, stabilization, and bridging behaviors are seen with the increasing strength ε_{pn} (units of kT) of the exponential interfacial attraction. Increasing attraction range (in units of monomer diameter, d) results in a thicker repulsive bound polymer layer and more bridging minima. Similar depletion, stabilization and bridging behavior of the spherical particle PMF in a branched star polymer melt of variable number of arms has been found based on PRISM theory (Burger et al., 2010; Nawani et al., 2009; Rudd & Broughton, 2000).

6.3.2 Concentrated Mixtures: Structure and Many Body Effects

Just crystal principle has been utilized to study the thick melt and non-dilute nanoparticle (volume portion U) states which is hugely important to nanocomposite materials. The distributed calculations accept conformational romanticizing. The most punctual investigations have focused on understanding a number muscle to impacts on the Polymer-Mediated molecule PMF, intermolecular combine correlations, aggregate diffusing structure factors and mass modulus to straight and star polymers in the window about miscibility (Harton, S. E., & Kumar, S. K. (2008); Jayaraman, A., & Schweizer, K. S. (2008b).). Recently, numerical progresses have permitted calculations of the full significant number of particular figure impacts on spinodal stage separation, from the weaken molecule to weaken polymer limits; specifically, investigations are performed towards consistent aggregate pressing portion about η_t= 0.4, alternately 0.5 fitting to solvent-free melts (Hall & Schweizer, 2008). Altogether late worth of effort infers that expanding ηt for molecule pressing fraction, which imitates the steady weight demand and the physical impact of polymers filling interstitial spaces between molecule surfaces, is only the tip of the iceberg faulty accounts to pressing impacts bringing about preferred concurrence about hypothetical structure factors for test dissipating examples (Iacovella et al., 2007).

Including nanoparticles dramatically modifies the aggregate polymer focus variances, as well as the diffusing capacity Spp(k). Nanoparticles ''imprint'' their spatial request on polymers through the adsorbed layer, bringing about a low point or ''micro phase-like'' top over Spp(k) on the period scale of the molecule beside its bound layer. This peak, absent in the immaculate polymer melt, intensifies for expanding interfacial fascination (Jayaraman & Schweizer, 2008a). The micro phase top disappears close to the exhaustion period separation, but it persists, as bridging stage detachment will be approached because of the proceeding vicinity of a determinedly held (now shared) certain polymer layer. The coming about Spp(k) is reminiscent of micro emulsions of substantial plentifulness centralization variances around both perceptible (k = 0) And mesoscopic ($k^{*1}0$) scales existing together (Hall & Schweizer, 2008). Proof of a particle-induced low point crest for Spp(k) has been seen in little point neutron dissipating analyses on a polystyrene-silica nanocomposite and compared favored with crystal calculations (Jayaraman & Schweizer, 2008b).

The elementary material property examined to date is the nanocomposites mass modulus. The addition of nanoparticles to a thick polymer melt seems to generically decrease the mass modulus (softening), or equivalently upgrade aggregate thickness fluctuations, for a monotonic design with filler volume portion (Iacovella et al., 2007; Jayaraman & Schweizer, 2008a). In the miscibility window of the period outline, this modulus diminishment is humble and not altogether touchy of the subtle elements of the interfacial union. The exhaustion and bridging prompted stage differentiated regimes offer a view of miscibility in intermediate interfacial union quality. The sort of polymer-mediated nanoparticle association is schematically shown. A lot of people form impacts with respect to period division have been investigated (Hall & Schweizer, 2008) for the hard sphere in a FJC polymer system at $\eta_t=0.4$.

6.4 NON-SPHERICAL NANOPARTICLES

Predictive minute hypotheses for nonspherical artificially homogeneous nanoparticles, for or without grafted chains, disintegrates over thick polymer results and melts in the thick, as initial phases of improvement.

6.4.1 Bare Non-Spherical Nanoparticles

SCMFT investigations of unbending rods to adsorbing homopolymer results are performed roused by carbon nanotube requisitions (Egorov, 2008). A Derjaguin close estimation to long parallel rods is utilized in consolidation with a portrayal and a geometrical close estimation of the introductory reliance of the polymer-mediated powerful possibility. A Flory-Onsager kind for dissection applied to low volume portion and high perspective proportion rods are connected to inspect stage solidness, including nematic fluid gem arrangement. At low Pole concentrations, exhaustion or bridging might happen bringing about decreased miscibility clinched alongside a way that relies on the level for polymer adsorption. Towards higher polymer focuses and adsorption lengths, miscibility expands. The permeation edge additionally exhibits an intricate reliance on degree of adsorption and polymer centralization. SCMFT has likewise been of late connected to an arrangement of parallel mud platelets of polymer melt in the vicinity of a little measure about grafted polymer and low sub-atomic weight added substance (Zhou et al., 2007). The PMF is computed for different species focuses in order to survey the plausibility to inter capitulars vein for polymer and shed platelets.

6.4.2 Non-Spherical Nanoparticles with Grafted Tethers

The rudimentary issue of the structure of a grafted layer made from the beginning in length chains joined to a solitary exceptionally bended barrel shaped molecule has been examined utilizing a moved forward adaptation for SCMFT (Profeta et al., 2005). The allowed chain wind avoidance zone is indicated to possess a limited portion of the brush in this raised geometry, clinched alongside both the dissolvable and melt states. A few investigations utilizing separate forms of DFT and a greater amount sub-atomic level form of SCMFT have been performed to nanorods functionalized with grafted chains. DFT has been utilized on figure thickness profiles and the PMF for a couple of interminably in length rods that copy carbon nanotubes broken down and watery adsorbing surfactant result (Zeng et al., 2008). In understanding late experiments, irregular adsorption of surfactants has been found and not barrel-shaped micelle structuring. A huge ghastly obstruction in the PMF is predicted which can prompt dynamic adjustment. Single chain SCMFT has been connected for a comparable model from the beginning single divider carbon nanotubes and settled for polyethylene oxide grafted chains (Buxton & Balazs, 2003). Polymer conformity and the inter-rod PMF are dictated. Under great dissolvable states of the fastened polymers, a noteworthy ghastly obstruction in the Pole PMF because of steric adjustment is predicted which could bring about handy scattering.

6.5 PHYSICAL PROCESS ANALYSIS

6.5.1 Small-Angles X-Ray Scattering SAXS Analysis of Polymer Nanocomposites

Small-Angles X-beam dissipating SAXS Investigation is conducted from beginning Polymer Nanocomposites. The properties of polymer nanocomposites material do not mainly rely on the properties of the individual parts with respect to their morphological tenet and interfacial cooperation. In lamellar nanocomposites, the interfacial associations are expanded because of the huge surface range of the filler particles laid open in the polymer matrix, which brings about exceptional anisotropic properties. This investigation displays the structural assessments of two distinctive sorts of polymer nanocomposites materials holding platelet-shaped nanofillers, eventually perusing X-beam diffusing. A standout amongst these is an engineered polymer–clay nanocomposite of polymer camwood intercalation under the galleries of organ clays (surfactant-decorated layered silicates). Alternate polymer nanocomposite regularly happening bone of collagen particles is a type of stringy polymer matrix filled with inorganic apatite crystals. Structure assessment is dependent upon synchrotron small-angle X-beam diffusing (SAXS) investigations (Burger et al., 2010).

6.5.2 SAXS Analysis of Collagen/Mineral Nanocomposites

Concerning the illustration stated earlier, bone tissues are created for mineralized collagen fibrils similar to the essential structural units, which are composed in distinctive routes and different bones. The mineralized collagen fibril comprises platform-shaped collagen particles and nanoscale platelet apatite crystals found towards particular positions of the platform (Jayaraman & Schweizer, 2008a). The collagen atoms are staggered in the longitudinal heading for a time like 67 nm, which corresponds to the periodicity watched, eventually perusing electron microscopy and SAXS. Since the atomic length (~300 nm) is something like 4.5D, a gap of roughly 0.5D is made between an atom and its next pivotal neighbor (Hall et al., 2009). The presence in the beginning gaps inside the collagen matrix brings about a thickness difference, keeping an exchanging entirety zone/overlap zone banding design along the fibril hub bearing. The mineral crystals are fundamentally placed in the entirety zones of the prompt phases of mineralization, yet all the camwood develops under the cover zones at late phases.

The SAXS outcomes indicate that the broadness of the "channels" over unmineralized fish bones are ca. 1.5nm. The development of mineral crystals in the thickness extent is compelled by the broadness of the "channels". Thus, the thickness of the mineral crystals' camwood is hazed and will have a limited appropriation. Therefore, we pick a gaussian appropriation to depict the thickness appropriation of the mineral crystals. The mineral crystals are haphazardly saved under the parallel "channels", thus the division between the closest neighbors, alternately the thickness of the intercalated natural layers, need further bolstering to have a moderately wide appropriation. The thickness circulation of the mineral crystals and the intercalated natural layers is plotted in fig. 5. The apatite platelets have a uniform thickness (~2.05 nm) with a limited dissemination (s. D. ~0.05 nm). On the other hand, the thickness of the natural layers has a more extensive dissemination (s. D. ~3.09 nm) with a normal quality of 5.47 nm. Such quantitative estimations of the 3d collagen/mineral superstructure are vital to a right examination of the mechanical properties, the physiological and living capacities of the mineralized collagen fibrils and the bone tissues.

6.6 NANO-DIELECTRICS TECHNOLOGY

The improvement of polymer nanocomposites depends generally on our examination of the structure–property relationship of the materials which requires a multi-scale model to foresee the material properties from the majority of the data about molecule properties, atomic structure, sub-atomic interaction, mesoscale morphological tenet. The present exploration on demonstrating and recreation for polymer nanocomposites are generally restricted to a distinctive period and time scale. However, it should be noted that few endeavors have as of late been made to create multiscale methodologies for foreseeing the multi-scale level for structure, properties and preparing execution from the beginning polymer nanocomposites in light of nanoparticle support. Balazs and co-workers (Buxton & Balazs, 2003), as of late joined DFT with SCF to ascertain the stage conduct from the beginning clay-based polymer nanocomposites and other polymer–nanoparticle mixtures. In this consolidated model, the thermodynamic conduct technique and collaboration of the majority of data "around different parts would be acquired starting with the SCF model which has served concerning illustration an information on a DFT to ascertain the stage conduct. Comparable multi-scale demonstrating methodologies have also been executed during Dow to anticipate the thermodynamic harmony morphologies and number of properties of nanocomposites starting with the compound structures and the relative sums of the plan parts and will assist in item improvement (Odegard et al., 2004).

On the other hand, Glotzer and co-workers (Iacovella et al., 2007) suggested a multi-scale methodology for nano-filled polymers which combines the coarse-grained MD, mesoscopic TDGL, and perceptible continuum limited component strategies. This methodology extends the period scales important to join atomic phenomena in perceptible properties. The connection between mesoscale structure and macro scale mechanical properties are finished towards utilizing the yield of the TDGL/CHC simulations as information for the limited component Investigation. The connection between the md simulations and the TDGL/CHC simulations are lesquerella direct and made just qualitatively through consolidation from the beginning filler–polymer interactional terms in the CHC free-energy work. Porter (Patel & Egorov, 2005b) suggested a multi-scale model that considers bone as a common mixture nanocomposite for secondary perspective proportion hydroxyapatite mineral particles over a polymer matrix for turned troop collagen to anticipate the mechanical properties and structure property relationship. The physical properties, for example, such as versatile modulus, vitality dissipation, stage move temperatures and disappointment start states are computed by a self-reliable mean-field system from their sub-atomic structure. These model parameters quantify the group Normal vitality parts of a material on an atomistic level and could be ascertained by atomistic simulations alternately basic experimental techniques.

Finally, this scrutiny has examined the vital improvement to attain the longstanding objective for foreseeing particle–structure–property connections up material outline and streamlining. The physical properties for example, such as versatile modulus, vitality dissipation, stage move temperatures and disappointment start states are ascertained by a self-reliable mean-field system starting with their atomic structure. The advancement of polymer nanocomposites at distinctive duration of the time and period scales has fundamentally invigorated the advancement of machine demonstrating and simulation, as a reciprocal alternately elective technobabble on experimentation of a significant number of accepted reproduction strategies (e.g., MC, MD, FEM), and a few novel recreation strategies have been produced on study polymer nanocomposites. Our Audit has been distinguished the most recent developments which given an accumulation from beginning explanatory devices inside person DFT package, CASTEP. A mix for demonstrating for analyze may be pointed to move forward the determination and affectability for heading in the end to purpose the structures and comprehending essential issues in materials science, chemistry, geochemistry, and so forth throughout this way. Observing and stock arrangement of all instrumentation may be new concepts, hypotheses and computational instruments which ought to make produced later on to settle on truly consistent multi-scale demonstrating actuality.

6.7 FILLERS IN COMMERCIAL POLYMERS

6.7.1 Acrylics

Water emulsions utilized within paints, coatings and sealants hold significant sums of fillers which are joined towards the routine of scattering. Frequently, grinding is utilized within the paint business. Prepared sealants and fillers are utilized for reinforcement, rheology and crosslinking. The rheology of a sealant can be controlled towards the consolidation for raged silica to amounts about 3%wt. The non-sag properties from the beginning sealant are incomplete because of raged silica, yet all these are controlled, eventually perusing ph change in the vicinity for different additives, such as extraordinary acrylic resins and polyurethane thickeners. The blending for both impacts provides for the sealant's last properties. The support of the sealant can be prepared through a mix from the beginning two processes: the connection from beginning silica particles and crosslinking through zinc oxide (Lee et al., 1996).

Fillers in a sealant are included in set amounts, since high loadings influence the elastomeric properties of acrylic pitch. In coatings, bigger amounts of fillers and pigments are utilized in case covering is not required, clinched alongside a large portion cases to endure expansive elongations. Outside coatings, which oblige split bridging capabilities, are a special case. In this case, prolongation in overabundance of

1000% is hazed. Here, the filler load is significantly decreased. A percentage of outside textured coatings, alternately stuccos, require surprising fillers for example, such as silica flours, glass beads and ceramic microspheres. These fillers are used to obtain separate enlivening impacts. Silica flour are included in extensive amounts assuming the part of the traditional filler included to decrease value.

6.7.2 Acrylonitrile-Butadiene-Styrene Copolymer (ABS)

Filler blending innovation organization is vital in ABS preparing. Carbon dark must be well scattered to obtain handy jet and sway quality. High jet is generally not difficult to acquire in the utilization of high surface range carbon dark and eventually perusing the change of its centralization with prerequisites. Similarly, as the carbon dark substance is increased, the secondary sway quality turns into additional which is troublesome to uphold in light sway quality declines, as undispersed surface range expands. The best effect maintenance is attained when an easier focus from the beginning carbon dark is subjected to two-phase blending. Initial carbon dark is scattered clinched alongside ABS, et cetera and the granulate obtained after dissolvable vanishing is utilized, as a color concentrate. The effect maintenance of carbon dark holding ABS scattered in this lifestyle is free of the surface zone for carbon dark abatements, as the fixation for carbon dark expands (Tsukuda et al., 1996).

Scattering of conductive fillers is indeed all the more basic. Here, two systems have been utilized. The first is the nickel covered carbon fibers which are included straightforwardly to the ABS. In the other, the covered fibers are pre-dispersed with a dissolvable over ABS, then the dissolvable is dissipated to structure a granulate. A covering for an ideal thickness about 0.2-0.5mm is considered a significant diminishment. Carbon fiber (as much as eventually perusing one half) necessary to acquire the conductance required. Nickel covering makes the fiber more conductive and provides for preferred mechanical properties of the fiber which serves it with withstanding transforming states.

6.7.3 Alkyd Resins

Alkyd resins are utilized within a number of universal provisions, as demonstrated previously, which have a well-established innovation organization. New investigations are rarely conducted. The inalienable secondary gloss of alkyd paints requires uncommon materials for complete camwood process matt surfaces. Rheological investigations demonstrate that some sorts of silica disturb the rheological networks of paints. Silica completes this hazed pore extent from the beginning of 100 to 400Å. Exceptional sorts from the beginning alkyd resins are required to take advantage of these rheological modifiers. Regardless of whether the pore extent of silica is substantial enough to catch the structure shaping segments of the alkyd resin, silica influences the execution of alkyd sap. Wax covering doesn't influence this procedure of rheological obstruction. The tangling impact relies on the amount of particles; the number of which protrude from the surface. These properties, consolidated with littler pore sizes, make a few silica handy tangling operators. An included preference is that they could be blended under the paint (Aldcroft, 1994).

6.7.4 Epoxy Resins

Epoxy mortars for sketchiness provisions are the exceedingly filled materials. For the best possible determination from the beginning silica sand mixture, as filler is demonstrated clinched alongside, the filler focus can be as secondary as 95%wt. These comparable materials handled from a sensitive polyurethane framework are presumably the vast majority filled plastics. An additional sample for profoundly filled epoxies can be found in the fluid metal exacerbates, which are epoxies excessively loaded with metals, such as steel, aluminum, titanium, bronze and copper. The stacking from the beginning metal powder in these materials can reach 70%wt. This level for iron powder is utilized within pipeline sealants. The incorporation of metal powders has a bearing on the erosion security of the beginning epoxy covering, as iron powder enhances erosion insurance in as much expansion as copper alternate nickel lessens the protective competencies of epoxy coatings. Boundary properties of epoxy coatings hold bond promoter, and glass flakes are assessed by electrochemical impedance spectroscopy (Akay et al., 1997; Gassan & Bledzki, 1997; Grady, 1997; Hshieh & Beson, 1997; Rebouillat et al., 1996; Wang & Ploehn, 1996).

Medication of glass globules for epoxy silane moves forward the bond from the beginning glass beads/epoxy composite which brings about a significantly easier water uptake and better maintenance of the beginning properties and then water drenching. A comparative change in composite properties can be obtained in a jute/epoxy framework. Glass globules blended with elastic particles progress the weariness imperviousness of an epoxy composite, lessening the confined stress in the split tip. The techniques of change in the bond between the carbon fiber and the matrix are reviewed elsewhere. The compressive quality for an epoxy composite can be moved forward towards glass beads, quartz and calcium carbonate. The state of the molecule has an impact, looking into crack conduct of filled epoxy composite. The best properties are obtained with round silica particles. Crack conduct of glass fabric/epoxy laminates are progressed, eventually perusing the addition of alumina. Mechanical properties can be dramatically expanded (tensile quality can be expanded 9 times), as the dirt is utilized as a filler in the nanocomposite manifestation. Filled epoxy resins does not grant camwood the opportunity to be grounded and utilized as a filler for different materials. Reused elastic particles are sometimes used for toughening epoxy resins.

6.7.5 Ethylene Vinyl Acetate Copolymer, EVA

The dynamic thickness and shear modulus of silica filled EVA are identified with the worth of effort for bond of filler particles. The worth of effort of bond relics on the molecule measure circulation and load of filler (Rothon & Hornsby, 1996). The increment for molecule span of filler causes an expansion of fill in for the bond.

6.7.6 Ethylene-Ethyl Acetate Copolymer, EEA

Calcium carbonate stabilizes EEA towards raised temperatures. The balancing out impact relies on molecule size, kind of calcium carbonate and covering. Calcite provides for better adjustment of whiting alternately precipitated calcium carbonate. Furthermore, a stearate covered review enhances warm strength. The filler keeps the arrangement of acidic and propionic acids, eventually perusing cooperation with carboxyl gatherings. Metal hydroxides (Mg and Al) also settle EEA, as they break down endo thermally. For this instrument to work, the metal hydroxide must be broken down at a much bring down temperature over the corruption temperature of the copolymer (Tang et al., 1996). Otherwise, the hydroxide cooperates with the sensitive gatherings of the copolymer and changes the debasement system of the polymer.

6.7.7 Ethylene-Propylene Copolymers, EPR and EPDM

Aspects of the beginning Ethylene-Propylene Copolymers will be demonstrated in this part. Clinched alongside EPR formulations, calcium borate is considered a chance for great reinstatement of the blending of antimony trioxide for a natural fire retardant. Calcium borate, influencing fire retardation, likewise strengthens the polymer. Another option is the huntite/hydromagnesite filler. Here, a portion of antimony trioxide and natural fire-retardant consolidation must be included. The huntite/magnesite filler mix cannot, by itself, end fire spread. In carbon dark loaded EPDM, the preparation from the beginning foamed materials is influenced, eventually perusing the filler. Cell thickness abatements for the measure from the beginning carbon dark builds with the measure of blowing agenize. The extent from the beginning unit's declines in the vicinity from the beginning carbon dark due to the basic surface of carbon dark (Datta, Bhattacharya, De et al, 1996; Datta, De, Kontos et al, 1996).

6.7.8 Liquid Crystalline Polymers, LCP

Thermo-tropic fluid crystalline polymers are figured with secondary centralization for glass fiber to withstand working temperatures exceeding 300°C. To prepare LCP, you quit offering regarding such issue. LCP orients itself in the bearing of shear alternate flow. In order to harmonize mechanical properties, it has been suggested that some amounts of short glass of alternate mineral fibers are included in LCP. The smallest build is because of the first mass of the magnesium carbonate, and the biggest is because of the vicinity of glass fibers. The mechanical properties of loaded composites are significantly moved forward, eventually perusing additions from the beginning magnesium carbonate, wollastonite and glass fiber (Otaigbe et al., 1997). The major essential change is crawling imperviousness.

6.7.9 Phenolic Resins

Shoot imperviousness is a paramount property of the beginning phenolic resins. The blending of phenolic resin for Expancel™ expandable microspheres prompts a large number of suitable items. Composites for secondary speed train interiors take advantage of the light weight, phenomenal fire rating and exact low warm conductivity. Polyester filled for aluminum hydroxide is an elective result for preparing inner part materials. The resin and filler camwood can be effortlessly transformed, when viscosity managing additives are included. The properties of novolac laminates camwood are enhanced, eventually perusing the addition of fillers. Erosion protective materials endure delamination due to various warm development rates. An overlay is produced starting with two layers - one holding carbon fiber and the other loaded with 15-20% graphite powder, the high temperature stream and temperature conveyance over the overlay to be moved forward, and delamination is wiped out (Akay et al., 1997; Datta, Bhattacharya, De et al, 1996; Datta, De, Kontos et al, 1996; Hshieh & Beson, 1997; Jakeman, 2000; Murayama & Min, 1997; Otaigbe et al., 1997; Rothon & Hornsby, 1996; Tang et al., 1996).

6.7.10 Polyamides, PA

In requisitions which require electric conductivity, polyamides are transformed, possibly with carbon fiber or for graphite. These requisitions incorporate benefits of the business machines (copying machines, workstation printers), electronic packaging, rug fiber and EMI protecting. Other fillers, such as nickel covered graphite, stainless steel fiber, aluminum flakes and metallized glass are utilized less frequently. Polyamide is a standout amongst the best EMI protecting materials. At the point when it is exacerbated for just 15% nickel covered glass fiber, it provides for a weakening of 50 db, eventually perusing comparison. Polyamide, exacerbated for 30% graphite fiber, provides for a weakening of main 30 db. The general standard of filled materials is that the higher the molecule measure of the conductive molecule is, the higher the conductivity of the resultant material becomes. In conductive materials filled with fibers, this relationship is all the more unpredictable and more dependent on filler kind. The utilization from the beginning magnesium hydroxide in polyamides is confined, eventually perusing the low corruption temperature and the low hydrolytic solidness of polyamides. Polyamide 6 and 6. 6 start is corrupted towards around 350°C, while magnesium hydroxide discharges water between 320 and 440oC (DellaCorte et al., 1993; Hamada et al., 1997; Jia & Kagan, 1997; Wagner et al., 1997).

Adding about 60%wt. magnesium hydroxide processes a fire-retardant polyamide 6 for its corruption temperature covering that of the filler. A comparable strategy is not as effective with polyamide 6. 6. Fillers assume a paramount part in powder covering for polyamide to structure articles of a metal-like gander (e.g., handles, mountings to radiators and pipes). In order for a material to be powder coated, it must withstand the stoving temperatures (170°C or more). It must additionally be electrically conductive and chargeable, and its sensitive bunches have the ability of connection with the covering framework (Jia & Kagan, 1997; Stejskal et al., 1996). Fillers, such as metal covered ceramic spheres and carbon fibers include polyamide for its quality and paint ability.

6.7.11 Polyaryletherketone, PAEK

The flexural weariness split draws on ductile properties of the composite. The yield quality of the matrix and the personal satisfaction of the interface influence the weariness properties of composites (Zhou et al., 1997).

6.7.12 Polycarbonate, PC

Fire retardant materials are generated starting with polycarbonate. A consolidation of PTFE fibers (2%wt.) and boric oxide (1%wt.) provides for v-1 rating. Handy outcomes are also obtained by utilizing a consolidation for alumina and silica, or a mix of first mass of the magnesium carbonate and calcium carbonate with zinc borate. Halogen-free fire-retardant evaluations are promptly prepared. The addition of zinc borate on polycarbonate significantly lessens the heat arrival and smoke era to the compound. The warm properties from the beginning polycarbonate are not remarkable, and the chances to move it forward are remote. Expansion of 30%wt. from the beginning carbon alternate glass fibers builds redirection temperature under load main, eventually perusing 20°C of the greatest feasible esteem for 150°C, provided that the coefficient of the warm extension is drastically diminished making it suitable for a number of its possibility requisitions (Tanaka et al., 1997). Polycarbonate is being utilized within provisions which require EMI protecting and static control. EMI protecting requires substantial amounts of conductive fillers. For example, 40%wt. aluminum chip provides for a weakening of 32 dB, 30%wt. Graphite fibers provide for 42 dB and 15%wt. nickel covered glass fiber provides for 45 db. Fiber support enhances mechanical properties by the sums relative to the sum of fiber utilized.

6.7.13 Polyetheretherketone, PEEK

The prologue of the beginning fillers to look causes a higher nucleation rate. The surface of carbon fibers and cores inside the look matrix seek crystallization development. Epitaxial transcrystalline development is habitually seen on the fiber surface in carbon fiber strengthened look composites. Supported with glass fiber or carbon fiber pairs, the elasticity and modulus and at the same time, the sway quality will be likewise expanded. Carbon fibers enhance properties, eventually perusing no less than half of look loaded with glass fibers. Warm properties are also enhanced in this reinforcement, once more, particularly by carbon fiber. Heat redirection temperature exceeds pairs because of the support to qualities at 300ºC. Dissolvable safety is typically resolved, eventually perusing the level from the beginning fiber support and the bond between the fiber and matrix. PEEK/carbon fiber composite has a phenomenal safety in water and high engineering (Selzer & Friedrich, 1997).

6.7.14 Polyethylene, PE

Mechanical properties rely on filler-matrix connection, and some portion qualities from the beginning fillers impact the mechanical conduct of a composite. Practically, fillers expand rigidity. However, calcinated kaolin increments elasticity about 3 times more than calcium carbonate alternately talc. The effect quality is progressed by calcinated kaolin. It can be brought down by the addition of whichever talc alternate calcium carbonate. 10^3 and these impacts can be changed by customizing the interface between the matrix and the filler. In one example, chlorinated PE holding carboxyl assemblies can be utilized similarly as compatibilizer. 140 and calcium carbonate can be altered towards grafting acryloamide for 0.2-1.8% amicus curiae short assemblies onto its surface. This expands its level of connection, and the elasticity of this composite is expanded, eventually perusing in half and its effect quality towards 120% compared with slick tar (When calcium carbonate is included without compatibilizer, the sway quality of the composite drops in the following 25% of flawless resin). 140 and similarly, the impact from the beginning kaolin is progressed by covering it with maleic grafted polyethylene which expands the sway vitality of the filled PE by an element of 4 compared with slick tar. Previously, blown films, calcium carbonate in the fixation range of 5-20%wt., have been discovered to increment dart effect quality. However, the rigidity and prolongation diminish as the fixation expands. Additions of talc in the same fixation range diminishes the greater part of mechanical properties (Bushko & Stokes, 1997; Gordienko & Dmitriev, 1996; Le Bras, 1997; Savadori et al., 1996; Wang et al., 1997).

The basic strategy to aggravate PE shoot retardant is the utilization of phosphoric esters regarding polyols. A substitute strategy employs magnesium hydroxide. Magnesium oxide provides for an insignificantly superior execution, since in both situations, an expansive measure from the beginning metal oxide will be necessary. If such expansive amounts are used, the sway safety of the material will be considerably lessened. The effect imperviousness can be progressed for these compositions, eventually perusing additions from the beginning silane-crosslinkable polyethylene. For polyethylene to be conductive, it must be loaded over a sure edge centralization for conductive filler. Beneath this edge level, conductivity remains very steady and it is generally autonomous from the beginning filler fixation. In this range, conductive filler particles are not in correct contact. At a certain edge, esteem will be reached and the conductivity will build quickly, since a greater amount of carbon dark is included. Eventually, a plateau of ultimacy conductivity is arrived at which relies on the matrix and on the kind from the beginning carbon dark utilized. Huge focuses (80-90%wt.) of ferromagnetic materials are required to arrive at threshold concentration.

Conductivity transforms quickly when 3%wt. stainless steel fiber is included in polyethylene. Similarly, attractive properties of camwood are transformed with additions from 10 to 30%wt. of nickel fibers. The fibers must be in the best possible introduction to create ideal attractive properties. Particular case investigation will attempt to acquire a material for low resistivity towards room temperature and secondary resistivity towards raised temperatures. It additionally endeavors to acquire material for which such change happens inside a couple of degrees Celsius. The composites which are created could switch quickly starting with a low to high resistivity. These composites are utilized within units which can break point electric issue ebbs and flows. The materials are produced towards selecting a fitting transform of consolidation (quality of mixing) and eventually perusing the decision from the beginning carbon dark. An arrangement which joins together coarse and fine carbon blacks provides for the hazed execution (Roychoudhury et al., 1995; Ventresca & Berard, 1997).

6.7.15 Poly (Ethylene Oxide), PEO & PEG

The poly (ethylene glycol)/graphite framework demonstrations are an exchanging component. A switching component is a polymer composite for the scattered conductive particles, such as carbon black and graphite alternate metal particles. The exchanging component has a low safety during low temperatures (switch-on) and a high safety towards high engineering (switch-off). A little distinction of temperature causes a fast change for imperviousness. The descriptions of this issue are two: the established illustration is in light of the premises that temperature expanding conductive particles are inaccessible on exchange electrons; a greater amount of later illustration qualities of the sudden passing progresses in resistivity of the change from the beginning dielectric consistent from beginning poly (ethylene glycol). A comparative framework is created in view of polyethylene. Poly (ethylene oxide) is, likewise, utilized in nanocomposites which hold numerous molybdenum disulfide or vanadium oxide. The inorganic filler and the natural matrix cooperate at a sub-atomic level in shaping xerogels, which are nanocomposites for regulated particle portability (Kimura et al., 1996).

6.7.16 Poly (Ethylene Terephthalate), PET

The treatment for glass globules for silane enhances their bond. This benefits the major properties of the composite, yet all of its rigidity will be marginally more level than that from the beginning flawless tar. The crystallization rate from the beginning pet is higher in the vicinity of filler. Glass globules result in a heterogeneous nucleating impact. Introduction of a modifier lowers the nucleation rate and acts as a compatibilizer. The blending from the beginning glass globules and the modifier builds pliable properties and the effect quality of the composite (Fomin & Guzeev, 2001).

6.7.17 Polyimide, PI

Polyimide is utilized as a model material in investigations of polymer metal interfaces. The place metal layers are framed towards metallization, plasma deposition, concoction vapor deposition, electrochemical deposition, and so forth in this way, observing and stock arranging of all instrumentation (Chen et al., 1994).

6.7.18 Polymethylmethacrylate, PMMA

NMR investigations show that hydroxyl bunches on the surface of silica are devoured through polymerization of PMMA. These gatherings are used for a response with polymer. Polymer adsorbed on the surface of alumina transforms conformity. Glass powder is used to fill bone bond. Manufactured material is additionally and promptly acknowledged, eventually perusing the figure if all the more inorganic filler is available. The certainty that calcium and silicate ions are expended starting with the bond demonstrates that silane coupling, despite being profitable for bond improvement, can degrade starting with the bioactive qualities of the bond. S has a little sum of titanium fiber (1.5%wt.) in another plan of bone bond, significantly expanding its imperviousness from the beginning split proliferation (Asai & Sumita, 1995; Visser, 1997). Phosphorus-containing PMMA is every now and then utilized for fire safe provisions. The imperviousness quickly drops the point when the fixation of aluminum to aluminum filled PMMA surpasses 20 Vol%. Marginally less (about 18-20 Vol%), nickel may be required to achieve the same safety.

6.7.19 Polyoxymethylene, POM

POM materials are for glass globules of lessened pliable strength, crack toughness and strain vitality discharge rate, yet enhanced flexural modulus. The elasticity is conversely proportional to the square root of the glass circle breadth. Reinforcing POM for glass fiber progresses constantly in its mechanical properties (Hashemi et al., 1996).

6.7.20 Polypropylene, PP

Fillers influence the nucleation rate as polypropylene crystalizes. The addition of 2.5%wt. titanium dioxide diminishes the measure for spherulite by an element of 3 because of an expanded number of locales which rival between developing spherulites. Spherulites doesn't develop in the surface skin. Indeed, an expansion from the beginning 40%wt. titanium dioxide doesn't cause spherulites to develop in the skin. An expansion from the beginning titanium dioxide influences scattering. Normally, 2.5 crystals for every bunch are found during 10%wt. titanium dioxide and 5 crystals for every group are distinguished in 40%wt. expansion. Trans crystallinity is alternately surprisingly seen on polypropylene. Trans crystallinity possibly develops on the surface of air pockets alternately on the surface of fibers, such as glass fibers. Trans crystalline structures are to a greater amount successfully produced at any level for mechanical anxiety made around the fiber. The impact on few fibers, for example, such as carbon fiber, E-glass fiber and Twaron are assessed utilizing polarizing light microscopy. Both the shear gradient at the interface and the temperature inclination are found with impact trans crystallinity. Filler particles turned during material stream influence the introduction of beginning polymer chains and crystallites, as they develop on the surface of the filler particles. A comparable introduction of talc particles and polymer chains are discovered in thermoformed and ruin formed polypropylene. The introduction of short glass fibers and related sub-atomic introduction of the matrix can be controlled by procedure parameters of infusion forming. Electron turn thunder investigations of calcium carbonate and talc filled polypropylene show that filler introduction throughout the infusion forming has depended on the filler load. The best introduction is obtained toward 15%Vol. filler (Curro, 1997; Nago & Mizutani, 1996; Schwab et al., 1997; Suri & Min, 1997; Thomason & Vlug, 1997; Zhao et al., 1996).

Mechanical properties of loaded polypropylene rely on few variables which are talked about beneath. The Izod sway quality of carbon dark filled polypropylene declines insignificantly, as the fixation from the beginning carbon dark abates. S is of 20%wt. carbon dark with polypropylene preparing a generous (400%) expand in flexural modulus and 40% expand in flexural quality. Pliable yield quality is an unpredictable property. Little additions from the beginning carbon dark (until 10%wt.) expand the ductile yield quality. However, progressively higher focuses in the end diminish ductile yield quality until it drops beneath that from the beginning slick tar. Talc and kaolin for focuses dependent upon 30%wt. do not increase the rigidity of polypropylene, yet move forward the flexural modulus towards an element about 2. Talc alone significantly diminishes sway quality. Calcium carbonate loaded polypropylene requires poorer mechanical properties and is more challenging to methodology than the flawless polymer. However, calcium carbonate is surface covered towards stearates prolongation, effect quality is administered and whiteness and preparing qualities are enhanced. S from the beginning glass globules holding rubbery inclusions brings forward sturdiness to polypropylene. The particles change the split development instrument, eventually perusing cavitation, shear yielding and molecule matrix de-bonding. Disappointment towards de-bonding starting with the surface of glass globules approached with amiidae silane happens at strain of 0.7%.

The engineering organization for processing microporous propylene sheets prefer filler debonding. A sheet for polypropylene which holds 65. 8%wt. calcium carbonate can be extruded. The base sheets are extended biaxially, clinched alongside an extending machine with 500 to 1500%. This causes debonding of the filler, which brings about a delicate microporous film with regulated gas and water vapor permeability. Microporous polypropylene empty fibers are arranged for a comparative methodology. Glass fibers influence the mechanical properties from the beginning polypropylene contingent upon the concentration, their length and their bond of the matrix. Filler mixtures can be chosen to streamline the properties of composite. Calcium carbonate and mica, joined with maleic anhydride, empower the result to meet hazed properties. The proportions of glass fiber and mica can be changed to control properties. For example, ductile and flexural quality rely on glass fiber content, and warpage and shrinkage can be controlled towards mica. Combinations of fillers for distinctive shapes (platelet, spherical, elongated) provide for an exceptional execution that can be acquired from a single filler.

6.7.21 Polypyrrole

In suspensions of carbon dark in pyrrole, anodic polymerization takes advantage of the certainty that carbon dark particles are contrarily charged with respect to their surface which makes it conceivable to them to keep moving to a positively charged anode to be inserted inside a developing polypyrrole matrix. 308 This preparation technique is suitable to processing materials to sensors, supercapacitors, fuel cells, and so on (Watson & Barteau, 1994).

6.7.22 Polystyrene and High Impact, PS and HIPS

Introduction of talc particles parallel to the divider of the mold, alternately the dies, are dictated for polystyrene filled for talc. S is of little sums (<1%wt.) for unbending particles (barium sulfide and cross-linked polystyrene globules are utilized within the experiment) dramatically enhancing the effect quality of polystyrene. A particular molecule breadth can be controlled and chosen to obtain the ideal change. The system of activity is identified with the shaping of voids and development from beginning crazes starting with these voids. Overfire retardant applications, such as a consolidation of zinc borate for ammonitic polyphosphate provides for V-0 rating. The utilization of zinc borate permits a diminishment in the measure of ammonitic polyphosphate. Red phosphorus alone alternately offers consolidation for ammonitic polyphosphate or melamine phosphate, likewise, prepared on a V-0 rating. The heat arrival rate can be adequately progressed, eventually perusing little additions (1-2%wt.) of silicone powder in consolidation with other fire retarding additives or toward higher focus (15%wt.), when utilized by itself (Meddad & Fisa, 1997).

6.7.23 Polytetrafluoroethylene, PTFE

Slick PTFE has few properties demonstrated which dispense with its compelling reason for additives. A high restricting oxygen list (>95%) implies that tar is nonflammable. It's warm and electric encasing properties and its low rubbing coefficient imply that further upgrades are infrequently required. A change of mechanical properties is frequently something to be required and this is typically achieved in the liability for compound imperviousness because the vast majority of fillers degrade starting with the compound safety of slick PTFE. S standing for NiZn ferrite powder is used to confer ferromagnetic properties of the polymer to the electronic business. The addition of graphite, molybdenum disulfide, boron nitride, metal powders, and glass fiber builds wear safety and warm conductivity. Graphite of over 40%wt. builds the porosity of the composite, and as a direct result, it generates a change in the morphological tenet. The wear rate for graphite/PTFE and molybdenum sulfide/PTFE composites transforms over 40%wt. of filler. Expansion of particulate fillers normally brings about crumbling mechanical properties. The standards of operation under high wear states with oil are mulled over (Zhang et al., 1997).

6.7.24 Polyurethanes, PU and TPU

Fire retardant polyurethanes are basically made for mixes from the beginning phosphorus, such as ammonitic phosphate alternately polyphosphate. Aluminum hydroxide is alternately in consolidation with melamine as a substitute methodology. On intumescent applications, graphite is utilized extensively. Calcium carbonate is helpful as a fire retarding additive in blending with different fire retarding materials, due to its huge endothermic top found as DTA curves (Chen et al., 1996; Torro-Palau et al., 1997).

6.7.25 Poly (Vinyl Chloride), PVC

Expansion from the beginning fillers affects mechanical properties of PVC. For example, calcium carbonate abatements yield stress and talc abatements sway quality. The purpose behind the examination of the PVC is to acquire compositions with adjusted properties. Ultrafine talc is utilized within requisition, provided that the sway and firmness must at the same time be moved forward. 360 Flexural moduli are enhanced towards expansion of ultrafine and general motivation talcs. However, all the expansion of the possible kinds of talc quickly abates sway quality. Ultrafine talc can be joined for an acrylic modifier with both firmness and sway enhanced (depending on the sums from the beginning talc and modifier, the flexural modulus can be enhanced, eventually perusing up to 60% and the sway quality by 1500%). In 360, the surface of fillers assumes a basic part. In spite of the fact that surface adjustment for filler can have advantages, the inalienable properties of the untreated filler are just as significant. The far-reaching utilization of calcium carbonate in PVC is, around other elements (mostly economic), identified with its basicity which allows it to greatly associate with PVC (which may be acidic). Whether surface change transforms this interaction or not, the execution of the composite is influenced. Fillers are included in PVC to render acceptable warm and uv adjustment. For example, calcium stearate can assume the part of a copartner warm stabilizer utilized within an arrangement of calcium salts about unsaturated fat acids. These stabilizers use combinations from the beginning two alternately additional metals - a standout amongst such (e. G. Zinc)- produces metal chlorides which quicken PVC corruption. The vicinity about extensive add up of calcium salts aides in changing over this chloride to calcium chloride which doesn't expand the corruption rate of PVC (Baggaley et al., 1997; Kovacevic et al., 1996; Wiebking, 1996).

Furthermore, calcium carbonate can react with hydrogen chloride which can be handled as PVC degrades. On the other hand, consideration from the beginning fillers which hold numerous admixtures of metals, such as iron, nickel, copper and so forth through this way, observing and stock arranging of all instrumentation can enhanced and PVC warm strength can be lessened. Fillers also influence UV adjustment towards adsorption from the beginning HALS stabilizers which immobilize them and keep them from performing as radical scavengers.

Antimony trioxide is a suitable fire retardant for PVC, taking into consideration that it needs chlorine to perform. Little additions (3-6%wt.) are additions for inflexible and semi-rigid compositions. Making plasticized PVC fire retardant requires higher focuses (15-20%wt.). Antimony trioxide is a white pigment that impacts the shade. Therefore, little molecule extent antimony pentoxide starting with a spread drying transform can be used to provide dim shades and transparent materials. Aluminum hydroxide is an added substance, although it's totally spread requisition is hampered by bigger focuses in which mechanical properties are influenced. Magnesium hydroxide is utilized similarly as a shoot retardant in order to restrict obstruction of stabilizers. Magnesium hydroxide is insignificantly superior to aluminum hydroxide for smoke concealment. Combinations of aluminum hydroxide and magnesium hydroxide with zinc borate are likewise examined as a possible supplant to antimony oxide. These combinations cause change to antimony oxide in the fire retarding properties from the beginning high temperature release, particular elimination area, smoke and co-emanation. The best performing formulations haze higher centralization of inorganic fire retardants (concentration expands eventually perusing ~250%), however, shockingly towards the same level from the beginning of loading. Formulations holding zinc borate clinched alongside blending with aluminum hydroxide offer exceptional execution over combinations of aluminum hydroxide for antimony oxide. The execution for aluminum and magnesium hydroxides can be improved, eventually perusing covering the powders for zinc hydroxystannate. Smoke decrease is obtained, eventually perusing different iron, copper, nickel and vanadium mixes which, in spite of the fact that they lessen the smoke, they influence warm strength and, frequently, the UV dependability from the beginning PVC. Different materials utilized incorporate molybdenum and boron mixes. In the range of conductive plastics, PVC is utilized within static control provisions and EMI protecting. Static control and directing, including tiling and sheeting for modern applications, can be attained by expansion of carbon dark. EMI protecting is a moderately new requisition for PVC in which metal and carbon fibers are utilized.

REFERENCES

Akay, M., Mun, S. K. A., & Stanley, A. (1997). Influence of moisture on the thermal and mechanical properties of autoclaved and oven-cured Kevlar-49/epoxy laminates. *Composites Science and Technology, 57*(5), 565–571. doi:10.1016/S0266-3538(97)00017-1

Akcora, P., Liu, H., Kumar, S. K., Moll, J., Li, Y., & Benicewicz, B. C. (2009). Anisotropi self-assembly of polymer-decorated spherical nanoparticles. *Nature Materials Journal, 8*, 354-359.

Aldcroft, D. (1994). Burning Time with Flame Retardant Coatings. *Polymers Paint Colour Journal, 184*(4366), 423–425.

Asai, S., & Sumita, M. (1995). Effect of interfacial energy and viscosity on percolation time of carbon black-filled poly(methyl methacrylate). *Journal of Macromolecular Science, Part B: Physics, 34*(3), 283–294. doi:10.1080/00222349508215536

Baggaley, R. G., Hornsby, P. R., Yahya, R., Cussak, P. A., & Monk, A. W. (1997). The influence of novel zinc hydroxystannate-coated fillers on the fire properties of flexible PVC. *Fire and Materials, 21*(4), 179–185. doi:10.1002/(SICI)1099-1018(199707/08)21:4<179::AID-FAM607>3.0.CO;2-O

Belyi, V. A. (2004). Exclusion zone of convex brushes in the strong-stretching limit. *The Journal of Chemical Physics, 121*(13), 6547–6554. doi:10.1063/1.1778153 PMID:15446956

Brown, S. P. (2009). Recent Advances in Solid-State MAS NMR Methodology for Probing Structure and Dynamics in Polymeric and Supramolecular Systems. *Macromolecular Rapid Communications, 30*(9-10), 688–716. doi:10.1002/marc.200800816 PMID:21706657

Burger, C., Hsiao, B. S., & Chu, B. (2010). Preferred Orientation in Polymer Fiber Scattering. *Polymer Reviews (Philadelphia, Pa.), 50*(1), 91–111. doi:10.1080/15583720903503494

Bushko, W. C., & Stokes, V. K. (1997). Estimates for Material Shrinkage in Molded Parts Caused by Time-Varying Cavity Pressure. In *Plastics saving planet earth: conference proceedings / ANTEC '97*. SPE, Society of Plastics Engineers.

Buxton, G. A., & Balazs, A. C. (2003). Simulating the morphology mechanical properties of filled diblock copolymers. *Physical Review E: Statistical, Nonlinear, and Soft Matter Physics, 67*(3), 031802–031814. doi:10.1103/PhysRevE.67.031802 PMID:12689091

Bymaster, A., Jain, S., & Chapman, W. G. (2008). Microstructure and depletion forces I polymer–colloid mixtures from an interfacial statistical associating fluid theory. *Journal of Chemical Physics, 128*(16), 164910-1:164910-13.

Chen, L., Liu, K., & Yang, C. Z. (1996). Preparation of ultrafine metal particles/polyurethane composites. *Polymer Bulletin, 37*(3), 377–383. doi:10.1007/BF00318071

Chen, X., Cai, J., Liu, H., & Hu, Y. (2006). Depletion interaction in colloid/polymer mixture application of density functional theory. *Proceedings of the Conference on Industrial Applications of Molecular Simulation*, 32, 877–885. 10.1080/08927020600935580

Chen, X., Gonsalves, K. E., Chow, G. M., & Xiao, T. D. (1994). Homogeneous dispersion of nano-structured aluminum nitride in a polyimide matrix. *Advanced Materials*, *6*(6), 481–484. doi:10.1002/adma.19940060608

Choi, W. J., Kim, S. H., Kim, Y. J., & Kimaj, S. C. (2004). Synthesis of chain-extended organifier and properties of polyurethane/clay nanocomposites. *Polymer*, *45*(17), 6045–6057. doi:10.1016/j.polymer.2004.06.033

Curro, J. G. (1997). Molecular Modeling of Amorphous Polymers in the Condensed Phase. In *Plastics saving planet earth: conference proceedings / ANTEC '97*. SPE, Society of Plastics Engineers.

Datta, S., Bhattacharya, A. K., De, S. K., Kontos, E. G., & Wefer, J. M. (1996). Reinforcement of EPDM-based ionic thermoplastic elastomer by precipitated silica filler. *Polymer*, *37*(12), 2581–2585. doi:10.1016/0032-3861(96)85376-6

Datta, S., De, S. K., Kontos, E. G., & Wefer, J. M. (1996). Ionic thermoplastic elastomer based on maleated epdm rubber. I. Effect of zinc stearate. *Journal of Applied Polymer Science*, *61*(1), 177–186. doi:10.1002/(SICI)1097-4628(19960705)61:1<177::AID-APP19>3.0.CO;2-4

DellaCorte, C., Pepper, S. V., & Honecy, F. S. (1993). Tribological properties of Ag/Ti films on Al2O3 ceramic substrates. *Surface and Coatings Technology*, *52*(1), 31–37. doi:10.1016/0257-8972(92)90368-K

Deng, D. (2008). Synthesis and structure analysis of zeolite AS-1 from HF-Al2O3-SiO2- ethylene-diamine-H2O. *Microporous and Mesoporous Materials Journal*, *116*(1-3), 491–497. doi:10.1016/j.micromeso.2008.05.021

Egorov, S. A. (2005). Interactions between nanoparticles in supercritical fluids: From repulsion to attraction. *Physical Review E: Statistical, Nonlinear, and Soft Matter Physics*, *72*(1), 401–405. doi:10.1103/PhysRevE.72.010401 PMID:16089926

Egorov, S. A. (2008). Interactions between polymer brushes in solvents of variable quality: A density functional theory study. *The Journal of Chemical Physics*, *129*(6), 064901–064909. doi:10.1063/1.2968545 PMID:18715103

Fomin, V. A., & Guzeev, V. V. (2001). Biodegradable Polymers, Their Present State and Future Prospects. *Progress in Rubber, Plastics and Recycling Technology*, *17*(3), 186–204. doi:10.1177/147776060101700303

Frischknecht, A. L. (2008). Forces between nanorods with end-adsorbed chains in a homopolymer melt. *The Journal of Chemical Physics*, *128*(22), 224902–224913. doi:10.1063/1.2929831 PMID:18554048

Fuchs, M., & Schweizer, K. S. (2002). Structure of colloid–polymer suspensions. *Journal of Physics Condensed Matter*, *14*(12), 239–269. doi:10.1088/0953-8984/14/12/201

Ganesan, V., Khounlavong, L., & Pryamitsyn, V. (2008). Equilibrium characteristics semi flexible polymer solutions near probe particles. *Physical Review E: Statistical, Nonlinear, and Soft Matter Physics*, *78*(5), 1804–1809. doi:10.1103/PhysRevE.78.051804

Gassan, J., & Bledzki, A. K. (1997). Effect of moisture content on the properties of silanized jute-epoxy composites. *Polymer Composites*, *18*(2), 179–184. doi:10.1002/pc.10272

Gordienko, V. P., & Dmitriev, Y. A. (1996). The degradation and stability of polyethylene with small additions of metal oxides under UV-irradiation. *Polymer Degradation & Stability*, *53*(1), 79–87. doi:10.1016/0141-3910(96)00033-X

Grady, B. P. (1997). Admicellar Polymerization as a Surface Modification Procedure to Improve Polymer-Filler Adhesion in Composites. In *Plastics saving planet earth: conference proceedings / ANTEC '97.* SPE, Society of Plastics Engineers.

Hall, L. M., Anderson, B. J., Zukoski, C. F., & Schweizer, K. S. (2009). Concentration fluctuations local order and the collective structure of polymer nanocomposites. *Macromolecules*, *42*(21), 8435–8442. doi:10.1021/ma901523w

Hall, L. M., & Schweizer, K. S. (2008). Many body effects on the phase separation and structure of dense polymer–particle melt. *The Journal of Chemical Physics*, *128*(23), 4901–4916. doi:10.1063/1.2938379 PMID:18570522

Hall, L. M., & Schweizer, K. S. (2010). Structure, scattering patterns and phase behavior o polymer nanocomposites with nonspherical fillers. *Soft Matter Journal*, *6*(5), 1015–1025. doi:10.1039/b919160g

Hamada, H., Hiragushi, M., Takahashi, K., & Machida, K. (1997). Article. In *Antec '97, Conference Proceedings*. Academic Press.

Harris, R. K., Hodgkinson, P., Pickard, C. J., Yates, J. R., & Zorin, V. (2007). Magnetic Resonance in Chemistry. *Chemistry Views Journal*, *45*, 174–186.

Harton, S. E., & Kumar, S. K. (2008). Mean-field theoretical analysis of brush-coated nanoparticle dispersion in polymer matrices. *Journal of Polymer Science. Part B, Polymer Physics*, *46*(4), 351–358. doi:10.1002/polb.21346

Hashemi, S., Din, K. J., & Low, P. (1996). Fracture behavior of glass bead-filled poly(oxymethylene) injection moldings. *Polymer Engineering and Science*, *36*(13), 1807–1820. doi:10.1002/pen.10576

Heine, D. R., Grest, G. S., & Curro, J. G. (2005). Structure of polymer melts and blends comparison of integral equation theory and computer simulations. *Advances in Polymer Science*, *173*, 209–249. doi:10.1007/b99431

Hooper, J. B., & Schweitzer, K. S. (2006). Theory of phase separation in polymer nanocomposites. *Macromolecules*, *39*(15), 5133–5142. doi:10.1021/ma060577m

Hooper, J. B., & Schweizer, K. S. (2005). Contact aggregation, bridging, and steric stabilization I dense polymer–particle mixtures. *Macromolecules Journal*, *38*(21), 8858–8869. doi:10.1021/ma051318k

Hooper, J. B., & Schweizer, K. S. (2007). Real space structure and scattering patterns o model polymer nanocomposites. *Macromolecules*, *40*(19), 6998–7008. doi:10.1021/ma071147e

Hshieh, F. Y., & Beson, H. D. (1997). Flammability testing of flame-retarded epoxy composites and phenolic composites. *Fire and Materials*, *21*(1), 41–49. doi:10.1002/(SICI)1099-1018(199701)21:1<41::AID-FAM595>3.0.CO;2-G

Iacovella, C. R., Keys, A. R., Horsch, M. A., & Glotzer, S. C. (2007). Icosahedral packing of polymer tethered nanospheres and stabilization of the gyroid phase. *Physical Review E: Statistical, Nonlinear, and Soft Matter Physics*, *75*(4), 801–809. doi:10.1103/PhysRevE.75.040801 PMID:17500854

Jakeman, R. (2000). Filling up on finer talcs cuts solvent content. *Polymers Paint Colour Journal*, *190*(4434), 18-20.

Jayaraman, A., & Schweizer, K. S. (2008). Effective interactions, structure, and phase behavior of lightly tethered nanoparticles in polymer melts. *Macromolecules*, *41*(23), 9430–9438. doi:10.1021/ma801722m

Jayaraman, A., & Schweizer, K. S. (2008a). Structure and assembly of dense solutions a melts of single tethered nanoparticles. *The Journal of Chemical Physics*, *128*(16), 4904–4917. doi:10.1063/1.2907717 PMID:18447497

Jayaraman, A., & Schweizer, K. S. (2008b). Effective Interactions, Structure, and Phase Behavior of Lightly Tethered Nanoparticles in Polymer Melts. *Macromolecules Journal*, *41*(23), 9430–9438. doi:10.1021/ma801722m

Jia, N., & Kagan, V. A. (1997). Effects of Time and Temperature Conditions on the Tensile-Tensile Fatigue Behavior of Short Fiber Reinforced Polyamides. In *Plastics saving planet earth: conference proceedings / ANTEC '97*. SPE, Society of Plastics Engineers.

Johnston, Iuliucci, Facelli, & Fitzgerald. (2009). Intermolecular shielding contributions studied by modeling the chemical-shift tensors of organic single crystals with plane waves. *The Journal of Chemical Physics*, *131*, 144503–144514. doi:10.1063/1.3225270 PMID:19831448

Kim, J. U., & Matsen, M. W. (2008). Interaction between polymer–grafted particles. *Macromolecules*, *41*(12), 4435–4443. doi:10.1021/ma8002856

Kim, K., Utracki, L. A., & Kamal, M. R. (2004). Numerical simulation of polymer nanocomposites using self-consistent mean-field model. *The Journal of Chemical Physics*, *121*(21), 10766–10778. doi:10.1063/1.1794636 PMID:15549962

Kim, S. C., & Lee, C. H. (2006). Depletion interactions between colloidal particles in polymer solutions: Density functional approach. *Molecular Physics*, *104*(9), 1487–1495. doi:10.1080/00268970600556618

Kimura, T., Asano, Y., & Yasuda, S. (1996). Self-temperature-control plane heaters by polyethylene glycol-graphite systems. *Polymer*, *37*(14), 2981–2987. doi:10.1016/0032-3861(96)89395-5

Kovacevic, V., Lucic, S., Hace, D., & Glasnovic, A. (1996). Rheology and morphology of poly(vinyl acetate) + calcite films. *Polymer Engineering and Science*, *36*(8), 1134–1139. doi:10.1002/pen.10507

Larina, L., & Lopyrev, V. (2008). Structure and Physical–Chemical Properties of Nitroazoles. *Chemical Physics Letters*, *463*, 195–200.

Lau, K. T., Chipara, M., Ling, H. Y., & Hui, D. (2004). On the effective elastic moduli of carbon nanotubes for nanocomposites structures. *Composites. Part B, Engineering*, *35*(2), 95–101. doi:10.1016/j.compositesb.2003.08.008

Le Bras, M. (1997). *Synergy in intumescence—application to β-cyclodextrin carbonisation agent in intumescent additives for fire retardant polyethylene formulations* (Vol. 56). Polymer Degradation Stability.

Lee, D. H., Condrate, R. A. Sr, & Reed, J. S. (1996). J., "Infrared spectral investigation of polyacrylate adsorption on alumina. *Journal of Materials Science*, *31*(2), 471–478. doi:10.1007/BF01139166

Li, Z., & Wu, J. (2007). Potential distribution theorem for the polymer-induced depletion between colloidal particles. *The Journal of Chemical Physics*, *126*(14), 4904–4912. doi:10.1063/1.2715595 PMID:17444740

Liu, J., Boo, W. J., Clearfield, A., & Sue, H. (2006). Applications of Synchrotron Light to Scattering and Diffraction in Materials. *Journal of Materials Processing & Manufacturing Science*, *2*, 143–151. doi:10.1080/AMP-200068646

Mark, J.E. (2006). Some Novel Polymeric Nanocomposites. *Accounts of Chemical Research - ACS Publications, 39*, 881–888.

Meddad, A., & Fisa, B. (1997). Filler–matrix debonding in glass bead-filled polystyrene. *Journal of Materials Science*, *32*(5), 1177–1185. doi:10.1023/A:1018575716563

Milman, V., Refson, K., & Clark, S. J. (2010). Electron and vibrational spectroscopies using DFT, plane waves and pseudo potentials: CASTEP implementation. *Journal of Molecular Structure THEOCHEM*, *954*, 22–35. doi:10.1016/j.theochem.2009.12.040

Milman, V., Refson, K., Clark, S. J., Pickard, C. J., Yates, J. R., Gao, S.-P., Hasnip, P. J., Probert, M. I. J., Perlov, A., & Segall, M. D. (2010). Electron and vibrational spectroscopies using DFT, plane waves and pseudopotentials: CASTEP implementation. *Journal of Molecular Structure*, *954*(1-3), 22–35. doi:10.1016/j.theochem.2009.12.040

Murayama, H., & Min, K. (1997). Effects of Glass Fibers on Kinetic Model and Rheology of Glass Fiber Reinforced Phenolic Resin. In *Plastics saving planet earth: conference proceedings / ANTEC '97*. SPE, Society of Plastics Engineers.

Nago, S., & Mizutani, Y. (1996). Microporous polypropylene fibers containing poly(methylsilsesquioxane) filler. *Journal of Applied Polymer Science*, *61*(13), 2355–2359. doi:10.1002/(SICI)1097-4628(19960926)61:13<2355::AID-APP14>3.0.CO;2-3

Nawani, P., Zhou, H., Chu, B., Burger, C., & Hsiao, B. S. (2009). *Structural Analysis of Biological and Technical Nanocomposites by X-Ray Scattering* (Vol. 776). Lect. Notes Phys. Springer-Verlag Berlin Heidelberg, Scattering.

Nawanit, P., Zhou, H., Chu, B., Burger, C., & Hsiao, B. S. (2009). *Structural Analysis of Biological and Technical Nanocomposites by X-Ray Scattering* (Vol. 776). Lecture Notes in Physics. doi:10.1007/978-3-540-95968-7_8

Odegard, G. M., Gates, T. S., Wise, K. E., Park, C., & Siochi, E. J. (2003). Constitutive modeling of nanotube-reinforced polymer composites. *Composites Science and Technology*, *63*(11), 1671–1687. doi:10.1016/S0266-3538(03)00063-0

Odegard, G. M., Pipes, R. B., & Hubert, P. (2004). Comparison of two models of SWCN polymer composites. *Composites Science and Technology*, *64*(7-8), 1011–1020. doi:10.1016/j.compscitech.2003.08.010

Olszta, M. J., Wang, J., & Dickey, E. C. (2006). Stoichiometry and valence measurements of niobium oxides using electron energy-loss spectroscopy. *Journal of Microscopy*, *224*(3), 233–241. doi:10.1111/j.1365-2818.2006.01709.x PMID:17210055

Otaigbe, J. U., Quinn, C. J., & Beall, G. H. (1997). Processibility and Properties of Novel Glass-Polymer Melt Blends. In *Plastics saving planet earth: conference proceedings / ANTEC '97*. SPE, Society of Plastics Engineers.

Patel, N., & Egorov, S. A. (2004). Interactions between colloidal particles in polymer solutions: A density functional theory study. *The Journal of Chemical Physics*, *121*(10), 4987–4997. doi:10.1063/1.1778671 PMID:15332935

Patel, N., & Egorov, S. A. (2005a). Interactions between nanocolloidal particles in polymer solutions: Effect of attractive interactions. *The Journal of Chemical Physics*, *123*(14), 4916–4926. doi:10.1063/1.2049275 PMID:16238433

Patel, N., & Egorov, S. A. (2005b). Dispersing nanotubes with surfactants: A microscopi statistical mechanical analysis. *Journal of the American Chemical Society*, *127*(41), 14124–14125. doi:10.1021/ja0530570 PMID:16218573

Patra, C. N., & Yethiraj, A. (2003). Density functional theory for nonuniform polymers: Accurate treatment of the effect of attractive interactions. *The Journal of Chemical Physics*, *118*(10), 4702–4706. doi:10.1063/1.1543141

Perdew, J. P. (2010). Electron and vibrational spectroscopies using DFT, plane waves and pseudopotentials: CASTEP implementation. *Journal of Molecular Structure THEOCHEM*, *954*(1-3), 22–35. doi:10.1016/j.theochem.2009.12.040

Pipes, R. B., & Hubert, P. (2002). Helical carbon nanotube arrays: Mechanical properties. *Composites Science and Technology*, *62*(3), 419–428. doi:10.1016/S0266-3538(02)00002-7

Profeta, Mauri, & Pickard. (2005). Article. *Journal of American Chemical Society, 125*, 541–548.

Rebouillat, S., Escoubes, M., & Gauthier, R. (1996). Interactions of Kevlar® fibers with selected model compounds: Water sorption and dynamic mechanical properties of fiber/matrix samples. *Journal of Adhesion Science and Technology*, *10*(7), 635–650. doi:10.1163/156856196X00689

Rothon, R. N., & Hornsby, P. R. (1996). Flame retardant effects of magnesium hydroxide. *Polymer Degradation Stability, 54*(2-3), 383–385.

Roychoudhury, A., De, P. P., Roychoudhury, N., & Vidal, A. (1995). Chemical Interaction between Chlorosulfonated Polyethylene and Silica—Effect of Surface Modifications of Silica. *Rubber Chemistry and Technology*, *68*(5), 815–823. doi:10.5254/1.3538777

Rudd, R.E., & Broughton, J.Q. (2000). Concurrent coupling of length scales in solid state systems. *Physical Status Solids, 217*, 251–291.

Savadori, A., Scapin, M., & Walter, R. (1996). Particle filled polyolefins with high stiffness and toughness, as used for load bearing components. *Macromolecular Symposia*, *108*(1), 183–202. doi:10.1002/masy.19961080116

Schadler, L. S., Kumar, S. K., Benicewicz, B. C., Lewis, S. L., & Harton, S. E. (2007). Designed Interfaces in Polymer Nanocomposites: A Fundamental Viewpoint. *MRS Bulletin, 32*(4), 335–340. doi:10.1557/mrs2007.232

Schwab, J. J., Haddad, T. S., Lichtenhan, J. D., & Mather, P. T. (1997). Property Enhancements of Common Thermoplastics via Incorporation of Silicon Based Monomers: Polyhedral Oligomeric Silsesquioxane Macromers and Polymers. In *Plastics saving planet earth: conference proceedings / ANTEC '97*. SPE, Society of Plastics Engineers.

Selzer, R., & Friedrich, K. (1997). Mechanical properties and failure behaviour of carbon fibre-reinforced polymer composites under the influence of moisture. *Composites. Part A, Applied Science and Manufacturing, 28*(6), 595–604. doi:10.1016/S1359-835X(96)00154-6

Sen, S., Xie, Y., Kumar, S. K., Yang, H., Bansal, A., Ho, D. L., Hall, L., Hooper, J. B., & Schweizer, K. S. (2007). Chain conformation and bound-layer correlations in polymer nanocomposites. *Physical Review Letters, 98*(12), 8302–8306. doi:10.1103/PhysRevLett.98.128302

Shao, L., & Jonathan, R. (2007). Carbon-13 Chemical Shift Tensors of Disaccharides: Measurement, Computation and Assignment. *The Journal of Physical Chemistry A, 111*(50), 13126–13132. doi:10.1021/jp075921b PMID:17999473

Sides, S. W., Kim, B. J., Kramer, E. J., & Fredrickson, G. H. (2006). Hybrid particle-field simulations of polymer nanocomposites. *Physical Review Letters, 96*(25), 250601–250605. doi:10.1103/PhysRevLett.96.250601 PMID:16907292

Stejskal, J., Kratochvil, P., Armes, S. P., Lascelles, S. F., Riede, A., Helmstedt, M., Prokes, J., & Krivka, I. (1996). Polyaniline Dispersions. 6. Stabilization by Colloidal Silica Particles. *Macromolecules, 29*(21), 6814–6819. doi:10.1021/ma9603903

Striolo, A., & Egorov, S. A. (2007). Steric stabilization of spherical colloidal particles implicit and explicit solvent. *The Journal of Chemical Physics, 126*(1), 4902–4907. doi:10.1063/1.2409710 PMID:17212514

Suri, A., & Min, K. (1997). Electrical Conductivity of Chemically Modified Polypropylene Filled with Copper. In *Plastics saving planet earth: conference proceedings / ANTEC '97*. SPE, Society of Plastics Engineers.

Surve, M., Pryamitsyn, V., & Ganesan, V. (2006). Nanoparticles in solutions of adsorbing polymers: Pair interactions, percolation, and phase behavior. *Langmuir, 22*(3), 969–981. doi:10.1021/la052422y PMID:16430256

Surve, M., Pryamitsyn, V., & Ganesan, V. (2007). Dispersion and percolation transitions of nanorods in polymer solutions. *Macromolecules, 40*(2), 344–354. doi:10.1021/ma061603j

Tanaka, A., Kitamura, M., Tokumitsu, K., & Nishimura, H. (1997). Ultrasonic Evaluation of Butt Fusion Welded Polyethylene Pipes 1. Preliminary Results. In *Plastics saving planet earth: conference proceedings / ANTEC '97*. SPE, Society of Plastics Engineers.

Tang, H., Chen, X., & Luo, Y. (1996). Electrical and dynamic mechanical behavior of carbon black filled polymer composites. *European Polymer Journal, 32*(8), 963–966. doi:10.1016/0014-3057(96)00026-2

Thomason, J. L., & Vlug, M. A. (1997). Influence of Fiber Length and Concentration on the Properties of Glass Fibre-Reinforced Polypropylene: 4. Impact Properties. *Composites. Part A, Applied Science and Manufacturing*, *28A*(3), 277–288. doi:10.1016/S1359-835X(96)00127-3

Torro-Palau, A., Fernandez-Garcia, J. C., Orgiles-Barcelo, A. C., Pastor-Blas, M. M., & Martin-Martinez, J. M. (1997). Structural modification of sepiolite (natural magnesium silicate) by thermal treatment: Effect on the properties of polyurethane adhesives. *International Journal of Adhesion and Adhesives*, *17*(2), 111–119. doi:10.1016/S0143-7496(96)00039-5

Tsukuda, R., Sumimoto, S., & Ozawa, T. (1996). Interaction between fillers and matrix in ABS resin composites observed by thermophysical measurements. *Journal of Applied Polymer Science*, *59*(6), 1043–1046. doi:10.1002/(SICI)1097-4628(19960207)59:6<1043::AID-APP19>3.0.CO;2-4

Uldry, A. C., Griffin, J. M., Yates, J. R., Perez-Torralba, M., Maria, M. D. S., Webber, A. L., Beaumont, M. L. L., Samoson, A., Claramunt, R. M., Pickard, C. J., & Brown, S. P. (2008). Quantifying Weak Hydrogen Bonding in Uracil and 4-Cyano-4'-ethynylbiphenyl: A Combined Computational and Experimental Investigation of NMR Chemical Shifts in the Solid State. *Journal of the American Chemical Society*, *130*(3), 945–954. doi:10.1021/ja075892i PMID:18166050

Ventresca, D. A., & Berard, M. T. (1997). Properties and Structure of Solid State Formed PET, PE and PP. In *Plastics saving planet earth: conference proceedings / ANTEC '97*. SPE, Society of Plastics Engineers.

Visser, S. A. (1997). Effect of filler type on the response of polysiloxane elastomers to cyclic stress at elevated temperatures. *Journal of Applied Polymer Science*, *63*(13), 1805–1820. doi:10.1002/(SICI)1097-4628(19970328)63:13<1805::AID-APP13>3.0.CO;2-X

Wagner, A. H., Kalyon, D. M., Yazici, R., & Fiske, T. J. (1997). Extensional Flow of Engineering Plastics with Glass Fibers. In *Plastics saving planet earth: conference proceedings / ANTEC '97*. SPE, Society of Plastics Engineers.

Wang, J. Y., & Ploehn, H. J. (1996). Dynamic mechanical analysis of the effect of water on glass bead–epoxy composites. *Journal of Applied Polymer Science*, *59*(2), 345–357. doi:10.1002/(SICI)1097-4628(19960110)59:2<345::AID-APP19>3.0.CO;2-V

Wang, X., Foltz, V. J., Rackaitis, M., & Bohm, G. A. G. (2008). Dispersing hairy nanoparticles I polymer melts. *Polymer*, *49*(26), 5683–5691. doi:10.1016/j.polymer.2008.10.019

Wang, Y., Lu, J., & Wang, G. (1997). Toughening and reinforcement of HDPE/CaCO3 blends by interfacial modification—Interfacial interaction. *Journal of Applied Polymer Science*, *64*(7), 1275–1281. doi:10.1002/(SICI)1097-4628(19970516)64:7<1275::AID-APP5>3.0.CO;2-G

Watson, B. A., & Barteau, M. A. (1994). Atomic Force Microscopy Imaging of TiO2 Surfaces Active for C-C Bond Formation Reactions in Ultrahigh Vacuum. *Chemistry of Materials*, *6*(6), 771–779. doi:10.1021/cm00042a012

Wiebking, H. E. (1996). The performance of ultrafine talc in rigid PVC. *Journal of Vinyl and Additive Technology*, *2*(3), 187–189. doi:10.1002/vnl.10121

Winey, K. I., & Vaia, R. A. (2007). Polymer Nanocomposites. *MRS Bulletin*, *32*(4), 314–319. doi:10.1557/mrs2007.229

Wu, J. (2006). Density functional theory for chemical engineering: From capillarity to soft materials. *AIChE Journal. American Institute of Chemical Engineers*, *52*(3), 1169–1193. doi:10.1002/aic.10713

Xu, J., Qiu, F., Zhang, H., & Yang, Y. (2006). Morphology and interactions of polymer brush-coate spheres in a polymer matrix. *Journal of Polymer Science. Part B, Polymer Physics*, *44*(19), 2811–2820. doi:10.1002/polb.20884

Zeng, Q. H., Yua, A. B., & Lu, G. Q. (2008). Multiscale modeling and simulation of polymer nanocomposites. *Progress in Polymer Science*, *33*(2), 191–269. doi:10.1016/j.progpolymsci.2007.09.002

Zhang, C.-Z., Liu, W.-M., Xue, Q.-J., & Shen, W.-C. (1997). Friction and wear characteristics of PTFE composites filled with metal oxides under lubrication by oil. *Journal of Applied Polymer Science*, *66*(1), 85–93. doi:10.1002/(SICI)1097-4628(19971003)66:1<85::AID-APP10>3.0.CO;2-6

Zhao, L., Li, Y. G., & Zhong, C. (2007). Integral equation theory study on the phase separation in star polymer nanocomposite melts. *The Journal of Chemical Physics*, *127*(15), 4909–4916. doi:10.1063/1.2795717 PMID:17949216

Zhao, W., Hasegawa, S., Fujita, J., Yoshii, F., Sasaki, T., Makuuchi, K., Sun, J., & Nishimoto, S. (1996). Effect of irradiation on pyrolysis of polypropylene in the presence of zeolites. *Polymer Degradation & Stability*, *53*(2), 199–206. doi:10.1016/0141-3910(96)00084-5

Zhou, H., Burger, C., Sics, I., Hsiao, B. S., Chu, B., Graham, L., & Glimcher, M. J. (2007). Small-angle X-ray study of the three-dimensional collagen/mineral superstructure in intramuscular fish bone. *Journal of Applied Crystallography*, *40*(s1), 666–668. doi:10.1107/S0021889806054409

Zhou, J., Li, G., Li, B., & He, T. (1997). Flexural fatigue behavior of injection-molded composites based on poly (phenylene ether ketone). *Journal of Applied Polymer Science*, *65*(10), 1857–1864. doi:10.1002/(SICI)1097-4628(19970906)65:10<1857::AID-APP2>3.0.CO;2-E

Zhu, J., Geris, A. J., & Wu, G. (2009). Solid-state (17) O NMR as a sensitive probe of keto and gem-diol forms of alpha-keto acid derivatives. *Physical Chemistry Chemical Physics*, *11*(32), 6972–6980. doi:10.1039/b906438a PMID:19652831

Section 2
Investment of High Voltage NanoDielectrics

In this section, degradation of nanodielectrics has been depicted to know the difference between traditional and innovative nanodielectric materials for high voltage engineering.

Chapter 7
Realistic NanoDielectrics Characterization

ABSTRACT

The utilization of polymers as electrical insulating materials has been developing quickly in recent decades. The build polymer properties have been created, eventually perusing the inclusion of a few diverse fillers if they are exorbitant of the polymer material. This chapter contains the realistic characterization of nanodielectrics that handled the polyethylene nanodielectrics characterization. The chapter contains also the polypropylene nanodielectrics, polyvinyl chloride nanodielectrics. Finally, this chapter focuses on new multi-nanocomposites insulation materials.

Recently, extraordinary desires have centered on costive nanoparticles. However, there are some concerns regarding the impact of the sorts costive nanoparticles in terms of electrical properties for polymeric nanocomposite. For a consistent advancement clinched alongside polymer nanocomposites, this examination depicts the impacts of sorts and fixation of controlling costive nanoparticles clinched alongside electrical properties about streamlined polymer material (Gouda & Thabet, 2014a; Thabet, 2011a; Thabet, 2017a; Thabet, Abdel-Moamen, & Abdelhady, 2016; Thabet, Al-Sharif, Abdel-Moamen et al, 2019; Thabet, Mobarak, & Kannan, 2019; Thabet et al., 2010; Thabet & Salem, 2017a; Thabet & Salem, 2017b; Thabet, Samir, & Mountasser, 2018; Thabet, Shaaban, & Allam, 2016). This investigation has been tentatively centered on the outcomes that identify the impacts of nanoparticles (clay and fumed silica) looking into electric and dielectric properties of polyethylene nanocomposites, low density polyethylene (LDPE) and high-density polyethylene (HDPE).

Polymer properties are experimentally tailored towards including few numbers of different nanoparticles for upgrading their mechanical, thermal and electrical properties. This chapter investigates upgrading the electric and dielectric properties of low-density polyethylene (LDPE), and high-density polyethylene (HDPE) polymer materials with cheap nanoparticles. Certain rates of clay and fumed silica nanoparticles are used to upgrade electric and dielectric properties of polyethylene nanocomposites films towards utilizing the dielectric Spectroscopy; the electric and dielectric properties of each polyethylene nanocomposites have been measured with and without nanoparticles in different frequencies dependent upon

DOI: 10.4018/978-1-7998-3829-6.ch007

1 kHz under different thermal states (20 °C and 60 °C). Moreover, we have attempted to specify the ideal nanoparticles sorts and their focuses in the control from controlling electric and dielectric characterization. Nanoparticles have proven significant advantage for upgrading electrical properties in controlling polymer industrial materials as structure nanocomposites, therefore, preparation of new Polyethylene nanocomposites support both the producers and clients in upgrading electrical execution of Polyethylene provisions. This chapter has improved electric and dielectric characterizations of controlling polyethylene for including costive nanoparticles to low density polyethylene (LDPE) and high-density polyethylene (HDPE) concerning illustration of a base matrix. Polyethylene trapping properties are exceptionally changed by the vicinity of costive nanoparticles (clay, and fumed silica) nanoparticles. In addition, there is focus tentatively on the impact of costive nanoparticles (clay, and fumed silica) and their concentrations looking into electric and dielectric properties for polyethylene materials. An experimental study has examined Polyethylene nanocomposites with respect to business polyethylene materials to illustrate the impact of sorts and focuses of nanoparticles on upgrading electric and dielectric Polyethylene qualities. Polymer nanocomposites guarantee high exhibitions as building materials, assuming that they are prepared and created legitimately. In this research, specimens from controlling nanocomposite polymers have been transformed, such as electrical insulating materials provided to the electric force cables towards utilizing the most recent systems of nanotechnology (Thabet & Mobarak, 2012; Thabet, 2011b; Thabet, 2011c; Thabet, 2012a; Thabet, 2012b; Thabet et al., 2012; Thabet & Hassan, 2011; Thabet et al., 2011). This single section has investigated improved dielectric and electrical properties of Polyvinyl chloride PVC whose matrix has demonstrated that trapping properties are profoundly altered towards the vicinity of costive nanoparticles clay and fumed silica. An experimental work for dielectric loss and capacitance of the new nanocomposite materials has been investigated and compared with unfilled modern materials. It can be discovered that a great connection exists between deference of capacitance and dielectric reduction qualities measured for rate of nanoparticles. Thus, the impact of costive nanoparticles material and its fixation ahead of dielectric properties of controlling modern polymers-based composite frameworks has been investigated. A similar examination will be performed between the unfilled build polymers, the frameworks holding person kind of nanoparticles clay or fumed silica inside the host polymer for different focuses. Nanostructured materials attract hobbies and applications, thus; physical and electrical properties of Polymethylmethacrylate (PMMA) nanocomposite materials under different thermal states are right now constantly focused on. The dielectric conduct for new nanocomposite materials for Polymethylmethacrylate loaded for nano-clay alternately nano-fumed silica has been measured in different recurrences (0.1 kHz - 1 kHz) and temperatures (20°C-60°C). Dielectric spectroscopy has been used to describe ionic conduction and to state the impact of filler centralization on the dielectric permittivity and dielectric losses, therefore; the relative permittivity and the reduction loss are measured towards dielectric spectroscopy for Polymethylmethacrylate with and without nanoparticles. Finally, a comparison is conducted between dielectric properties of new Polymethylmethacrylate nanocomposites which are prepared to include nanoparticles of clay or fumed silica for different focuses under different temperatures. In this study, the upgrade of dielectric characterization has been investigated in Polypropylene (PP) as matrix that has exceedingly changed, eventually perusing vicinity for costive nanoparticles clay and fumed silica. In addition, the dielectric quality of nanocomposites has been moved forward altogether regarding unfilled materials under secondary voltage exchanging current (HVAC) electric fields. Filling nanoparticles under polymers give acceptable preferences over unfilled polymers in view that they expand imperviousness with corruption, as stated by their sorts and focuses. Therefore, a test worth of effort to dielectric passing and capacitance of the new polypropylene nanocomposite materials has been

investigated and compared with unfilled polypropylene modern materials (Gouda et al., 2014; Gouda & Thabet, 2014b; Gouda & Thabet, 2014c; Thabet, 2013a; Thabet, 2013b; Thabet & Repetto, 2013; Thabet & Repetto, 2014; Thabet, 2012c; Thabet, 2014; Thabet & Mubarak, 2012). A streamlined HVAC breakdown model test has been utilized for tests and measurements, whenever the rate for dielectric quality of tried nanocomposite materials precedes the customary polypropylene encasing materials under AC electric fields. Therefore, the dielectric quality is measured for a few new polypropylene nanocomposite examples. The test estimations portray that the consolidation from controlling clay or fumed silica nanoparticles under Polypropylene (PP) has regulated the dielectric quality and voltage persistence, fundamentally contrasted with routine materials concerning their sorts and focuses. Finally, a thermal impact once suggested about nanocomposites under uniform and non-uniform electric fields has been investigated that energizes modern creation preparation for the suggested cost-fewer nanoparticles. This Section transforms and portrays cost-fewer polypropylene (PP) nanocomposite films; a test fill that has been investigated for considering the electric properties of the new nanocomposite materials and comparing unfilled streamlined materials in a recurrence go dependent upon 1 kHz. A little expansion from controlling nanoparticles (clay, and fumed silica) on polypropylene demonstrated calculable change in the electric reactance and conductance during different recurrences up to 1kHz. In addition, an electric spectroscopy has measured the electric properties of polypropylene for and without nanoparticles under variant temperatures (20°C, and 60°C). Cambridge Engineering Selector (CES) program has conveyed the electrical/mechanical predictable models for the recommended materials. Finally, this part prompts integrate electrical insulating polypropylene nanocomposite films in which the electrical properties are appropriately upheld, so as to attain a greater amount of cost-effective, energy-effective and subsequently naturally exceptional materials in electrical encasing engineering.

7.1 POLYETHYLENE NANO-DIELECTRIC CHARACTERIZATION

Nano-clay is a circular particle in shape, and it may be largely regarded as significant mark in controlling nanoclay for polymer provisions. Costless of clay impetus, it can be the best filler of nanoparticles industrial materials. Nano-fumed silica is generally utilized with a rheology modifier, conferring exceptionally thixotropic properties towards generally low rates. Fumed silica powders are utilized in paints and coatings, silicone elastic and silicone sealants, adhesives, link mixes and gels, printing inks and toner, and plant security. On the other hand, low-density polyethylene (LDPE) is a thermoplastic aggravated from petroleum, and it may be characterized by a thickness range of (0.910 - 0.940) g/cm^3. LDPE holds the concoction components of carbon and hydrogen. High-Density Polyethylene (HDPE) is a polyethylene thermoplastic aggravated from petroleum. HDPE has minimal branching, providing for it stronger intermolecular drives and rigidity over lower-density polyethylene.

Polyethylene is a thermoplastic constructed from petroleum, inert during space temperatures and with everything; except solid oxidizing agents and a percentage of solvents creating swelling. It can withstand temperatures of 80°C and 95°C for a short to a long haul. This polymer is a business material that is utilized in the manufacturing of high-voltage mechanical items. Polyethylene nanocomposites films are made, eventually perusing the utilization of dissolving polyethylene (LDPE, and HDPE), then, blending and infiltrating nanoparticles inside the base matrix polyethylene towards current ultra-nationalistic gadgets. Clay and fumed silica nanoparticles are cheap impetuses that transform the properties of industrial materials with respect to the physical fabricate procedure. SEM pictures for polyethylene nanocomposites films delineate the infiltration from the controlling nanoparticles inside low-density polyethylene and high-density polyethylene, as demonstrated in Fig. 1. Table 1 depicts the measured electric and dielectric properties for polyethylene nanocomposites materials

Figure 1. SEM images for polyethylene nanocomposite films

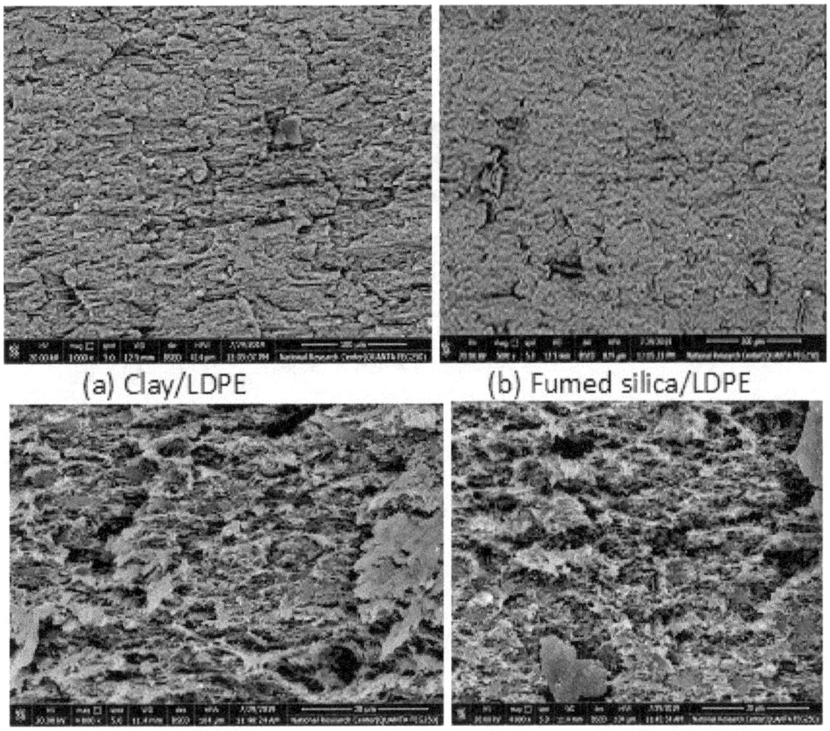

(a) Clay/LDPE (b) Fumed silica/LDPE (c) Clay/HDPE (d) Fumed silica/HDPE

Finally, it can measure the greater part of dielectric properties to immaculate nanocomposite streamlined materials, eventually perusing the utilization of HIOKI 3522-50 LCR Hi-tester gadget and allowing it to be distinguished. The mulled over modern materials in this investigation have been figured using nano particulates. Electric and dielectric properties of the examined materials are nitty gritty in Table (1).

Measurement Setup: HIOKI 3522-50 LCR Hi-tester gadget has measured electrical parameters of nano-metric robust dielectric encasing examples during different frequencies. Determination of LCR may be force supply: 100, 120, 220 alternately 240 v (±10%) AC (selectable), 50/60 Hz, and Frequency: DC, 1 mHz should 100 kHz, show Screen: LCD with backlight / 99999 (full 5 digits), essential Accuracy: Z: ± 0.08% rdg. θ: ± 0.05þ, and outside dc inclination ± 40 v max. (option) (3522-50 utilized alone ± 10 v max. / utilizing 9268 ± 40 Vmax.). It can measure the greater part of dielectric properties to immaculate nanocomposite industrial materials, eventually perusing the utilization of HIOKI 3522-50 LCR Hi-tester gadget. Figure (2) indicates HIOKI 3522-50 LCR Hi-tester gadget for measuring characterization of nanocomposite encasing industrial materials.

Figure 2. HIOKI 3522-50 LCR Hi-tester device

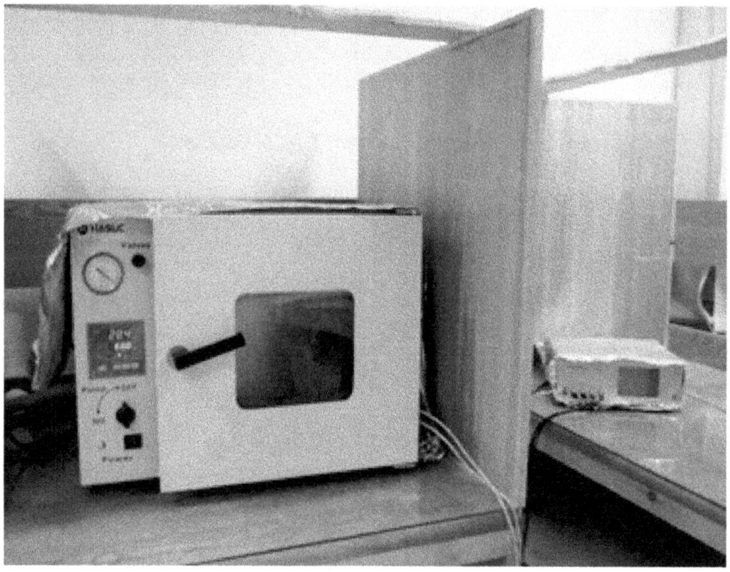

Table 1. Electric and dielectric properties of pure and nanocomposite materials

Characteristics	Dielectric constant		Resistivity (Ω.m)	
Materials	LDPE	HDPE	LDPE	HDPE
Pure	2.3	2.3	10^{14}	10^{15}
1 wt.% Clay	2.23	2.23	10^{15}	10^{16}
5 wt.% Clay	1.99	1.99	10^{15}-10^{18}	10^{16}-10^{19}
10 wt.% Clay	1.76	1.76	10^{18}-10^{20}	10^{19}-10^{21}
1 wt.% SiO_2	2.32	2.32	10^{13}	10^{14}
5 wt.% SiO_2	2.39	2.39	10^{13}-10^{11}	10^{14}-10^{12}
10 wt.% SiO_2	2.49	2.49	10^{11}-10^{9}	10^{12}-10^{10}

Dielectric spectroscopy is a capable test strategy to explore the dynamic electric and dielectric conduct of the polymeric test through recurrence reaction dissection. This system is in the view of estimation of resistance, conductance and susceptance as a work for recurrence to an example sandwiched between pin-plate electrodes. Thus, the conductance and susceptance are measured for the sum tests concerning illustration of the capacity of recurrence up to 1 kHz under variant temperatures of (20oC Also 60oC). Moreover, the estimations are produced by utilizing high determination dielectric spectroscopy, and this strategy is dependent upon the estimation of the capacitance of a work from controlling the recurrence of a test sandwiched between two electrodes. Dielectric reduction (tan δ) and capacitance (C) are measured concerning illustration of a capacity of controlling recurrence up to 100 kHz at 25 °C for the greater part test specimens.

Figure 3. Measured conductance of clay/LDPE nanocomposite films

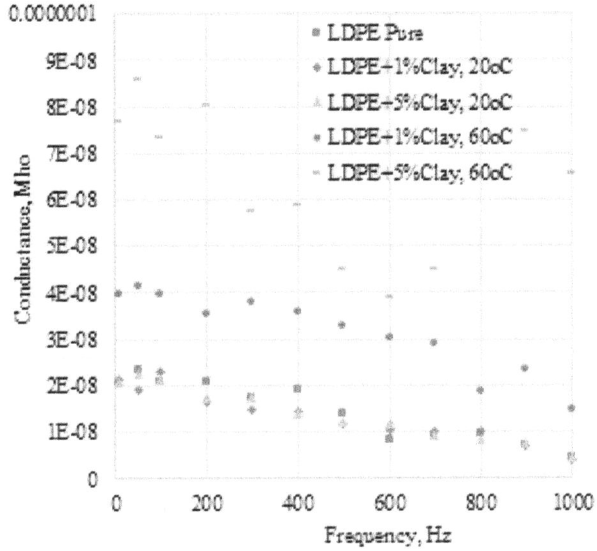

Figure 4. Measured conductance of SiO_2/LDPE nanocomposite films

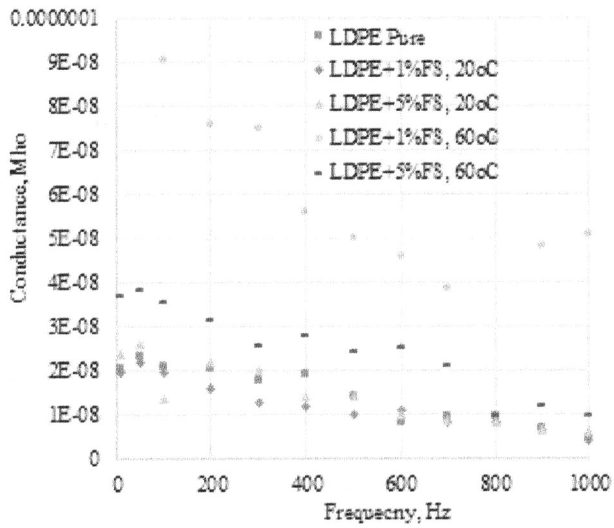

7.1.1 Effect of Nanoparticles Inside Low Density Polyethylene

Figure 3 depicts the conductance of clay/LDPE nanocomposites films that declines with expanding focus of clay nanoparticles in the nanocomposites up to 5 wt. % at room temperature (20°C). However, during high temperature (60°C), the conductance execution of clay/LDPE nanocomposites films is turned around inside the same fixation go of nanoparticles. Therefore, expanding the temperature of nanocomposite materials transforms temperature degrees of nanoparticles that evolve the electric conductance conduct technique against typical states. On the other hand, fig. 4 indicates the conductance of SiO_2/LDPE nanocomposites films concerning illustration of a capacity of controlling recurrence. It is to be noted that the measured conductance abatements with expanding fixation about fumed silica nanoparticles are up to 1 wt. %. It expands because of the expanding centralization of fumed silica nanoparticles up to 5 wt. % without arriving at values from the controlling low-density polyethylene.

Figure 5. Measured susceptance of clay/LDPE nanocomposite films

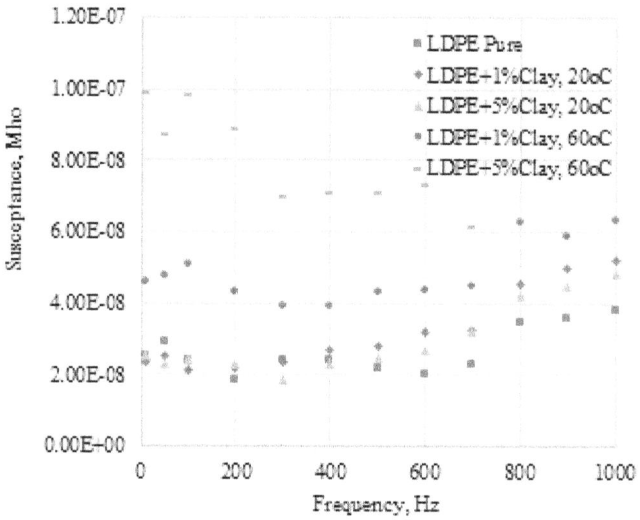

Under high temperature (60°C), the measured conductance for fumed silica/LDPE nanocomposites films builds for expanding fixation from controlling fumed silica nanoparticles in the nanocomposites up to 1 wt. %. Then, it declines with the expanding rate of fumed silica nanoparticles in the nanocomposites up to 5 wt. %. Therefore, there is reasonableness for conductance property conduct to utilize fumed silica nanoparticles in low-density polyethylene that has opposite conductance property conduct under high temperature (60°C).

Figures (5, 6), indicate the results of the estimations for susceptance, as a work for recurrence to clay/ LDPE, and SiO_2/LDPE nanocomposite films tests under changing thermal temperatures. It is noted that fig. 5 demonstrates that the susceptance from controlling clay/LDPE nanocomposites films increments with the expanding centralization from the controlling clay nanoparticles in the nanocomposites up to 5 wt. % under changing thermal states (low and high). However, fig. 6 indicates the measured susceptance of SiO_2/LDPE nanocomposites films that show the same execution about conductance with expanding fumed silica nanoparticles to low-density polyethylene under fluctuating thermal states (low and high). Therefore, the rising temperature of nanocomposites materials raises the temperature of nanoparticles that advance the electric conduct technique against the ordinary states. Thus, vicinity from controlling clay nanoparticles in low-density polyethylene causes unsteadiness about susceptance property conduct technique in the event of high temperatures with respect to room temperature.

Figure 6. Measured susceptance of SiO2/LDPE nanocomposite films

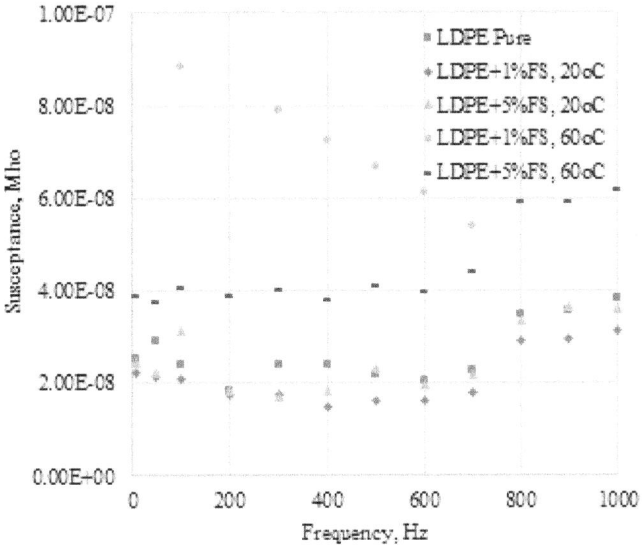

Figure 7. Measured loss tangent of clay/LDPE nanocomposites at room temperature (20°C)

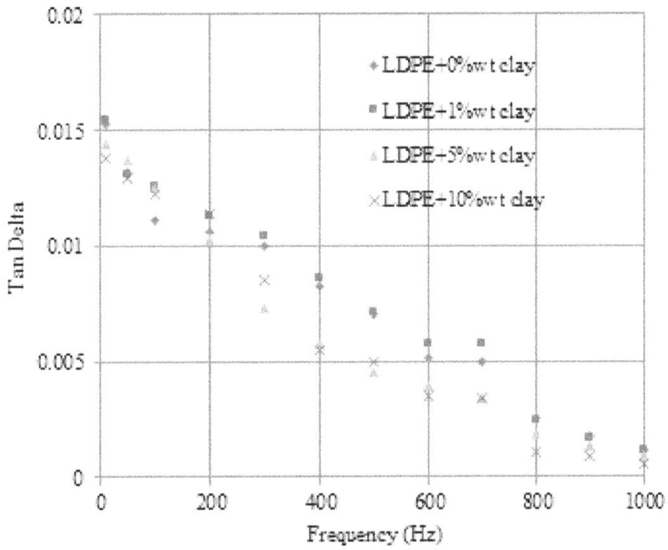

In case of studying the performance of dielectric losses in nanocomposites, Fig. 7 demonstrates reduction loss (Tan Delta) of the tried tests, as a capacity from controlling recurrence to Clay/ low-density polyethylene nanocomposites in room temperature (20°C). The measured loss tangent contrasts looking into expanding the misfortune loss for expanding the rate of clay nanoparticles is dependent upon 1%wt, especially, during low frequencies and it diminishes with expanding clay nanoparticles rate up to 10%wt, specially, at high frequencies. In any case, fig. 8 contrasts the measured loss tangent that establishes for the expanding rate of fumed silica nanoparticles in the nanocomposite dependent upon

10%wt, particularly during high frequencies. On the other hand, fig. 9 contrasts the capacitance of Clay/ LDPE nanocomposites specimens versus recurrence towards room temperature (20°C). The measured capacitance declines for the expanding rate of clay nanoparticles in the nanocomposite dependent upon 5%wt, yet the measured capacitance for low-density polyethylene nanocomposites increments for expanding the clay rate of nanoparticles up to the rate of 10%wt. In any case, fig.8 demonstrates that the

Figure 8. Measured loss tangent of Fumed Silica/LDPE nanocomposites at room temperature (20°C)

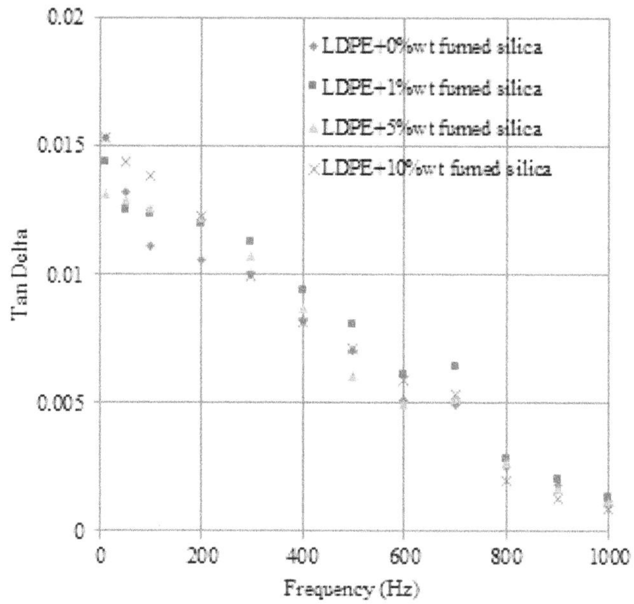

Figure 9. Measured Capacitance of Clay/LDPE nanocomposites at room temperature (20°C)

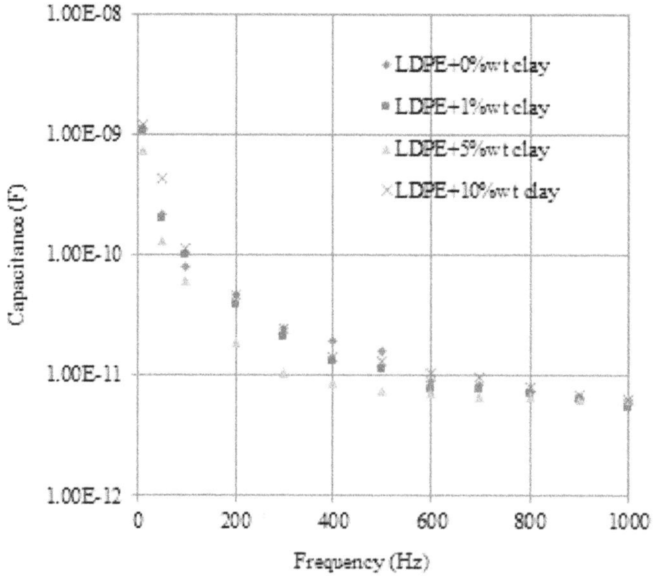

measured capacitance is shut for expanding the rate of fumed silica nanoparticles in the nanocomposite, particularly at high frequencies. It is clear that nanoparticles have transformed the electric and dielectric polymer properties. The dielectric properties of insulating polymer nanocomposites have been investigated in the recurrence space from 0.1Hz to 1kHz. Additionally, it has been found that the dielectric properties have an end relationship with the interfacial conduct technique between the fillers and the polymer matrix on such composites.

Figure 10.Measured Capacitance of Fumed Silica/LDPE nanocomposites at room temperature (20°C)

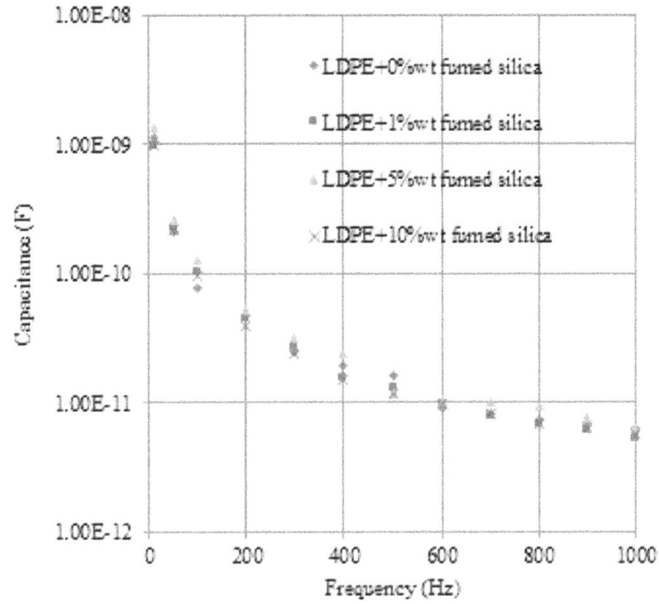

Figure 11. Measured loss tangent of Clay/LDPE nanocomposites at certain temperature (40oC)

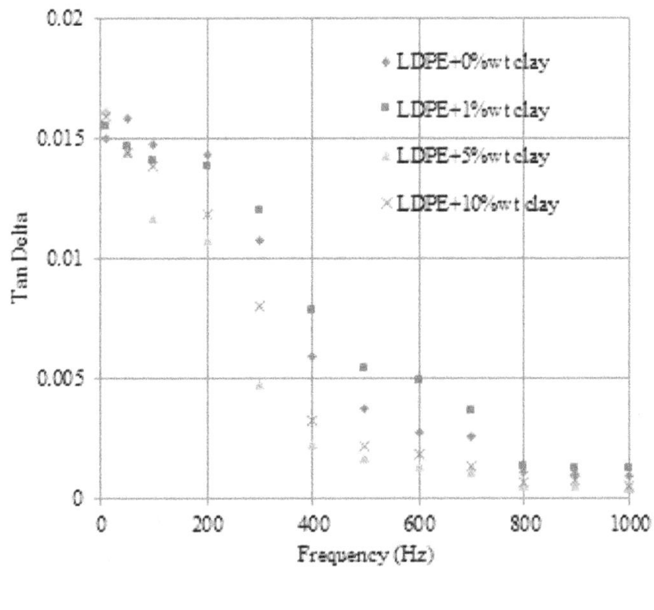

Figure 12. Measured loss tangent of Fumed Silica/LDPE nanocomposites at certain temperature (40°C)

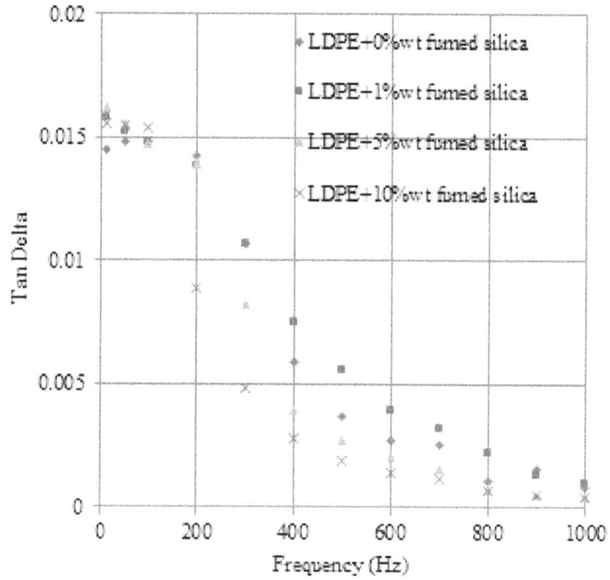

Figure 13. Measured Capacitance of Clay/LDPE nanocomposites at certain temperature (40°C)

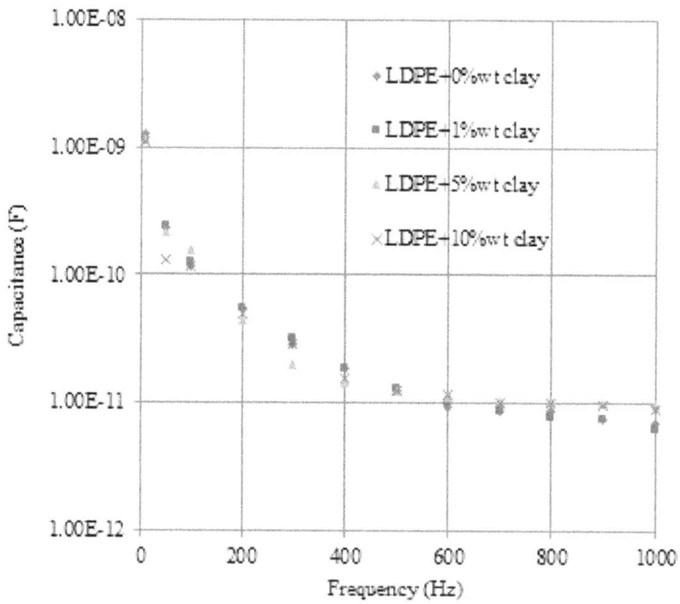

There are impacts from controlling rising temperatures on nanoparticles inside the proposed nanocomposites. It is clear that the rising temperature for nanocomposite materials transforms nanoparticles temperatures that evolve dielectric conduct technique through the typical states. Fig. 11 provides for the loss tangent concerning illustration of a capacity of recurrence for Clay/ LDPE nanocomposites tests under trial at temperature (40°C). It is recognized that the loss tangent of Clay/ LDPE nanocomposites expands for the expanding rate of clay nanoparticles in the nanocomposite dependent upon 1%wt., following that the misfortune loss of Clay/ LDPE nanocomposites declines with the expanding rate of clay nanoparticles in the nanocomposite dependent upon 10%wt.

Figure 14. Measured Capacitance of Fumed Silica/LDPE nanocomposites at certain temperature (40°C)

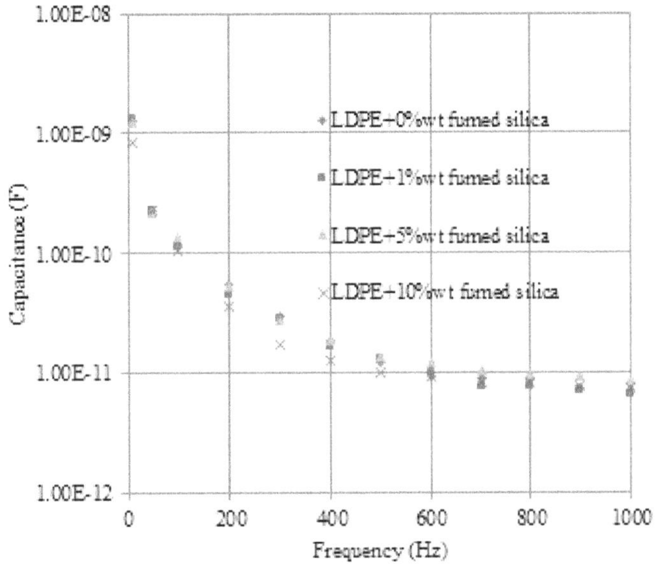

Figure 15. Measured loss tangent of Clay/LDPE nanocomposites at certain temperature (60°C)

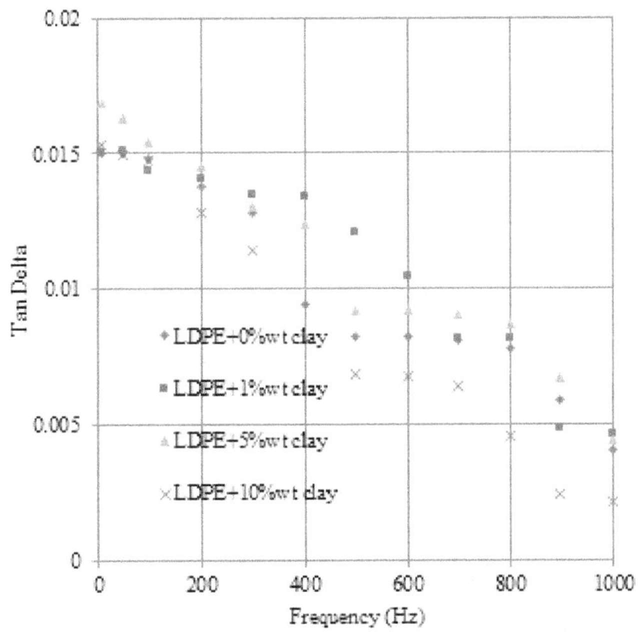

Figure 16. Measured loss tangent of Fumed Silica/LDPE nanocomposites at certain temperature (60°C)

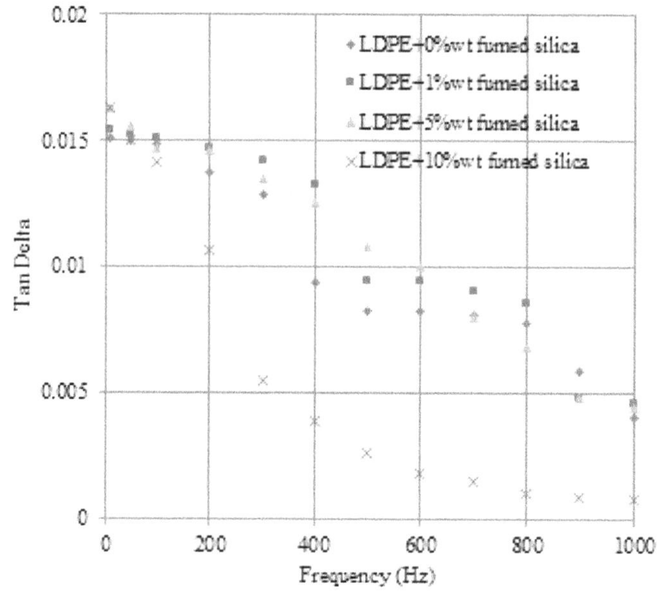

Similarly, Fig. 12 depicts that the same conduct of clay nanoparticles inside low-density polyethylene is clear in the expanding rate of fumed silica nanoparticles in the low-density polyethylene nanocomposite dependent upon 10%wt. On the other hand, Fig. 13 indicates the tried specimen's capacitance, such as capacity for recurrence of Clay/LDPE nanocomposites during temperature (40°C); the measured capacitance builds for expanding clay nanoparticles rate up to 1%wt. However, it abates for expanding clay nanoparticles rate up to 10%wt. It is to be noted that fig. 14 indicates the measured capacitance of fumed silica/LDPE nanocomposites resembling the conduct from controlling Clay/LDPE nanocomposites.

In view of the impacts of high-temperature qualities on nanoparticles inside the nanocomposites, fig. 15 indicates the connection between loss tangent and the connected voltage recurrence for Clay/LDPE nanocomposites towards temperature (60°C). The measured loss tangent for low-density polyethylene nanocomposite builds for expanding clay rate nanoparticles dependent upon 5%wt., especially, during low frequencies. Following that, the misfortune loss from controlling Clay/LDPE nanocomposite abatements for expanding clay rate nanoparticles dependent upon 10% wt, especially towards high frequencies is demonstrated.

Additionally, Fig. 16 demonstrates that the measured reduction loss of fumed silica/LDPE expands with expanding fumed silica nanoparticles rate up to 5%wt, especially during low frequencies. However, it declines with expanding fumed silica nanoparticles rate dependent upon 10%wt. Fig. 17 depicts capacitance versus the connected voltage recurrence to Clay/ LDPE nanocomposites at temperature (60°C); the measured capacitance abatements for expanding clay nanoparticles rate up to 1%wt, then, the measured capacitance expands for expanding clay rate nanoparticles dependent upon 10%wt. Although Fig. 18 demonstrates that the measured capacitance expands for expanding fumed silica rate nanoparticles up to 5%wt, the capacitance for fumed silica/LDPE nanocomposites abates for expanding fumed silica nanoparticles rate (5%wt -10%wt). Hence, it is self-evident that the rising temperature of nanocomposite materials raises nanoparticles temperatures that evolve dielectric conduct with respect to typical states.

Figure 17. Measured Capacitance of Clay/LDPE nanocomposites at certain temperature (60°C)

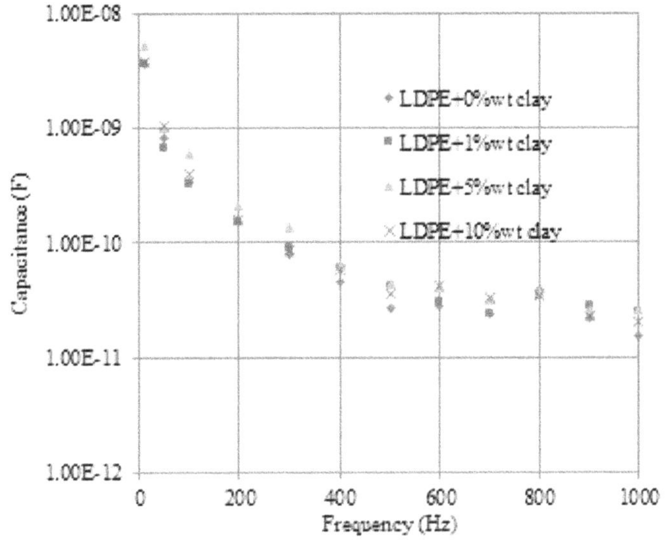

Figure 18. Measured Capacitance of Fumed Silica/LDPE nanocomposites at certain temperature (60°C)

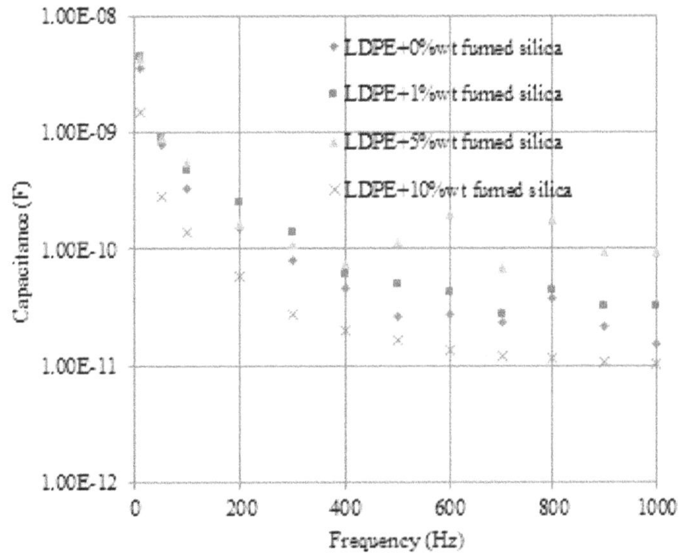

In case of high thickness polyethylene, fig. 19 indicates the measured conductance of the tried specimens of clay/HDPE nanocomposites films as a capacity from controlling recurrence towards temperatures from controlling (20°C and 60°C). It is self-evident that the measured qualities of conductance are focalized and expand for the expansion of the focus of clay nanoparticles up to 5 wt. %. However, there is no merging between measured qualities from controlling conductance of high-density polyethylene nanocomposites in high temperature (60°C). On the other hand, fig. 20 indicates the combination of measured values of conductance for SiO2/HDPE nanocomposite films for expanding centralization from controlling fumed silica nanoparticles up to 5 wt. % at room temperature (20°C). Thus, the measured conductance builds with expanding fixation from controlling fumed silica nanoparticles in the nanocomposites dependent upon 5wt. %, bit by bit under high thermal states.

Figure 19. Measured conductance of clay/HDPE nanocomposite films

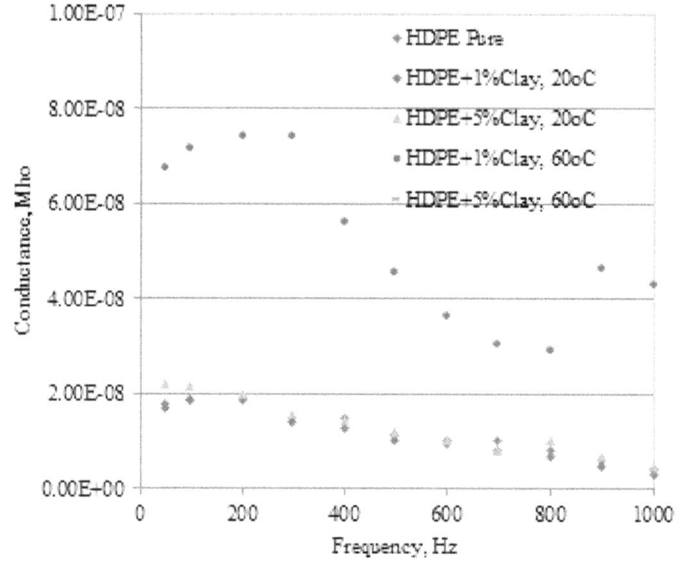

Figure 20.Measured conductance of SiO2/HDPE nanocomposite films

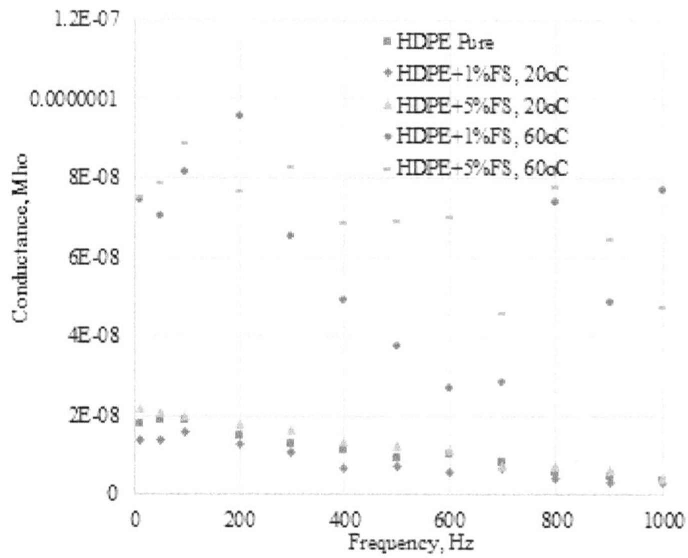

Figure 21. Measured susceptance of clay/HDPE nanocomposite films

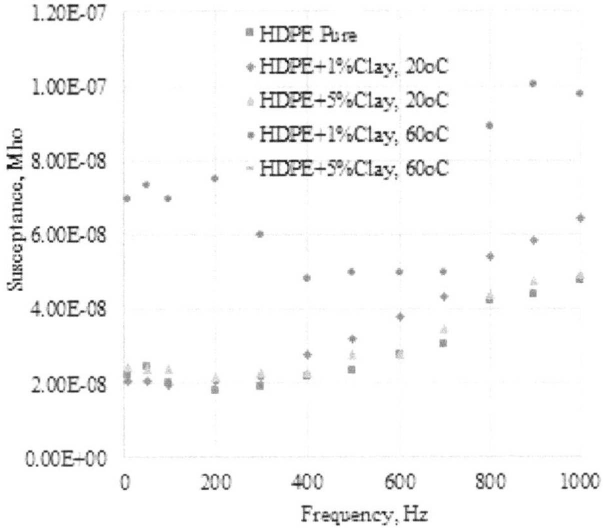

Figure 22. Measured susceptance of SiO2/HDPE nanocomposite films

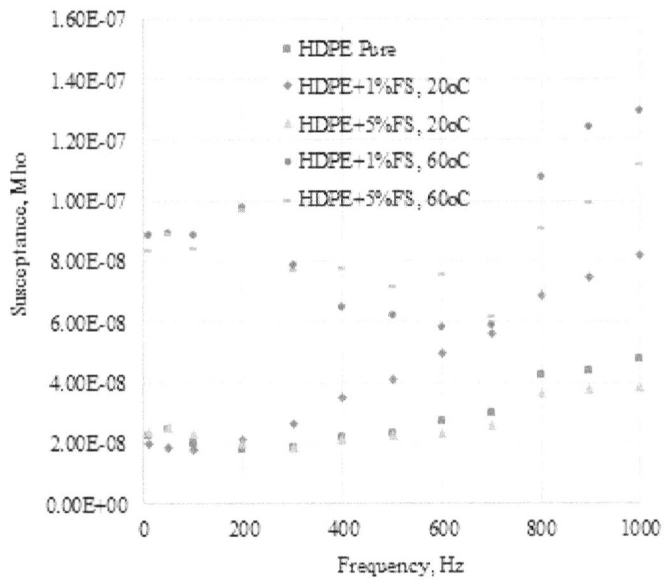

Figures (22, 23) provide for the outcomes of the estimations about the susceptance which are similar to those of a work about recurrence of clay/HDPE and SiO_2/HDPE nanocomposite films specimens at temperatures of (20°C and 60°C). It is clear that Fig. 22 concentrates on expanding susceptance with expanding focus from controlling clay nanoparticles in nanocomposites dependent upon 1wt.%, then, the measured susceptance abates for expanding centralization of clay nanoparticles dependent upon 5wt. %. On the other hand, the susceptance of clay/HDPE nanocomposites films builds with expanding centralization of clay nanoparticles up to 5wt. % during high temperature (60°C). Fig. 23, in particular, indicates the measured susceptance of SiO_2/HDPE nanocomposites films tests versus recurrence in temperatures about (20°C and 60°C), the susceptance of SiO_2/HDPE nanocomposites films builds with expanding centralization from controlling fumed silica nanoparticles in high-density polyethylene nanocomposites dependent upon 1 wt. % during room temperature (20°C). Accordingly, it abates for the expanding focus of fumed silica nanoparticles dependent upon 5wt. % at high temperature (60°C). It is clear that the dielectric properties of insulating polymer nanocomposites films have been investigated in the recurrence area from 0.1 Hz to 1 kHz, and there may be a merging between the measured values of electric and dielectric polymer properties at room temperature (20°C).

Figure 23. Measured loss tangent for clay/ HDPE nanocomposites at T=20 oC

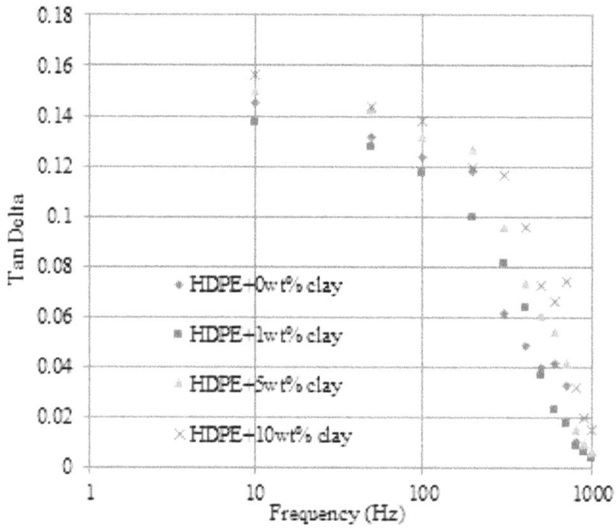

Figure 24. Measured loss tangent for SiO2/HDPE nanocomposites at T=20 oC

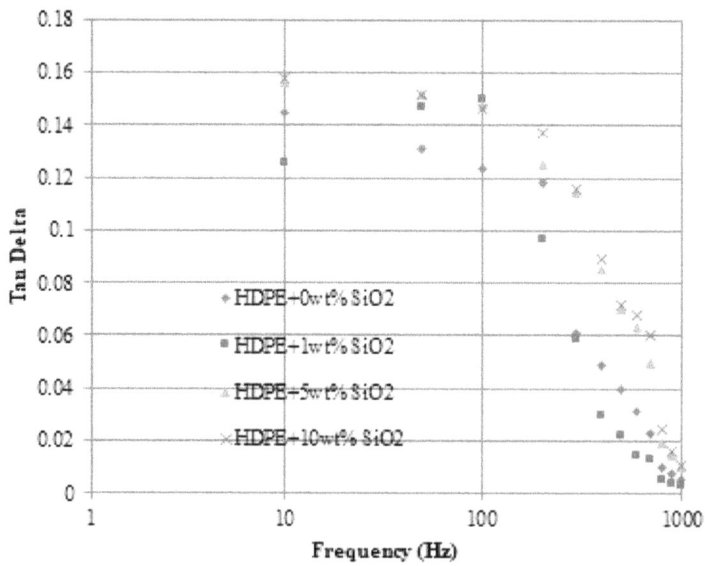

Figure 25. Measured capacitance clay/ HDPE nanocomposites T=20 oC

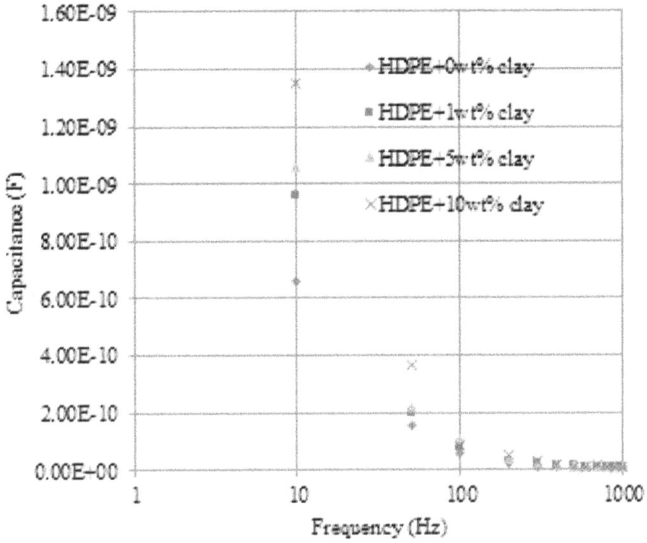

Figure 26. Measured capacitance SiO2/HDPE nanocomposites T=20 oC

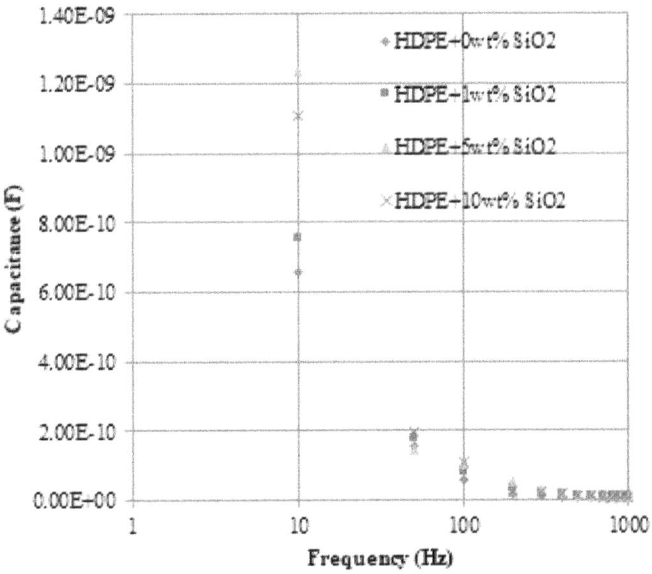

For more experimental results, Figure 23 indicates the loss tangent of Clay/HDPE nanocomposites at room temperature (20°C). This figure illustrates the reduction loss of clay/HDPE nanocomposites which increments with expanding clay nanoparticles fixation up to 1%wt., especially towards low frequencies, where it abates with expanding clay nanoparticles fixation up to 10%wt. In addition, Fig. 24 reveals the loss tangent for SiO_2/HDPE nanocomposites at room temperature (20°C). The misfortune loss for SiO_2/HDPE nanocomposites declines with expanding fumed silica fixation nanoparticles up to 1%wt., especially in high frequencies and it increments with expanding fumed silica centralization nanoparticles (1%wt.-10%wt.). Figure 25 demonstrates capacitance of clay/HDPE nanocomposites at room temperature (20°C). It is clear that the measured capacitance of clay/HDPE nanocomposites tends to expand clay centralization nanoparticles up to10%wt. On the other hand, Fig 26 demonstrates capacitance similarity as SiO_2/HDPE nanocomposites at room temperature (20°C), whenever the measured capacitance of SiO_2/HDPE nanocomposites increments with expanding fumed silica centralization nanoparticles dependent upon 5%wt. It also declines with expanding fumed silica centralization nanoparticles dependent upon 10%wt.

Figure 27. Measured loss tangent for clay/ HDPE nanocomposites at T=40 °C

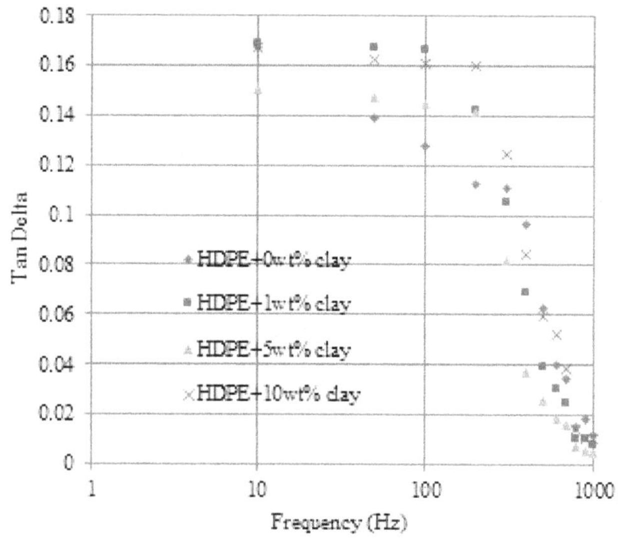

Figure 28. Measured loss tangent for SiO2/HDPE nanocomposites at T=40 °C

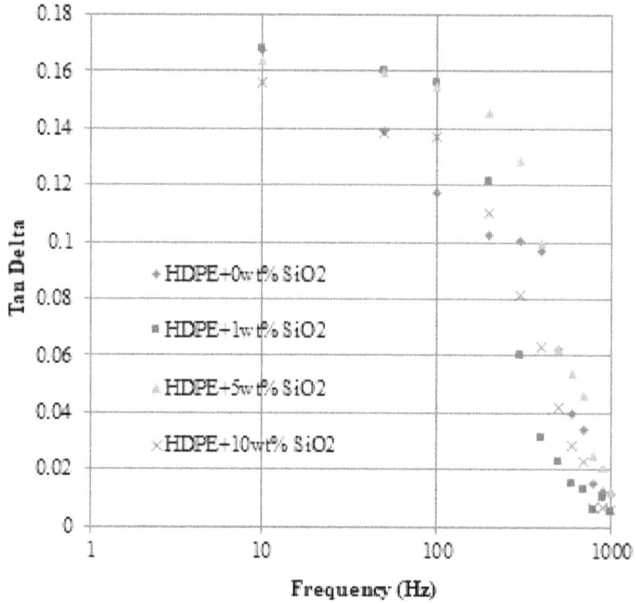

Figure 29. Measured capacitance for clay/ HDPE nanocomposites at T=40 °C

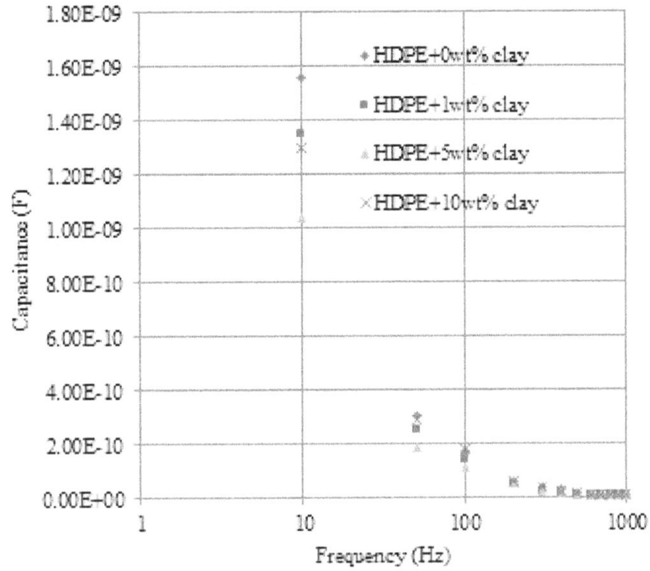

Figure 30.Measured capacitance for SiO2/HDPE nanocomposites at T=40 °C

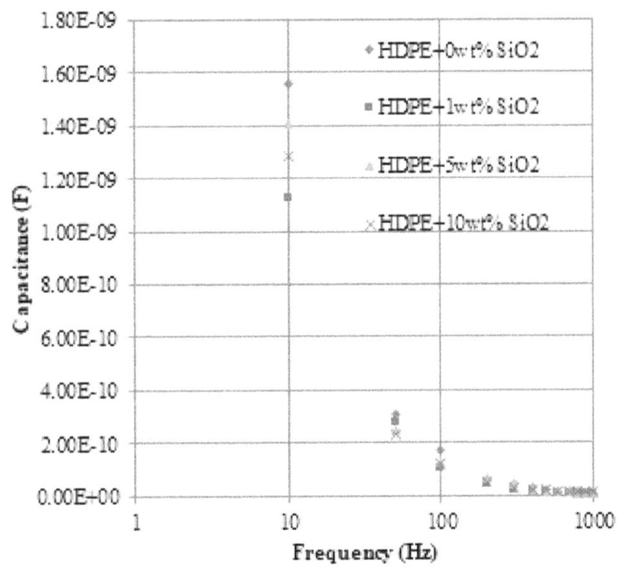

Figure 27 indicates reduction of capacity for clay/HDPE nanocomposites at temperature (40°C). The misfortune loss of clay/HDPE nanocomposites builds for expanding clay nanoparticles centralization dependent upon 10%wt., especially at low frequencies. In any case, Fig. 28 demonstrates reduction loss as a capacity of recurrence to SiO_2/HDPE nanocomposites at temperature (40°C). The measured misfortune loss of SiO_2/HDPE nanocomposites expands with expanding fumed silica fixation nanoparticles up to 1%wt., especially during low frequencies. Noteworthy, the loss tangent of SiO_2/HDPE nanocomposites declines with expanding fumed silica fixation nanoparticles (1%wt.- 10%wt.). Figure 28 reveals the capacitance concerning illustration the recurrence of clay/HDPE nanocomposites at temperature (40°C). The measured capacitance from controlling clay/HDPE nanocomposites declines for expanding clay focus nanoparticles. Similarly, Fig. 29 indicates the measured capacitance of SiO_2/HDPE nanocomposites declines for expanding fumed silica focus nanoparticles capacitance, as a work for recurrence towards temperature (40°C).

Figure 31. Measured loss tangent for clay/ HDPE nanocomposites at T=60 °C

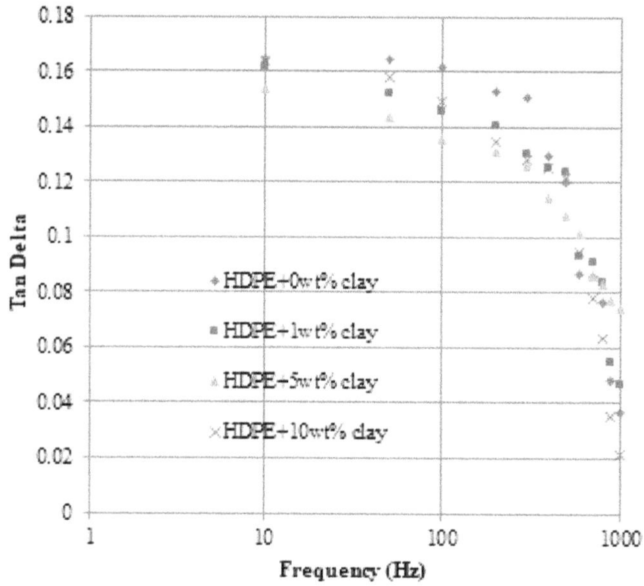

Figure 32. Measured loss tangent for SiO_2/ HDPE nanocomposites at T=60 °C

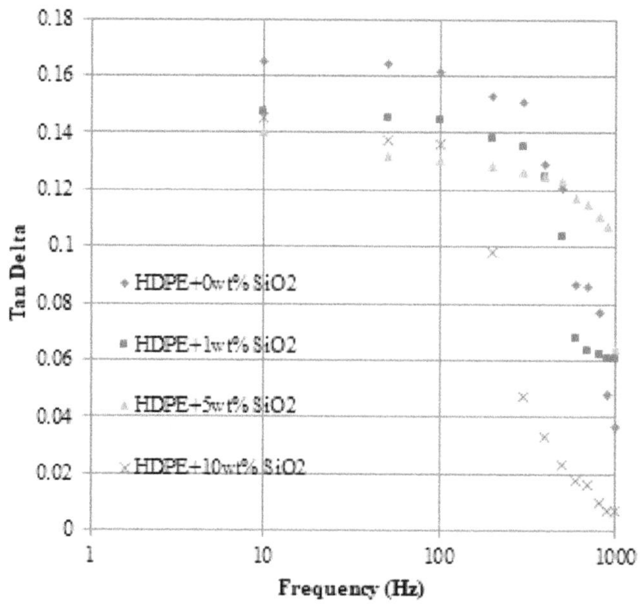

Figure 33. Measured capacitance for clay/ HDPE nanocomposites at T=60 oC

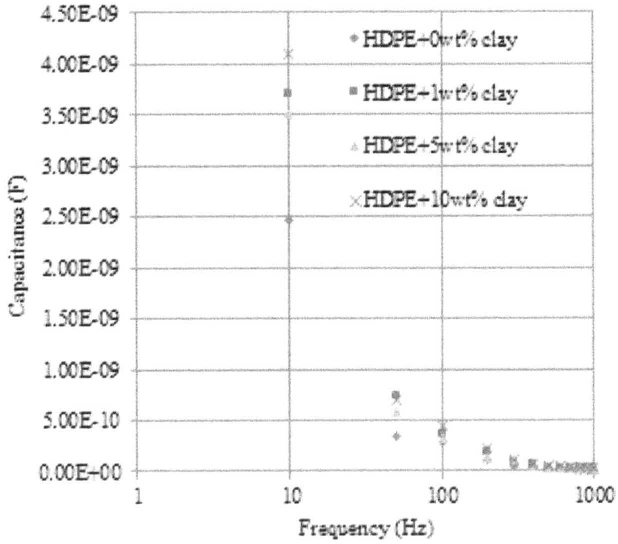

Figure 34. Measured capacitance for SiO2/ HDPE nanocomposites at T=60 oC

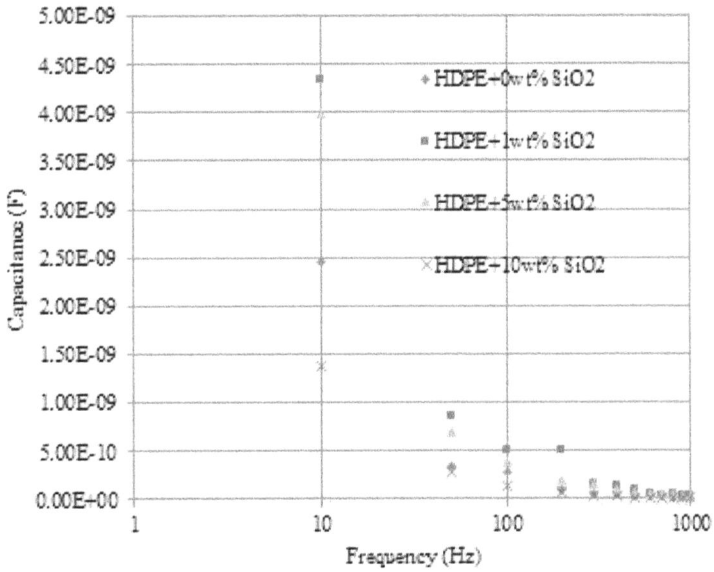

Figure 31 demonstrates reduction loss as a capacity from controlling recurrence to clay/HDPE nanocomposites at temperature (60°C). The misfortune loss of clay/HDPE nanocomposites declines with expanding clay nanoparticles fixation dependent upon 10%wt., especially at low frequencies. On the other hand, Fig. 32 demonstrates reduction loss similarly as a work for recurrence for SiO_2/HDPE nanocomposites at temperature (60°C). In any case, it is noticed that the reduction loss for SiO_2/HDPE nanocomposites declines for expanding fumed silica fixation nanoparticles up to 10%wt., especially during low frequencies.

Figure 34 indicates capacitance, as a work for recurrence to clay/HDPE nanocomposites at temperature (60°C). It illustrates the measured capacitance of clay/HDPE nanocomposites expands with expanding clay centralization nanoparticles dependent upon 10%wt. On the other hand, fig. 34 reveals the capacitance concerning illustration of a capacity about recurrence to SiO_2/HDPE nanocomposites towards temperature (60°C). Anyhow, it is illustrated that the capacitance of SiO_2/HDPE nanocomposites expands for expanding fumed silica centralization nanoparticles dependent upon 1%wt and abates for expanding fumed silica centralization nanoparticles (5%wt. -10%wt.).

7.1.2 Thermal Stability of Polyethylene Nanodielectrics

Polymer properties are tentatively custom-made, eventually perusing the inclusion of few numbers of controlling distinctive fillers. However, they are unreasonable with respect to the traditional polymer materials. This single section mulls over the upgrade and control of electric and dielectric properties of low-density polyethylene (LDPE) polymer materials, eventually perusing cost-fewer nanoparticles. Specific rates of clay and fumed silica nanoparticles have improved electric and dielectric properties from controlling low-density polyethylene nanocomposite. Dielectric spectroscopy has measured the electric and dielectric properties from controlling low-density polyethylene with and without nanoparticles towards different frequencies (10Hz-100kHz) and temperatures (20°C, 40°C and 60°C). Moreover, the ideal rates from controlling nanoparticles with respect to nanoparticles type, filler focus and temperature for upgrading electric and dielectric characterization of low-density polyethylene are investigated. Test estimations have been built on dielectric breakdown quality of new polyethylene nanocomposite materials under variant electric fields (uniform and non-uniform) and variant temperatures. The compelling nanoparticles factors on dielectric breakdown quality of polyethylene nanocomposites materials are also addressed.

Nanoparticles over-regulating and over-physical properties from controlling polymeric nanocomposite materials drives us to investigate the impacts of the nanoparticles with respect to electric and dielectric properties for polymeric commercial enterprises. In this research, the dielectric conduct of high-density polyethylene (HDPE) nanocomposite materials loaded nanoparticles of clay or fumed silica has been investigated at Different frequencies (10Hz-1kHz) and temperatures (20°C-60°C). Dielectric spectroscopy has been used to describe ionic conduction, the impacts of nanoparticles centralization on the dielectric loss and capacitive charge of the new nanocomposites. Capacitive charge and loss tangent over high density polyethylene nanocomposites measure dielectric spectroscopy. Separate dielectric self-destructive considerations and conduct have been watched, contingent upon type and fixation about nanoparticles under variant thermal states. In this work, low-density polyethylene (LDPE) has been created for inorganic nanoparticles additives for producing new nanocomposites. Material morphologies for low density polyethylene have utilized examining electron magnifying instrument (SEM) and have discovered that the blends kept the qualities demonstrated in the new low-density polyethylene nanocomposites. The reliance on controlling dielectric consistent of nanocomposites looking into recurrence can be investigated over a moderate total recurrence range, and the parts of the size and stacking level of the filler are talked about. The dielectric conduct technique for alumina nanoparticles loaded low density polyethylene is investigated through a recurrence reach of (0.01Hz-1MHz) and temperatures (20°C - 80°C). Test outcomes from controlling new polymer nanocomposites has been compared with traditional sing-phase insulation, which has progressed dielectric properties. In addition, the facts, sorts and focuses from controlling nano-additives in the successful dielectric steady in a total assortment for recurrence and thermal states have been examined. Extraordinary desires have concentrated on costive nanoparticles. However, this part has been dedicated to the impact of the sorts from controlling cost-fewer nanoparticles on electrical properties from controlling polymeric nanocomposite. With a consistent advancement in polymer nanocomposites, this examination depicts the impacts about the sorts and centralization from controlling cost-fewer nanoparticles to electrical properties of mechanical polymer material. Test effects have addressed the impacts of clay and fumed silica nanoparticles with different volume portions and temperatures on electric and dielectric properties for low-density polyethylene (LDPE). Cost-fewer nanoparticles (clay and fumed silica) have exceptionally low cost and secondary capacity for evolving polymer matrix characterization, therefore, a test examination of thermal impacts of cost-fewer nanoparticles on electric and dielectric properties of Polypropylene nanocomposites is exhibited in this research. This test has been conveyed out to portray and state the impact of the type's centralization of nanoparticles on the electric and dielectric nanocomposites materials. Namely, dielectric spectroscopy has measured the relative permittivity and the reduction loss from controlling Polypropylene with and without nanoparticles. The greater part estimations have been conveyed out at variant frequencies and temperatures (20°C, 40°C and 60°C). Distinctive dielectric conduct technique has been observed contingent upon nanoparticle type, nanoparticle centralization and nanocomposite temperature. The test reaches some results on the impacts of nanoparticles on electric characterization under variant thermal states. In the beginning, including fumed silica nanoparticles expanded permittivity of the created polyethylene nanocomposites materials, however, including clay has diminished permittivity of the new nanocomposite materials, as indicated in table 1. Expanding centralization of clay and fumed silica nanoparticles at room temperature (20°C) influences conduct about conductance and susceptance of polyethylene nanocomposites films and is relied upon in evolving the fixation of nanoparticles inside polyethylene materials under low and high frequencies. Sorts and concentrations of nanoparticles present the relationship between electric properties and interfacial medium conduct technique between nanoparticles and the polymer matrix of nanocomposite dainty films. The point behind including nanoparticles of clay alternately fumed silica will be to regulate the dielectric quality of business polyethylene, eventually perusing utilization of nanotechnology systems. The variety of conductance and susceptance qualities of polyethylene nanocomposite films can be controlled towards evolving the sorts and focuses for nanoparticles. Expanding fixation of clay nanoparticles in polyethylene declines the viable permittivity. However, expanding fixation from controlling fumed silica nanoparticles builds compelling permittivity from controlling polyethylene nanocomposites films. The vicinity of exceptional sorts of nanoparticles inside polyethylene is confined to the chain mobility. Then, the outcome expands electric encasing and restricts the era of versatile charge for the development of charge transporters for polymer dielectrics. Therefore, the number of controlling charge transporters and connected recurrence is the ruling of the electrical encasing about polyethylene nanocomposite films. New created polyethylene nanocomposites films have high thermal dependability during little focuses of clay or fumed silica nanoparticles. Including a lot from controlling the nanoparticles, the polyethylene may be opposite to the electric and dielectric conduct technique qualities bit by bit. For addition, rising thermal states from controlling nanocomposite materials influence temperatures from controlling nanoparticles, and Subsequently transform the electric characterization.

In view of the effects of the impact of temperature and sorts of nanoparticles on the electric characterization in the beginning, including fumed silica expanded permittivity of the new nanocomposite materials, including clay which has diminished permittivity of the new nanocomposite materials similarly as included in Table 1. While comparing the results conveying the impact of raising centralization from controlling clay and fumed silica nanoparticles are pointed out in Fig's (3-6) at room temperature (20°C), the measured loss tangent relies on expanding the specific rates for nanoparticles, especially the measured reduction loss aspects that vary between low and high frequencies. On the other hand, the measured capacitance varies for the expanding rate of clay nanoparticles inside the nanocomposite. In any case, the measured capacitance is shut for expanding the rate of fumed silica nanoparticles in the nanocomposite, particularly in high frequencies.

For high thermal conditions, towards the temperature of 40°C, as indicated in Fig's (7-10), the misfortune loss of Clay/ LDPE nanocomposites builds with expanding rate of clay nanoparticles in the nanocomposite up to 1%wt. Then, afterwards, the misfortune loss of Clay/ LDPE nanocomposites abates with the expanding rate of clay nanoparticles in the nanocomposite dependent upon 10%wt. Thus, fumed silica nanoparticles take after the same conduct of clay nanoparticles inside low-density polyethylene up to 10%wt. For rising specimen's temperature up to 60oC, the impacts high temperature values on nanoparticles inside the nanocomposites and the impact of rising fixation of nanoparticles will be pointed out in Fig's (11-14). The measured capacitance declines and increments in a specific quality about clay or fumed silica nanoparticles rate dependent upon 10%wt. Hence, it is clear that the rising temperature for nanocomposite materials impacts nanoparticles temperatures which evolve the dielectric conduct again around the typical states. Finally, the vitality of including nanoparticles from the controlling clay or fumed silica can be inferred for controlling, expanding and alternately diminishing the dielectric quality of immaculate LDPE, eventually perusing the utilization of nanotechnology systems and expanding nature's domain temperature from controlling nanocomposite materials that make nanoparticles temperatures evolve the dielectric conduct in the typical states.

Great hopes are there for the impacts and the benefit of costive nanoparticles (Gouda et al., 2014; Thabet, 2013a; Thabet, 2013b; Thabet & Repetto, 2013; Thabet, 2012c; Thabet & Mubarak, 2012). However, one of the concerns in this section is the impact of sorts of costive nanoparticles, looking into electrical properties for polymeric nanocomposite. For a consistent advancement in polymer nanocomposites, this examination depicts the impacts of sorts and centralization of costive nanoparticles on electrical properties of streamlined polymer material. Known test outcomes for dielectric spectroscopy are investigated and examined, while recognizing the greater impacts of controlling nanoparticles on electrical properties from controlling the nanocomposite modern material fabricated, similar to high thickness Polyethylene (HDPE) with different nanoparticles from controlling clay and fumed silica.

Every delineated outcome has clarified that including fumed silica expands permittivity of high-density polyethylene encasing materials, including clay which has abatements permittivity of high thickness polyethylene encasing materials concerning the illustration indicated in tab. 1. Physical interface between high thickness polyethylene and nanoparticles has influenced capacitance and dielectric passing point curves under typical thermal states (T=20°C) that are pointed out in Figs (3-6). Therefore, it is clear that the loss tangent of clay/HDPE nanocomposites establishes for expanding clay nanoparticles fixation dependent upon 1%wt., especially in low frequencies, However it declines for expanding clay nanoparticles focus dependent upon 10%wt. In any case, the loss tangent from controlling SiO2/HDPE nanocomposites declines with expanding fumed silica focus nanoparticles up to 1%wt., especially towards high frequencies and expands with expanding fumed silica fixation nanoparticles (1%wt. -10%wt.). The measured capacitance of clay/HDPE nanocomposites establishes expanding clay focus nanoparticles up to10%wt. The measured capacitance for SiO_2/HDPE nanocomposites establishes expanding fumed silica focus nanoparticles dependent upon 5%wt, in any case, it abates with expanding fumed silica centralization nanoparticles dependent upon 10%wt.

Evolving thermal states can be influenced by the physical interface between high thickness polyethylene and nanoparticles, and others will influence capacitance and dielectric loss point curves, as pointed out in Figs (7-10) for high thermal state (T=40oC). Thus, the reduction loss of clay/HDPE nanocomposites expands with expanding clay nanoparticles focus up to 10%wt., especially, at low frequencies. In any case, the measured reduction loss of SiO2/HDPE nanocomposites increments for expanding fumed silica centralization nanoparticles dependent upon 1%wt., especially during low frequencies. It is noted the reduction loss of SiO_2/HDPE nanocomposites abatements with expanding fumed silica centralization nanoparticles (1%wt. - 10wt %). In addition, the measured capacitance of clay/HDPE nanocomposites declines for expanding clay fixation nanoparticles up to 10%wt. Consequently, it is clear that the measured capacitance for SiO2/HDPE nanocomposites declines for expanding fumed silica centralization nanoparticles dependent upon 1%wt., then it expands with expanding fumed silica fixation nanoparticles up to 10%wt.

Finally, it concentrates on the impact of raising thermal states (T=60°C) for physical interface between high thickness polyethylene and nanoparticles that is pointed out in Figs (11-14); instead the misfortune loss and capacitance from controlling new nanocomposite materials have news person for distinctive focuses of weight about altered nanoparticles centralization towards temperature. Thus, the misfortune loss of clay/HDPE nanocomposites abates with expanding clay nanoparticles focus dependent upon 10%wt., especially, under low frequencies. Anyway, it may be noticed that the loss tangent on SiO_2/HDPE nanocomposites declines with expanding fumed silica focus of nanoparticles dependent upon 10%wt., especially in low frequencies. The measured capacitance of clay/HDPE nanocomposites expands for expanding clay centralization nanoparticles dependent upon 10%wt. In any case, the capacitance for SiO_2/HDPE nanocomposites is established with expanding fumed silica centralization nanoparticles up to 1%wt, but it declines with expanding fumed silica centralization nanoparticles (5%wt. -10%wt.).

7.1.3 Comparative Characterizations of Commercial Polyethylene Nanocomposite

In the beginning, including fumed silica has expanded permittivity of the new nanocomposite materials, including clay which has a diminished permittivity of the new nanocomposite materials, as indicated in table (1).

In thinking about the effects of portraying the impact of raising fixation from controlling clay and fumed silica nanoparticles to reduction loss. Moreover, capacitance about new low-density polyethylene nanocomposite materials is relied upon in separate weight focuses about changed nanoparticles centralization at room temperature (T=25 °C). For example, the reduction loss expands for the clay nanoparticles rate up to 5%wt, specially, at low frequencies, yet all the declines in expanding clay nanoparticles rate dependent upon 10%wt, especially towards high frequencies. In any case, the measured loss tangent declines with the rising rate of fumed silica nanoparticles in the nanocomposite up to 5%wt, especially, towards low frequencies, and increments for expanding fumed silica nanoparticles rate up to 10%wt, especially, during high frequencies. With respect to the measured capacitance of new low-density polyethylene nanocomposites, it is clear that the measured capacitance declines for the rising rate of clay nanoparticles in the nanocomposite dependent upon 5%wt, yet the measured capacitance from controlling low-density polyethylene nanocomposites establishes for expanding clay rate of nanoparticles up to

10%wt. In all cases, the measured capacitance declines for climbing rate of fumed silica nanoparticles in the low-density polyethylene nanocomposite dependent upon 1%wt and it establishes with expanding fumed silica rate nanoparticles at 5%wt - 10%wt.

Table 2. Electric and Dielectric Properties of Pure and Nanocomposite Materials

Materials	**Dielectric Constant at 1kHz**	**Resistivity (Ω.m)**
Pure PP	2.28	108
PP + 1%wt Clay	2.21	109
PP + 5%wt Clay	1.97	109-1010
PP + 10%wt Clay	1.75	1010-1012
PP + 1%wt Fumed Silica	2.29	107
PP + 5%wt Fumed Silica	2.37	107-105
PP + 10%wt Fumed Silica	2.47	105-104

Additionally, the impact on raising centralization of clay nanoparticles is pointed out in Figures (4,5). The reduction loss and capacitance of new high-density polyethylene nanocomposite materials to news person for diverse weight focuses on changed nanoparticles centralization. Hence, the reduction loss of Clay/ high density polyethylene nanocomposite abates with expanding clay rate nanoparticles dependent upon 1%wt clay, but establishes expanding clay nanoparticles starting with 5% wt. up to 10%wt. In addition, the measured reduction loss of fumed Silica/High thickness Polyethylene nanocomposites declines with expanding fumed silica rate nanoparticles dependent upon 1%wt fumed silica but, it increments for expanding fumed silica rate nanoparticles dependent upon 10%wt fumed silica nanoparticles. In any case, the measured capacitance from controlling high thickness Polyethylene nanocomposites expands for expanding clay and fumed silica nanoparticles rate dependent upon 10%wt. Including clay will diminish the permittivity of new nanocomposite materials and the impacts for including little measure for clay nanoparticles rate to low density polyethylene declines in capacitance and loss tangent. In any case, including little measure of clay nanoparticles rate in high density polyethylene builds capacitance and misfortune tangent, especially towards low frequencies. Including clay nanoparticles increments capacitance for high density polyethylene more than expanding fumed silica nanoparticles in high density polyethylene in the same rates. Including fumed silica expands the permittivity of new nanocomposite materials, few numbers of fumed silica nanoparticles with respect to low density polyethylene establish capacitance and reduction tangent, especially during low frequencies. Whatever, little measure about fumed silica nanoparticles in high density polyethylene expands capacitance in any case with abatements reduction tangent, especially at low frequencies. Including a lot of clay or fumed silica nanoparticles with polyethylene will allow for the reverse dielectric conduct technique aspects bit by bit which rely on nanoparticles structure for polymer matrix.

Figure 35. Measured loss tangent of Polypropylene nanocomposites at room temperature (25oC)

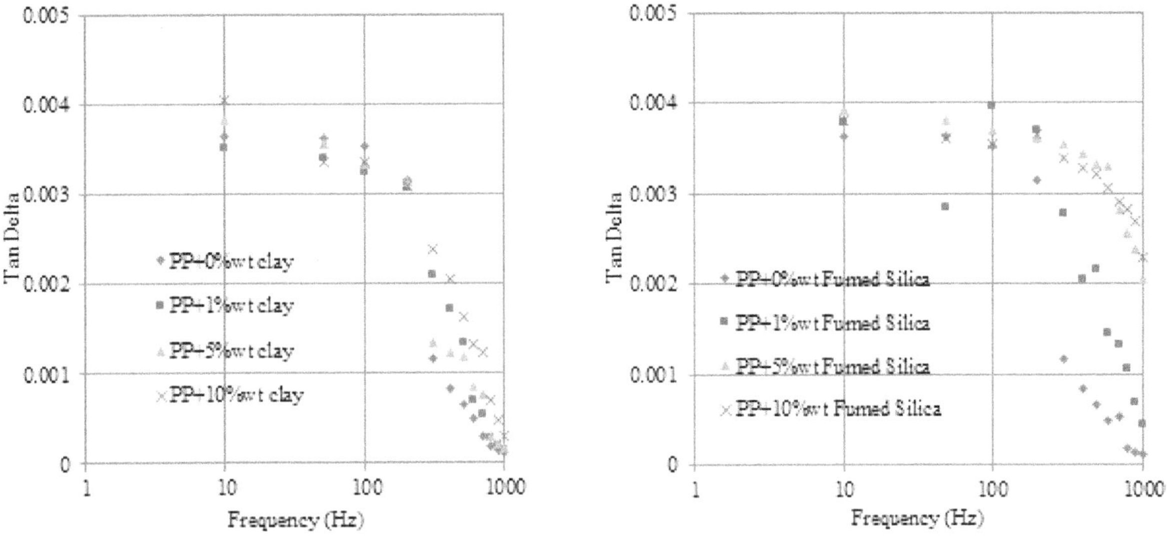

(a) Clay/PP nanocomposites (b) Fumed Silica/PP nanocomposites

Figure 36. Measured capacitance of Polypropylene nanocomposites at room temperature (25oC)

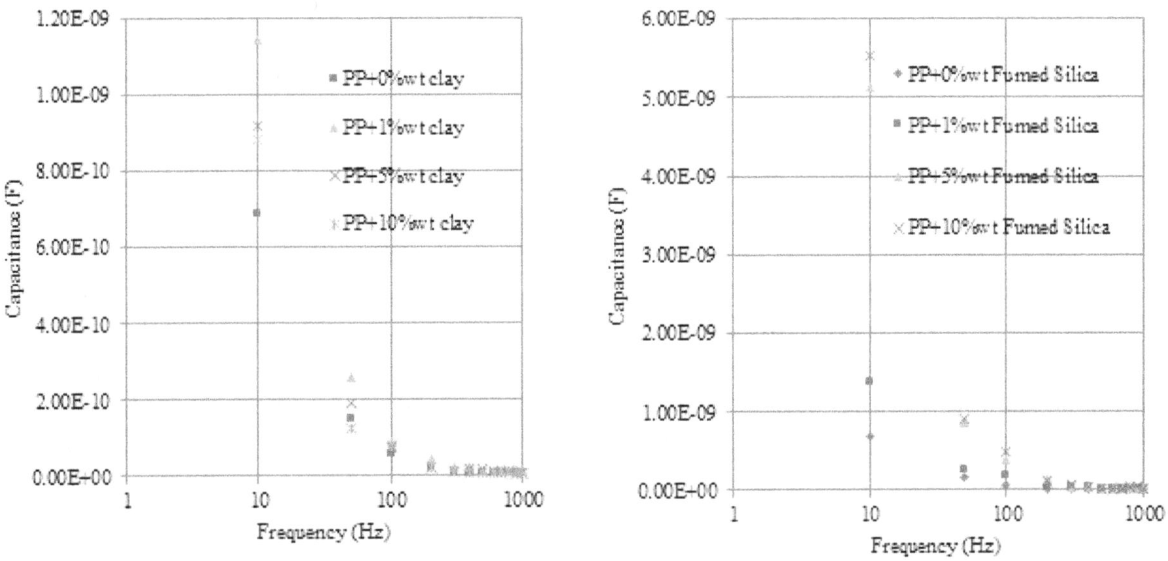

(a) Clay/PP nanocomposites (b) Fumed Silica/PP nanocomposites

7.2 POLYPROPYLENE NANO-DIELECTRICS

The build of polymer materials may be an economically accessible material now being used in the manufacturing of high voltage (HV) streamlined results and their properties point by point in Table (2).

Figure 37. Measured loss tangent of Polypropylene nanocomposites at a certain temperature (T=40oC)

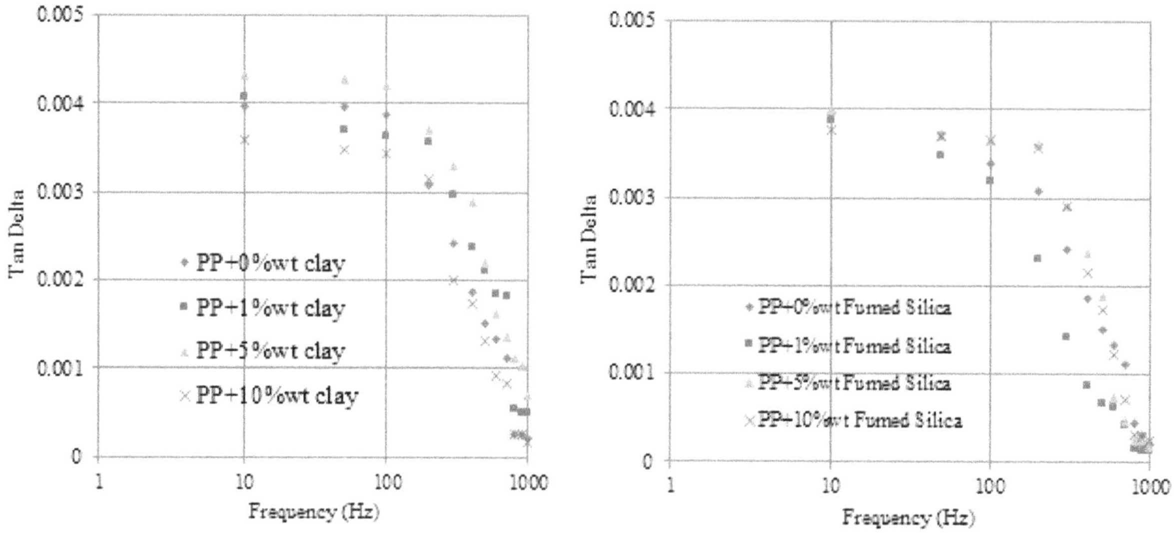

(a) Clay/PP nanocomposites (b) Fumed Silica/PP nanocomposites

Dielectric spectroscopy is capable of testing the strategy of research in dynamic conduct technique of through the investigation of its recurrence subordinate dielectric reaction. This technobabble is in light of the estimation of the capacitance, as a work of recurrence for a test sandwiched between two electrodes. The tan δ, and capacitance (C) are measured as a work of recurrence in the range of 10 Hz to 50 kHz in 25°C for every last test. The estimations are committed to utilizing secondary determination dielectric spectroscopy.

7.2.1 Effect of Cost-Fewer Nanoparticles Inside Polypropylene Characterization

Figure 35. (a) indicates misfortune loss as a capacity of recurrence for Clay/Polypropylene nanocomposites towards room temperature (25°C). The misfortune loss of polypropylene declines for expanding clay nanoparticles rate dependent upon 1%wt, especially at low frequencies. However, it establishes expanding clay nanoparticles rate up to 10%wt, especially during high frequencies. In any case, figure 35. (b) indicates the measured reduction loss with the rising rate of fumed silica nanoparticles in the nanocomposite at room temperature (25°C). It is clear that the loss tangent of fumed silica/Polypropylene nanocomposite establishes expanding fumed silica rate nanoparticles dependent upon 10%wt, especially at high frequencies. Figure 36 (a) indicates capacitance of the work about recurrence to Clay/Polypropylene nanocomposites at room temperature (25°C). The capacitance of Clay/Polypropylene nanocomposite establishes expanding clay rate nanoparticles dependent upon 1%wt. In any case, it falls down with expanding nanoparticle rate up to 10%wt. Figure 36. (b) contrasts capacitance from the controlling fumed silica/Polypropylene nanocomposites at room temperature (25°C). The measured capacitance from controlling fumed silica/Polypropylene establishes expanding fumed silica rate nanoparticles up to 10% wt.

Figure 38. Measured capacitance of Polypropylene nanocomposites at a certain temperature (T=40oC)

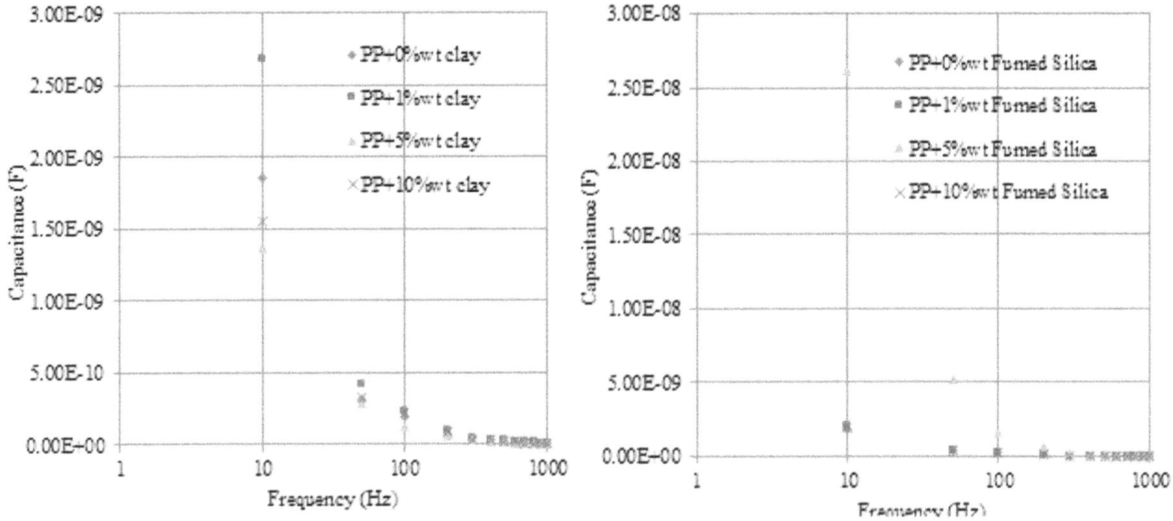

(a) Clay/PP nanocomposites (b) Fumed Silica/PP nanocomposites

Figure 39. Measured loss tangent of Polypropylene nanocomposites at a certain temperature (T=60oC)

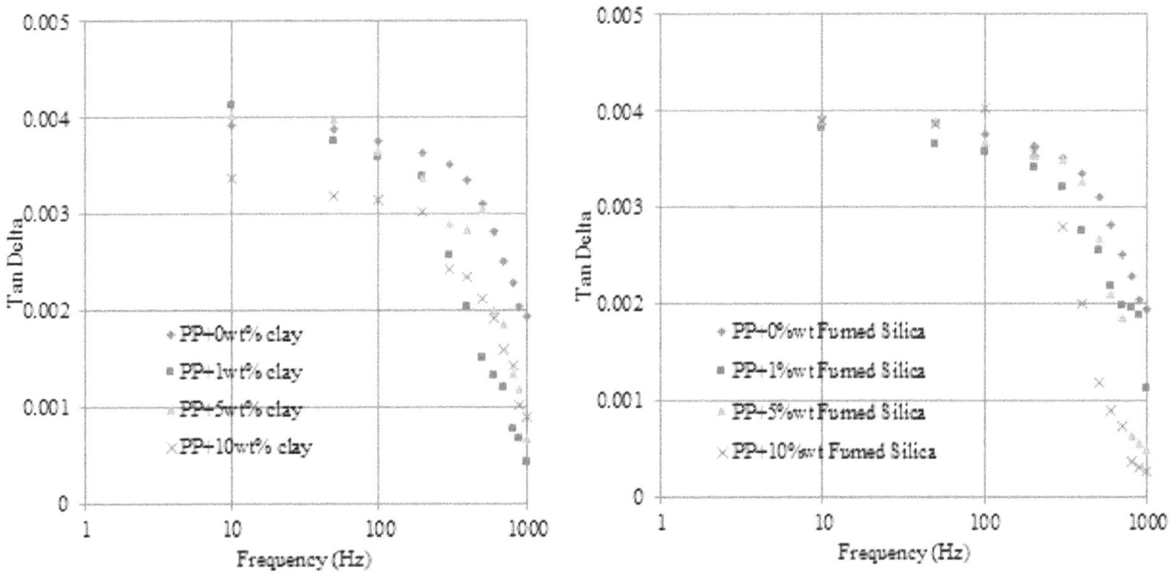

(a) Clay/PP nanocomposites (b) Fumed Silica/PP nanocomposites

Figure 37 (a) indicates reduction loss as a work of recurrence for Clay/Polypropylene nanocomposites towards temperature (40°C). The reduction loss of Clay/Polypropylene increments with expanding clay nanoparticles rate dependent upon 1%wt, especially in low frequencies. Then, the reduction loss abates with expanding clay nanoparticles rate up to 10%wt, especially at high frequencies. Figure 37. (b) demonstrates loss tangent as a capacity of recurrence for fumed silica/Polypropylene nanocomposites at temperature (40°C). The loss tangent of fumed silica/Polypropylene nanocomposite abates for expanding fumed silica nanoparticles rate dependent upon 1%wt, especially towards low frequencies, but it builds with expanding fumed silica nanoparticles rate (1%wt -10%wt).

Figure 40.Measured capacitance of Polypropylene nanocomposites at a certain temperature (T=60oC)

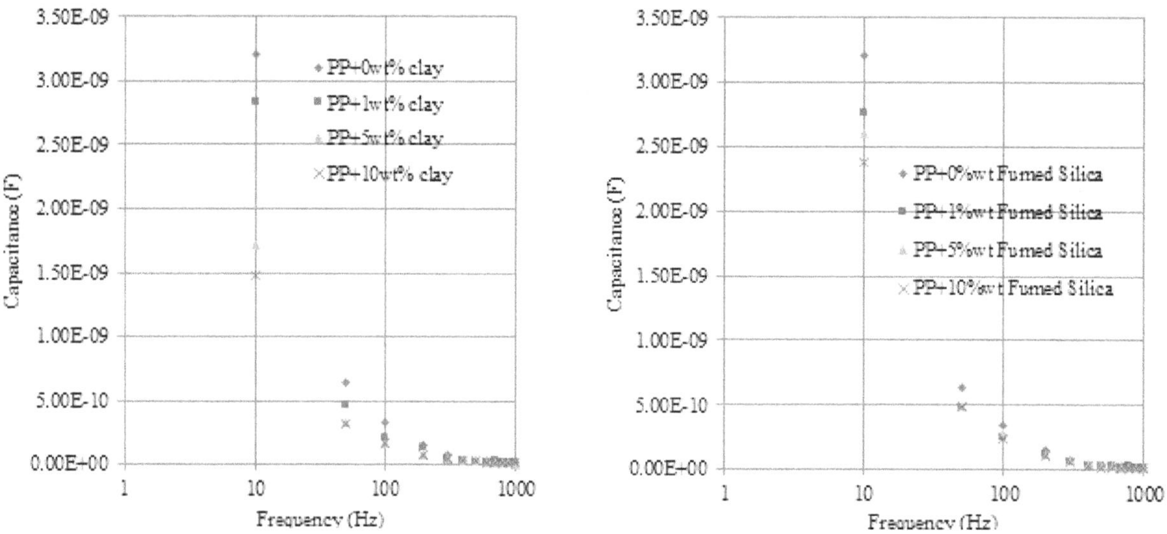

(b) Clay/PP nanocomposites (b) Fumed Silica/PP nanocomposites

Figure 38. (a) reveals capacitance of the work about recurrence to Clay/Polypropylene nanocomposites at temperature (40°C). The measured capacitance of Clay/Polypropylene nanocomposites increments for expanding clay nanoparticles rate up to 1%wt. Then, it declines with expanding clay nanoparticles rate dependent upon 10%wt. Figure 38. (b) demonstrates capacitance concerning illustration of the work of recurrence for fumed silica/Polypropylene nanocomposites at temperature (40°C). The measured capacitance of fumed silica /Polypropylene nanocomposites establishes expanding fumed silica nanoparticles rate dependent upon 10%wt. Figure 39. (a) indicates the reduction loss of the work about recurrence for Clay/Polypropylene nanocomposites at temperature (60°C). The loss tangent for Caly/Polypropylene nanocomposites abates with expanding clay nanoparticles rates dependent upon 10%wt, especially at low frequencies. Figure 39. (b) reveals loss tangent of the work about recurrence of fumed silica/Polypropylene nanocomposites in temperature (60°C). The misfortune loss for fumed silica/Polypropylene nanocomposite declines with expanding fumed silica rate nanoparticles up to 10%wt, especially towards low frequencies. However, Figure 40.(a) demonstrates capacitance as a capacity of recurrence for Clay/Polypropylene nanocomposites at temperature (60°C). The capacitance for Clay/Polypropylene declines for expanding clay nanoparticles rate up to 10%wt. Figure 40.(b) demonstrates capacitance as a capacity of recurrence of fumed silica/Polypropylene nanocomposites at temperature (60°C). The capacitance for fumed silica/Polypropylene abates with expanding fumed silica rate nanoparticles dependent upon 10%wt.

Table 3. Electric and dielectric properties of polyvinyl chloride nanocomposites

Materials	Dielectric Constant At 1kHz	Resistivity (Ω.m)
Pure PVC	3.3	1013
1wt. % Clay/PVC	3.20	1013-1014
1wt. % ZnO/PVC	2.63	1014-1017
1wt. % SiO2/PVC	3.42	1012-1010
1wt. % TiO2/PVC	1.85	1022-1020
5wt. % Clay/PVC	2.83	1014-1017
5wt. % ZnO/PVC	2.39	1017-1020
5wt. % SiO2/PVC	3.53	1010-108
5wt. % TiO2/PVC	3.01	1012-1011
10wt. % Clay/PVC	2.49	1020
10wt. % SiO2/PVC	3.60	108
10wt. % ZnO/PVC	2.15	1020
10wt. % TiO2/PVC	3.07	1011

7.2.2 Thermal Stability of Nanodielectrics

In view of the effects of conveying the impact from controlling sorts and controlling nanoparticles, clay and fumed silica have expanded permittivity of the new nanocomposite materials. The electric and dielectric characterization from controlling immaculate and new nanocomposites about polyvinyl chloride materials has been measured towards utilizing HIOKI 3522-50 LCR Hi-tester gadget towards room temperature, as demonstrated clinched alongside Table (3). The principle determinations of LCR are energy supply: 220 v (±10%) AC (selectable), 50 Hz and Frequency: DC, 1 mHz on 100 kHz, essential Accuracy: Z: ±0.08% RDG. H: ±0.05.

Thus, in contrast with the greater outcomes on portraying the impact from controlling raising focus from controlling nanoparticles at room temperature, it can be demonstrated that the misfortune loss for polypropylene declines with expanding clay nanoparticles rate up to 1%wt, especially during low frequencies, but it expands with expanding clay nanoparticles rate up to 10%wt, especially towards high frequencies. Moreover, the misfortune loss of fumed silica/Polypropylene nanocomposite increases with expanding fumed silica rate nanoparticles up to 10%wt, especially at high frequencies. The capacitance of Clay/Polypropylene nanocomposite increases with expanding clay rate nanoparticles up to 1%wt, provided that it tumbles down with expanding nanoparticle rate up to 10%wt. Therefore, the measured capacitance of fumed silica/Polypropylene increases with expanding fumed silica rate nanoparticles dependent upon 10%wt. Furthermore, the effects of portraying the impact of raising centralization about nanoparticles towards 40°C, and it is clear that the misfortune loss from controlling Clay/Polypropylene expands with expanding clay nanoparticles rate up to 1%wt, especially towards low frequencies. Then, the misfortune loss abates for expanding clay nanoparticles rate up to 10%wt, especially in high frequencies. Moreover, there is a reduction loss of fumed silica/Polypropylene nanocomposite abatements with expanding fumed silica nanoparticles rate up to 1%wt., especially in low frequencies, but it increments for expanding fumed silica nanoparticles rate (1%wt-10%wt). In any case, the measure capacitance for Clay/Polypropylene nanocomposites increments for expanding clay nanoparticles rate up to 1%wt, then, it abates for expanding clay nanoparticles rate dependent upon 10%wt. Therefore, the measured capacitance of fumed silica /Polypropylene nanocomposites increments with expanding fumed silica nanoparticles rate up to 10%wt.

Figure 41. SEM images for PVC nanocomposites

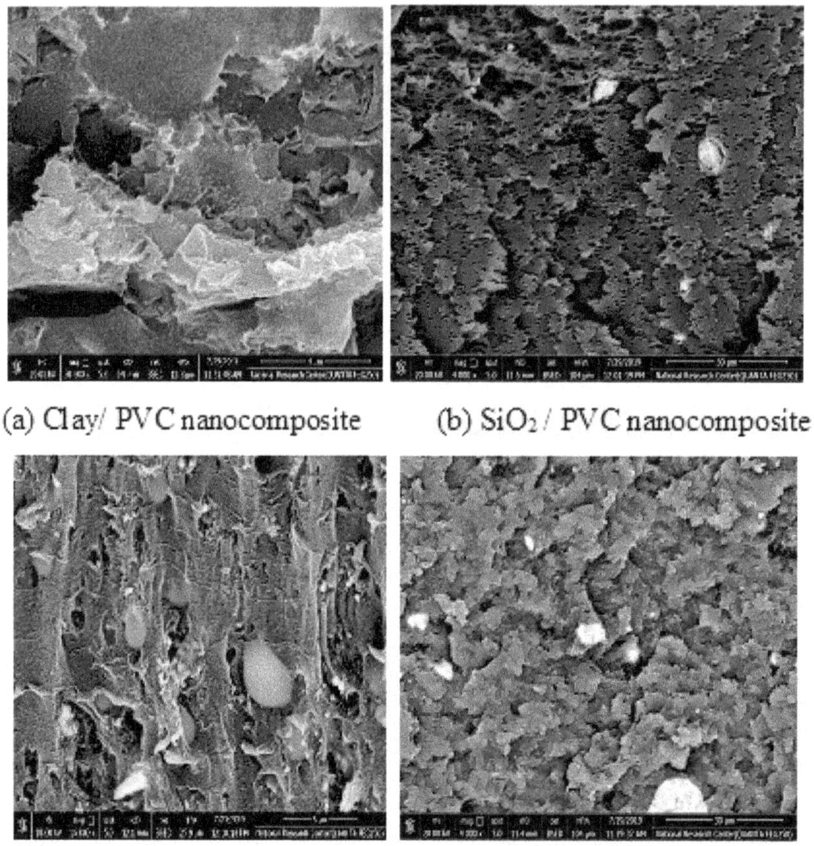

(a) Clay/ PVC nanocomposite (b) SiO_2 / PVC nanocomposite

(c) ZnO /PVC nanocomposite (d) TiO_2/ PVC nanocomposite

Finally, regarding the outcomes for portraying the impact of raising centralization of nanoparticles towards 60°C, the misfortune loss of Caly/Polypropylene nanocomposites declines with expanding clay nanoparticles rate up to 10%wt, especially in low frequencies. In addition, the reduction loss of fumed silica/Polypropylene nanocomposite abates with expanding fumed silica rate nanoparticles dependent upon 10%wt, especially during low frequencies. In any case, the capacitance of Clay/Polypropylene declines with expanding clay nanoparticles rate up to 10%wt. Furthermore, the capacitance of fumed silica/Polypropylene declines with expanding fumed silica rate nanoparticles up to 10%wt.

7.3 POLYVINYL CHLORIDE NANODIELECTRICS

SOLARTRON dielectric reaction can measure recurrence reaction of dielectric reduction conduct technique for polyvinyl chloride towards variant thermal states (20-80°C). However, the primary determinations of controlling SOLARTRON gadget have detail limits: 10 to 30°C. Humidity, non-condensing: 95% to 40°C. Safety: intended to consent with IEC384 (BS4743). This book focuses on recurrence reaction examination of dielectric reduction of the inalienable dispersal from controlling electric vitality in polyvinyl chloride (10^2Hz will 10^3Hz). In variant temperatures (20°C-80°C), for now nanocomposites have made specimens. SEM images of PVC nanocomposites are in Fig. 41.

7.3.1 Effect of Nanoparticles Inside Dielectric Constant of Polyvinyl Chloride

In this study, the vicinity of a certain kind and focus of nanoparticles inside polyvinyl chloride has essential factors for confining and alternately permitting chain portability and bringing about expanding or diminishing of the electric encasing of all things considered confinement, as claimed about versatile accuse and the development for charge transporters in polymer dielectrics. Thus, the variety for safety and susceptance values towards low recurrence reach is expected from the impact of inorganic nanoparticles (SiO_2) inside electrical encasing. It is discovered that trapping properties of matrix are exceptionally altered towards the vicinity for clay nanoparticles (cost-less) scattered homogenously clinched alongside polyvinyl chloride dependent upon 10wt. %. Little additives of clay nanoparticles in polyvinyl chloride indicated calculable change in the electric resistivity of diverse recurrence.

Likewise, the electrical encasing of polypropylene composites commitment to its reactance and conductance esteem in polypropylene nanocomposite films clinched alongside easier recurrence extent can bring about the electrical encasing of the nanocomposite films hosting influenced by the vicinity of nanoparticles. There is no straight execution knowledge from controlling the expanding rate of clay nanoparticles for polypropylene electrical properties. In any case, the expanding rate of fumed silica nanoparticles increments polypropylene conductance but abates its reactance in high thermal states. Thermal dependability of the new polypropylene nanocomposite films has been confined and identified with the most reduced softening point for both nanoparticles and polymer matrix for receiving modern securely requisitions. Under high thermal conditions, the impact of the unwinding time of the accuse transporters on the electrical encasing of polypropylene nanocomposite films can be disregarded. Therefore, the number of charge transporters and connected recurrence end up ruling elements of the electrical encasing of polypropylene nanocomposite films. The vicinity of nanoparticles inside polypropylene will limit the chain versatility and bring about expanding electric encasing to all things considered confinement for portable accuse and the development of charge transporters over polymer dielectrics, particularly at an easier recurrence extent, if the encasing will assume a more essential part. Consequently, the variety of reactance and conductance worth towards low recurrence extent can be expected of the impact of inorganic fillers' electrical encasing. Electrical Strength for new nanocomposite films happens at few numbers of clay or fumed silica nanoparticles, in any case including a lot of the nanoparticles on polypropylene to opposite electrical conduct qualities bit by bit. A high thermal condition for polypropylene nanocomposite materials may change electrical conduct technique in the typical states (Ebnalwaled et al., 2016; Thabet, 2014; Thabet, 2015a; Thabet, 2015b; Thabet, 2015c; Thabet, 2015d; Thabet, 2016e; Thabet, 2016; Thabet & Mubarak, 2015; Thabet & Mubarak, 2016a).

Figure 42. Measured real permittivity of 1%wt clay/PVC nanocomposite

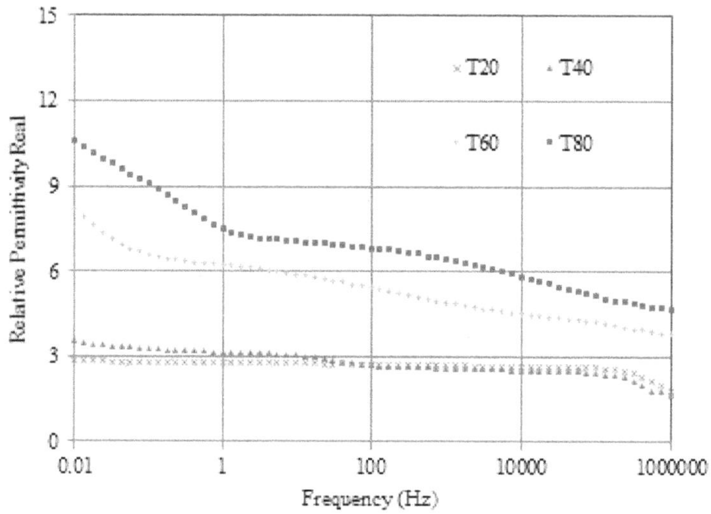

Figure 43. Measured real permittivity of 5%wt clay/PVC nanocomposite

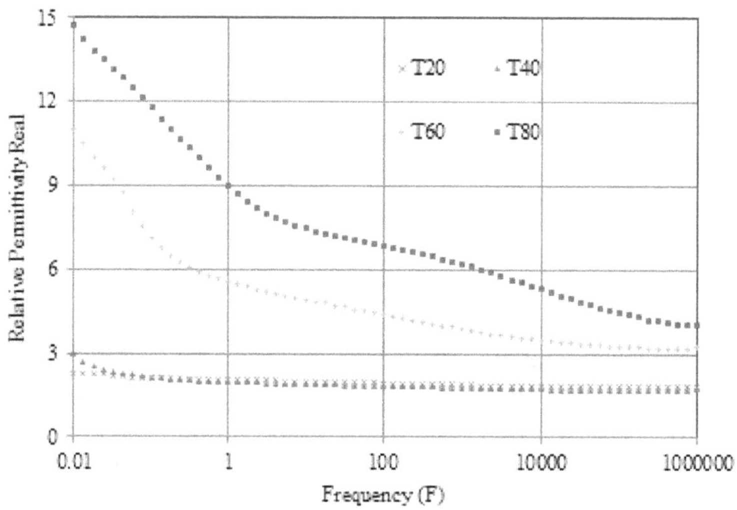

Expanding fixation of Clay, ZnO, and silica concerning illustration of a second kind of nanoparticle causes diminishing in the compelling dielectric constants for any multi-nanocomposites. In any case, expanding focus of alumina and TiO_2 as a second kind of nanoparticle expands the viable dielectric constant from controlling any multi-nanocomposites. Utilizing SiO_2 as a second sort of nanoparticle causes expanding in the successful dielectric constant for (Clay, ZnO, and Silica)/polymer for multi-nanocomposites, yet it all causes diminishing in the powerful dielectric constant of (MgO, TiO2, and Alumina)/polymer for multi-nanocomposites towards expanding the volume portion of SiO2 nanoparticles. In any case, utilizing MgO similarly as a second kind of nanoparticles causes expanding in the compelling dielectric constant for (Clay, ZnO, Silica, SiO_2, and TiO_2)/polymer to multi-nanocomposites.

It causes diminishing in the compelling dielectric constant from controlling Alumina/polymer for multi-nanocomposites by expanding the volume portion from controlling MgO nanoparticles.

Figure 44. Measured real permittivity of 10%wt clay/PVC nanocomposites

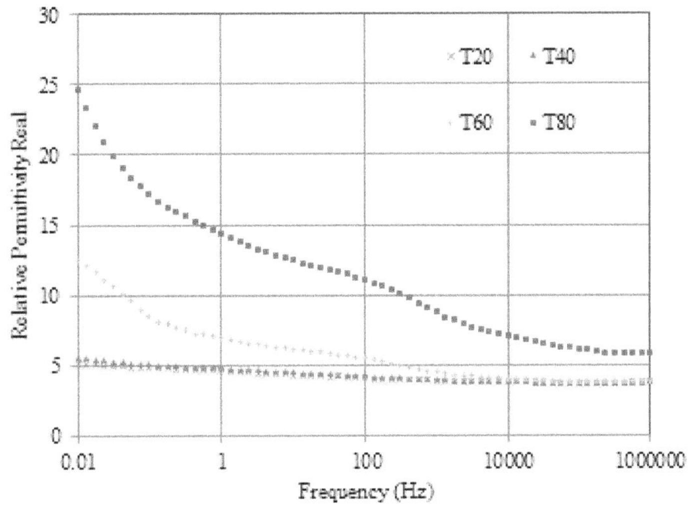

Figure 45. Measured real permittivity of 1%wt. fumed silica/PVC nanocomposite

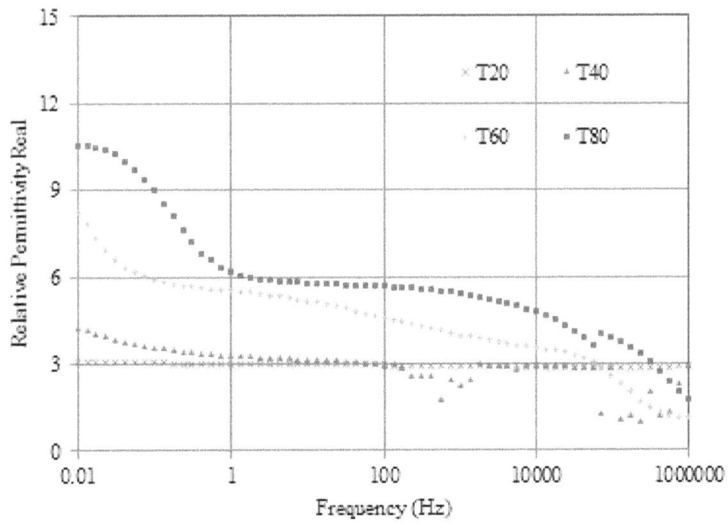

Figure 46. Measured real permittivity of 5%wt. fumed silica/PVC nanocomposite

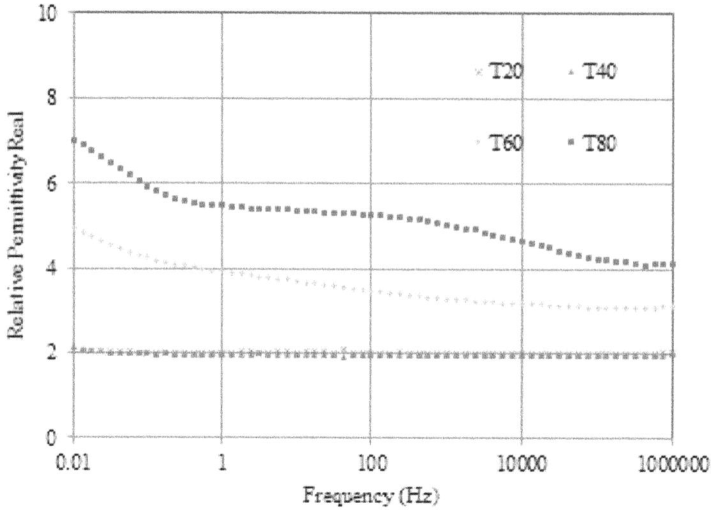

Figure 47. Measured real permittivity of 10%wt. fumed silica/PVC nanocomposite

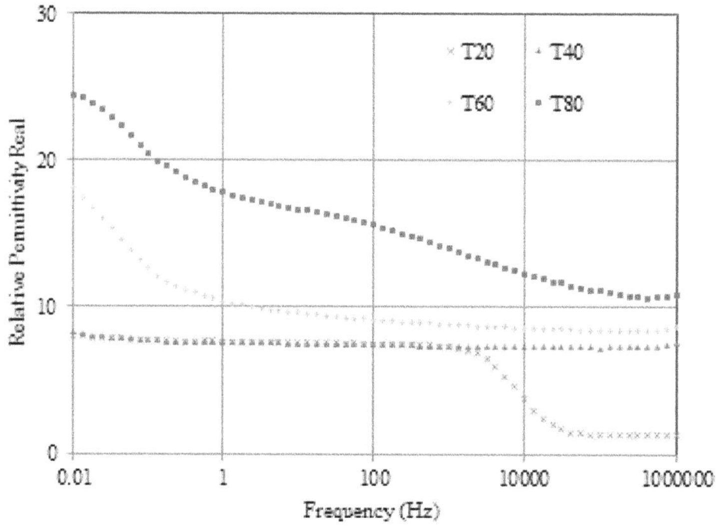

Figures (42-44) indicate the impacts of natural nanoparticles looking into dielectric constant for polyvinyl chloride specimens as a work of recurrence to clay/PVC nanocomposites at variant temperatures. In fig. 41, it is self-evident the conduct technique for dielectric constant difference abatements genuine relative permittivity for expanding the rate of clay nanoparticles dependent upon 1%wt in view of immaculate polyvinyl characterization, especially in high frequencies. The genuine relative permittivity increments with expanding thermal temperatures, especially during low frequencies; however, there may be a semi-matching between dielectric estimations under low thermal temperatures (20°C - 40°C). On the other hand, fig. 43 is opposite with respect to expanding true relative permittivity for expanding the rate of clay nanoparticles dependent upon 5%wt, especially in secondary temperatures (60°C - 80°C).

Expanding genuine relative permittivity has begun and Johnson has proceeded to expand the rate of clay nanoparticles dependent upon 10%wt, as indicated in fig 44. Fig. 's (45-47) show dielectric constant characterization as a work of recurrence for fumed silica/ PVC nanocomposites specimens at variant temperatures. As shown in Fig. 's (45, 46), it is shown the dielectric constant from controlling fumed silica/ PVC nanocomposites abatements with expanding rate of fumed silica nanoparticles up to 5%wt with respect to immaculate polyvinyl chloride execution. Previously, in fig. 46, the true relative permittivity increments for expanding thermal temperatures, especially towards low frequencies, however, there will be a total matching between dielectric estimations under low thermal temperatures (20°C - 40°C).

Figure 48. Measured real permittivity of 1%wt. ZnO/PVC nanocomposite

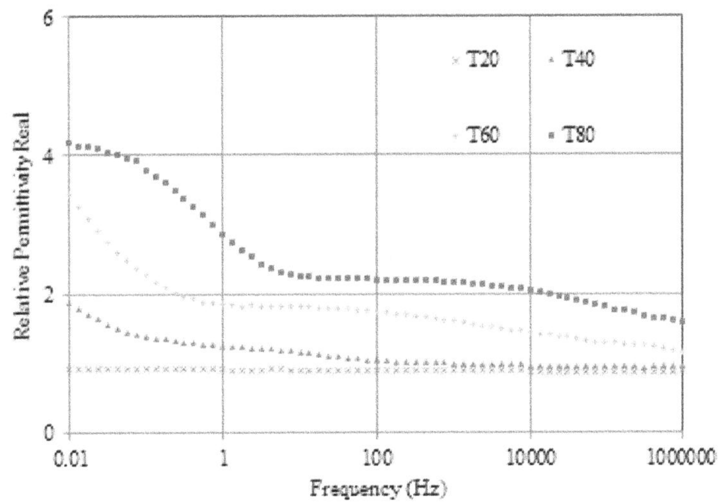

Figure 49. Measured real permittivity of 5%wt. ZnO/PVC nanocomposite

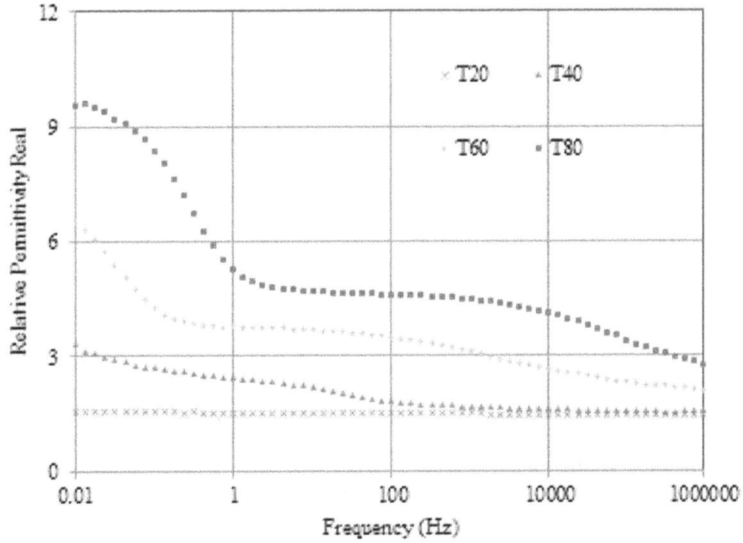

Figure 50.Measured real permittivity of 10%wt. ZnO/PVC nanocomposite

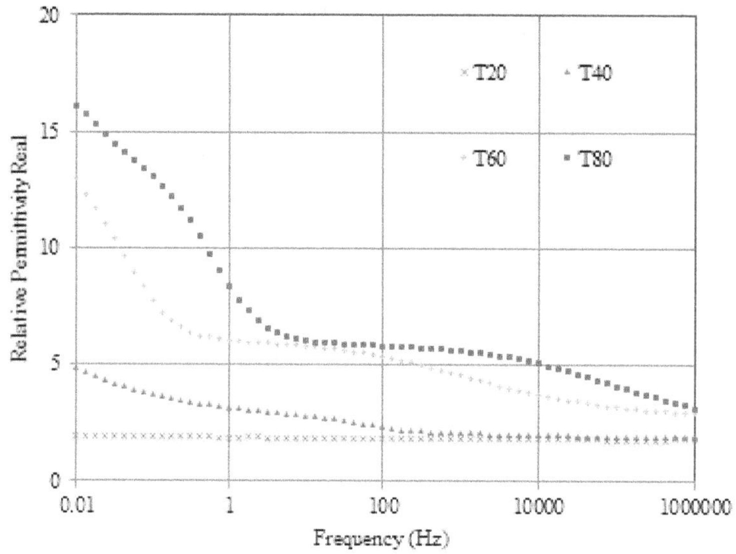

On the other hand, fig. 47 illustrates expanding dielectric constant characterization with expanding the rate of fumed silica nanoparticles up to 10%wt.

Figure 51. Measured real permittivity of 1%wt. TiO_2/PVC nanocomposite

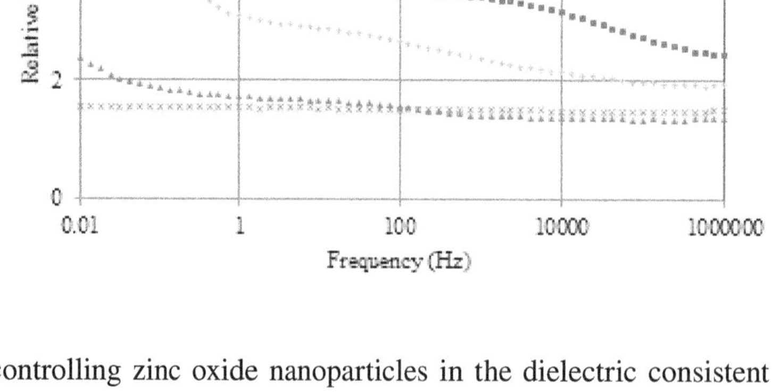

Impacts from controlling zinc oxide nanoparticles in the dielectric consistent characterization of polyvinyl chloride is demonstrated in Fig. 's (48-50) as a work of controlling recurrence and variant temperatures. Figure 47 shows the dielectric constant of ZnO/PVC nanocomposites abatements with expanding rate of zinc oxide nanoparticles in the nanocomposite dependent upon 1%wt. The genuine relative permittivity expands with the expanding thermal temperatures, especially towards low frequencies; however, there is a semi-matching between dielectric estimations under high thermal temperatures (20°C - 40°C). The diminishment of true relative permittivity by utilizing 1%wt. ZnO nanoparticles is the most elevated diminishment that can happen about the suggested different nanoparticles. However, Fig. 's (49, 50) indicate that the dielectric constant of polyvinyl chloride nanocomposites expands with expanding zinc oxide rate dependent upon 10%wt towards variant temperatures. It is self-evident that the genuine relative permittivity expands with expanding thermal temperatures, especially, towards low frequencies; however, there may be a semi-matching between dielectric estimations under low thermal temperatures (20°C - 40°C).

Figure 52. Measured real permittivity of 5%wt. TiO2/PVC nanocomposite

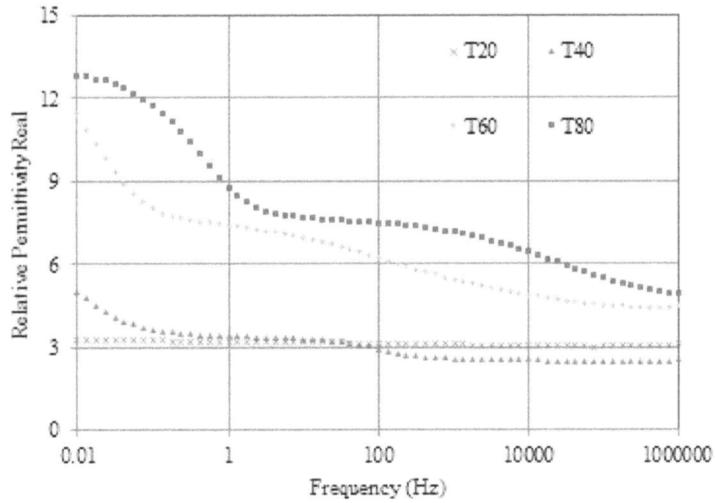

Figure 53. Measured real permittivity of 10%wt. TiO2/PVC nanocomposite

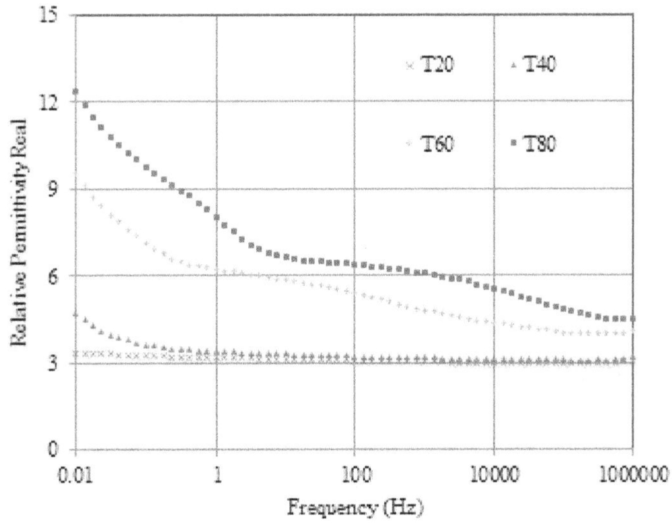

Figure 54. Measured real permittivity of pure polyvinyl chloride

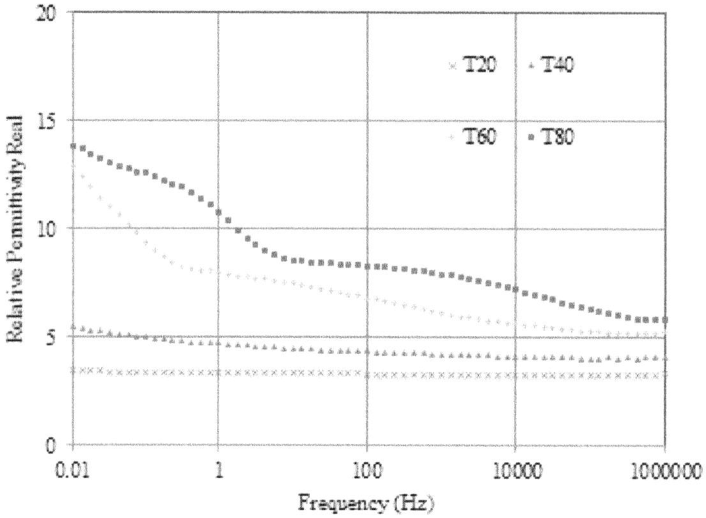

Figure 55. Measured loss tangent of pure PVC nanocomposites

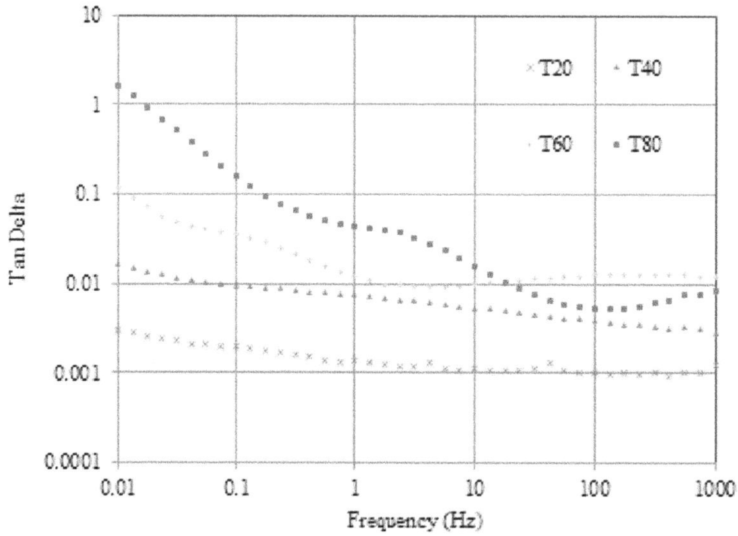

Impacts for titanium dioxide nanoparticles are indicated in fig. 's (51-53) that demonstrate the dielectric constant as a capacity from the controlling recurrence of TiO_2/PVC nanocomposite tests during variant temperatures. Fig. 51 shows that the dielectric steady for TiO_2/PVC nanocomposites declines for the expanding rate of titanium dioxide nanoparticles in the nanocomposite up to 1%wt, with respect to immaculate polyvinyl chloride characterization. On the other hand, the genuine relative permittivity increases with expanding thermal temperatures, especially at low frequencies; however, there may be a semi-matching between the dielectric estimations at high thermal temperatures (20°C - 40°C). Figure 51 depicts that the dielectric constant for polyvinyl chloride nanocomposites expands with expanding titanium dioxide rate dependent upon 5%wt in variant temperatures. However, concerning the illustration demonstrated in Fig. 53, there is a little diminishing of the dielectric constant from controlling polyvinyl chloride with expanding titanium dioxide nanoparticles to 10%wt. TiO_2/PVC nanocomposites during variant temperatures. Thus, there is a nonlinear dielectric characterization from controlling polyvinyl chloride for including variant TiO_2 nanoparticles rates.

Figure 56. Measured loss tangent of 1wt. % Clay/PVC nanocomposites

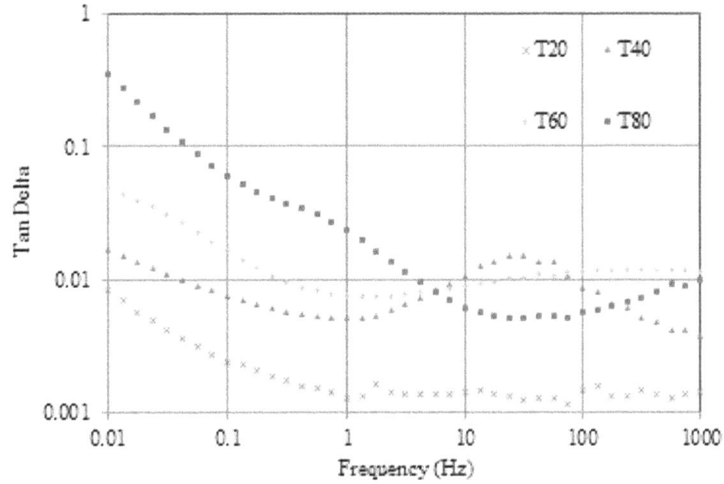

Figure 57. Measured loss tangent of 10wt. % Clay/PVC nanocomposites

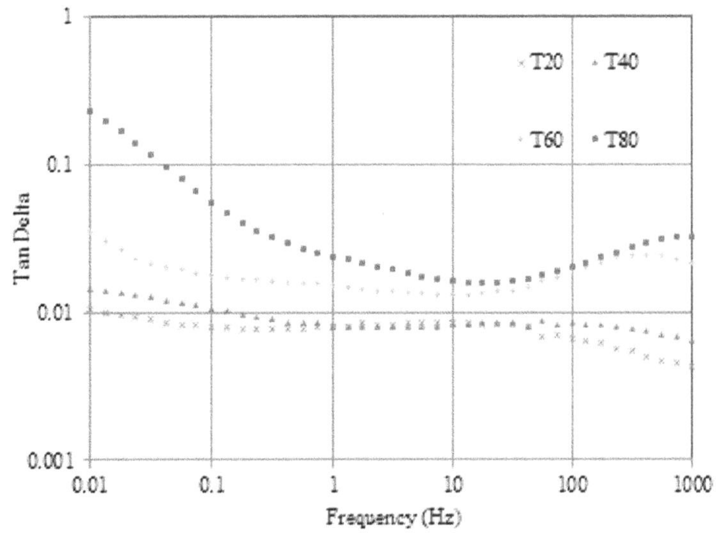

Figure 58. Measured loss tangent of 1wt. % SiO2/PVC nanocomposites

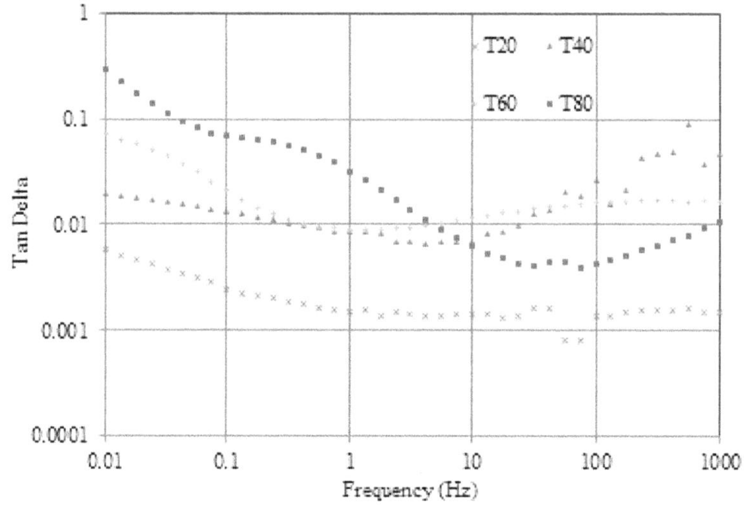

In view of the outcomes, the impact of temperature and sorts of nanoparticles on the dielectric characterization becomes clear. It has been recognized that including variant nanoparticles increments and declines electrical properties of the new polyvinyl chloride nanocomposite materials, as tabulated previously in table 3. The impact on clay nanoparticles is pointed out in Fig's (42-44); then the measured dielectric constant contrasts by examination of the diminishing genuine relative permittivity with expanding the rate of clay nanoparticles up to 1%wt regarding immaculate polyvinyl characterization, especially, towards high frequencies. On the other hand, the true relative permittivity expands with expanding thermal temperatures, especially, towards low frequencies; however, the measured dielectric constant increases with expanding the rate of clay nanoparticles up to 10%wt.

Figure 59. Measured loss tangent of 10wt. % SiO_2/PVC nanocomposites

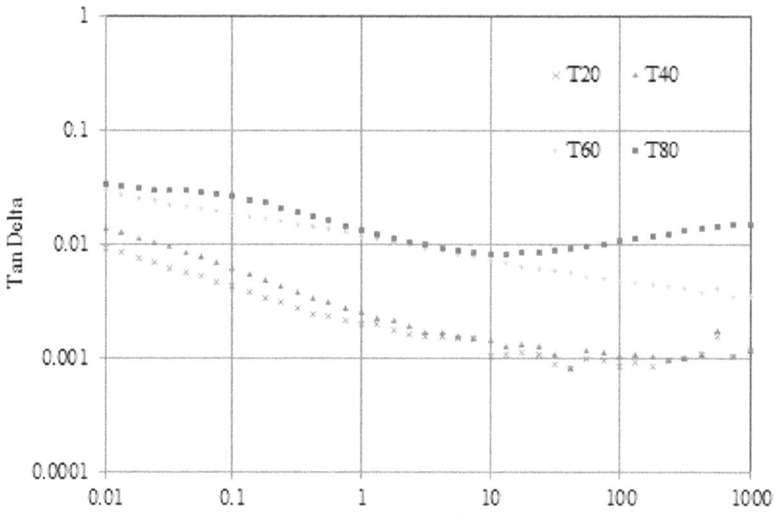

Figure 60.Measured loss tangent of 1wt. % ZnO/PVC nanocomposites

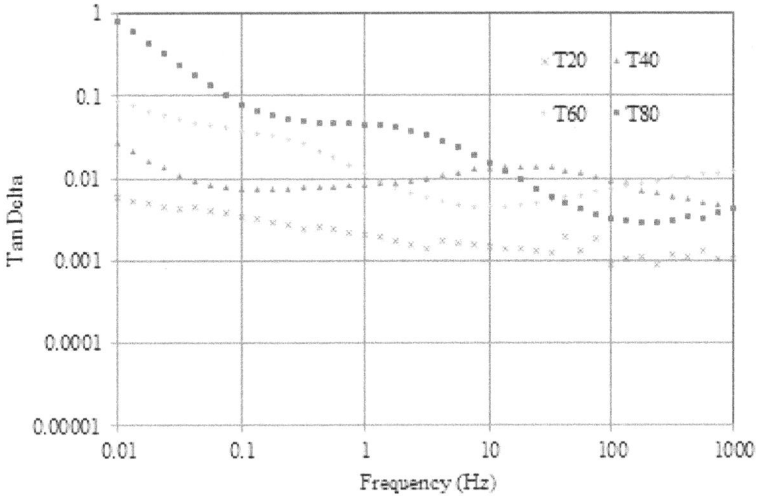

Figure 61. Measured loss tangent of 10wt. % ZnO/PVC nanocomposites

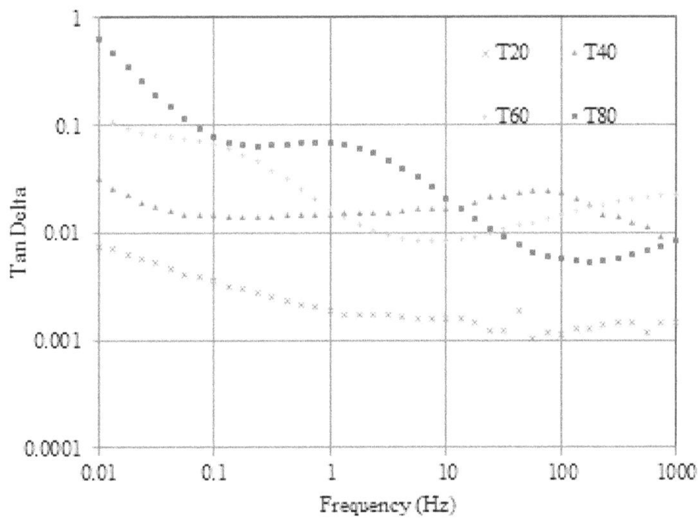

For inorganic nanoparticles, as demonstrated clinched alongside Fig's (47-49), they include fumed silica nanoparticles (up to max. 5%wt.) of polyvinyl chloride abatements in the genuine relative permittivity. The true relative permittivity increases with the expanding thermal temperatures, especially, in low frequencies. For zinc oxide nanoparticles, the impact of expanding zinc oxide nanoparticle rates inside polyvinyl chloride is pointed out in Fig's (48-50), i.e., zinc oxide nanoparticles are successful nanoparticles for diminishing the dielectric constant characterization between the suggested nanoparticles. The diminishment of true relative permittivity by utilizing zinc oxide nanoparticles is the most elevated diminishment that can happen across the proposed different nanoparticles. This similar investigation has specified a nonlinear dielectric characterization at a point including the variant TiO_2 nanoparticles rates clinched alongside polyvinyl chloride, as indicated for Fig's (51-53), i.e. dielectric steady characterization declines and expands with expanding rate of titanium dioxide nanoparticles over a nonlinear conduct. This view is shared regarding controlling dielectric characterization from controlling polyvinyl chloride tentatively under variant recurrence bands and thermal states, taking into account late researches.

Figure 62. Measured loss tangent of 1wt. % TiO_2/PVC nanocomposites

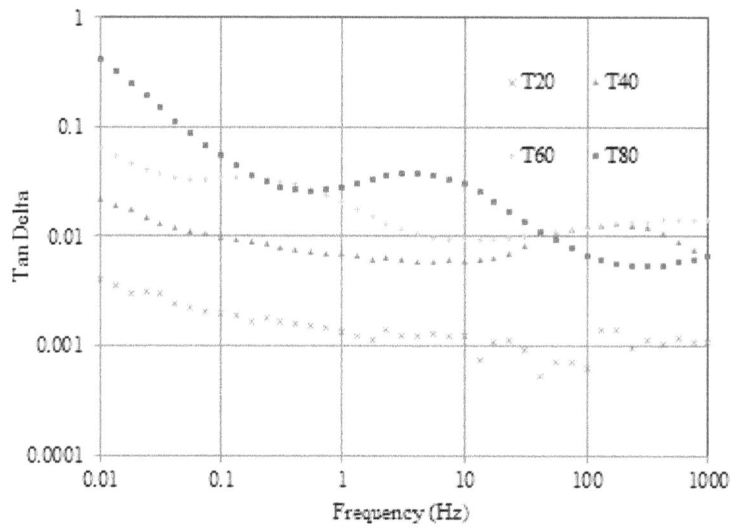

Figure 63. Measured loss tangent of 10wt. % TiO2/PVC nanocomposites

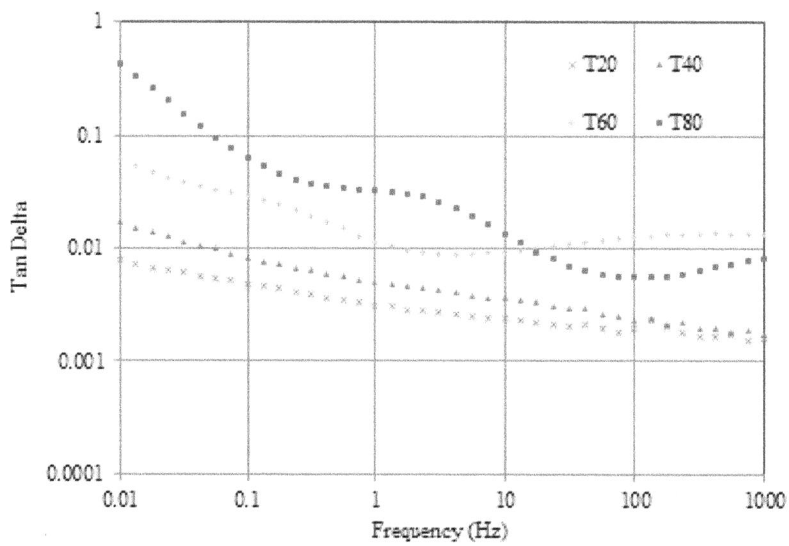

Table 4. Dielectric losses of pure and nanocomposite materials under high frequency [1mhz]

Materials	20oC	40oC	60oC	80oC
Pure PVC	0.01496	0.01777	0.0218	0.0336
1wt. % Clay/PVC	0.09166	0.09172	0.0295	0.0392
1wt. % ZnO/PVC	0.01922	0.01908	0.0500	0.0339
1wt. % SiO2/PVC	0.01106	0.01583	0.0734	0.1398
1wt. % TiO2/PVC	0.01066	0.01070	0.0170	0.02708
5wt. % Clay/PVC	0.00743	0.00872	0.0159	0.0303
5wt. % ZnO/PVC	0.01228	0.01731	0.0479	0.0465
5wt. % SiO2/PVC	0.01087	0.01283	0.0164	0.0209
5wt. % TiO2/PVC	0.01270	0.01135	0.0226	0.03497
10wt. % Clay/PVC	0.01267	0.01331	0.0149	0.0236
10wt. % ZnO/PVC	0.01176	0.01598	0.0340	0.0687
10wt. % SiO2/PVC	0.02266	0.02636	0.0287	0.0376
10wt. % TiO2/PVC	0.01504	0.02011	0.0199	0.02760

Frequency-domain dielectric estimations are performed in the reach between 10^{-2} Hz Also 10^{6} Hz throughout a run starting with 20°C to 80°C, utilizing SOLARTRON gadget that has Vibration: tried understanding of IEC68 (BS2011), Safety: outlined with consent for IEC348 (BS4743). The produced information can be checked towards measuring few examples. Dielectric spectroscopy outcomes demonstrate that for proposed nanocomposites, a nearby "quasi-conductive" district is exhibited as proof by the manifestation of established quasi-DC scattering during sub-Hz frequencies. Figure 54 reveals the dielectric constant characterization, as a capacity of recurrence for immaculate polyvinyl chloride in variant temperatures. The dielectric consistent conduct declines for expanding frequencies, especially at high frequencies. However, the true relative permittivity of immaculate polyvinyl chloride expands with expanding thermal temperatures, especially at low frequencies.

7.3.2 Effect of Nanoparticles on Dielectric Loss of Polyvinyl Chloride

Figure 55 reveals dielectric loss characterization, as a work from controlling recurrence of immaculate polyvinyl chloride in variant temperatures (20°C-80°C). The diminish in loss tangent conduct happens with expanding frequencies up to 1 kHz. In any case, the increase of misfortune loss conduct technique happens with expanding temperatures, especially in instances about low frequencies.

Figures (56, 57) demonstrate the impact of clay nanoparticles on the dielectric loss of polyvinyl chloride tests as a capacity of recurrence and temperature. The conduct technique from controlling misfortune loss declines for expanding clay nanoparticles focuses dependent upon 10wt. % with respect to traditional polyvinyl chloride characterization; especially clinched alongside situations about high thermal states. Note that clay nanoparticles are more proficient in diminishing the dielectric loss for polyvinyl chloride during low frequencies and secondary temperatures (60°C - 80°C). Figures (58, 59) indicate the impact of fumed silica nanoparticles with respect to dielectric reduction of polyvinyl chloride tests, as a capacity of recurrence and temperature. The diminishing clinched alongside loss tangent conduct technique happens with expanding frequencies up to 1 kHz. Moreover, the decline of misfortune loss is clear with expanding centralization of fumed silica nanoparticles dependent upon 10wt. % identified with routine polyvinyl conduct technique in situations for high thermal states of low frequencies. Thus, fumed silica nanoparticles are skilled in diminishing the reduction loss of polyvinyl chloride, especially in case of controlling secondary temperatures (60°C - 80°C).

Figures (60, 61) show the reduction loss execution abatements with expanding frequencies dependent upon 1 kHz. A minor diminishment, or reduction loss conduct happens with expanding focus for zinc oxide nanoparticles dependent upon 10wt. % identified with accepted polyvinyl chloride characterization about low frequencies. Utilizing few numbers of zinc oxide nanoparticles (1wt.%) is the only tip of the iceberg capable of diminishing the reduction loss towards main high temperature levels (60°C - 80°C). Titanium di-oxide nanoparticles diminish the dielectric reduction conduct technique for expanding frequencies dependent upon 1 kHz, as indicated in Fig. (62, 63). It is recognized that the expansion of loss tangent happens for expanding thermal temperatures, especially in low frequencies. Furthermore, little focuses from controlling titanium dioxide nanoparticles are additionally beneficial for diminishing the dielectric loss.

Many factors can influence the choice in controlling ideal nanoparticles identified with economic, modern application, dielectric characterization and thermal stability, reliability…etc. Physical framing of new nanocomposites doesn't progress with the essential compound bonds. It rather advances with the first electric industrial materials without deformations. For upgrading control of modern encasing utilization of nanotechnology techniques (Ebnalwaled & Thabet, 2016; Gouda & Thabet, 2016; Thabet, 2017b; Thabet & Ebnalwaled, 2017; Thabet & Mubarak, 2016b; Thabet & Mubarak, 2017a; Thabet & Mubarak, 2017b), it is paramount to point out the sort and the focus quality in controlling nanoparticles at a sure temperature go. The impact of expanding the fixation of the clay nanoparticles on the dielectric loss for polyvinyl chloride is indicated in Fig. 's (56, 57). i.e. clay nanoparticles are more successful for diminishing loss tangent from controlling polyvinyl chloride with expanding force frequencies at high temperatures (60°C - 80°C) and low frequencies. Moreover, fumed silica nanoparticles are additionally powerful nanoparticles for diminishing dielectric reduction of polyvinyl chloride, especially, in the event of controlling high thermal states (60°C - 80°C), as demonstrated clinched alongside Fig. 's (58, 59).

The impact of controlling expanding centralization for zinc oxide nanoparticles is demonstrated in Fig's (60, 61); i. e, the diminish from controlling misfortune loss happens at little centralization of zinc oxide nanoparticles to ideal worth dependent upon 1wt. % at high temperatures. In any case, the expansion of loss tangent happens with expanding the fixation for zinc oxide nanoparticles more than 1wt.%, especially, at low temperatures. The misfortune loss abates with expanding the fixation for titanium dioxide nanoparticles up to 1wt. % at low temperature levels, as demonstrated clinched alongside Fig. 's (62, 63), and it increments with expanding temperatures, especially at low frequencies. On the other hand, titanium dioxide nanoparticles impacts examine diminishing the dielectric loss particularly towards high temperature levels (60°C - 80°C). Undertaking electronic items and provisions, such as SOLARTRON estimation gadget, has intrigued contemplation of dielectric characterization under high frequencies and variant thermal conditions. It is all the more profitable to portray numerically the impacts of nanoparticles on dielectric passing for polyvinyl chloride under variant thermal states toward 1MHz, as indicated in Table 4. Moreover, it is clear that titanium di-oxide is the best nanoparticle to diminish dielectric reduction of polyvinyl chloride.

7.3.3 Thermal Stability of Nanodielecric Characterization

Including clay diminishes the permittivity for new nanocomposite materials and the impacts of control, including little measurement from the controlling clay nanoparticles rate to low-density polyethylene abatements capacitance and misfortune loss relying upon thermal nanoparticles temperatures. Otherwise, including fumed silica will expand the permittivity from controlling new nanocomposite materials and the impacts of few numbers of fumed silica nanoparticles rate to low-density polyethylene increments capacitance and reduction loss pertaining to thermal nanoparticle temperatures. Thermal solidness for new nanocomposites happens in few numbers of clay or fumed silica nanoparticles. However, including a lot of nanoparticles in low-density polyethylene will make reverse dielectric conduct qualities bit by bit. It is clear that rising temperature from controlling nanocomposites materials impacts nanoparticles temperatures which evolve the dielectric conduct again in the ordinary states. Including fumed silica has expanded permittivity of the new high thickness polyethylene nanocomposite materials, but including clay has diminished permittivity of the new high thickness polyethylene nanocomposite materials. It can be regulated in the reduction loss and capacitance of controlling new high thickness polyethylene nanocomposites relying on kind and focus of nanoparticles in nanocomposites, according to a physical interface between high density polyethylene and nanoparticles. Thermal surroundings are a viable parameter to expand and diminishing the loss tangent and capacitance of new high thickness polyethylene nanocomposites concerning kind and centralization of nanoparticles. Thermal nature's domain states are influenced once the physical interface made between high thickness polyethylene and nanoparticles due to clay nanoparticles is more productive than fumed silica nanoparticles in diminishing charging capacitance and reduction of loss execution under rom thermal states. However, under high thermal conditions, execution of charging capacitance and misfortune loss has turned around gradually. Altered polypropylene provisions, eventually perusing nanotechnology composites, rely upon, focus for nanoparticles and encompassed temperatures. Including fumed silica expands permittivity of the new Polypropylene nanocomposite materials, but including clay abates permittivity of the new Polypropylene nanocomposite materials. By including clay nanoparticles at room temperature (25°C), the loss tangent for polypropylene declines with expanding clay nanoparticles rate up to 1%wt, especially, in low frequencies, but it increments with expanding clay nanoparticles rate up to 10%wt, especially at high frequencies. In any case, the capacitance of Clay/Polypropylene nanocomposite increments expanding clay rate nanoparticles up to 1%wt, yet it decreases for expanding nanoparticle rate up to 10% wt. However, at moderate and high temperatures (40°C:60°C), the reduction loss and capacitance of Clay/Polypropylene abates with expanding clay nanoparticles rate dependent upon 10%wt, especially during high frequencies. Including fumed silica nanoparticles towards space and moderate temperatures (25°C:40°C), the reduction loss and capacitance for fumed silica/Polypropylene nanocomposite increase with the expanding fumed silica rate nanoparticles dependent upon 10%wt, especially at high frequencies. However, during high temperatures (60°C), the loss tangent and capacitance for fumed silica/Polypropylene nanocomposite diminish for expanding fumed silica rate nanoparticles up to 10%wt, especially, at low frequencies. The chosen nanoparticles have succeeded in processing high electrical encasing materials to electrical force frameworks utilizing nanotechnology strategies under separate recurrence and thermal states. The electric and dielectric characterization of the new electrical encasing relies on the type, concentration and dissemination of nanoparticles inside immaculate encasing materials. Clay nanoparticles are costfewer nanoparticles and powerful to diminishing the dielectric passing about polyvinyl chloride at high temperature levels (60°C - 80°C), In any case, the dielectric passing will be fluctuating for diminishing and expanding the rate of clay nanoparticles at low temperatures. On the other hand, fumed silica nanoparticles are more viable nanoparticles over others concentrated on nanoparticles for diminishing the dielectric passing of polyvinyl chloride, especially high temperature levels (60°C - 80°C). The best qualities from controlling zinc oxide nanoparticles for diminishing the dielectric passing are 1wt. %, especially at secondary temperatures. The dielectric passing increments with expanding zinc oxide fixation up to10wt. % especially during low temperature levels. On the other hand, titanium dioxide nanoparticles have the best impacts for diminishing the dielectric passing particularly at bring down focuses (1wt.%) and high temperature levels (60°C - 80°C).

7.4 FOCUS AND RECOMMENDATIONS

The prediction of dielectric constant of nanocomposite materials has been exhibited hypothetically and in the examination in view of exponential control low model. Including high dielectric constant nanoparticles in the polymer progresses the dielectric constant of the matrix and prepares nanocomposites for secondary dielectric constant to be utilized clinched alongside secondary voltage up to date capacitors, such as PP-Al_2O_3, and PP-SiO_2 composites. Furthermore, including nanoparticles for low dielectric constant decreases the dielectric constant of matrix and generates composite for low dielectric constant for utilizing insulations like Epoxy-SiO_2, and PP-ZnO composites. Including clay can diminish the permittivity from controlling new nanocomposite materials and the impacts of including little measure for clay nanoparticles rate on low density polyethylene abatements capacitance and reduction loss (Thabet, Abdelhady, & Abdel-Moamen, 2018; Thabet, Abdelhady, Ebnalwaled et al, 2019a; Thabet, Abdelhady, Ebnalwaled et al, 2019b; Thabet et al., 2020; Thabet, Allam, & Shaaban, 2018; Thabet, Allam, & Shaaban, 2019a; Thabet, Allam, & Shaaban, 2019b; Thabet & Ebnalwaled, 2018; Thabet, Salem, & Essam, 2018).

On the other hand, the little measure about clay nanoparticles rate in high density polyethylene builds capacitance and misfortune tangent, especially at low frequencies. Including clay nanoparticles builds capacitance from controlling high density polyethylene more than expanding fumed silica nanoparticles over high density polyethylene during the same rates. In addition, including fumed silica will expand the permittivity from controlling new nanocomposite materials, few numbers of fumed silica nanoparticles for low density polyethylene expands capacitance and misfortune tangent, especially during low frequencies. Nevertheless, the few number of fumed silica nanoparticles by examining high thickness polyethylene increments capacitance, However, abatements pass tangent, especially during low frequencies. In any case, a lot of clay alternately fumed silica nanoparticles of polyethylene can be opposite dielectric conduct aspects bit by bit regarding the nanoparticles structure for polymer matrix.

The dielectric qualities of the interphase locale help the generally compelling dielectric aspects of the nanocomposite framework. The volume portion of the interphase district progresses non-linearly for the volume portion of the controlling filler, and the permittivity of the nanocomposite progresses non-linearly, as a capacity for filler stacking. Expanding the nanoparticle molecule surface range in the composite framework holds an interphase area with permittivity higher than that of the matrix yields to the expansion of the compelling permittivity. Moreover, the impact of expanding the interphase thickness in the nanocomposite framework has a tendency to increase the successful permittivity. While, the impact of the state parameters on the successful permittivity of the nanocomposite framework has a tendency to decline for the viable permittivity. The interphase permittivity inside is roughly 10% of the matrix period.

The electrical resistivity from the controlling polymers has been measured and indicated as a move from insulating polymers to conducting polymers. It has been a change in the electrical resistivity to the extent that that there are eight requests of extent. The electrical and thermal conductivities attained towards the exfoliated graphite loaded polymers end up being higher than carbon nanofiber and HHT nanofiber- built polymer composites. Directing materials has separate degrees with be the ideal filler, as silver has high conductivity and secondary cost, yet copper has beneficial properties for low expense, but not beneficial properties concerning illustration silver in conductivity. Conductive adhesives are made, eventually perusing the inclusion of nano measured particles for diverse sorts of polymers. Film resistivity is a point that is measured as a work for nanoparticles volume alternate centralization. Micro-sized silver chip for PVA is close to the permeation threshold, the addition of few numbers for nano-sized silver colloids can assist in increasing the conductive way, and subsequently more level of resistivity. However, at the micro-sized silver chip, there is an addition to the structure of the conductive way toward itself, expansion of nano-sized silver colloids can reveal a negative impact on conductivity because of commitment to contact resistances. The film with a higher temperature camwood allays has this negative effect, most likely because of the secondary movement connected with nanosized particles. The conduct of the powerful dielectric constant for every parameter in nanocomposites relies on the quality of the dielectric constant of the inclusion, the quality of interphase dielectric constant and the profile that best depicts the inhomogeneous dielectric properties of the interphase area. Nanoparticles from controlling clay and fumed silica has secondary powerful parameters in the dielectric composite constant and expand the encasing modern materials. Low expense of Clay and fumed silica fillers are utilized within adhesives, protective coatings, joint compounds, plastics, cables, a lot of other different applications and low dielectric constant.

Clay has a useful record conduct technique as a nanoparticle utilized for encasing materials and has a powerful dielectric distinguishing mark to separate polymer modern materials for nanocomposite encasing materials eventually perusing IPL. Nanoparticle clay makes novel costive encasing industrial materials, eventually perusing getting easily diminished compelling dielectric constant with the variety of filler volume portion. Nanocomposite materials for clay and chosen mechanical polymers, as ABS, PE, Epoxy, and PVC are smoothness, diminished for the variety of filler volume portion. Expanding clay filler surface range and its interphase thickness additional smoothness diminishes the viable dielectric constant of the nanocomposite. Corruption levels from controlling permittivity execution vary as stated by kind and focus on nanoparticles that can be pulled in the mechanical polymers to their dielectric properties. Kind and centralization of nanoparticles inside polymers can be used alternately to limit the chain portability and bring about diminishing alternate expanding electric encasing to get alternately the farthest point from controlling versatile charge and the development of accuse transporters in polymer dielectrics.

In this chapter, the vicinity of a certain kind and fixation for nanoparticles inside polyvinyl chloride are the essential factors for confining and alternately permitting the chain portability to cause expanding and alternately diminishing of the electric encasing in that capacity to confine and constrain portable accuse and the development of charge transporters for polymer dielectrics. Thus, the variety of safety and susceptance values towards low recurrence can be expected of the impact of inorganic nanoparticles (SiO_2) inside electrical encasing. It is discovered that trapping properties of matrix is exceptionally altered, eventually perusing the vicinity of clay nanoparticles (cost-less) that are scattered homogenously in polyvinyl chloride up to 10wt. %. Little additives of clay nanoparticles with polyvinyl chloride have demonstrated calculable change in the electric resistivity at separate recurrence. Likewise, the electrical encasing of polypropylene composites commitment to its reactance and conductance worth in polypropylene nanocomposite films in easier recurrence extent brings about the electrical encasing of the nanocomposite films hosting to be influenced, eventually perusing the vicinity of nanoparticles. There is no straight execution knowledge from controlling the expanding rate of clay nanoparticles with respect to polypropylene electrical properties. In any case, the expanding rate of fumed silica nanoparticles increases polypropylene conductance but declines its reactance during high thermal states. Thermal soundness of the new polypropylene nanocomposite films has been confined and with the least softening point about both nanoparticles and polymer matrix for getting streamlined securely provisions. Under high thermal conditions, the impact of the unwinding time of the accuse transporters on the electrical encasing of polypropylene nanocomposite films may be overlooked. Along the lines, the amount of accuse transporters and connected recurrence turns into ruling factors of the electrical encasing about polypropylene nanocomposite films. The vicinity of nanoparticles inside polypropylene will confine the chain portability and bring about expanding electric encasing in which the capacity confinement will be set to versatile accuse and the development of accuse transporters in polymer dielectrics, particularly towards an easier recurrence extent, if the encasing will assume a more significant part. Therefore, the variety from controlling reactance and conductance quality towards low recurrence go can be expected as one of the impacts of inorganic fillers' electrical encasing. Electrical soundness from controlling new nanocomposite films happens in little amounts of clay or fumed silica nanoparticles. However, including a lot of nanoparticles in polypropylene will reverse electrical conduct technique aspects bit by bit. A high thermal state of polypropylene nanocomposite materials will change the electrical conduct again in the typical states.

Expanding centralization of Clay, ZnO, and silica as a second kind of nanoparticle causes diminishing in the successful dielectric consistent of any multi-nanocomposites. The expanding focus of alumina and TiO_2 as a second sort of nanoparticles expands the compelling dielectric constant from controlling any multi-nanocomposites. Utilizing SiO_2 as a second sort of nanoparticle expands the successful dielectric steady from controlling (Clay, ZnO, and Silica)/polymer to multi-nanocomposites. However, it diminishes the successful dielectric consistent of (MgO, TiO_2, and Alumina)/polymer to multi-nanocomposites, eventually perusing the expansion of the volume portion of SiO_2 nanoparticles. Utilizing MgO concerning illustration as a second kind of nanoparticles expands the successful dielectric constant in controlling (Clay, ZnO, Silica, SiO_2, and TiO_2)/polymer for multi-nanocomposites and it diminishes the viable dielectric constant of Alumina/polymer for multi-nanocomposites towards expanding the volume portion in controlling MgO nanoparticles.

REFERENCES

Ebnalwaled, A. A., & Thabet, A. (2016, July). Controlling the optical constants of PVC nanocomposite films for optoelectronic applications. *Synthetic Metals Journal*, *220*, 374–383. doi:10.1016/j.synthmet.2016.07.006

Ebnalwaled, A. A., Yousef, A., Gerges, M. K., & Thabet, A. (2016, January). Synthesis of Nano-Polyimide for Microelectronic Applications. *Journal of Applied Chemical Science International*, *6*(1), 18–30.

Gouda, Thabet, Mubarak, & Samir. (2014). Nanotechnology Effects on Space Charge Relaxation Measurements for Polyvinyl Chloride Thin Films. *International Journal on Electrical Engineering and Informatics, 6*(1), 1-12.

Gouda, O., & Thabet, A. (2014a). *Experimental Measurements for HVDC Breakdown Voltage in Polyvinyl Chloride Insulation Nanocomposite Materials.* IEEE-16th International Middle East Power System Conference, Ain Shams, Egypt.

Gouda, O. E., & Thabet, A. (2014b). Frequency Response Analysis for New Magnetic Power Transformer Composite Crystalline Core. *International Electrical Engineering Journal, 5*(8), 1519–1525.

Gouda, O. E., & Thabet, A. (2014c, December). Thermal Experimental Dielectric Characterization of Cost-Fewer Low-density Polyethylene Nanocomposites. *Advances in Electrical and Electronic Engineering Journal, 12*(5), 537–546. doi:10.15598/aeee.v12i5.1179

Gouda, O., & Thabet, A. (2016). Experimental Verification on Enhancing Electric and Dielectric Phenomena of Transformer Nanofluids. *Materials Research Forum, Materials Sciences and Engineering, International Symposium on Dielectric Materials and Applications (ISyDMA'2016).*

Thabet. (2013a). Influence of Cost-Less Nanoparticles on Electric and Dielectric Characteristics of Polyethylene Industrial Materials. *International Journal of Electrical Engineering and Technology, 4*(1), 58-67.

Thabet. (2013b). Experimental Investigation on Thermal Electric and Dielectric Characterization for Polypropylene Nanocomposites Using Cost-fewer Nanoparticles. *International Journal of Electrical Engineering and Technology, 4*(2), 1-12.

Thabet. (2014). A Study on Space Charge Distribution in LDPE Nanocomposites for Future Electric Power Applications. *European Journal of Electrical Engineering, 17*(3-4), 189-202.

Thabet & Ebnalwaled. (2017). Improvement of surface energy properties of PVC nanocomposites for enhancing electrical applications. *Journal of the International Measurement Confederation, 110*, 78–83.

Thabet & Mobarak. (2012). Experimental Study for Dielectric Strength of New Nanocomposite Polyethylene Industrial Materials. *International Journal of Electrical Engineering and Technology, 3*(1), 353-364.

Thabet & Repetto. (2013). Predicting Effective Permeability of Nanodielectric Composites Bonded by Soft Magnetic Nanoparticles. *International Journal of Chemical, Materials Science and Engineering, World Academy of Science, Engineering and Technology, 7*(11), 354-359.

Thabet & Repetto. (2014). A Theoretical Investigation on Effective Permeability of New Magnetic Composite Materials. *International Journal on Electrical Engineering and Informatics, 6*(3), 521-531.

Thabet, A. (2011a, January). Advanced Simulation Models for Predicting Electric Response and Aging of Nanocomposite Industrial Insulation Materials. *Journal of Engineering Sciences, Assiut University, 39*(1), 73–86.

Thabet, A. (2011b, July). A Model for Electrical Treeing in Epoxy Nano-Composite Insulating Materials. *Journal of Engineering Sciences, Assiut University, 39*(4), 853–869.

Thabet, A. (2011c, September). Advanced Model for Predicting Dielectric Properties of Nanocomposite Industrial Materials. *Journal of Engineering Sciences, Assiut University, 39*(5), 1055–1068.

Thabet, A. (2012a, January). Effect of Nanoparticles on Water Treeing Characteristics in XLPE Industrial Insulating Materials. *Journal of Engineering Sciences, Assiut University, 40*(1), 191–208.

Thabet, A. (2012b, May). Enhancing Performance of High Voltage Metallized Film Capacitors by Using New Industrial Nano-Composites. *Journal of Engineering Sciences, Assiut University, 40*(3), 799–818.

Thabet, A. (2012c, July). Experimental Measurements for Space Charge Relaxation in Novel PVC Thin Film Nano-Composites. *Journal of Engineering Sciences, Assiut University, 40*(4), 1105–1120.

Thabet, A. (2014, December). Experimental Investigation on Thermal Dielectric Characterization for New Cost-fewer PMMA Nanocomposites. *International Journal on Electrical Engineering and Informatics, 6*(4), 631–643. doi:10.15676/ijeei.2014.6.4.1

Thabet, A. (2015a, January). Experimental enhancement for dielectric strength of polyethylene insulation materials using cost-fewer nanoparticles. *International Journal of Electrical Power & Energy Systems, 64*, 469–475. doi:10.1016/j.ijepes.2014.06.075

Thabet, A. (2015b, March). Experimental Study of Space Charge Characteristics in Thin Films of Polyvinyl Chloride Nanocomposites. *International Journal on Electrical Engineering and Informatics, 7*(1), 1–11. doi:10.15676/ijeei.2015.7.1.1

Thabet, A. (2015c, June). Experimental Verification for Improving Dielectric Strength of Polymers by Using Clay Nanoparticles. *Advances in Electrical and Electronic Engineering Journal, 13*(2), 182–190.

Thabet, A. (2015d, December). Percolation Phenomena for New Magnetic Composites and TIM Nanocomposites Materials. *Advances in Electrical and Electronic Engineering Journal, 13*(5), 558–566.

Thabet, A. (2016). Thermal experimental verification on effects of nanoparticles for enhancing electric and dielectric performance of polyvinyl chloride. *Journal of the International Measurement Confederation, 89*, 28–33. doi:10.1016/j.measurement.2016.04.002

Thabet, A. (2016e, April). Experimental Study for Effects of Cost-fewer Nanoparticles on Dielectric Performance of Polypropylene Nanocomposites. *International Journal of Electronics and Electrical Engineering, 4*(2), 134–139. doi:10.18178/ijeee.4.2.134-139

Thabet, A. (2017a). Experimental Control of Dielectric Loss Behavior of Polyvinyl Chloride Nanocomposites under Thermal Conditions. *IEEE, International Middle East Power System Conference*, 12-17.

Thabet, A. (2017b, June). Theoretical Analysis for effects of nanoparticles on dielectric characterization of electrical industrial materials. *Electrical Engineering (ELEN). Journal, 99*(2), 487–493.

Thabet, A., Abdel-Moamen, M. A., & Abdelhady, S. (2016). Effective Magnetic Characterization for New Nanocomposites Industrial Materials Using Multi-Nanoparticles Technique. *IEEE, International Middle East Power System Conference*, 52-57.

Thabet, A., Abdelhady, S., & Abdel-Moamen, M. A. (2018, June). Design of multi-nanoparticles technique for enhancing magnetic characterization of power transformers cores. *Advances in Electrical and Electronic Engineering Journal, 16*(2), 167–177. doi:10.15598/aeee.v16i2.2354

Thabet, A., Abdelhady, S., Ebnalwaled, A. A., & Ibrahim, A. A. (2019a, June). Innovative industrial Cu(In,Ga)Se2 thin film solar cell with high characterization using nanoparticles structure. *Indonesian Journal of Electrical Engineering and Informatics, 7*(2), 382–392.

Thabet, A., Abdelhady, S., Ebnalwaled, A. A., & Ibrahim, A. A. (2019b, October). Improvement Optical and Electrical Characteristics of Thin Film Solar Cells Using Nanotechnology Techniques. *International Journal of Electronics and Telecommunications, 65*(4), 625–634.

Thabet, A., Abdelhady, S., & Mobarak, Y. (2020). Design Modern Structure for Heterojunction Quantum Dot Solar Cells. *Iranian Journal of Electrical and Computer Engineering, 10*(2), 2918. doi:10.11591/ijece.v10i3.pp2918-2925

Thabet, A., Al-Sharif, M. R., Abdel-Moamen, M. A., & El-Nobi, A. (2019). Improvement of Electrical Field Distribution on Non-Ceramic Insulators String Using Nanoparticles. *IEEE, International Middle East Power System Conference.*

Thabet, A., Allam, M., & Shaaban, S. A. (2018, March). Investigation on enhancing breakdown voltages of transformer oil nanofluids using multi-nanoparticles technique. *IET Generation Transmission and Distribution Journal, 12*(5), 1171–1176. doi:10.1049/iet-gtd.2017.1183

Thabet, A., Allam, M., & Shaaban, S. A. (2019a, January). Slowing Positive Streamer Propagation in Silicon and Ester Transformer Oil Using Multi-Nanoparticles Technique. *International Journal of Applied Energy Systems, 1*(1), 15–20.

Thabet, A., Allam, M., & Shaaban, S. A. (2019b). Assessment of Individual and Multiple Nanoparticles on Electric Insulation of Power Transformers Nanofluids. *Electric Power Components and Systems Journal, 47*(4-5), 420–430. doi:10.1080/15325008.2019.1609624

Thabet, A., & Ebnalwaled, A. A. (2018, June). Controlling on attraction forces of water droplets on surfaces of polypropylene nanocomposites coatings. *Transactions on Electrical and Electronic Materials Journal, 19*(3), 1–9.

Thabet, A., El-Dein, A. Z., & Youssef, N. M. (2012). Frequency Modulation Response for Effective Dielectric Response of New Nanocomposite Liquids. International Review on Modelling and Simulation, 5(3), 1348-1355.

Thabet, A., & Hassan, A. (2011, November). Design of Compact Microstrip Patch Antenna with and Without Ground Plane Slot using New Nano-Composite Materials. *Journal of Engineering Sciences, Assiut University, 39*(6), 1375–1385.

Thabet, A., Mobarak, Y., & Kannan, N. (2019). Investigation on optimal nanoparticles for enhancing performance of suspension insulators string. *International Conference on Inventive Material Science applications (ICIMA 2019).* 10.1063/1.5131593

Thabet, A., & Mubarak, Y. A. (2012, September). A Model for Dielectric Characterization of New Nanocomposite Polymeric Industrial Materials. *Journal of Engineering Sciences, Assiut University, 40*(5), 1375–1388.

Thabet, A., & Mubarak, Y. A. (2015, February). Experimental Dielectric Measurements for Cost-fewer Polyvinyl Chloride Nanocomposites. *Iranian Journal of Electrical and Computer Engineering*, *5*(1), 13–22. doi:10.11591/ijece.v5i1.pp13-22

Thabet, A., & Mubarak, Y. A. (2016a, February). Predictable Models and Experimental Measurements for Electric Properties of Polypropylene Nanocomposite Films. *Iranian Journal of Electrical and Computer Engineering*, *6*(1), 120–129.

Thabet, A., & Mubarak, Y. A. (2016b, September). Thermal Experiment Analysis for Dielectric Characterization of High-Density Polyethylene Nanocomposites. *Advances in Electrical and Electronic Engineering Journal*, *14*(3), 295–303.

Thabet, A., & Mubarak, Y. A. (2017a, March). Experimental Enhancement for Electric Properties of Polyethylene Nanocomposites under Thermal Conditions. *Advances in Electrical and Electronic Engineering Journal*, *15*(1), 55–62. doi:10.15598/aeee.v15i1.1727

Thabet, A., & Mubarak, Y. A. (2017b, June). The Effect of Cost-Fewer Nanoparticles on the Electrical Properties of Polyvinyl Chloride. *Electrical Engineering Journal, Springer*, *99*(2), 625–631. doi:10.100700202-016-0392-3

Thabet, A., Mubarak, Y. A., & Bakry, M. A. (2010, September). Developing Characterization of Industrial Polymer Conducting Materials by using Nano-Metric Fillers. *Journal of Engineering Sciences, Assiut University*, *38*(5), 1227–1246.

Thabet, A., Mubarak, Y. A., & Bakry, M. A. (2011, March). A Review of Nano –Fillers effects on Industrial and their characteristics. *Journal of Engineering Sciences, Assiut University*, *39*(2), 377–403.

Thabet, A., & Salem, N. (2017a). Optimizing Dielectric Characteristics of Electrical Materials Using Multi-Nanoparticles Technique. *IEEE, International Middle East Power System Conference*, 716-721.

Thabet, A., & Salem, N. (2017b). Thermal Experimental Verification on Electric and Dielectric Properties for Alumina/LDPE Nanocomposites. *The 4th International Conference on Energy Engineering (ICEE-4)*.

Thabet, A., Salem, N., & Essam, E. M. (2018, June). Mohamed, "Modern Insulations for Power Cables Using Multi-Nanoparticles Technique. *International Journal on Electrical Engineering and Informatics*, *10*(2), 271–279. doi:10.15676/ijeei.2018.10.2.6

Thabet, A., Samir, M., & Mountasser, M. M. (2018). New Design for polypropylene thin film of HV metallized power capacitor using multiple nanoparticles technique. *IEEE, International Middle East Power System Conference.*

Thabet, A., Shaaban, S. A., & Allam, M. (2016). Enhancing Dielectric Constant of Transformer Oils Using Multi-Nanoparticles Technique under Thermal Conditions. *IEEE, International Middle East Power System Conference*, 220-225.

Chapter 8
Space Charge in NanoDielectrics

ABSTRACT

This chapter describes the space charge degradation of nanodielectrics that handled the space charge measurements, thin films nanodielectrics materials. This chapter expects to offer acceptable energetic space charge accumulation to control the new nanocomposite thin films. This chapter draws attention also to characterization of thin films nanodielectrics. This chapter sheds light on the effects of nanoparticles on space charge characterization for nanodielectrics degradation. The forecasting and recommendations of the work is presented in this chapter.

Nanotechnology techniques are utilized for evolving properties of polymers to add up to their reliability in claiming electrical power applications. This chapter displays the impact of small amount of nanometer measure fillers (Clay, ZnO or Al_2O_3) that are homogeneously scattered towards the best of few weight rates in low density polyethylene (LDPE) for space charge characterization. Pulsed electroacoustic (PEA) framework is used to consider the evolving in the accumulation conduction phenomena, the amount of stored charges and the trap density distribution in new low-density polyethylene nanocomposite insulating materials with depolarization present. It has regulated control in electric and dielectric characterization of the new low-density polyethylene nanocomposite insulating materials for upgrading electrical power applications. Space charge framing which is a nanocomposite thin film under dc stress towards different temperatures has been watched utilizing the pulsed electro-acoustic (PEA) system. The Polyvinyl chloride (PVC) / (Clay, fumed Silica, and ZnO) nanocomposite materials, which are aggravated from claiming Polyvinyl chloride (PVC) blended for each standout amongst Clay, fumed Silica, or ZnO nano-size filler, have changed their electrical properties over immaculate Polyvinyl chloride. Therefore, the materials have have been relied upon for a chance to be utilized with more diminutive thickness as an insulation in the same dc applications. However, it is necessary to study the space charge characteristics. Therefore, we have attempted to investigate the space charge formation in thin film of Polyvinyl Chloride (PVC) / (Clay, Fumed Silica, and ZnO) nanocomposite materials under dc electric field at various concentrations of nanoparticles and dc voltages using PEA system. From the lab-test results, it is found out that the space charge shaping is determinedly influenced, eventually perusing the Clay, fumed Silica, and ZnO substance in their nanocomposites. Space charge aggregation control in insulation materials dem-

DOI: 10.4018/978-1-7998-3829-6.ch008

onstrates if conduction is present through a dielectric material, and it may be vital to be recognized and to conceive the measure of accuse put away in the dielectric. In this chapter, space charges lab-tests are carried and applied the neighborhood electric fields, high-field conduction and breakdown phenomena in thin films for Polyvinyl chloride nanocomposite insulation materials. Therefore, space charge characteristics have been investigated, eventually perusing pulsed electro-acoustic (PEA) framework to slim thickness for Polyvinyl chloride nanocomposite materials, filled with separate different nanoparticles similar to Clay and fumed Silica. Starting with the lab-test, the measurements including clay and fumed silica are recognized in Polyvinyl chloride which accelerates their breakdown phenomena. This chapter investigates the impacts of sorts and concentrations of nanoparticles with respect to space charge accumulation control in Polyvinyl chloride nanocomposites and illustrates which are fillers that upgrade Polyvinyl chloride streamlined applications.

Space charges in slim films of polyvinyl chloride nanocomposite insulation materials are made in applied electric fields and influence high-field conduction and breakdown phenomena, including nanoparticles in which polyvinyl chloride transforms their electrical ages, and thus space charge characteristics have been investigated, eventually perusing pulsed electroacoustic (PEA) estimation framework to slim thickness from claiming polyvinyl chloride nanocomposite materials filled with different independent nanoparticles in ZnO, TiO_2, and Al_2O_3. This chapter depicts that space charge accumulation for polyvinyl chloride nanocomposites contrasts with respect to its kind and fixation about nanoparticles.

8.1 SPACE CHARGE MEASUREMENTS

Until now, there has been reporting about weight dielectric properties of polymeric nanocomposite, from claiming which space charge distribution is investigated in detail, such as space charge at different electric stresses, its inception field and so on. Since the majority of engineering dielectrics are concerned with different electrode materials, it is inexorable to recognize if space charge distribution on polymer nanocomposite is touchy with electrode materials. It fortifies us to follow step space charge characterization in the group of dielectric materials. For the pulsed electro-acoustic strategy (PEA), space charge over a nanocomposite is investigated with diverse metal electrode pairs and space charge. However, the concealment instrument for space charge accumulation by including nanoparticles to polymers under high electric field during high engineering has not been clarified yet. Therefore, it is recommended to investigate the charge trapping impact of great prompted possibility towards the interface between polymer and nanoparticles in view of the distinction between their permittivities (Chen et al., 2010; Guastavino et al., 2010; Li et al., 2010; Sarathi et al., 2007; Singha & Thomas, 2008). However, there are few researches concerning the impact of sorts of nanoparticles on space charge conveyance in polymeric nanocomposite. For a consistent advancement in polymer nanocomposites, this investigation depicts the impacts of sorts and concentrations from claiming nanoparticles on new nanocomposite industrial polymer material. All lab-tests conducted have been investigated and examined to identify the impacts of nanoparticles on space charge dynamics in PVC/(Clay, fumed Silica, and ZnO) nanocomposite materials under dc anxiety. Created and lab-testing nanocomposite streamlined materials have been completed by utilizing all lab-test setup and supplies in nanotechnology exploration focus. However, constantly in PEA framework, trial outcomes have been performed. An introductory estimation is conveyed out utilizing two semiconductor electrodes. Electrical breakdown measurements are conveyed by setting examples of a mineral oil vessel, held towards 20°C, and utilizing barrel shaped stainless-steel electrodes of breadth 50 mm and breakdown voltage up to 7.5 kV. Lab-tests are acknowledged applying an expanding dc voltage. The figures displayed hint at space charge distribution at different dc voltage pushed and in variant times. This a chance to contemplate the sure and negative space charge peaks identified with the cathode and anode electrodes discovered, therefore utilizing variant connected voltages and times. It is recognized that our measurements have been acquired by utilizing silicone oil to support an acoustic contact.

Figure 1. PEA system lab-testing equipment's

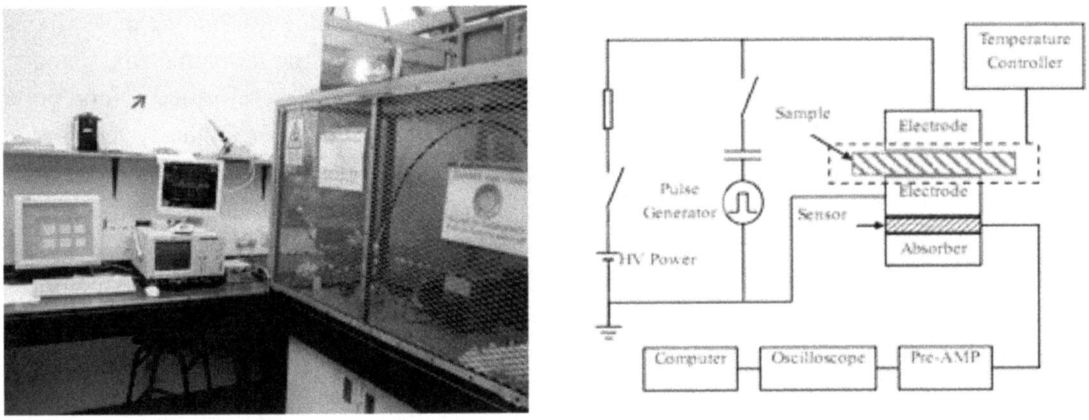

(a) Life photo for experimental PEA system (b) Schematic diagram of PEA system

Acoustic wave proliferation in the thick lab-tests is altogether lessened and scattered because of the passing and dispersive properties of the polymer and guidelines from claiming space charge distribution utilizing the PEA technique. The determinations from claiming every module in PEA framework is undoubtedly as follows: The oscilloscope (LECROY, LC334AM, 500MHz), impulse generator (TECHMP, 10nsec), VHF settled attenuator (30W, 50Hz, 40dB) and balanced out voltage supply (Type L30). Measurements are performed in silicon oil to avoid halfway discharges and keep the temperature steady. The connected voltage is monitored all the while by utilizing an imperviousness voltage divider. Hence, the impact of pre-stressing on impulse breakdown can be recognized with an impulse generator and a coupling capacitor (2000 pF). Figure (1) demonstrates the term photograph and schematic outline of the pea framework utilized in measurement.

8.2 THIN FILMS NANODIELECTRIC MATERIALS

Nanocomposites can, in principle, have a chance to be shaped starting with clays and organoclays to an amount of claiming approaches, including different in situ polymerization (Vaugh 2008; Tanaka, T., Matsuo, Y., & Uchida, K. (2008), result (Bamji et al., 2005; Vaughan et al., 2006) and latex (Chen et al., 2010; Fabiani et al., 2009) techniques. However, the best investment that must be seized is melt transforming (Andritsch et al. 2010) which is by and large the most economical and all the more adaptable for formulation. It also includes aggravating and creation offices regularly utilized for business acts. For most purposes, complete exfoliation of the clay platelets, i.e., detachment of platelets from each other, scattered separately in the polymer matrix, will be the aspired objective of the arrangement procedure. There are three general systems utilized to prepare organic–inorganic nanocomposites: the natural part which can be acquainted. (i) a precursor, which can be a monomer or an oligomer, (ii) a preformed linear polymer (in molten, solution, or emulsion states), or. (iii) a polymer network, physically (e.g., semi-crystalline straight polymer) or artificially (e.g., thermosets, elastomers) cross-linked.

The mineral part can be presented as follows: (i) A precursor or (ii) Preformed nanoparticles. Natural or inorganic polymerization is by and large essential in case a standout amongst the beginning moieties is a forerunner. This prompts the accompanying three strategies for the preparation of organic–inorganic nanocomposites as stated by the beginning materials and preparing techniques: mixing or mixing, sol–gel courses and in situ polymerization. Mixing is basically blending of the inorganic fillers under the polymers; A sol–gel procedure can be performed in-situ in the vicinity of a preformed natural polymer or completed at the same time throughout the polymerization of the monomer(s); the system for in situ polymerization primary includes the scattering of inorganic fillers in the monomer(s) taken after by polymerization. In addition, significant exertions have been dedicated to the plan and regulated creation of mixture nanocomposite particles for custom-made morphologies recently.

Generally, there are three preparative routines to integrate polymer/inorganic filler nanocomposites. The primary is immediate blending or mixing of the polymer and the inorganic fillers, as discrete stages (known as melt mixing) or in result (solution mixing). The second is sol–gel process, which begins with a sub-atomic forerunner at encompassing temperature and manifestations of metal oxide schema towards hydrolysis and buildup. The third is in situ polymerization from claiming monomers in the vicinity from claiming fillers, as indicated in Fig. (2). Preparation from claiming contemplated polymers has utilized SOL-GEL technique. The sol-gel preparation of the nanoparticles inside the polymer broken down control in non-aqueous or watery result is the perfect gas system for the structuring from claiming interpenetrating networks between inorganic and natural moieties during the milder temperature control in enhancing handy similarity and engineering solid interfacial cooperation between two stages. This methodology has been utilized effectively to prepare nanocomposites with nanoparticles in a range of polymer matrices. Few methodologies of the sol-gel transform are related to the arrangement of the mixture materials. You quit offering on that one technique including the polymerization from claiming natural utilitarian gatherings from a preformed sol-gel system. The sol- gel methodology is a rich chemistry, which has been reviewed elsewhere in the transforming of materials from glass to polymers. The organic–inorganic mixture nanocomposites including polymer and nanoparticles are synthesized through sol–gel strategy at encompassing temperature. The inorganic period is created in situ towards hydrolysis–condensation of tetraethoxysilane (TEOS) in distinctive concentrations under corrosive catalysis in vicinity of the natural phase polymer and disintegrated control in formic corrosive (Fabiani et al., 2009).

Figure 2. Scheme shows the three general approaches for preparation of polymer/inorganic filler nanocomposites in ScCO2.

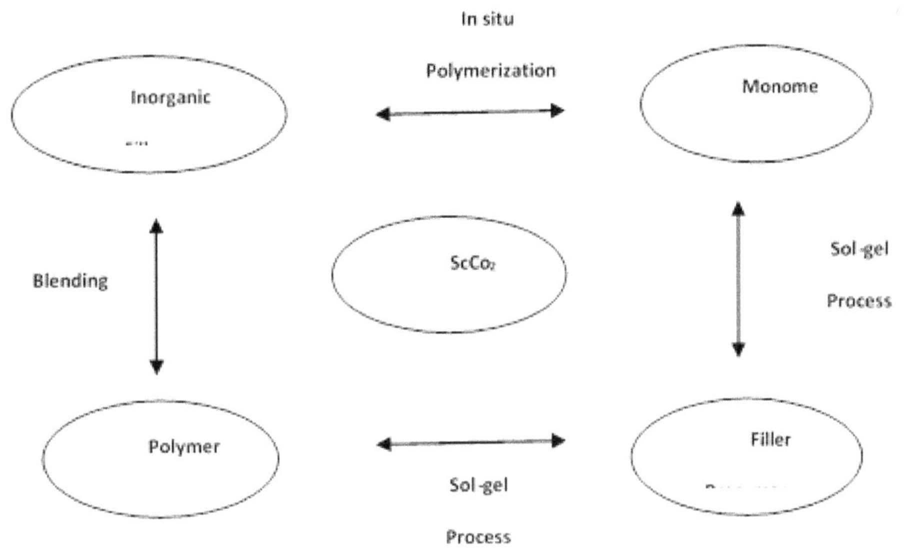

The scattering of nanoparticles is a vital element for the last properties of nanocomposites. Nanoparticles have a tendency to structure agglomerates and groups over a polymer matrix because of their high surface vitality (Chen, Y., Imai, T., Ohki, Y., & Tanaka, T. (2010); Friedrich, K., & Schlarb, A. K. 2008). The agglomeration may prompt a crumbling of the pointed properties of the last results. Therefore, as with handy and stable scattering of nanoparticles in polymer matrices, few strategies are connected to split the agglomerates. There are two approaches: Mechanical scattering methods, including ultranationalistic vibration (Chisholm et al., 2005; Lam & Lau, 2006), exceptional sol-gel systems (Baraton, 2003; Matějka et al., 2000), high shear vitality scattering blending (Wetzel et al., 2003) and surface change of nanoparticles (Sun et al., 2004; Zhu et al., 2004). The surface adjustment is a concoction technique for enhancing the similarity and interaction between polymer and joined filler, prompting improved scattering (Geng et al., 2008).

The filler geometry is an enter element that impacts the scattering of nanoparticles. In general, low-dimensional fillers are additional and troublesome in scatter over three-dimensional. The distinction arises starting with the reality that three-dimensional quasi-spherical particles show best point-to-point contacts, while one-dimensional rods or tubes can have contact along the full length of the cylinder, which builds the particle-particle connection, two-dimensional sheets are moving to a bigger contact territory. The expanded molecule contact territory and association considerably settle on a homogeneous scattering, that's only the tip of the iceberg challenging. Therefore, it has been decided to investigate round particles as much as possible to scatter them over rods or sheets (Sun, 2010). The surface functionalization from claiming particles can be acknowledged for a surfactant or coupling agenize. A surfactant is compound that lowers the interfacial strain of the middle of a polymer and robust filler. Indeed, an exceptional option is the utilization from claiming coupling operators.

Coupling Agent

Stable scattering for filler in the last composite is vital for dispensing with filler agglomerates that might go about as feeble concentrations to actuate electrical or mechanical disappointment. A coupling agenize is a compound that is connected to the surface of a material that should be changed to make it perfect with an alternate material of a separate nature (Özdilek, 2006). The atomic structure empowers the coupling agenize on fill in as a go-between control in holding natural and inorganic materials (Yung et al., 2008). The initial coupling operators were connected as late as 1940s, when glass fibers saw their principal use for upgrading the properties of natural polymers. An assortment for coupling agents, for example, such as silanes, zirconates, titanates and zircoaluminates has been acquainted to the showcase from that point forward, so as on enhance the interface between the polymer and the filler (Levering, 1995). The significant parts of surface functionalization are: • Settling the nanoparticles inside a polymer matrix (Obtaining a great scattering of the particles) through compound bonding, • Acquiring thermodynamical and concoction similarity between polymer matrix and joined filler to stay with the divided particles starting with each other; • Moving forward the bond during the interfaces between matrix and particles (Reed, 2010).

Fabrication Precautions

The nanoparticles are not risky concerning illustration. Concerning illustration, they are blended under a polymer host, due to the solid holding between the filler and matrix. Safety measures are necessary to keep the arrival of nanoparticles under air. Fume hoods and respiratory masks are necessary to be utilized in the factories, which process nanocomposites, so as to secure lungs against airborne nanoparticles. A little under-pressure in the attempting territory is prescribed to forestall get away from nanoparticles to different working regions. Ventilation frameworks must be prepared for nano confirmed filters which prevent outpouring of nanoparticles of the surroundings (Cherkasova, 2009; Schulte & Salamanca-Buentello, 2007). Toxicological viewpoints must be taken into account, when the lifetime of a composite will be reused or arranged. A particular case which needs caution is the likelihood of nano-sized particles being discharged in the air, as an aftereffect of burning and being continuously breathed by people (Agarwal et al., 2005). The determination of the poisonous quality of nanoparticles is a truly expensive, confounded and drawn out investigation (Oberdörster, 2005; Thomas & Sayre, 2005). The expense of a toxicological contemplate is high, yet inconsequential compared to the fines and punishments paid to the organizations for asbestos victimized people trials. More fill in is necessary for the new nano materials to evaluate their poisonous quality and wellbeing dangers (Hurt et al., 2006; Maeon et al., 1988; Oyegoke, 2001).

An additional issue is the right understanding of the information obtained utilizing these complex bio/nanomaterial frameworks. A deserving objective to toxicologists and material researchers is the joint improvement of 'green' nanomaterial formulations – the individuals optimized for capacity and negligible wellbeing sway. The European requisition recognizes security issues and precautionary standards to nanotechnology requisition about everyday support. Warm medication from claiming nanoparticles in vacuum in front of the surface of functionalization might free up the hydroxyl accumulations on the surface of the nanoparticles to make them accessible to hydrogen holding for the polymer matrix. The impact from claiming temperature medication can be acknowledged concerning illustration of an extra venture to enhance the contact between the polymer matrix and presented filler.

Electron Microscopes (EMs) Experimental Setup

Sort supplies utilized in the estimation from claiming specimens and lab-test measurements in this postulation, such as electron microscopes (EMs), work precisely as their optical counterparts but they utilize a concentrated shaft for electrons as opposed to light with the "image" of the example and increased data concerning illustration of its structure and creation. The essential steps included in the whole electron microscopes EMs are the following: A stream about electrons will be framed over high vacuum (by electron guns). This stream may be accelerated towards the example (with a certain electrical potential). At the same time, there will be limited and kept tabs utilizing metal apertures and attractive lenses under a thin, focused, monochromatic shaft. The example is Lighted, eventually perusing the shaft and collaborations happen inside the Lighted sample, influencing the electron shaft. These associations' impacts are distinguished and changed under a picture. The oversteps are conveyed out at EMs in any case about the kind. For the purpose of point by point materials characterization, two intense instruments are used: The filtering electron magnifying instrument (SEM) and the Transmission Electron Microscope instrument (TEM) (Brundle et al., 1992; Chen et al., ; Ciuprina & Plesa, ; Gouda et al., 2013; Masuda et al., 2007; Mayer, 2007; Muhamad et al., 2009; Murakami et al., 2008; Saha & Purkait, 2004).

Nanoclay: It is a nanomer 1. 30E, clay surface altered for 25-30wt.% octadecylamine. The principle constituent of nanoparticle clays critical amounts of other nanoclays is frequently introduced. Circular molecule shape is the major essential trademark of nanoclay for polymer applications. The platy way implies that clay fillers have more amazing impact on properties, such as viscosity, firmness and strength. Utilizing clay as nanoparticle provides for large amounts of fire retardancy of the generated composite and it is chosen in this study. Cosset les about clay impetus can be the best filler of nanoparticles modern materials. Fumed silica (SiO_2) is feathery white powder with a greatly low density, showcased under professional names, such as Aerosol and Cab-o-sil. For both hydrophobic and hydrophilic evaluations available, it may be broadly utilized concerning illustration of a rheology modifier, conferring exceedingly thixotropic properties at moderately low rates. Likewise, it can give expanded track, finer solidness in suspensions, keeps "sagging" and settling of solids for a fluid framework. For this reason, it will be especially suitable to coatings, inks, adhesives, resins, sealants and greases. Fumed silica, or fumed silicon dioxide, will be generated, eventually perusing the vapor-phase hydrolysis from claiming silicon tetrachloride in an H2/O2 fire. Hydrophilic fumed silica bearing hydroxyl gatherings by looking into its surface will be generated towards this transform. Fumed silica powders are utilized within paints and coatings, silicone elastic and silicone sealants, adhesives, link mixes and gels, printing inks and toner and plant insurance. Zinc oxide (ZnO): Is a prevalent cross-linker for elastic and for different resins. It is likewise utilized as a UV stabilizer, and it has a moderately high refractive file which makes it a proficient white pigment. Zinc oxide is an inorganic compound with the equation ZnO. It has high refractive index, high warm conductivity, non-toxic and perfect with skin, making it a suitably added substance to materials and surfaces that come in contact with people. Zinc oxide is utilized as an impetus for methanol union. The increment control in surface range for Nano scale zinc oxide contrasted with bigger powders has the possibility of enhancing the effectiveness of these methods.

Polyvinyl chloride (PVC) is the major generally utilized thermoplastic, polymerized vinyl chloride. It is transformed starting with ethylene and anhydrous hydrochloric corrosive. PVC is stronger and more unbending over other general reason thermoplastic materials. It has a high elasticity and modulus of flexibility. Additives are used for particular wind uses, such as warm stabilizers, lubricity, sway modifiers and pigmentation. There are two fundamental types of PVC unbending and plasticized. Unbending

PVC, as its name suggests, is an unmodified polymer that exhibits high unbending nature. Unmodified PVC is stronger and stiffer than PE and PP. Plasticized PVC is changed by s, as a low sub-atomic weight species to flexibilize the polymer. Plasticized PVC can be figured to provide for items with rubbery self-destructive considerations and conduct. It may be altered by s for styrene butadiene elastic which enhances score sturdiness and effect quality. PVC's are essentially extreme and strong, oppose water and abrasion and are phenomenal electrical insulators.

Sol-gel process has been reviewed elsewhere in the preparation of materials from glass to polymers. The organic–inorganic mixture nanocomposites, including polymer and nanoparticles, are synthesized through sol–gel technique in encompassing temperature. The inorganic period is produced in situ towards hydrolysis–condensation of tetraethoxysilane (TEOS) in distinctive concentrations, under corrosive catalysis, over vicinity of the natural phase, polymer broken down in formic corrosive (Murugaraj et al., 2005; Okuzumi et al., 2008; Vella et al., 2000; Zha et al., 2010). The sol-gel preparing of the nanoparticles inside the polymer disintegrated in non-aqueous or watery result will be the Perfect methodology to the arrangement about interpenetrating networks between inorganic and natural moieties towards the milder temperature for moving forward to great similarity and engineering solid interfacial communication between the two stages. This methodology has been utilized effectively to obtain ready nanocomposites with nanoparticles for an extent from claiming polymer matrices. Few methodologies for the sol-gel procedure are related to the creation of the mixture materials. One system includes the polymerization of natural utilitarian gatherings starting with a preformed sol–gel organize. The sol- gel methodology is a rich science which has been reviewed elsewhere on transforming from claiming materials starting with glass to polymers. The organic–inorganic mixture nanocomposites, including polymer and nanoparticles, are synthesized through sol–gel method in encompassing temperature. The inorganic stage is produced in situ towards hydrolysis–condensation of tetraethoxysilane (TEOS) in distinctive concentrations, under corrosive catalysis, to vicinity of the natural phase, polymer broken down in formic corrosive (Bois et al., 2009; Huang et al., 2009).

Low-thickness polyethylene is a thermoplastic made from petroleum. Particles of LDPE are less hard stuffed and less crystalline due to the side branches and low thickness. Low-thickness polyethylene holds the concoction components of carbon and hydrogen. Infiltration of nanoparticles within polyethylene and polyvinyl chloride is demonstrated by SEM measurements, as demonstrated in Figure 3.

Figure 3. SEM images for polyvinyl chloride and low-density polyethylene nanocomposite films

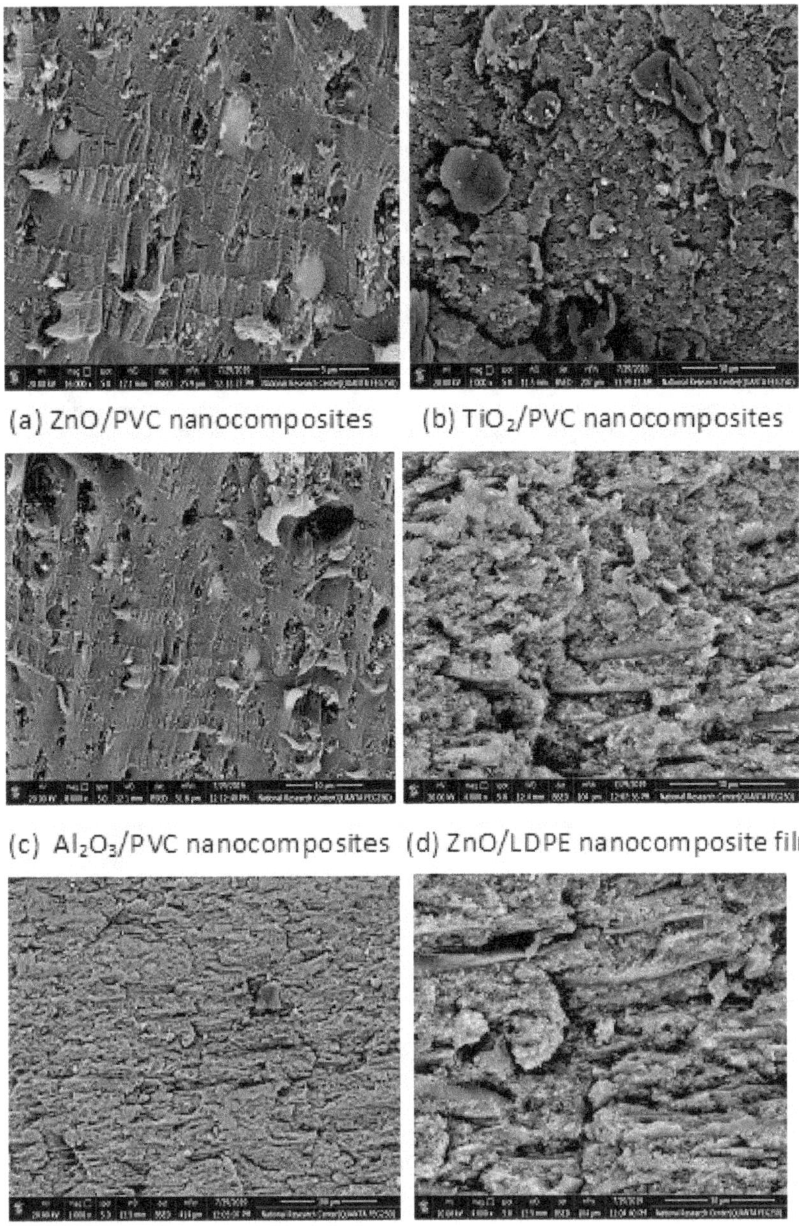

(a) ZnO/PVC nanocomposites (b) TiO_2/PVC nanocomposites

(c) Al_2O_3/PVC nanocomposites (d) ZnO/LDPE nanocomposite film

(e) Clay/LDPE nanocomposite film (f) Al_2O_3/LDPE nanocomposite film

Figure 4. Photo for lab-tested samples under DC electric field breakdown for nanocomposite of Polyvinyl Chloride PVC with 10%Fumed Silica

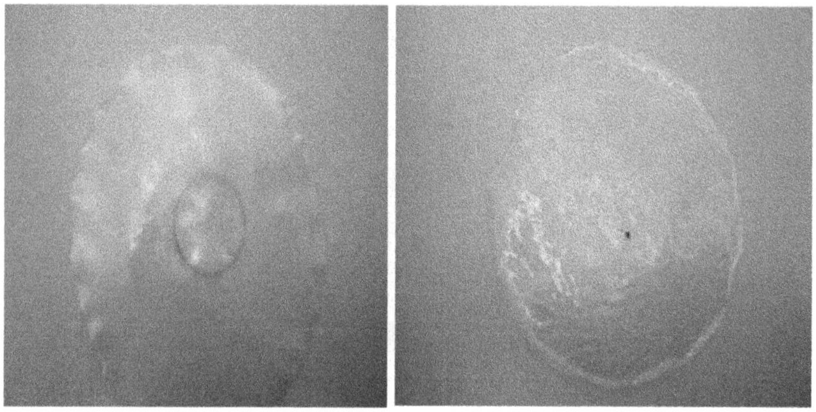

Table 1. Electric and Dielectric Properties of LDPE Nanocomposite films

Materials	Dielectric Constant at 1kHz	Resistivity (Ω.m)
Pure LDPE	2.3	10^{14}
LDPE + 1%wt Clay	2.23	10^{15}
LDPE + 5%wt Clay	1.99	10^{15}-10^{18}
LDPE + 10%wt Clay	1.76	10^{18}-10^{20}
LDPE + 1%wt ZnO	2.28	10^{14}
LDPE + 5%wt ZnO	2.27	10^{14} -10^{15}
LDPE + 10%wt ZnO	2.25	10^{15}
LDPE + 1%wt Al2O3	2.42	10^{13}
LDPE + 5%wt Al2O3	2.59	10^{13}-10^{11}
LDPE + 10%wt Al2O3	2.97	10^{11}-10^{9}

8.3 CHARACTERIZATION OF THIN FILMS NANODIELECTRICS

HIOKI 3522-50 LCR Hi-lab-tester device measures electrical parameters of nano-metric strong dielectric insulation examples at different frequencies, therefore; it can constantly measure dielectric properties for immaculate nanocomposite industrial materials. The contemplated modern materials here are low-thickness polyethylene nanocomposite films that have been figured towards utilizing nanotechnology strategies and nanoparticles (Dia.: 10nm) of clay (montmorillonite-type), ZnO and Al_2O_3. The base ofevery last bit of these materials is an economically accessible material recently used in the manufacture of high-voltage (HV) modern results and their properties point by point in Table 1. The streamlined material contemplated here is Polyvinyl Chloride which is figured by using nano particulates of clay. The base of all these polymer materials is an economically accessible material generally used in the manufacture of high-voltage (HV) modern results and their properties point by point in Table (2). Additives from claiming clay, ZnO, TiO_2, Al_2O_3 and fumed silica nanoparticles of the base mechanical polymers are created by utilizing mixing, ultrasonic, additives of clay, zinc oxide and fumed silica nanoparticles of the built modern polymers. Polyvinyl Chloride has been created towards utilizing mixing, ultrasonic and warming forms. Arrangements for contemplated Polyvinyl chloride nanocomposites utilize the SOL-GEL technique. The sol-gel transforming of the nanoparticles inside the polymer disintegrated in non-aqueous or watery result is the Perfect methodology for the structuring of interpenetrating networks between inorganic and natural moieties during the milder temperature control in moving forward with handy similarity and engineering solid interfacial collaboration between the two periods. This procedure has been utilized effectively to obtain ready nanocomposites for nanoparticles and for an extent of polymer matrices.

Table 2. Dielectric properties of pure and nanocomposite materials

Materials	Dielectric Constant at 1kHz	Resistivity (Ω.m)
Pure PVC	3.3	10^{13}
PVC + 1%wt Clay	3.25	10^{13}
PVC + 5%wt Clay	3.04	10^{13}-10^{14}
PVC +10%wt Clay	2.85	10^{14}-10^{15}
PVC + 1%wt Fumed Silica	3.35	10^{13}-10^{12}
PVC + 5%wt Fumed Silica	3.42	10^{13}-10^{12}
PVC + 10%wt Fumed Silica	3.5	10^{12}-10^{11}
PVC + 1%wt ZnO	2.99	10^{13}-10^{14}
PVC + 5%wt ZnO	2.75	10^{14}-10^{15}
PVC + 10%wt ZnO	2.5	10^{15}-10^{16}
PVC + 1%wt ZnO	3.24	10^{14}
PVC + 5%wt ZnO	3.19	10^{14}-10^{15}
PVC +10%wt ZnO	3.04	10^{15}-10^{16}
PVC + 1%wt TiO_2	3.94	10^{12}
PVC + 5%wt TiO_2	4.64	10^{12}-10^{11}
PVC +10%wt TiO_2	5.17	10^{11}-10^{10}
PVC + 1%wt Al_2O_3	3.39	10^{13}
PVC + 5%wt Al_2O_3	3.59	10^{12}-10^{11}
PVC + 10%wt Al_2O_3	3.99	10^{12}-10^{11}

8.4 EFFECTS OF NANOPARTICLES ON SPACE CHARGE CHARACTERIZATION

PEA system measurements are conveyed out utilizing two semiconductor barrel-shaped stainless steel electrodes with breadth 50 mm, putting examples in a mineral oil vessel kept at 20°C. Lab-tests are understood by applying an expanding dc breakdown voltage dependent upon 7.5kV. Every measurement shows space charge distribution towards different dc voltage pushed and variant time. It is a chance to see the certain and negative space charge peaks identified with the cathode and anode electrodes that are discovered, hence utilizing variant connected voltages and times.

Figure 5. Space charge profiles for pure Low- density polyethylene with varying DC electric field

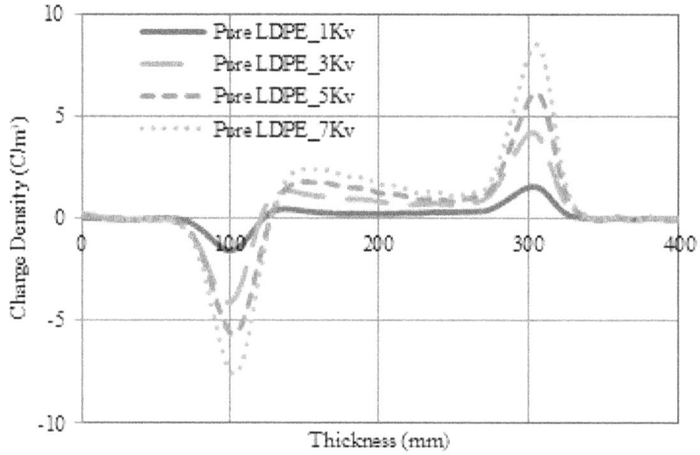

Figure 6. Space charge profiles for 1%wt. ZnO /LDPE nanocomposite films

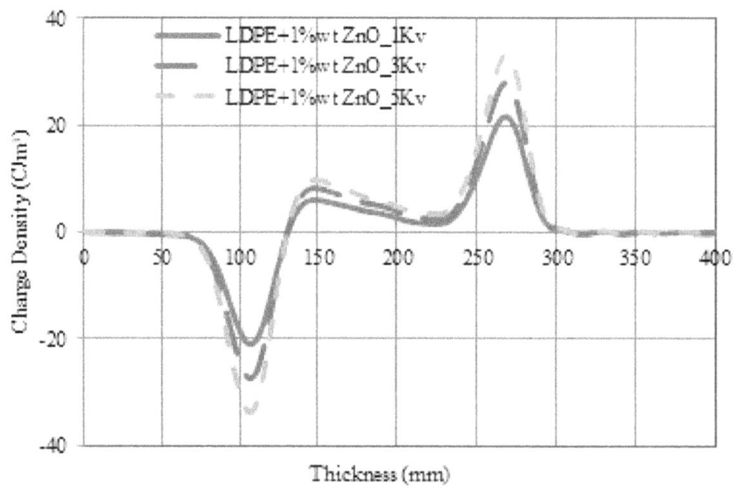

Figure 5 indicates the measurements that infer the infusion of impulse high voltages on the accuse thickness for immaculate PVC dielectric materials; the electrode will be towards the cleared out and the anode will be towards the good and the charge thickness builds expanding impulse voltages and -charge is gathered close the electrode. On the other hand, Fig. 6 reveals space charge distribution important ZnO/LDPE nanocomposite films with 1%wt rate of zinc oxide nanoparticles under changing dc electric fields; the accuse thickness towards both electrodes expand for expanding impulse connected voltages, and there may be high hetero-charge gathered towards the electrode. However, Fig. 7 contrasts space charge circulation applicable ZnO/LDPE nanocomposite films with 5%wt rate of zinc oxide nanoparticles under fluctuating dc electric fields. The charge thickness at both electrodes expands with expanding impulse connected voltages and there will be high hetero-charge gathered between electrodes. Figure 8 reveals the space charge distribution important ZnO/LDPE nanocomposite films with 10%wt. rate of zinc oxide nanoparticles under changing dc electric fields; the charge thickness at both electrodes expands with expanding impulse connected voltages, and there may be high hetero-charge gathered in electrode. Expanding zinc oxide nanoparticles starting with 5%wt up to 10%wt declines accuse thickness in the nanocomposite particles.

Figure 7. Space charge profiles for 5%wt. ZnO / LDPE nanocomposite

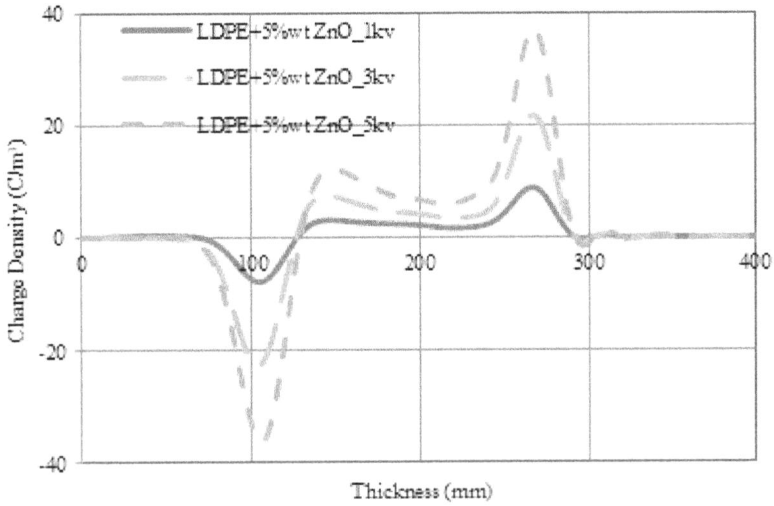

Figure 8. Space charge profiles for 10%wt. ZnO / LDPE nanocomposite films

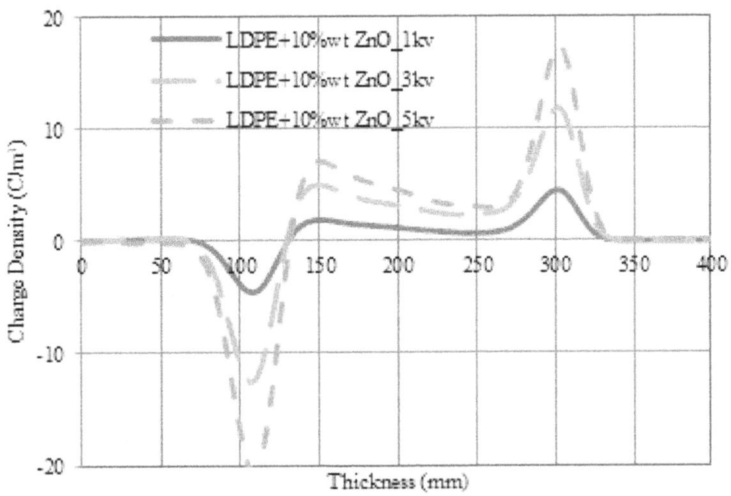

Figure 9. Space charge profiles for 1%wt. Clay / LDPE nanocomposite films

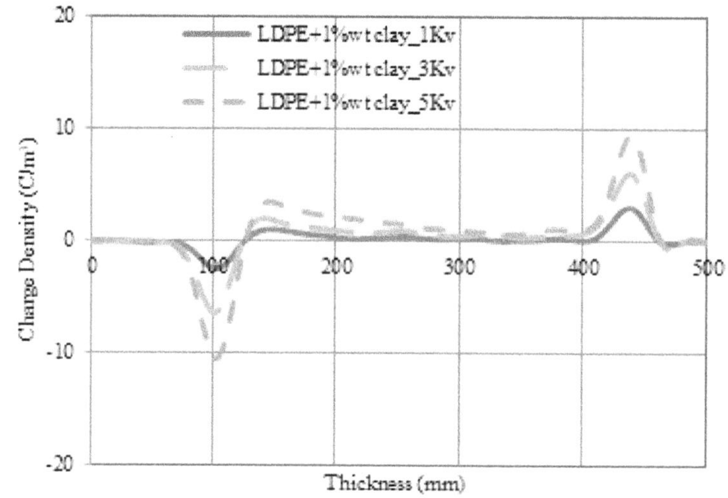

Figure 10.Space charge profiles for 5%wt. Clay / LDPE nanocomposite films

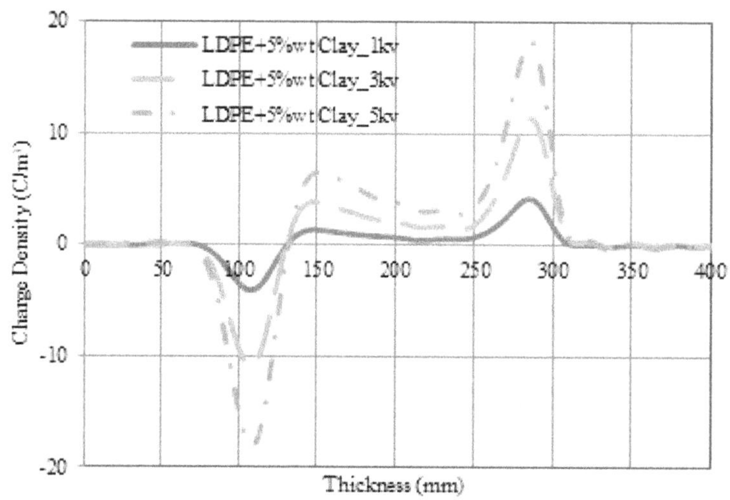

8.4.1 Effect of Clay Nanoparticles on Space Charge Distribution

Figure 9 demonstrates space charge distributions important on Clay/LDPE nanocomposite films for 1%wt. rate of clay nanoparticles under different dc electric fields. It is recognized that charging thickness during the anode is more than charging thickness in the electrode for expanding impulse connected voltages, and there is a hetero-charge gathered close the electrode. However, figure 10 demonstrates space charge distributions pertinent to Clay/LDPE nanocomposite films with 5%wt rate of clay nanoparticles under different dc electric fields. The measurements suggest the infusion of impulse high voltages that depict accuse thickness during the anode and cathode that may be expanded and there are hetero-charge gathered charges between both electrodes. It is obvious that expanding clay nanoparticles starting with

1%wt up to 5%wt expands accuse thickness in the nanocomposite particles. Figure 11 demonstrates space charge distributions important on Clay/LDPE nanocomposite films for 10%wt rate of clay nanoparticles under different dc electric fields. The measurements infer the infusion of impulse high voltages that depict charge thickness during the anode and cathode which can be expanded, and there are homo-charge gathered charges between the anode or cathode electrodes. It is acknowledged that expanding clay nanoparticles starting with 5%wt up to 10%wt increments charge thickness in the nanocomposite molecules between electrodes.

Figure 11. Space charge profiles for 10%wt. Clay/ LDPE nanocomposite films

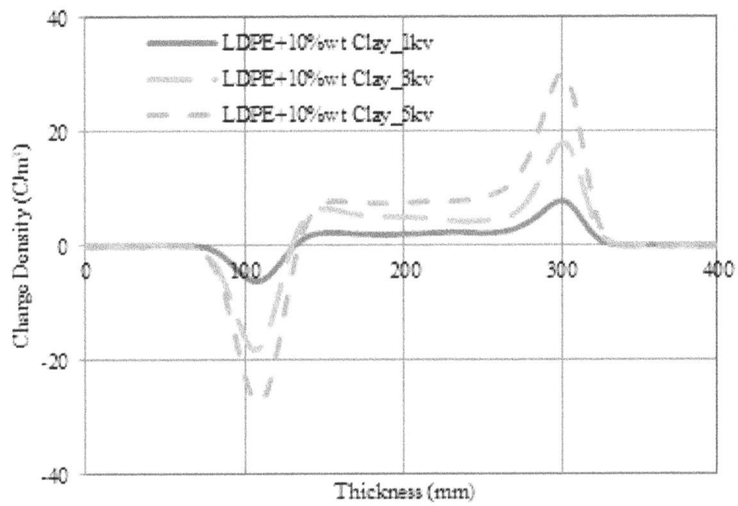

Figure 12. Space charge profiles for 1%wt. Al2O3 / LDPE nanocomposite films

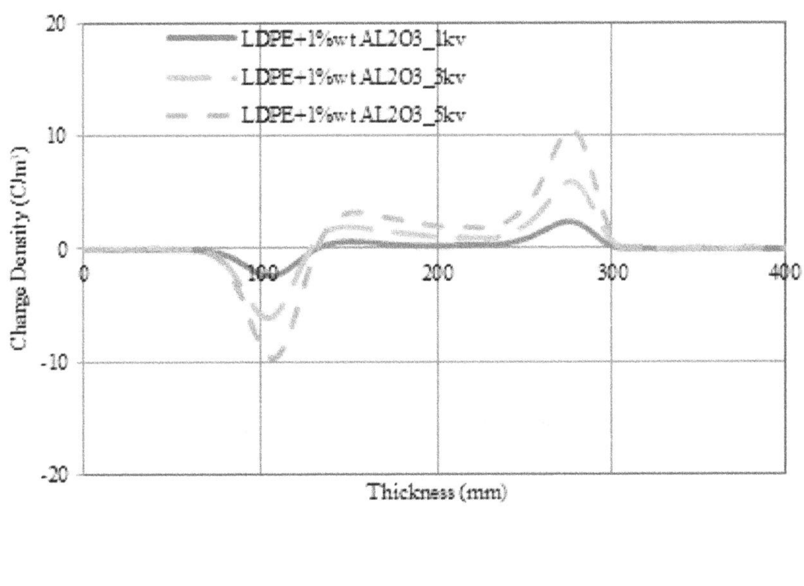

Figure 13. Space charge profiles for 5%wt. Al2O3 / LDPE nanocomposite films

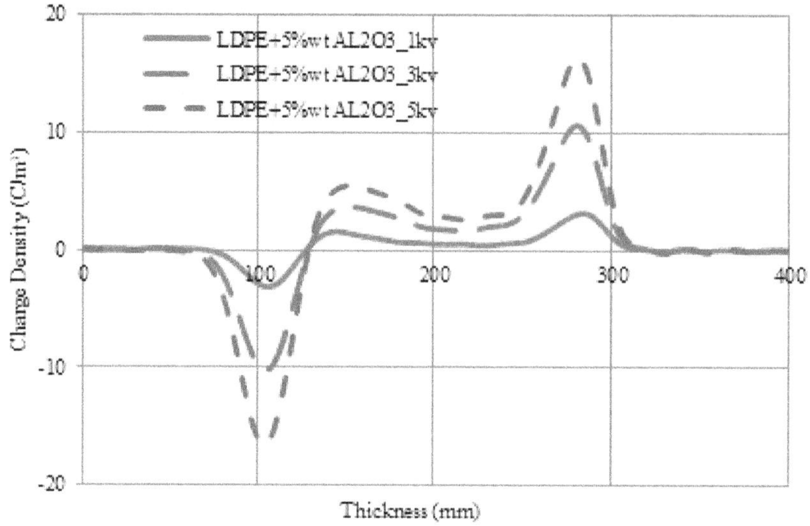

Figure 14. Space charge profiles for 10%wt Al2O3 / LDPE nanocomposite

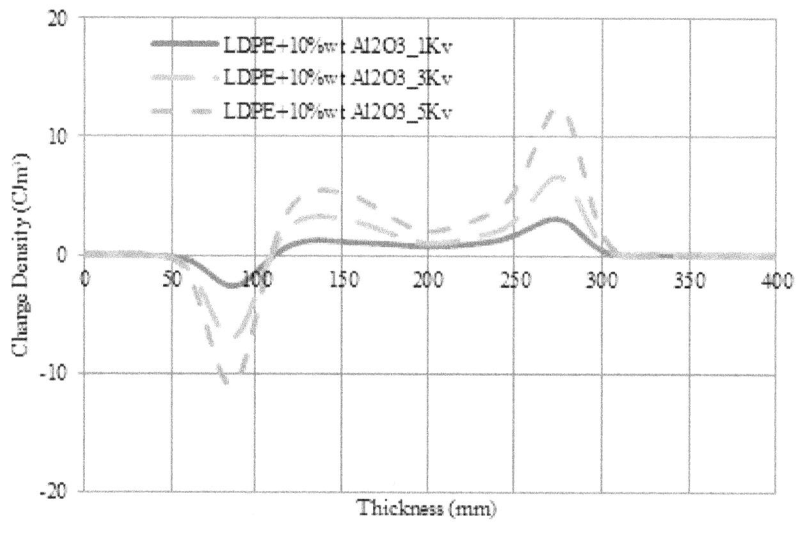

8.4.2 Effect of Aluminum Oxide Nanoparticles on Space Charge Distribution

Figure 12 demonstrates space charge circulation important Al_2O_3/LDPE nanocomposite films for 1%wt. rate of aluminum oxide nanoparticles under fluctuating dc electric fields; the accuse thickness towards both electrodes expands with expanding impulse connected voltages, and there is high hetero-charge gathered between anode and cathode. Furthermore, Fig. 13 contrasts space charge distributions important to Al_2O_3/LDPE nanocomposite films for 5%wt rate of aluminum oxide nanoparticles under different dc electric fields. It is clear that expanding clay nanoparticles starting with 1%wt up to 5%wt. increments accuse thickness in the nanocomposite molecules. Hence, the accuse thickness towards the anode and cathode will be expanded and there will be hetero-charge gathered charges between both electrodes. Figure 14 reveals space charge conveyance pertinent Al_2O_3/LDPE nanocomposite films with 10%wt rate of aluminum oxide nanoparticles under changing dc electric fields; the accuse thickness in both electrodes builds for expanding impulse connected voltages. There is also high hetero-charge gathered between the anode and cathode. It is recognized that expanding Al_2O_3 nanoparticles starting with 5%wt. dependent upon 10%wt. increments charge thickness in the nanocomposite molecules towards both electrodes.

Figure 15. Space charge profiles for Pure Polyvinyl Chloride under DC electric field

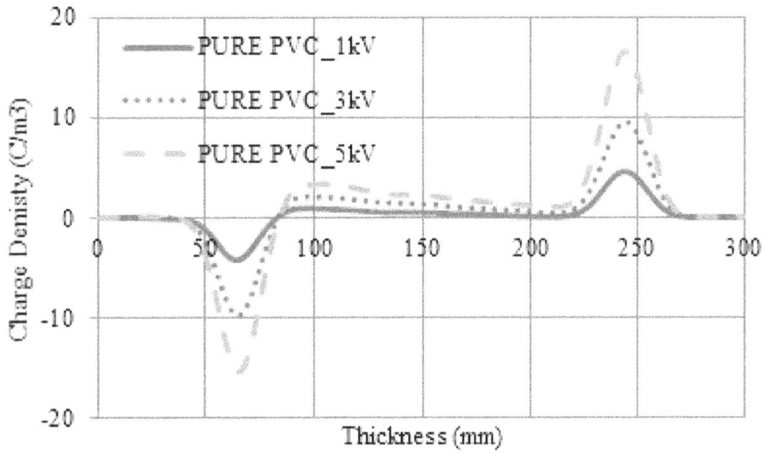

Figure 16. Space charge profiles for Pure Polyvinyl Chloride under DC electric field

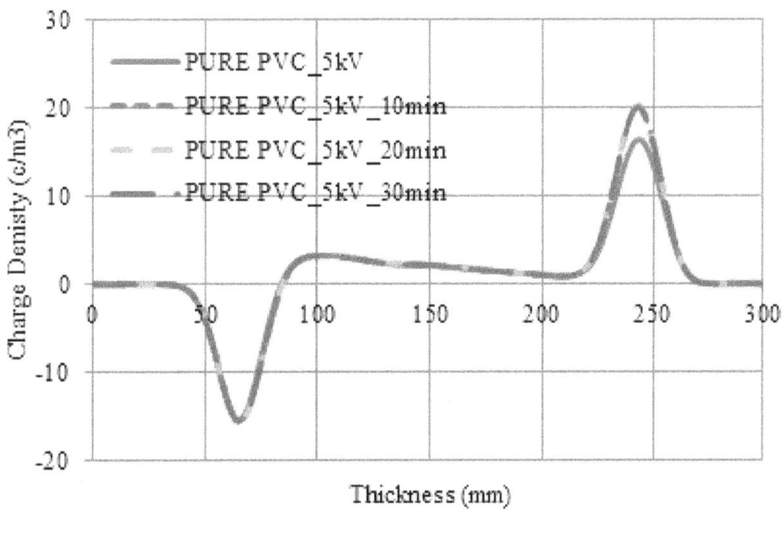

Figure 17 Space charge profiles for nanocomposite of PVC with 5%Clay under DC electric field

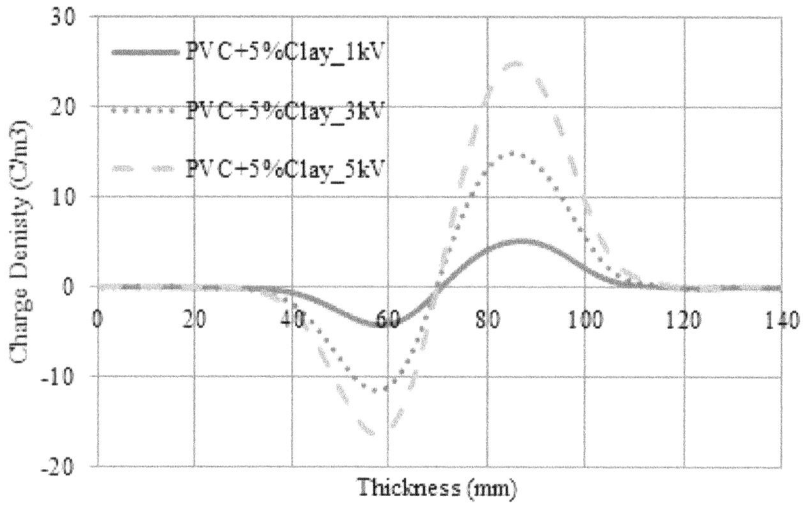

Figure 18. Space charge profiles for nanocomposite of PVC with 5%Clay under DC electric field

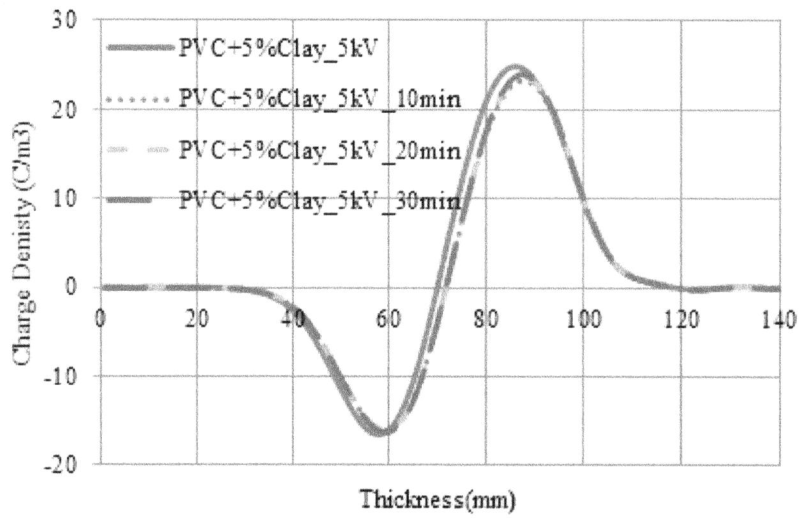

Figure 15 demonstrates space charge profiles applicable on PVC pure, up to dc electric field for 70 kV/mm. A lot of charge can definitely be seen in the example of PVC. Thus, figure 16 contrasts the space charge conveyance for climbing voltage. The measurements suggest the infusion of impulse high voltages on the charge thickness for immaculate PVC dielectric materials; the electrode is cleared out and the anode may be during the straight, so that the charge thickness increments dependent upon ±17 C/m^3 for expanding impulse voltages up to 5kV and hetero-charge is gathered close to the electrode. Figure 16 illustrates the space charge distribution, and its rot happened control in the remaining voltage, watched in the lab-tests after half an hour space charge electrodes. Thus, figure 16 contrasts the impacts of the space charge distribution, in the remaining 5kV voltage around immaculate Polyvinyl chloride PVC materials. Accuse thickness increments starting with $\pm 15\ C/m^3$ up to$\pm\ 20\ C/m^3$ with remaining impulse voltages towards 5kV best towards the anode and the hetero-charge gathered close to the electrode is not expanded. The information starting with measurement will be evaluated to be related to a rot duration of the time from claiming 6 secs and must hence be treated with certain alert. The measurements consequently infer the infusion for negative charge starting with the electrode, which remains trapped near the injecting electrode when the duration is 1/2hr.

Figure 19. Space charge profiles for nanocomposite of PVC with 10%Clay under DC electric field

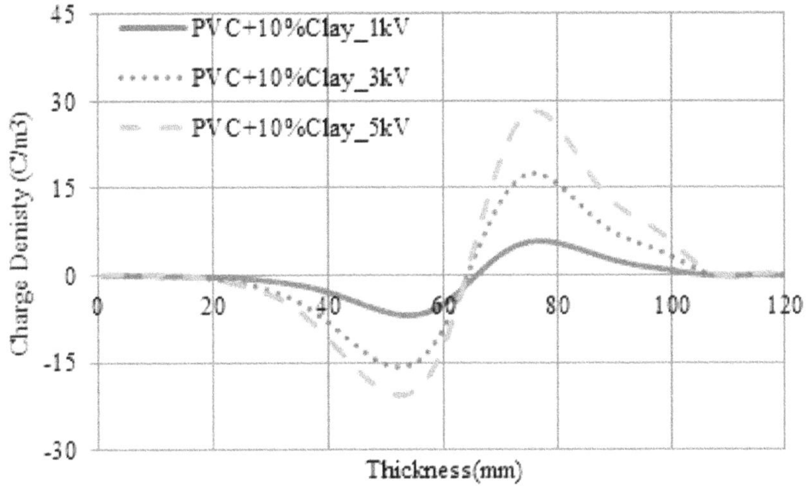

Figure 20.Space charge profiles for nanocomposite of Polyvinyl Chloride PVC with 10%Clay under DC electric field

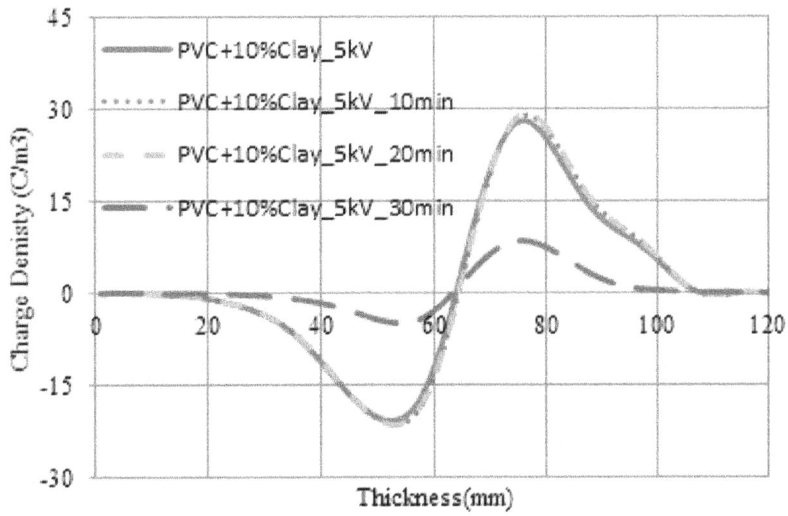

Figure 21. Space charge profiles for nanocomposite of PVC with 5%Fumed Silica under DC electric field

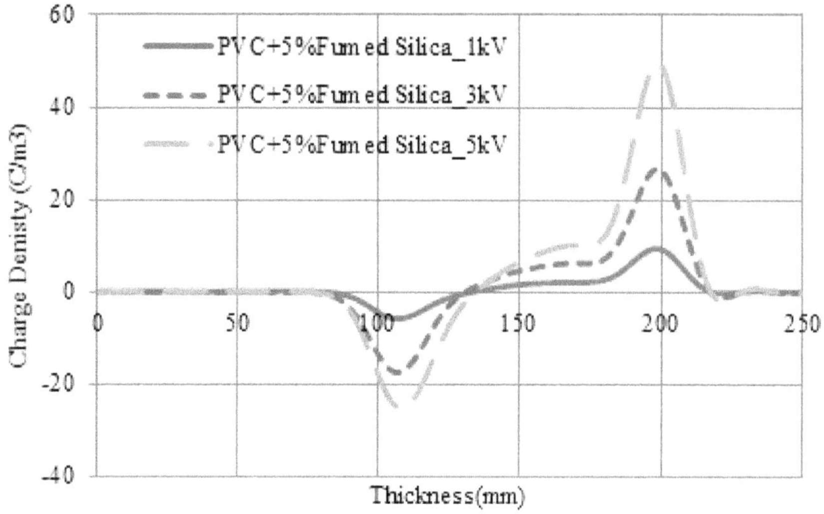

Figure 22. Space charge profiles for nanocomposite of PVC with 5%Fumed Silica under DC electric field

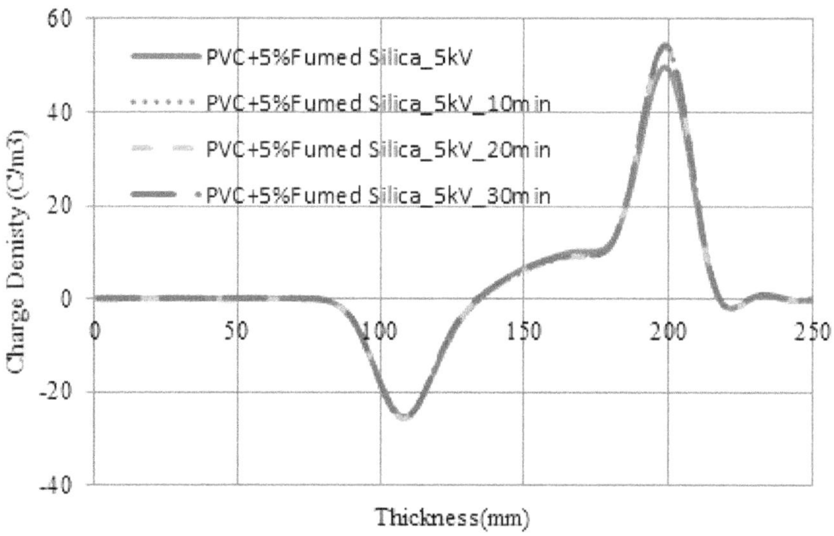

Figure 17 reveals space charge profiles important on PVC for 5% clay nanoparticles, up to dc electric field of 70 kV/mm. A lot about charge is certainly seen in the nanocomposite specimen, and so, figure 18 contrasts the space charge distribution for climbing voltage dependent upon 5kV. The measurements suggest the infusion of impulse high voltages on the charge thickness of Polyvinyl chloride PVC for 5% Clay nanocomposite materials; charging thickness during the anode may be more than charging thickness at the electrode for expanding impulse connected voltages dependent upon 5kV, and there will be hetero-charge gathered close to the electrode. Figure 19 illustrates the space charge distribution, and its rot which will have control in the remaining 5kV dc voltage and will be watched in the specimens, then after half an hour between space charge electrodes. Thus, figure 6 contrasts the impacts of the space charge distribution, during the remaining 5kV voltage about Polyvinyl chloride PVC for 5% clay nanocomposite material. There is no evolving in control thickness for the remaining impulse voltages, yet there will be a little decrease during the anode.

Figure 23. Space charge profiles for nanocomposite of PVC with 10% Fumed Silica under DC electric field

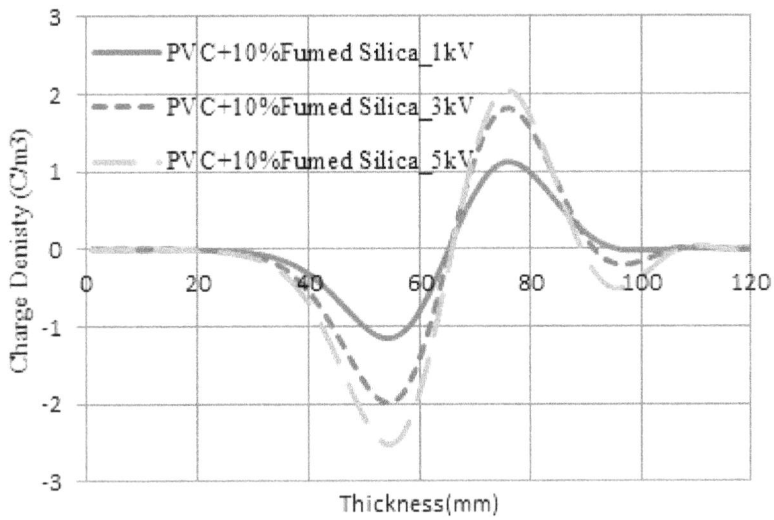

Figure 24. Space charge profiles for nanocomposite of PVC with 10% Fumed Silica under DC electric field

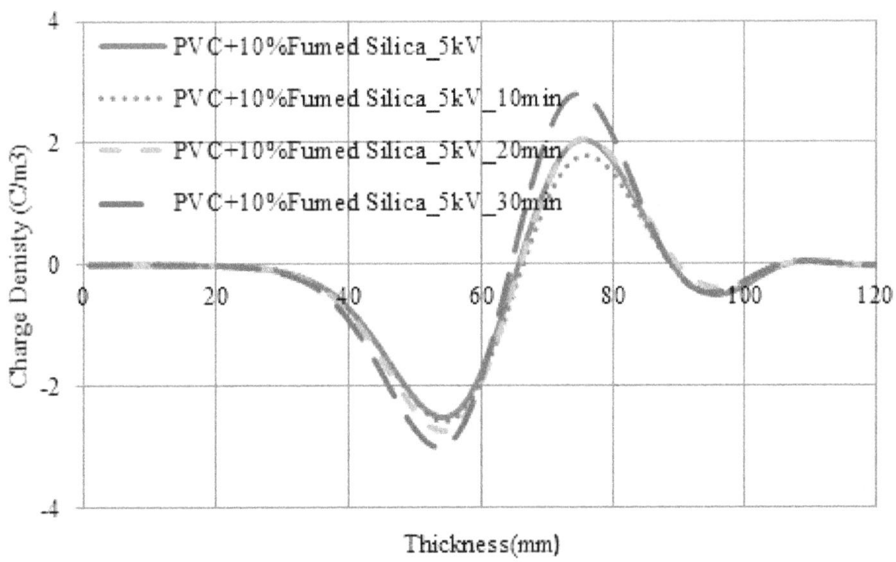

Figure 25. Space charge profiles for nanocomposite of PVC with 10%ZnO under DC electric field

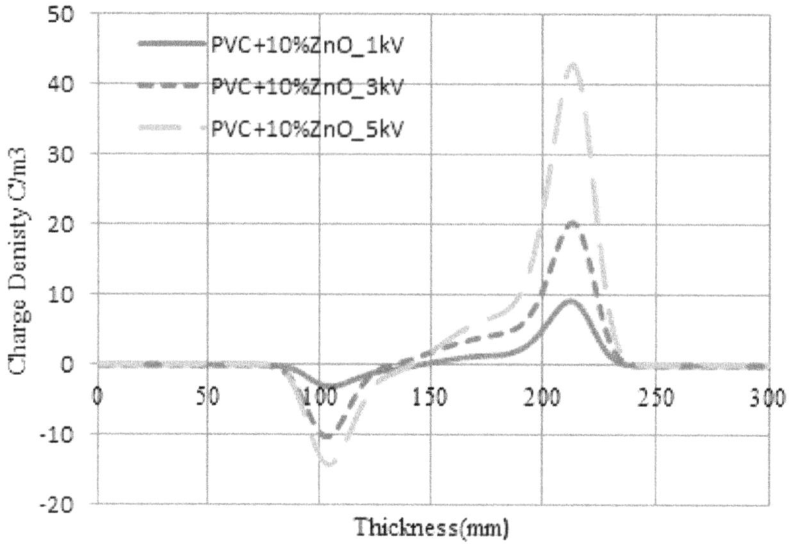

Figure 19 indicates the space charge conveyance for climbing voltage up to 5kV. The measurements suggest the infusion from claiming impulse high voltages on the charge thickness from claiming PVC for 10% clay nanocomposite materials; accuse thickness in the anode and cathode will be expanded to nanocomposite PVC for 10%clay and there are hetero-charge gathered charges between the anode and cathode electrodes. Figure 20 contrasts the space charge distribution and its rot happened control in the remaining voltage, watched in the lab-tests after half an hour between space charge electrodes. Thus, figure 8 illustrates the impacts of the space charge distribution, in the remaining 5kV voltage over PVC with 10%clay nanocomposite material. There is a high decrease in control thickness during anode and cathode with remaining impulse voltages during 5kV. The impact of raising concentrations of nanoparticle will be pointed out in Figures. The space charge profiles under connected dc impulse voltages are accounted for to diverse weight concentrations of altered nanoparticles focus. In other words, expanding rates for clay nanoparticles from 0% clay in immaculate PVC, dependent upon 5%clay and 10% clay builds accuse thickness about 20%. Likewise, a progressive for the greater part space charge in the electrode field level is watched as the nanoparticles concentrations builds, as alluded with charge in the greater part and not the electrode peaks. With maturing under 5kV, charge thickness at the anode and cathode is diminished daintily in PVC with 5%clay in any case there is more decrease in control thickness that occur in nanocomposite PVC with 10% clay and there are hetero-charge gathered charges between the anode and cathode electrodes. An outline judgment from claiming space charge distributions is given by the edge aspects of Figures 5-8, applicable with pure, 5%clay and10% clay separately.

Figure 21 demonstrates space charge profiles pertinent to PVC for 5% fumed silica nanoparticles, for a dc electric field from claiming 70 kV/mm. A lot about charge can be certainly seen in the nanocomposite specimen. Hence, figure 9 contrasts the space charge circulation with climbing voltage dependent upon 5kV. The measurements suggest the infusion of impulse high voltages on the charge thickness for Polyvinyl chloride PVC for 5%fumed silica nanocomposite materials; charging thickness during the anode will be more than charging thickness towards the electrode for expanding impulse connected voltages dependent upon 5kV, and there is hetero-charge gathered close to the anode.

Figure 26. Space charge profiles for nanocomposite of PVC with 10%ZnO under DC electric field

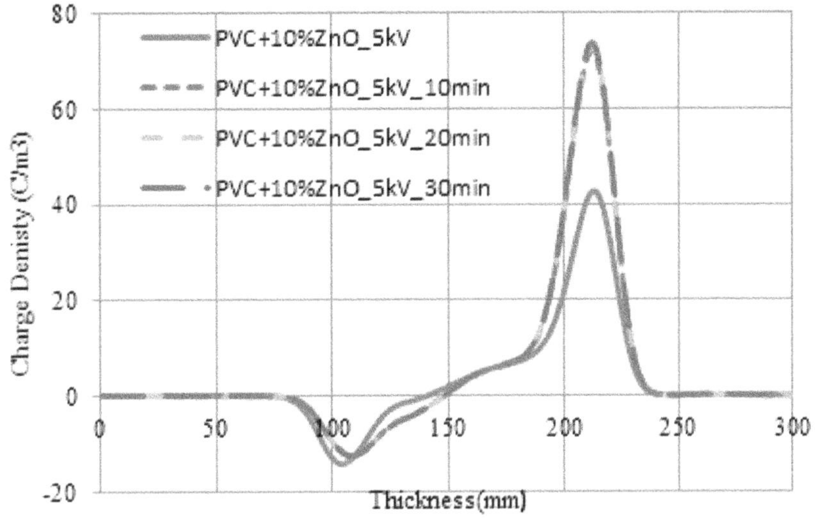

Figure 22 illustrates the space charge distribution and its rot which may happen in the remaining 5kV dc voltage, and is watched in the lab-tests, following half an hour between space charge electrodes. In this way, figure 10 contrasts the impacts of the space charge distribution, at the remaining 5kV voltage around PVC with 5%fumed silica nanocomposite material. There is no evolving in control thickness for the remaining impulse voltages, yet there is a little climbing at the anode. Figure 23 illustrates the space charge distribution for climbing voltage dependent upon 5kV. The measurements suggest the infusion of impulse high voltages on the accuse thickness from claiming PVC for 10%wt. fumed silica nanocomposite materials; accuse thickness in the anode and cathode will be diminished with least characteristics in nanocomposite PVC for 10%wt. clay and there are no hetero-charge gathered charges between the anode or cathode electrodes. Figure 25 indicates the space charge distribution and its rot happened control in the remaining voltage, watched in the specimens after half an hour between space charge electrodes. Thus, Fig. 24 illustrates the impacts of the space charge distribution, towards the remaining 5kV voltage over Polyvinyl chloride PVC for 10%wt. fumed silica nanocomposite material. There are little progressions in control thickness at anode and cathode for the remaining impulse voltages during 5kV.

A similar investigation of the outcomes for portraying the impact for raising fixation of fumed silica nanoparticles can be pointed out in Fig. (22-26). The place space charge profiles under connected dc impulse voltages are news person for distinctive weight concentrations of altered nanoparticles focus. i.e. expanding rates from claiming fumed silica nanoparticles from 0%fumed silica, up to 5%wt. Fumed silica builds charge thickness of 200%. Similarly, as can be seen, a progressive for greater part space charge in the electrode field level is seen as the nanoparticle concentrations expand to 5%wt. fumed silica and there will be high decrease in control density reaching 80%, assuming that fumed silica rate expands from 5%wt. up to 10%wt, as alluded to charge in the mass and not the electrode peaks. For maturing under 5kV, accuse thickness at the anode and cathode is diminished daintily in PVC with 5%wt. fumed silica, in any case there is recognized high decrease in control thickness that occur for nanocomposite PVC for 10%fumed silica and there are no hetero-charge gathered charges between the anode and cathode electrodes. An outline of space charge distributions is furnished, eventually perusing the edge aspects of figures pertinent to pure, 5%wt. fumed silica and10% fumed silica separately. Figure 14 indicates space charge profiles applicable to PVC for 5%wt. ZnO nanoparticles, for a dc electric field about 70 kV/mm. A lot of charge can certainly be seen in the nanocomposite specimen, and so, figure 26 contrasts the space charge conveyance with climbing voltage dependent upon 5kV. The measurements suggest the infusion from claiming impulse high voltages on the charge thickness of Polyvinyl chloride PVC with 5%ZnO nanocomposite materials; charging thickness in the anode is more than charging thickness in the electrode with expanding impulse connected voltages up to 5kV and there are hetero-charges gathered close to the anode. Figure 27 contrasts the space charge distribution and its rot which occur in the over remaining 5kV dc voltage and will be watched in the lab-tests after half an hour between space charge electrodes. Thus, figure 27 illustrates the impacts of the space charge distribution towards the remaining 5kV voltage for Polyvinyl chloride PVC with 5%wt. ZnO nanocomposite material. There is no evolv-

Figure 27. Space charge profiles for Pure Polyvinyl Chloride PVC

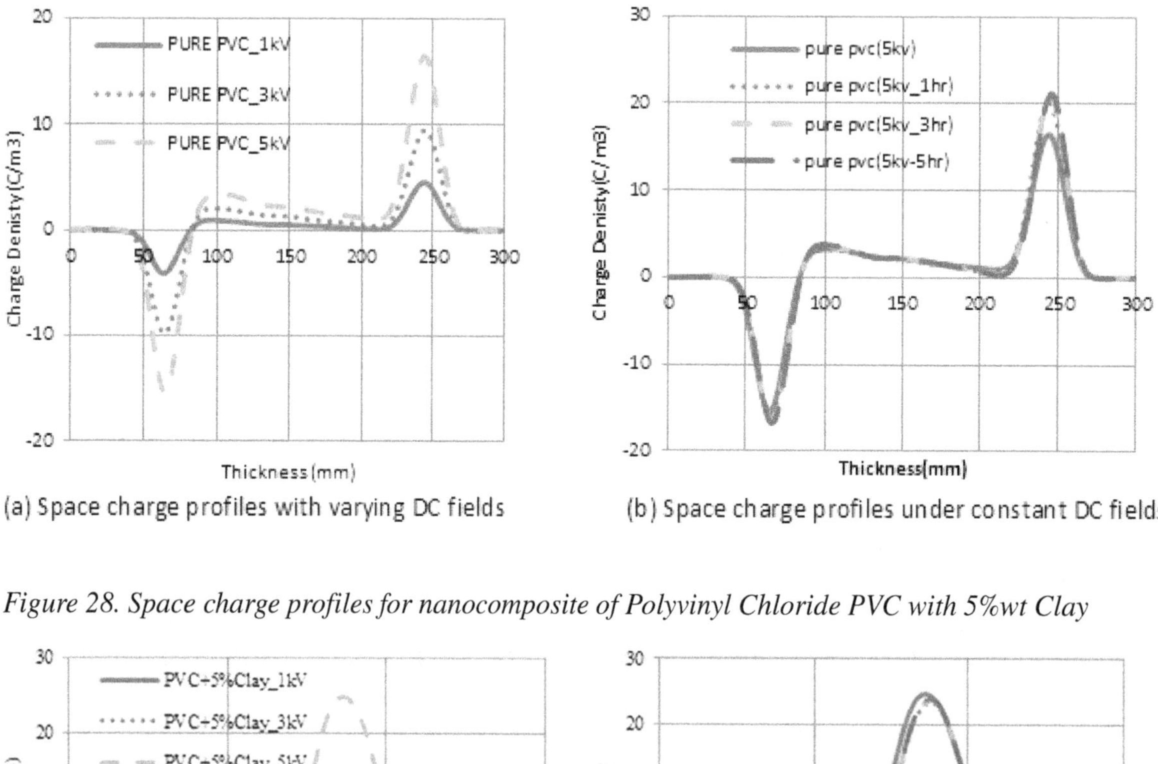

(a) Space charge profiles with varying DC fields (b) Space charge profiles under constant DC fields

Figure 28. Space charge profiles for nanocomposite of Polyvinyl Chloride PVC with 5%wt Clay

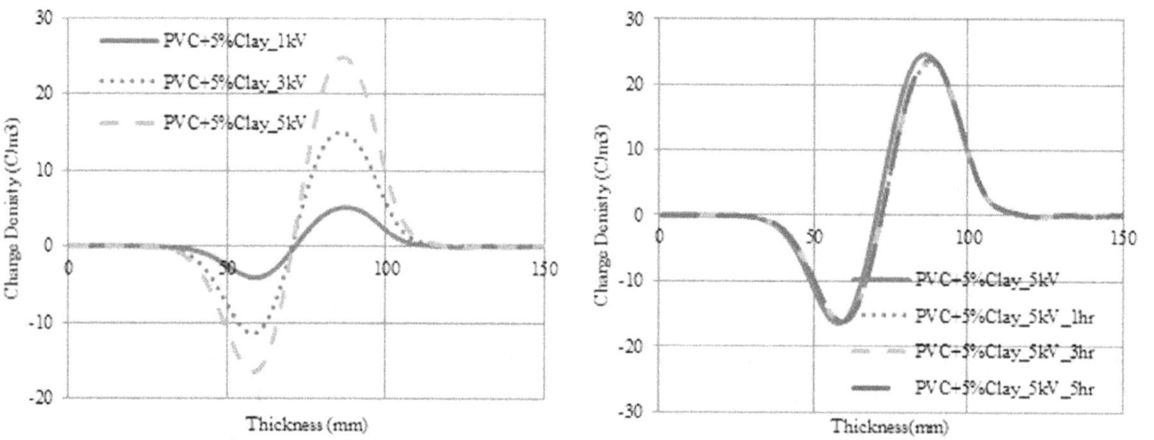

(a) Space charge profiles with varying DC electric fields (b) Space charge profiles under constant DC electric fields ing in control thickness with the remaining impulse voltages, yet there is a little climbing in the anode dependent upon 250% of immaculate PVC.

Figure 29. Space charge profiles for nanocomposite of Polyvinyl Chloride PVC with 10%wt Clay

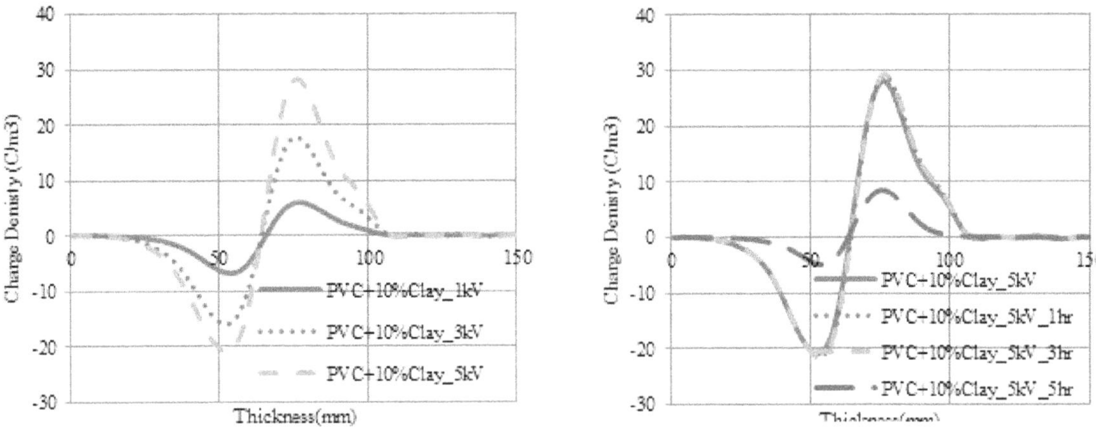

(a) Space charge profiles with varying DC electric fields (b) Space charge profiles under constant DC electric fields

Figure 30.Space charge profiles for nanocomposite of Polyvinyl Chloride PVC with 5%wt fumed silica

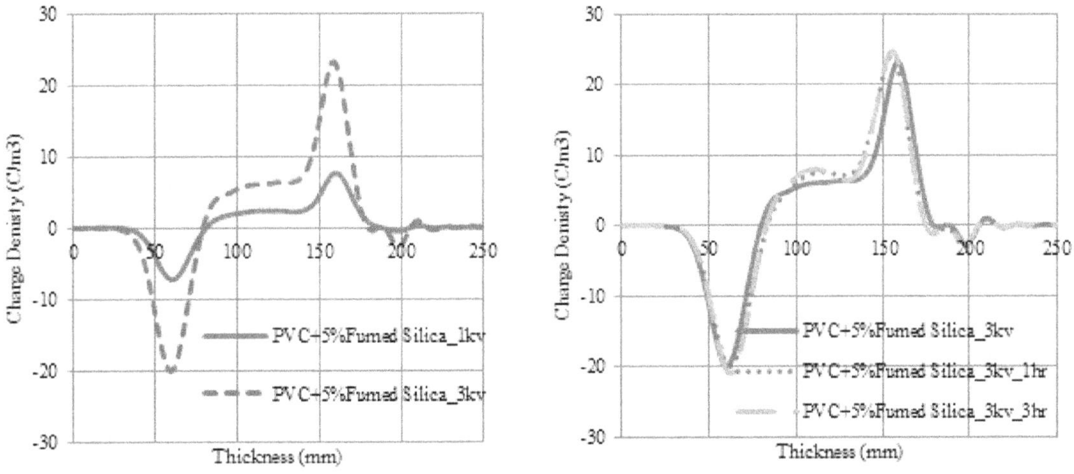

(a) Space charge profiles with varying DC electric fields (b) Space charge profiles under constant DC electric fields

Accordingly, comparing all outcomes to portray the impact of raising focus from claiming ZnO nanoparticles is pointed out in Fig.'s (25, 26). The space charge profiles under connected dc impulse voltages are attributed to expanding ZnO concentrations for altered nanoparticles focus. i.e. expanding rates of ZnO nanoparticles starting with 0%wt. ZnO, up to 10%wt. ZnO increments accuse thickness by 250% starting with the happened control in pure PVC. As can be seen, a progressive of heft space charge at the electrode field level watched concerning illustration of the nanoparticles focus expands at

10%wt. ZnO. This once concerns illustration alluded to accuse in the mass and not the electrode peaks, for maturing under 5kV, accuse thickness in the anode and cathode will be expanded in PVC at 10%wt. ZnO and there are hetero-charge gathered charges between the anode and cathode electrodes. An outline of space charge distributions is given by the edge characteristics of Fig.'s (26,27), applicable to immaculate PVC and 10%wt. ZnO nanocomposite individually. It is recognized that charge thickness increments with raising ZnO nanoparticles rate to 10%wt. ZnO in PVC. At aging under 5kV, charge thickness has been expanded towards the anode and cathode more than that happened in immaculate PVC and other contemplated nanocomposites.

Figure 31. Space charge profiles for nanocomposite of Polyvinyl Chloride PVC with 10%wt Fumed silica

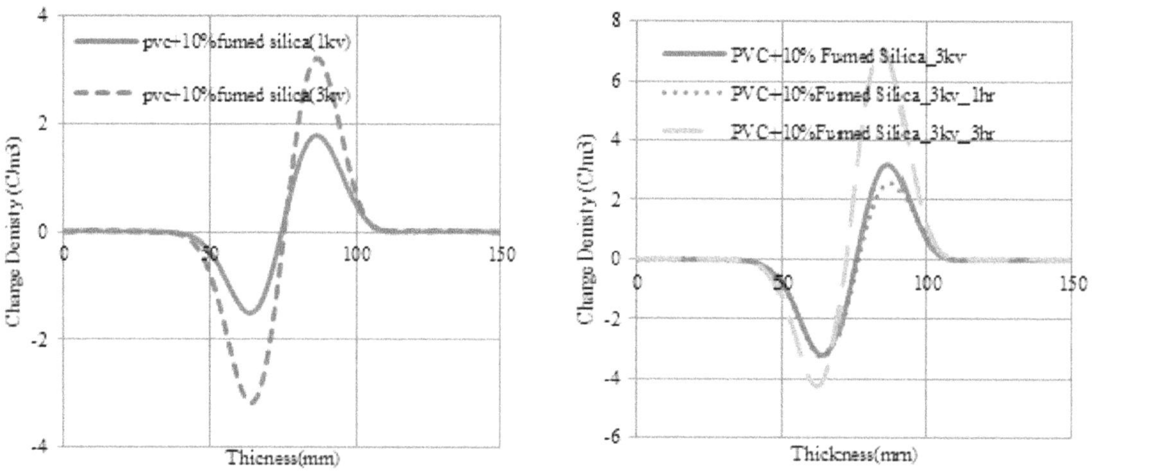

(a) Space charge profiles with varying DC electric fields (b) Space charge profiles under constant DC electric fields

Figure 27 (a) indicates space charge distribution important to immaculate Polyvinyl chloride (PVC) under different dc electric field 70 kV/mm. The measurements infer the infusion of impulse high voltages on the charge thickness from claiming immaculate PVC dielectric materials; the electrode is towards the cleared out and the anode is toward the correct, so that the accuse thickness expands by expanding impulse voltages and hetero-charge is gathered close the electrode. Fig. 27. (b) contrasts the space charge distribution and its rot happened for the remaining 5kV, watched in the lab-tests following three-hour space charge electrodes. Accuse thickness builds for the remaining impulse voltages at 5kV main towards the anode and the hetero-charge gathered close to the electrode will be not expanded. The measurements consequently suggest the infusion of negative charge starting with the electrode, which remains trapped near the injecting electrode at a point when the duration is three hours. Figure 28. (a) demonstrates space charge profiles important for Polyvinyl Chloride (PVC) with 5%wt. clay nanoparticles dependent upon 70 kV/mm dc electric fields. This figure contrasts the space charge conveyance for the climbing voltage up to 5kV. The measurements suggest the infusion of impulse high voltages on the charge thickness for Polyvinyl chloride PVC with 5%wt. clay nanocomposite materials; charging thickness at the anode will be more than charging thickness towards the electrode for the expanding impulse connected to voltages up to 5kV and there are no hetero-charge gathered close to the electrode. Figure 28. (b) contrasts the space charge distribution and its rot which has happened over the remaining

5kV dc voltage and will be watched in the lab-tests after half an hour between space charge electrodes. There will be no evolving in control thickness with the remaining impulse voltages. However, there is a little diminishment at the anode.

Figure 32. Space charge measurement profiles for pure polyvinyl chloride

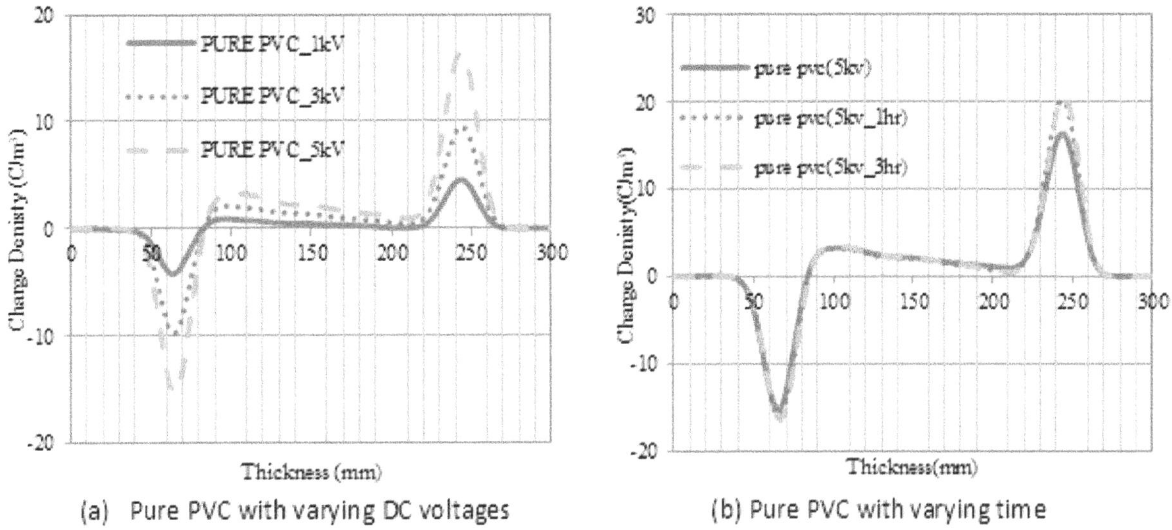

Figure 29. (a) contrasts the space charge distribution for climbing dc connected voltage dependent upon 5kV. The measurements suggest the infusion of impulse high voltages on the charge thickness for Polyvinyl chloride PVC for 10%wt. clay nanocomposite materials; accuse thickness during the anode and cathode will be expanded to nanocomposite PVC with 10%wt. clay and there are no hetero-charge gathered charges between the anode and cathode electrodes. Figure 29. (b) contrasts the space charge distribution and its rot happened in remaining voltage and watched in the specimens, but after half an hour between space charge electrodes. Thus, figure 8 contrasts the impacts of the space charge distribution, at the remaining 5kV voltage for Polyvinyl chloride PVC with 10%clay nanocomposite material. There is high diminishment in control thickness in anode and cathode with the remaining impulse voltages at 5kV. Figure 30.(a) demonstrates space charge distributions important to fumed silica/Polyvinyl chloride nanocomposites with 5%wt nanoparticles rate under different dc electric field 70 kV/mm. The measurements suggest the infusion of impulse high voltages on the accuse thickness for fumed silica/Polyvinyl chloride nanocomposites; charging thickness during the anode will be more than charging thickness at the electrode for expanding impulse connected voltages dependent upon 5kV, and there will be hetero-charge gathered close to the anode. Fig. 30.(b) contrasts the space charge distribution and its rot which connected 5kV dc voltage through three hours in fumed silica/Polyvinyl chloride nanocomposites for 5%wt. nanoparticles rate. Thus, this figure illustrates that there will not be any evolving in control thickness with the remaining impulse voltages, in any case there is a deviation and a little climbing in the anode.

Figure 31 shows Space charge profiles of nanocomposite of Polyvinyl chloride PVC with 5%wt fumed silica. Figure 31. (a) reveals space charge distributions important to fumed silica/Polyvinyl chloride nanocomposites with 10%wt nanoparticles rate under different dc electric field 70 kV/mm. The measurements infer the infusion for impulse high voltages that depict that charge thickness in the anode and cathode will be diminished to the least characteristics in fumed silica/Polyvinyl chloride nanocomposites and there will not be hetero-charge gathered charges between the anode or cathode electrodes. Moreover, expanding fumed silica nanoparticles from 5%wt. dependent upon 10%wt diminishes accuse thickness in the nanocomposite molecules. Fig. 31. (b) contrasts the space charge distribution in fumed silica/Polyvinyl chloride nanocomposites for 10%wt. nanoparticles percentage, and its rot that happened in the lab-tests through three hours remaining voltage. Charge thickness abates before the time for remaining impulse voltages at 3kV right away hour, as it expands for expanding the remaining time dependent upon three hours, particularly at the anode.

Figure 33. Space charge profiles for ZnO/PVC nanocomposite with varying DC electric fields

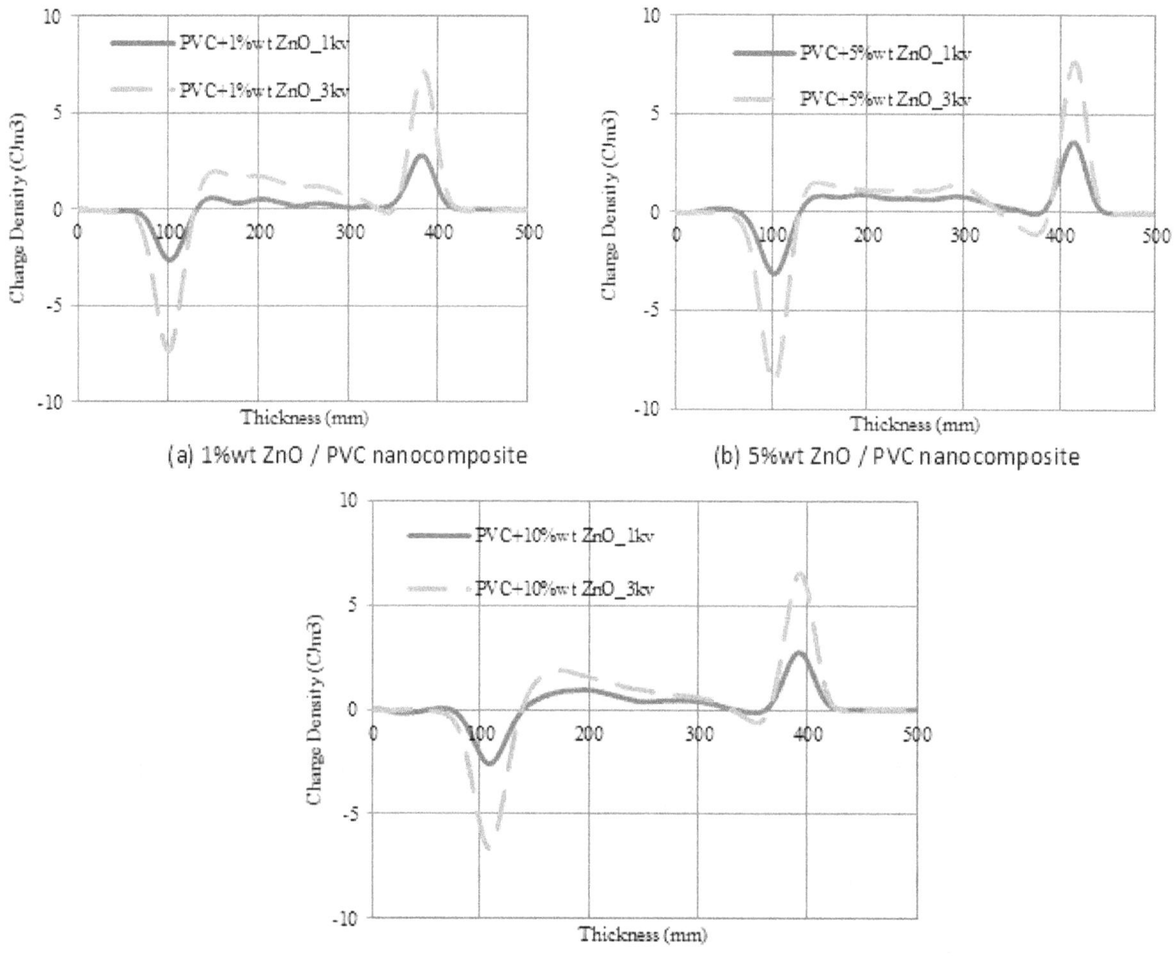

(a) 1%wt ZnO / PVC nanocomposite

(b) 5%wt ZnO / PVC nanocomposite

(c) 10%wt ZnO / PVC nanocomposite

Figure 34. Space charge profiles for 10%wt ZnO/PVC nanocomposite with varying time

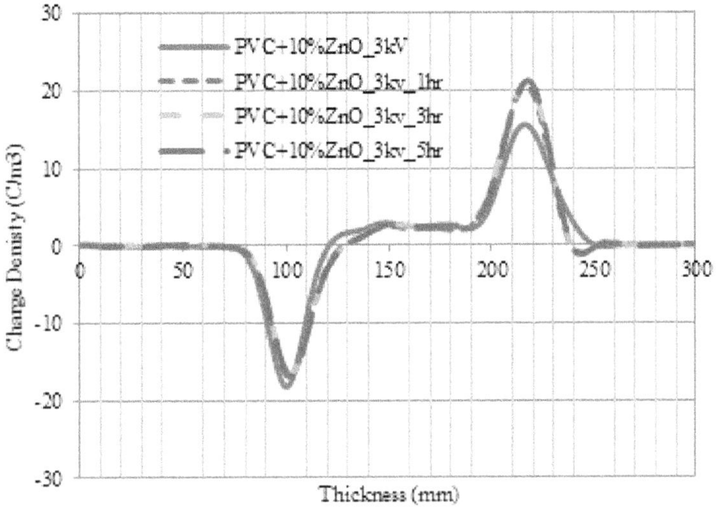

Figure 35. Space charge profiles for TiO_2/PVC nanocomposite with varying DC electric fields

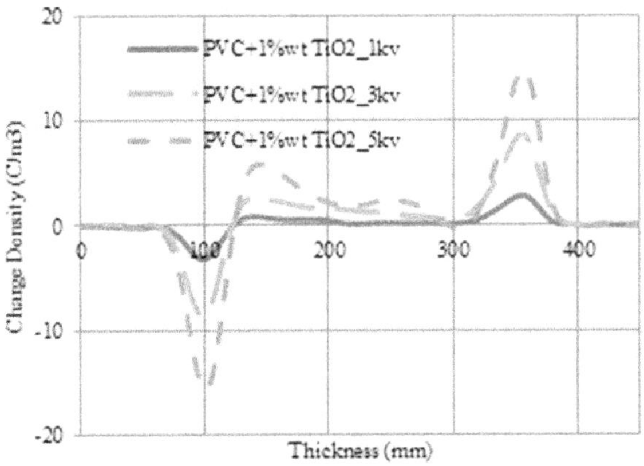

(a) 1%wt TiO_2 /PVC nanocomposite

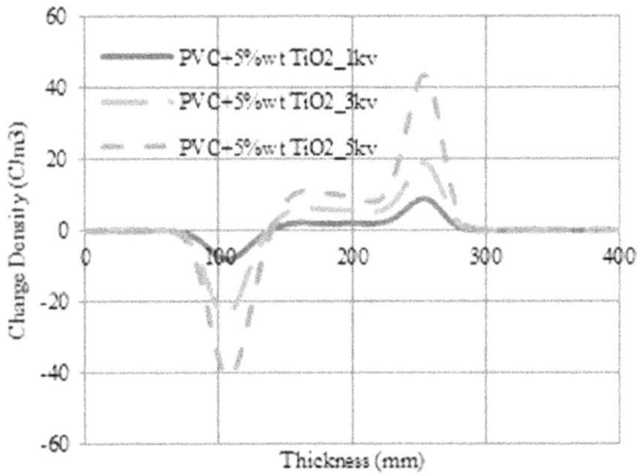

(b) 5%wt TiO_2/PVC nanocomposites

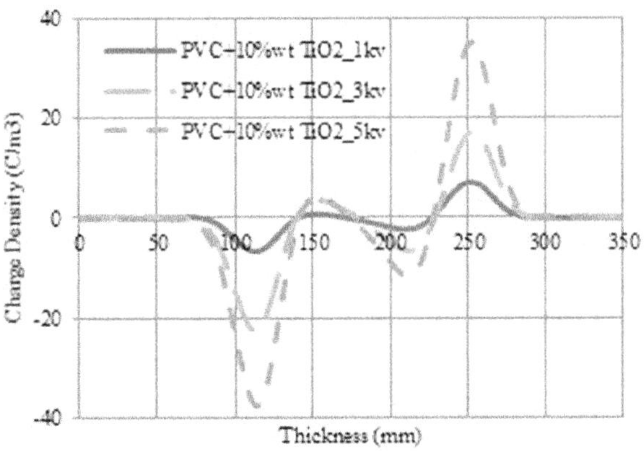

(c) 10%wt TiO_2/PVC nanocomposite

Figure 36. Space charge profiles for 10%wt TiO_2/PVC nanocomposite with varying time

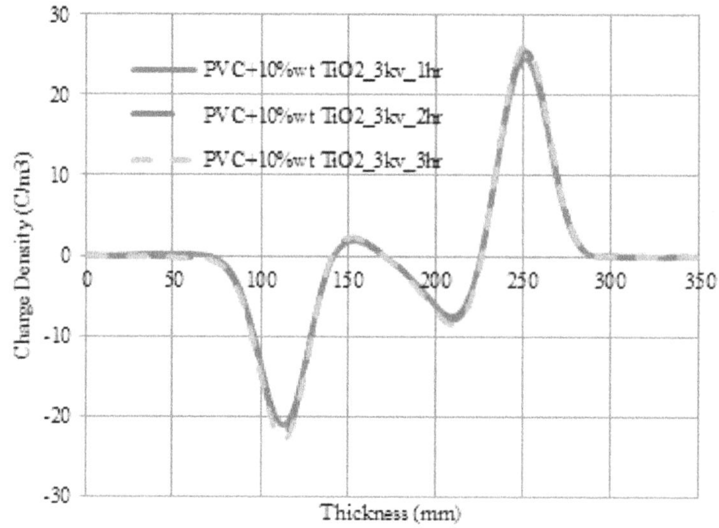

Characterization from claiming Polyvinyl chloride indicated in Fig. 32 is pointed towards the measurements inferring the infusion from claiming impulse high voltages on the accuse thickness of immaculate PVC dielectric materials; the accuse thickness expands with expanding impulse voltages and hetero-charge gathers close to the electrode. Furthermore, accuse thickness builds for the remaining impulse voltages during 5kV just towards the anode, and the hetero-charge gathered close to the electrode will not be expanded. The impact from the claiming raising focus about clay nanoparticles control in Polyvinyl chloride is pointed out in Fig.'s (33, 34). These figures depict clay nanoparticles effects, i.e. expanding rates of clay nanoparticles dependent upon 10%wt clay increments accuse thickness over 20%. With maturing under 5kV, charge thickness at the anode and cathode may be diminished daintily for PVC for 5%wt clay, that's only the tip of the iceberg diminishment in control thickness that happened to nanocomposite PVC for 10%wt clay, and there are no hetero-charge gathered charges between the anode and cathode electrodes. The impact from claiming raising fixation about fumed silica nanoparticles in Polyvinyl chloride are pointed out in Figures (35, 36), i. e.: it depicts that at 5%wt fumed silica, the measurements infer the infusion of impulse high voltages on the charge thickness of fumed silica/ Polyvinyl chloride nanocomposites; charging thickness during the anode will be more than charging thickness in the electrode for expanding impulse connected voltages up to 5kV, and there will be hetero-charge gathered close to the anode. However, in 10%wt fumed silica, there are no hetero-charge gathered charges between the anode or cathode electrodes. Moreover, expanding fumed silica nanoparticles starts with 5%wt. up to 10%wt. diminishing charge thickness in the nanocomposite molecules.

Figure 32 (a) reveals to space charge circulation control in immaculate polyvinyl chloride under different dc electric fields; the accuse thickness increments with expanding impulse voltages and hetero-charge gathered close to the electrode. On the other hand, Fig. 32. (b) contrasts the space charge distribution, and its rot happened in the remaining 5kV and watched in the specimens following three-hour space in charge electrodes. Accuse thickness increments with the remaining impulse voltages in 5kV, particularly towards the anode, yet the hetero-charge gathering happens close to the electrode. The measurements along these lines suggest the infusion from claiming negative charge from the electrode, which remains trapped near the injecting electrode at the duration of three hours.

Figure 37. Space charge profiles for Al_2O_3/ PVC nanocomposite with varying DC electric fields

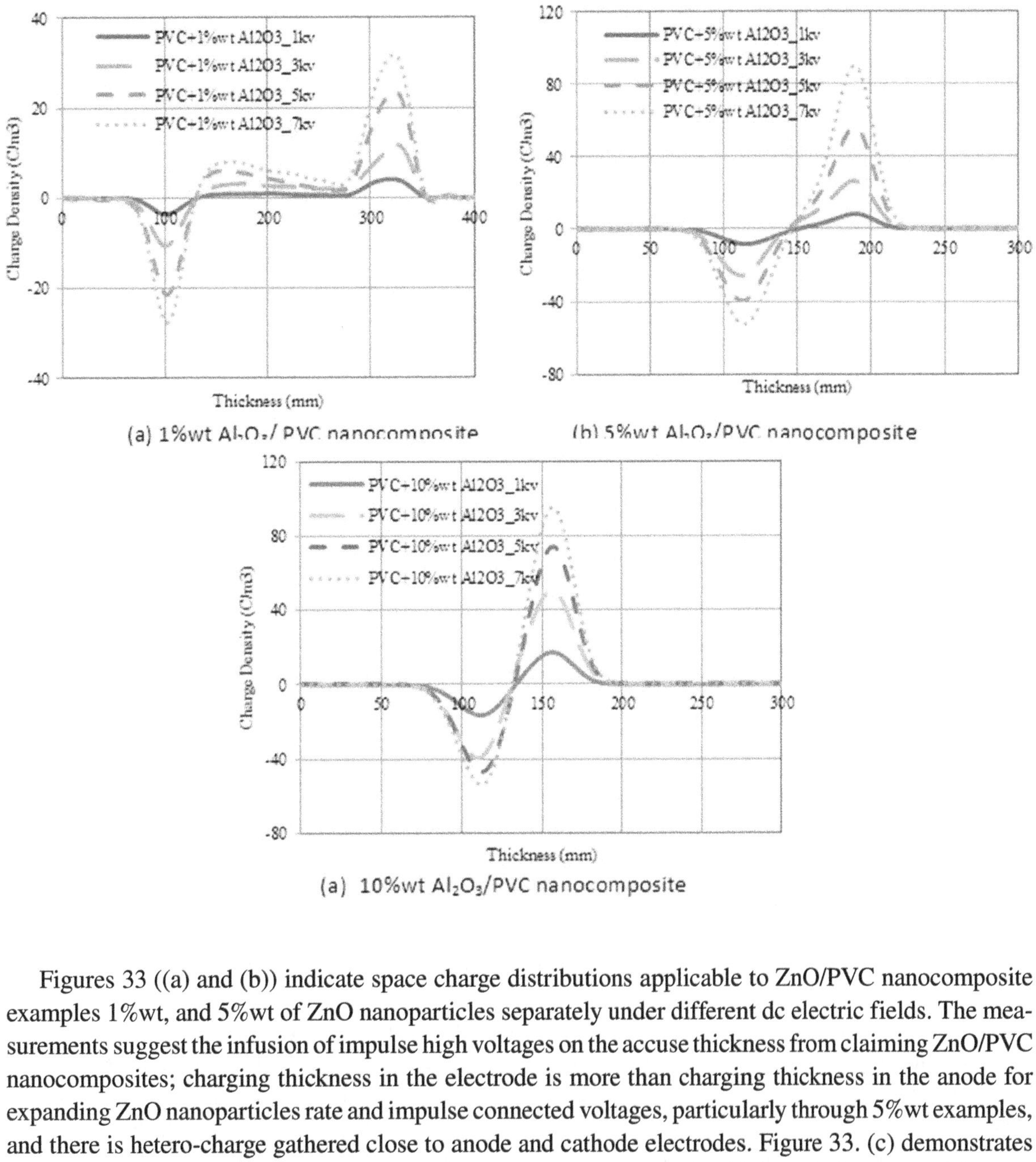

(a) 1%wt Al_2O_3/ PVC nanocomposite (b) 5%wt Al_2O_3/PVC nanocomposite

(a) 10%wt Al_2O_3/PVC nanocomposite

Figures 33 ((a) and (b)) indicate space charge distributions applicable to ZnO/PVC nanocomposite examples 1%wt, and 5%wt of ZnO nanoparticles separately under different dc electric fields. The measurements suggest the infusion of impulse high voltages on the accuse thickness from claiming ZnO/PVC nanocomposites; charging thickness in the electrode is more than charging thickness in the anode for expanding ZnO nanoparticles rate and impulse connected voltages, particularly through 5%wt examples, and there is hetero-charge gathered close to anode and cathode electrodes. Figure 33. (c) demonstrates space charge distributions pertinent to ZnO/PVC nanocomposites with 10%wt. ZnO nanoparticles rate under different dc electric fields. The measurements infer the infusion from claiming impulse high voltages on the accuse thickness for ZnO/PVC nanocomposites; charging thickness during the electrode will be more than charging thickness in the anode for expanding impulse connected voltages, and there may be hetero-charge gathered close to the anode. It has been recognized that expanding ZnO nanoparticles from 5%wt. up to 10%wt. diminishes charge thickness in the nanocomposite molecules in electrodes. On the other hand, Fig. 34 contrasts the space charge distribution in ZnO/PVC nanocomposites for 10%wt nanoparticles percentage and its rot that happened in the specimens through a three-hour remaining voltage. Charge thickness expands for the expanding remaining duration of the time up to three hours, particularly during the anode.

Figure 38 Space charge profiles for 10%wt Al2O3/ PVC nanocomposite under constant DC electric field

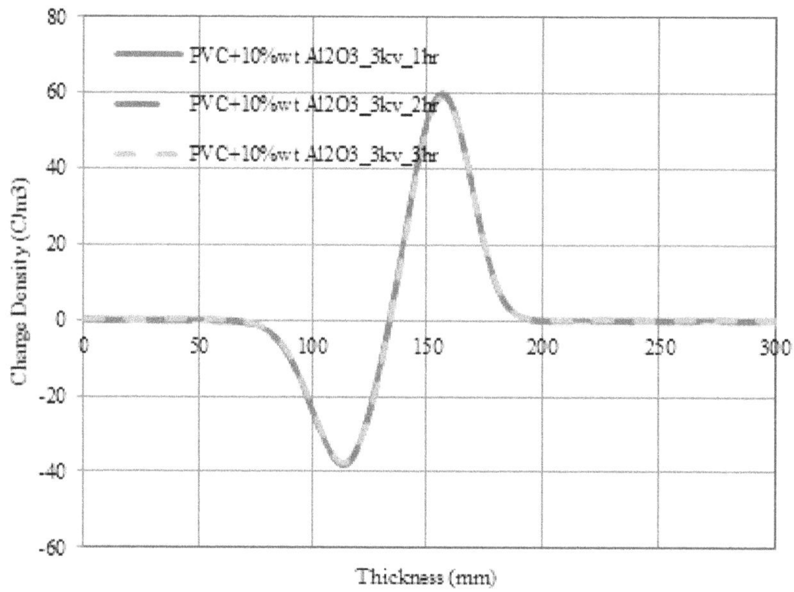

Figure 35 ((a) and (b)) demonstrate space charge distributions pertinent to TiO_2/PVC nanocomposite examples 1%wt, and 5%wt about TiO_2 nanoparticles, individually under different dc electric fields. The measurements infer the infusion from claiming impulse high voltages on the charge thickness of TiO_2/ PVC nanocomposites; charging thickness between the electrode increments for the expanding TiO_2 nanoparticles rate and impulse connected voltages, particularly through 5%wt examples, and there is hetero-charge gathered between electrodes. Figure 35. (C) reveals space charge distribution applicable TiO_2/PVC nanocomposites for 10%wt nanoparticles rate with changing dc electric fields. The measurements suggest the infusion of impulse high voltages on the accuse thickness pertinent to TiO_2/PVC nanocomposites with 10%wt nanoparticles percentage; the charge thickness towards both electrodes expands with expanding impulse connected voltages, and there may be high hetero-charge gathered between anode and cathode electrodes. It is recognized that expanding TiO_2 nanoparticles from 5%wt. dependent upon 10%wt diminishes charge thickness in the nanocomposite particles at electrodes. Moreover, Fig. 36 contrasts the space charge distribution for the remaining 3 kV in TiO_2/PVC nanocomposites for 10%wt nanoparticles rate through three hours. There will be no evolving in control thickness with the remaining impulse voltages between both anode and cathode electrodes.

Figure 37 demonstrates space charge distribution pertinent Al_2O_3/Polyvinyl chloride nanocomposites aluminum oxide nanoparticles rates (1%wt., 5%wt. and 10%wt.) under fluctuating dc electric field 70kV/mm. The measurements suggest the infusion from claiming impulse high voltages on the accuse thickness from claiming applicable Al_2O_3/Polyvinyl chloride nanocomposites with variant nanoparticles percentage; the charge thickness at both electrodes builds with expanding impulse connected voltages up to 5kV, and there may be high hetero-charge gathered between the anode and cathode, especially, in the event including 1%wt rate. It has been noticed the space charge distribution for the remaining 3 kV over Al_2O_3/Polyvinyl chloride nanocomposites with 10%wt nanoparticles under fluctuating dc electric field. There may be no evolving in control thickness for the remaining impulse voltages towards electrode, in case there is a little climbing during the anode. Figure 38 reveals the space charge conveyance pertinent Al_2O_3/Polyvinyl chloride nanocomposites with 10%wt nanoparticles rate with the remaining 3 kV over examples. The measurements infer the infusion from claiming impulse high voltages on the accuse thickness of pertinent Al_2O_3/Polyvinyl chloride nanocomposites for 10%wt nanoparticles percentage; there may be no expanding in control thickness towards both electrodes for the remaining impulse voltages at electrode, and there may be no hetero-charge gathered between the anode and cathode.

Space charge accumulation over immaculate polyvinyl chloride is indicated in figure 3. The charge thickness builds with expanding impulse voltages and hetero-charge gathered close to the electrode. Furthermore, charge thickness increments for the remaining impulse voltages in the anode and the hetero-charge gathered close to the electrode. The impact of raising fixation of ZnO nanoparticles over polyvinyl chloride declines its dielectric constant, as indicated and tabulated in this chapter. It has been depicted that the charging thickness during the electrode has more charging thickness at the anode with expanding ZnO nanoparticles rate and impulse connected voltages; particularly through 5%wt examples, and there may be hetero-charge gathered close to the electrode. Expanding ZnO nanoparticles starting with 5%wt dependent upon 10%wt abates charge thickness in the nanocomposite particles towards electrodes. The impact of claiming raising fixation from claiming TiO_2 nanoparticles to polyvinyl chloride increments its dielectric constant, as indicated and tabulated in this chapter. The charging thickness during the electrode are more than the charging thickness during the anode with expanding TiO_2 nanoparticles rate and impulse connected voltages; particularly through 5%wt examples, and there are hetero-charge gathered between electrodes as demonstrated. It is recognized that expanding TiO_2 nanoparticles from 5%wt up to 10%wt diminishes the charge thickness in the nanocomposite molecules at electrodes. On the other hand, the impact of raising concentrations of Al_2O_3 nanoparticles in polyvinyl chloride builds its dielectric constant similarly, as demonstrated in table I. The charging thickness during the electrode will be more than the charging thickness during the anode with expanding Al_2O_3 nanoparticles rate and impulse-connected voltages; particularly through 1%wt examples, and there will be no hetero-charge gathered between the electrodes through 5%wt and 10%wt from claiming aluminum oxide. Additionally, it is recognized that there may be no expanding in control thickness during both electrodes with remaining impulse voltages at electrode, and there is no hetero-charge gathered between anode and cathode.

8.5 FORECAST AND RECOMMENDATIONS

Trends of Nanoparticles for Space Charge Distribution

A critical stage certainly starts with homogeneity infiltration of nanoparticles over base matrix polymer (LDPE). The impact from claiming expanding focus of ZnO, clay and Al_2O_3 nanoparticles on low density polyethylene transforms the dielectric constant properties concerning illustration, as tabulated in this chapter. Third, the impacts of nanoparticles on the conduct technique for space charge accumulation have been contemplated in immaculate and nanocomposites with respect to impulse voltages and gathered charges close to the poles. Thus, it is depicted that the charging thickness towards the electrode is more than the charging thickness during the anode for expanding ZnO nanoparticles rate and impulse connected voltages; particularly through 5%wt examples, and there will be hetero-charge gathered close the electrode. Expanding ZnO nanoparticles from 5%wt up to 10%wt abates the charge thickness of nanocomposite molecules in electrodes. It is to be noted that the charging thickness during the electrode is more than the charging thickness towards the anode for expanding clay nanoparticles rate and impulse connected voltages; particularly through 5%wt examples, and there may be hetero-charge gathered between both electrodes. In any case, expanding clay nanoparticles starting with 5%wt dependent upon 10%wt increments charge thickness in the nanocomposite particles during both electrodes. Although the impact for raising concentrations of Al_2O_3 nanoparticles control in low density polyethylene increments its dielectric constant concerning illustration as tabulated in this chapter, whatever similarity is demonstrated, the charging thickness at the electrode is more than the charging thickness towards the anode for expanding Al_2O_3 nanoparticles rate and impulse connected voltages; particularly through 5%wt examples, and there is hetero-charge gathered towards electrodes. In addition, it is recognized that expanding Al_2O_3 nanoparticles from 5%wt dependent upon 10%wt diminishes charge thickness in the nanocomposite molecules during electrodes.

Sort and rate of nanoparticles are indispensable parameters in specifying electrical maturing and the dissemination of space charge thickness accumulation in the nanocomposite molecules between electrodes in dc electric fields. Therefore, including clay and ZnO nanoparticles over Polyvinyl chloride abates its dielectric constant. Hence, ZnO is more compelling than clay nanoparticles up to 5%wt control in expanding the charge thickness during both electrodes with expanding dc impulse connected voltages and other expands of hetero-charge gathered electrodes. Including Al_2O_3 nanoparticles in Polyvinyl chloride expands its dielectric constant, and the accuse thickness in both electrodes increases with expanding Al_2O_3 rates up to 5%wt. and impulse connected voltages. Thus, it expands hetero-charge gathered between electrodes.

Including clay nanoparticles builds accuse thickness of 20%wt. and others for aging under 5kV. Accuse thickness towards the anode and cathode is diminished daintily in PVC for 5%wt. clay, yet there is a greater amount of decrease in control thickness that happened for nanocomposite PVC with 10%wt. clay, and there are hetero-charge gathered charges between the anode and cathode electrodes.

Including fumed silica nanoparticles from up to 5%wt. increments accuse thickness to 200%. Furthermore, a progressive of greater part space charge at the electrode field level is watched, as the nanoparticles fixation expands to 5%wt. fumed silica, yet all the three of them is high diminishment in control density bt 80%, assuming that fumed silica rate has expanded starting with 5%wt. up to 10%wt. With aging under 5kV, accuse thickness towards the anode and cathode will be diminished daintily over PVC for 5% fumed silica, in case there is a recognized high decrease in control thickness that happens to nanocomposite PVC with 10%wt. fumed silica, and there are no hetero-charge gathered charges between the anode and cathode electrodes. Including ZnO nanoparticles up to 10%wt. ZnO expands accuse thickness that has happened over immaculate PVC by 250%. Moreover, a progressive for heft space charge in the electrode field level is seen, as the nanoparticles focus builds with 10%wt. ZnO. For maturing under 5kV, charge thickness towards the anode and cathode will be expanded in PVC for 10%wt. ZnO and there are hetero-charge gathered charges between the anode and cathode electrodes. At aging under 5kV, accuse thickness is expanded towards the anode and cathode more than that which happened in immaculate PVC and different mulled over nanocomposites. The accuse thickness in Polyvinyl chloride increments for expanding impulse voltages and hetero-charge gathered close to the particular case electrode. However, including clay nanoparticles in Polyvinyl chloride builds the charge thickness towards both electrodes for expanding impulse connected voltages and builds hetero-charge gathered between both electrodes. Including fumed silica nanoparticles from up to 5%wt. expands accuse thickness for expanding impulse connected voltages, and there may be hetero-charge gathered close to the electrodes. Nevertheless, including fumed silica nanoparticles from up to 10% the measurements suggests the infusion of impulse high voltages that depict accuse thickness towards the anode and cathode and it is diminished to the least characteristics over fumed silica/Polyvinyl chloride nanocomposites, and there are no hetero-charge gathered charges between the anode or cathode electrodes in nanocomposite molecules. Including ZnO nanoparticles in polyvinyl chloride declines its dielectric constant, but the accuse thickness towards both electrodes expands with expanding ZnO rates up to 5%wt and impulse connected voltages. Then, it increments hetero-charge gathered during electrode. In any case, including TiO_2 nanoparticles for polyvinyl chloride expands its dielectric constant, and the accuse thickness in both electrodes increments with expanding TiO_2 rates dependent upon 5%wt and impulse connected voltages. Then, it increments hetero-charge gathered between electrodes. On the other hand, including Al_2O_3 nanoparticles over polyvinyl chloride increments its dielectric constant, and the charging thickness at the electrode is more than charging thickness towards the anode for expanding Al_2O_3 nanoparticles rate and impulse connected voltages; particularly through 1%wt examples. Moreover, there is no hetero-charge gathered between electrodes through 5%wt and 10%wt from claiming aluminum oxide. Subsequently, there will be no expanding in control thickness in both electrodes for the remaining impulse voltages in electrode, and there will be no hetero-charge gathered between the anode and cathode. Kind and rate of nanoparticles are crucial parameters in specifying electrical maturing and the circulation of space charge thickness accumulation in the nanocomposite particles between electrodes in dc electric fields.

REFERENCES

Agarwal, S., Tatli, E., Clark, N. N., & Gupta, R. (2005). Potential Health Effects of Manufactured Nanomaterials: Nanoparticle Emission Arising From Incineration of Polymer Nanocomposites. *International Symposium on Polymer Nanocomposites Science and Technology*.

Andritsch, T., Kochetov, R., Gebrekiros, Y. T., Lafont, U., Morshuis, P. H. F., & Smit, J. J. (2009). Synthesis and Dielectric Properties of Epoxy Based Nanocomposites. *IEEE Conference on Electrical Insulation and Dielectric Phenomena*, 523-526. 10.1109/CEIDP.2009.5377771

Andritsch, T., Kochetov, R., Gebrekiros, Y. T., Morshuis, P. H. F., & Smit, J. J. (2010). Short Term DC Breakdown Strength In Epoxy Based BN Nano- And Microcomposites. *IEEE International Conference on Solid Dielectrics*, 179-182. 10.1109/ICSD.2010.5568098

Bamji, S. S., Abou-Dakka, M., Bulinski, A. T., Utracki, L., & Cole, K. (2005). Dielectric Properties of Propylene Containing Nano-Particles. *IEEE Conference on Electrical Insulation and Dielectric Phenomena*, 166-170.

Baraton. (2003). *Synthesis, Functionalization and Surface Treatment of Nanoparticles*. American Scientific Publisher.

Bois, L., Chassagneux, F., Parola, S., Bessueille, F., Battie, Y., Destouches, N., Boukenter, A., Moncoffre, N., & Toulhoat, N. (2009). Growth of ordered silver nanoparticles in silica film mesostructured with a triblock copolymer PEO–PPO–PEO. *Journal of Solid State Chemistry*, *182*(7), 1700–1707. doi:10.1016/j.jssc.2009.01.044

Brundle, Evans, & Wilson. (1992). Encyclopedia of Materials Characterization. Butterworth-Heinemann Publications.

Chen, G., Sadipe, J. T., Zhuang, Y., Zhang, C., & Stevens, G. C. (2009). Conduction In Linear Low Density Polyethylene Nanodielectric Materials. *Properties and Applications of Dielectric Materials, ICPADM. IEEE 9th International Conference*, 845-848. 10.1109/ICPADM.2009.5252208

Chen, X., Gonsalves, K. E., Chow, G.-M., & Xiao, T. D. (1994). Homogeneous Dispersion of Nanostructured Aluminum Nitride In A Polyimide Matrix. *Advanced Materials*, *6*(6), 481–484. doi:10.1002/adma.19940060608

Chen, Y., Imai, T., Ohki, Y., & Tanaka, T. (2010). Tree initiation phenomena in nanostructured epoxy composites. *IEEE Transactions on Dielectrics and Electrical Insulation*, *17*(5), 1509–1515. doi:10.1109/TDEI.2010.5595552

Cherkasova, A. S. (2009). *Thermal Conductivity Enhancement In Micro- and Nano-Particle Suspensions* (PhD thesis). State University of New Jersey.

Chisholm, N., Mahfuz, H., Rangari, V. K., Ashfaq, A., & Jeelani, S. (2005). Fabrication and Mechanical Characterization of Carbon/Sic Nanocomposites. *Composite Structures*, *67*(1), 115–124. doi:10.1016/j.compstruct.2004.01.010

Ciuprina, F., & Plesa, I. (2011). DC and AC Conductivity of LDPE Nanocomposites. *Advanced Topics in Electrical Engineering (ATEE), 7th International Symposium*, 1-6.

Fabiani, D., Cavallini, A., Montanari, G. C., & Lab-testa, L. (2009). Extraction of Aging Markers For Nanostructured Epoxy Resin Under Surface Discharges. *IEEE Electrical Insulation Conference*. 10.1109/EIC.2009.5166376

Friedrich, K., & Schlarb, A. K. (2008). *Tribology of Polymeric Nanocomposites. Friction And Wear of Bulk Materials And Coatings* (Vol. 55). Elsevier.

Geng, Y., Liu, M. Y., Li, J., Shi, X. M., & Kim, J. K. (2008). Effects of Surfactant Treatment On Mechanical and Electrical Properties Of CNT/Epoxy Nanocomposites. *Composites. Part A, Applied Science and Manufacturing*, *39*(12), 1876–1883. doi:10.1016/j.compositesa.2008.09.009

Gouda, O., Thabet, A., Mobarak, Y. A., & Samir, M. (2013). *Nanotechnology Effects on Space Charge Relaxation Measurements for Polyvinyl Chloride Thin Films*. IET Nanobiotechnology Journal.

Guastavino, F., Coletti, G., Dardano, A., Fina, A., & Thelakkadan, A. S. (2010). Thermo-Mechanical and Electrical Characterization of Epoxy/Nanoclay Composites. *IEEE Conference on Electrical Insulation and Dielectric Phenomena*, 549-552. 10.1109/CEIDP.2010.5723997

Henk, P. O., Korsten, T. W., & Kvarts, T. (1999). Increasing The Electrical Discharge Endurance of Acid Anhydride Cured DGEBA Epoxy Resin By Dispersion of Nanoparticle Silica. *High Performance Polymers*, *11*(3), 281–296. doi:10.1088/0954-0083/11/3/304

Henk, P. O., Korsten, T. W., Kvarts, T., & Saeidi, A. (2001). Increasing The PD-Endurance of Epoxy And XLPE Insulation By Nanoparticles Silica Dispersion In The Polymer. *Nordic Insulation Symposium*.

Huang, Jiang, & Yin. (2009). Nanoparticle surface modification induced space charge suppression in linear low density polyethylene. *Applied Physics Letters, 95*(24), 242905 - 242905-3.

Hurt, R. H., Monthioux, M., & Kane, A. (2006). Toxicology of Carbon Nanomaterials: Status, Trends, and Perspectives On The Special Issue. *Carbon*, *44*(6), 1028–1033. doi:10.1016/j.carbon.2005.12.023

Imai, T., Sawa, F., Nakano, T., Ozaki, T., Shimizu, T., Kozako, M., & Tanaka, T. (2006). Effects of Nano- And Micro-Filler Mixture On Electrical Insulation Properties of Epoxy Based Composites. *IEEE Transactions on Dielectrics and Electrical Insulation*, *13*(1), 319–326. doi:10.1109/TDEI.2006.1624276

Imai, T., Sawa, F., Ozaki, T., Shimizu, T., Kido, R., Kozako, M., & Tanaka, T. (2006). Influence of Temperature On Mechanical And Insulation Properties of Epoxy-Layered Silicate Nanocomposite. *IEEE Transactions on Dielectrics and Electrical Insulation*, *13*(1), 445–452. doi:10.1109/TDEI.2006.1624291

Imai, T., Sawa, F., Ozaki, T., Shimuzu, T., Kido, R., Kozako, M., & Tanaka, T. (2005). Evaluation of Insulation Properties of Epoxy Resin With Nano-Scale Silica Particles. *International Symposium on Electrical Insulating Materials*, 239-242. 10.1109/ISEIM.2005.193387

Kozako, M., Kuge, S., Imai, T., Ozaki, T., Shimizu, T., & Tanaka, T. (2005). Surface Erosion Due To Partial Discharges On Several Kinds of Epoxy Nanocomposites. *IEEE Conference on Electrical Insulation and Dielectric Phenomena*, 162-165. 10.1109/CEIDP.2005.1560646

Lam, C. K., & Lau, K. T. (2006). Localized Elastic Modulus Distribution of Nanoclay / Epoxy Composites By Using Nanoindentation. *Composite Structures*, *75*(1), 553–558. doi:10.1016/j.compstruct.2006.04.045

Levering, A. W. (1995). *Interphases In Zirconium Silicate Filled High Density Polyethylene And Polypropylene* (PhD thesis). Delft University of Technology.

Li, S., Yin, G., Chen, G., Li, J., Bai, S., Zhong, L., Zhang, Y., & Lei, Q. (2010). Short-Term Breakdown and Long-Term Failure In Nanodielectrics: A Review. *IEEE Transactions on Dielectrics and Electrical Insulation*, *17*(5), 1523–1535. doi:10.1109/TDEI.2010.5595554

Maeon, T., Futami, T., & Kushibe, H. (1988). T. Takada And C. M. Cooke, "Measurement of Space Charge Distribution In Thick Dielectric Using The PEA Method. *IEEE Transactions on Electrical Insulation, EI-23*, 433–439. doi:10.1109/14.2384

Masuda, S., Okuzumi, S., Kurniant, R., Murakami, Y., Nagao, M., Murata, Y., & Sekiguchi, Y. (2007). *DC Conduction and Electrical Breakdown of Mgo/LDPE Nanocomposite. In Electrical Insulation and Dielectric Phenomena, CEIDP*. Conference.

Matějka, L., Dukh, O., & Kolařík, J. (2000). Reinforcement of Crosslinked Rubbery Epoxies By In Situ Formed Silica. *Polymer, 41*(4), 1449–1459. doi:10.1016/S0032-3861(99)00317-1

Mayer, A. (2007). TEM Sample Preparation And FIB-Induced Damage. *Materials Research Society Bulletin, 32*(5), 400–407. doi:10.1557/mrs2007.63

Muhamad, N. A., Phung, B. T., Blackburn, T. R., & Lai, K. X. (2009). Polarization and Depolarization Current (PDC) Lab-tests On Biodegradable and Mineral Transformer Oils At Different Moisture Levels. In *Power Engineering Conference, AUPEC*. Australasian Universities.

Murakami, Y., Nemoto, M., Okuzumi, S., Masuda, S., Nagao, M., Hozumi, N., Sekiguchi, Y., & Murata, Y. (2008). DC Conduction and Electrical Breakdown of Mgo/LDPE Nanocomposite. *IEEE Transactions on Dielectrics and Electrical Insulation, 15*(1), 33–39. doi:10.1109/T-DEI.2008.4446734

Murugaraj, Mainwaring, & Mora-Huertas. (2005). Dielectric enhancement in polymer-nanoparticle composites through interphase polarizability. *Journal of Applied Physics, 98*(5), 054304 - 054304-6.

Oberdörster. (2005). Principles For Characterizing The Potential Human Health Effects From Exposure To Nanomaterials: Elements of A Screening Strategy. *Particle and Fibre Toxicology, 2*(8), 1-35. PMID:15813962

Okuzumi, Sh., Murakami, Y., Nagao, M., Sekiguchi, Y., Reddy, Ch. Ch., & Murata, Y. (2008). DC Breakdown Strength and Conduction Current of MgO/LDPE Composite Influenced by by Filler Size. *IEEE, Annual Report Conference on Electrical Insulation Dielectric Phenomena*, 722-725. 10.1109/CEIDP.2008.4772933

Oyegoke. (2001). Space Charge Measurement As Possible Diagnostic Tool For High-Voltage DC Cable Systems. In *Future*. Helsinki University of Technology Book.

Özdilek, C. (2006). *Colloidal Liquid Crystalline Reinforced Nanocomposites* (PhD thesis). Delft University of Technology.

Reed, C. W. (2010). Functionalization of Nanocomposite Dielectrics. *IEEE International Symposium on Electrical Insulation*, 353-356.

Saha, T. K., & Purkait, P. (2004). Investigation of Polarization and Depolarization Current Measurements For The Assessment of Oil-Chapter Insulation of Aged Transformers. *IEEE Transactions on Dielectrics and Electrical Insulation, 11*(1), 144–154. doi:10.1109/TDEI.2004.1266329

Sarathi, R., Sahu, R. K., & Rajeshkumar, P. (2007). Understanding The Thermal, Mechanical And Electrical Properties of Epoxy Nanocomposites. *Materials Science and Engineering A, 445-446*, 567–578. doi:10.1016/j.msea.2006.09.077

Schulte, P. A., & Salamanca-Buentello, F. (2007). Ethical And Scientific Issues of Nanotechnology In The Workplace. *Environmental Health Perspectives*, *115*(1), 5–12. doi:10.1289/ehp.9456 PMID:17366812

Singha, S., & Thomas, M. J. (2006). Polymer Composite/Nanocomposite Processing and Its Effect On The Electrical Properties. *IEEE Conference on Electrical Insulation and Dielectric Phenomena*, 557-560.10.1109/CEIDP.2006.311993

Singha, S., & Thomas, M. J. (2008). Dielectric Properties of Epoxy Nanocomposites. *IEEE Transactions on Dielectrics and Electrical Insulation*, *15*(1), 12–23. doi:10.1109/T-DEI.2008.4446732

Sun, C. (2010). *Controlling The Rheology of Polymers/Silica Nanocomposites* (PhD thesis). Eindhoven University of Technology.

Sun, Y., Zhang, Z., & Wong, C. P. (2004). Fundamental Research On Surface Modification of Nano-Size Silica For Underfill Applications. *IEEE 9th Symposium on Advanced Packaging Materials*, 132-138.

Tanaka, T., Matsuo, Y., & Uchida, K. (2008). Partial Discharge Endurance of Epoxy/Sic Nanocomposite. *IEEE Conference on Electrical Insulation and Dielectric Phenomena*, 13-16. 10.1109/CEIDP.2008.4772782

Tanaka, T., Yazagawa, T., Ohki, Y., Ochi, M., Harada, M., & Imai, T. (2007). Frequency Accelerated Partial Discharge Resistance of Epoxy/ Clay Nanocomposite Prepared By Newly Developed Organic Modification and Solubilization Methods. *IEEE International Conference on Solid Dielectrics*, 337-340.10.1109/ICSD.2007.4290821

Thomas, K., & Sayre, P. (2005). Research Strategies For Safety Evaluation of Nanomaterials, Part I: Evaluating He Human Health Implications of Exposure To Nanoscale Materials. *Toxicological Sciences*, *87*(2), 316–321. doi:10.1093/toxsci/kfi270 PMID:16049265

Vaughan, A. S. (2008). Raman Nanotechnology – The Lycurgus Cup. *IEEE Electrical Insulation Magazine*, *24*(6), 4. doi:10.1109/MEI.2008.4665344

Vaughan, A. S., Swingler, S. G., & Zhang, Y. (2006). Polyethylene Nanodielectrics: The Influence of Nanoclays On Structure Formation and Dielectric Breakdown. *IEEJ Transactions on Fundamentals and Materials*, *126*(11), 1057–1063. doi:10.1541/ieejfms.126.1057

Vella, N., Toureille, A., Guarrotxena, N., & Millan, J. L. (2000, June). Thermal Step and TSDC Measurement in PVC. *IEEE Transactions on Dielectrics and Electrical Insulation*, *7*(3), 329–333. doi:10.1109/94.848909

Wang, Q., Curtis, P., & Chen, G. (2010). Effect of Nanoparticles On Electrical Breakdown Behavior of Epoxy Resin. *IEEE Conference on Electrical Insulation and Dielectric Phenomena*, 393-396.

Wetzel, B., Haupert, F., & Zhang, M. Q. (2003). Epoxy Nanocomposites With High Mechanical and Tribological Performance. *Composites Science and Technology*, *63*(14), 2055–2067. doi:10.1016/S0266-3538(03)00115-5

Wiesner, M. R. (2007). *Bottero, Environmental Nanotechnology*. McGraw-Hill.

Yung, K. C., Wang, J., & Yue, T. M. (2008). Thermal Management For Boron Nitride Filled Metal Core Printed Circuit Board. *Journal of Composite Materials*, *42*(24), 2615–2627. doi:10.1177/0021998308096326

Zha, Dang, Song, Yin, & Chen. (2010). Dielectric properties and effect of electrical aging on space charge accumulation in polyimide/TiO2 nanocomposite films. *Journal of Applied Physics, 108*(9), 094113 - 094113-6.

Zhu, Y. W., Shen, X. Q., Wang, B. C., Xu, X. Y., & Feng, Z. J. (2004). Chemical Mechanical Modification of Nanodiamond In An Aqueous System. *Physics of the Solid State*, *46*(4), 681–684. doi:10.1134/1.1711451

Chapter 9
NanoDielectrics Surfaces and Barriers

ABSTRACT

This chapter describes the nanodielectrics surfaces and barriers. Thermal interface materials (TIM) used for industrial applications have been mentioned. Wetting nanodielectrics surfaces, packaging, and battery applications have been chosen to present the effect of nanotechnology on industrial applications. This chapter draws attention on the suggested investment procedures for industrial insulation materials in the future.

This chapter reviews explanatory models for estimating successful conductivity of new recommended thermal interface polymeric composite modern materials whose characterization reaction has been improved in terms of particle sorts and their concentrations, as well as forces applied to these materials. A detailed examination of the nanostructure aspects and the impact of the effective thermal conductivity from claiming TIMs is included in the objectives of this chapter. Thus, their characterization response with respect to types and concentrations of selected nanoparticles has been enhanced. In this work, sol gel system has been used to prepare polypropylene nanocomposites; a percent of diverse sorts of the nanoparticles (clay, ZnO, SiO_2 and TiO_2) and diverse centralization (1%wt., 5%wt. and 10%wt.) are used to control the attraction forces of water droplets on surfaces for polypropylene nanocomposites. The fabricated polypropylene nanocomposites are characterized eventually perusing FTIR, SEM, dielectric constant, contact angle, wetting energy, spreading coefficient and fill in of bond measurements. Lab-test measurements have deduced that clay and ZnO nanoparticles decrease the dielectric constant's ability for claiming polypropylene, while at the same time, SiO_2 and TiO_2 increment this value. It has been reported that the wettability of the arranged nanocomposites is lessened, eventually perusing the expansion of sure nanoparticles ratio, which demonstrates the ability of the obtained nanocomposites for bundling and battery cases applications.

Nanotechnology strategy is utilized in this effort to enhance the surface energy properties for polyvinyl chloride (PVC). Diverse sorts (clay, ZnO, SiO_2 and Al_2O_3) and concentrations (1%wt., 5%wt. and 10%wt.) about nanoparticles are investigated. The morphology, dielectric constant, contact angle, wet-

DOI: 10.4018/978-1-7998-3829-6.ch009

ting energy, spreading coefficient and worth of effort of bond are investigated to immaculate PVC and PVC nanocomposites. Tap and salt water are used to study the surface energy properties. The outcomes have uncovered that the sort and concentration of the utilized nanoparticles influence the properties of the acquired nanocomposites. Changing the sort and concentration of the utilized nanoparticles is the principle purpose for changing surface roughness, hydrophilic cites arrangement and dipole/dipole interactions, and hence, for enhancing the surface energy properties of PVC nanocomposites.

9.1 THERMAL INTERFACE MATERIALS (TIM)

Predictive modelling based on fundamental physical principles is critical to developing new TIMs, since it can be used to quantify the impact of particle volume fractions and plans on the effective thermal conductivity. Such models will be empowering to streamline the structure and plan of the material (Dan et al., 2010; Kanuparthi, Subbarayan, Sammakia et al, 2008; Kanuparthi, Subbarayan, Siegmund et al, 2008; Renukappa & Rashmi, 2012).

9.1.1 Theory and Effective Parameters

Thermal interface materials (TIMs) have been generally embraced to minimize the thermal interface resistance between the rough surfaces of heat generating components and the heat dissipation devices. The greater part of TIMs are committed to polymers with thermally conductive particles conveyed inside to upgrade the thermal conductivity. There are different sorts of methodologies to discover the effective thermal conductivity of the two-phase composite frameworks. The successful thermal conductivity of TIMs influences large fractions factors; for example, the thermal conductivity of filler particles, the thermal conductivity of the matrix, the volume fraction for filler particles and the particle size conveyance of filler particles, et cetera (Dan et al., 2012; Kanuparthi et al., 2009; Ouchetto et al., 2006; Yue et al., 2008; Zanden et al., 2013). This chapter likewise looks into connected hypothetical models for estimating thermal conductivity of nanocomposite modern materials comprising nano-sized nano-crystalline particles installed in distinctive matrices. Dependent upon the presented theoretical model (Thabet, 2015b; Thabet, 2016; Thabet, 2017) that has been utilized for foreseeing an effective thermal conductivity for thermal interface materials nanocomposites; fig. 1 indicates the barrel-shaped area between the two round particles. There are recommendations for claiming contact between the two particles in a thermal flux thickness crosswise over the surface for particles in the irregular course of action of the matrix. Thus, thermal conductance between round particles can be predicted without shifting attention to the volume fraction from claiming particles in the matrix (Φ), the thermal conductivity from claiming particles (K_p), the thermal conductivity from claiming matrix (K_m), radii from claiming two neighborhood circle surfaces (R_1, And R_2) and factors of the ability from claiming framing nonstop organization from claiming fillers in the matrix (C_1, Also C_2) and through (0:1) concerning illustration as follows (Dan et al., 2010):.

$$K_{eff} = \left[\frac{5.2933}{2\grave{A}k_p R_1} + \frac{1}{\grave{A}k_m a \ln\left(1+\frac{R^2}{ha}\right)} + \frac{5.2933}{2\grave{A}k_p R_2} \right]^{-1} \quad (1)$$

Where,

$R = \alpha.a$ (1.i)

$a = \frac{2R_1R_2}{R_1 + R_2}$ (1.ii)

$\alpha = 5.0539\varphi^2 - 7.3994\varphi + 3.203$ (1.iii)

$$\log K_{eff} = \varphi C_2 \log k_p + (1 - \varphi) \log (C_1 k_m) \quad (2)$$

Figure 1. Schematic diagram for cylindrical region between two spherical particles (Thabet, 2016).

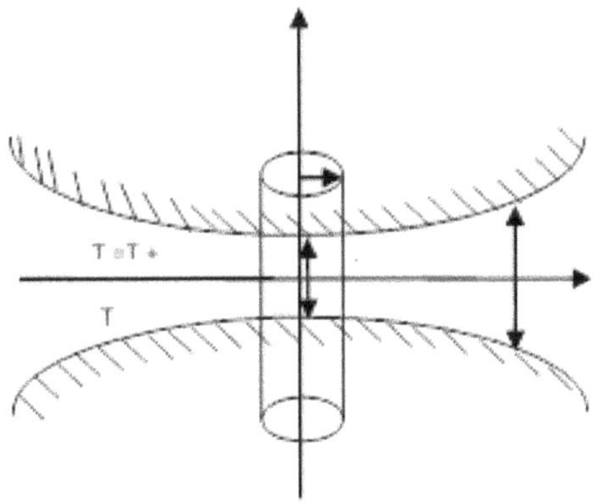

The effective thermal conductivity K_{eff} from claiming nanoparticles/nonmagnetic matrix composite can offer a chance to ascertain the use of the over hypotheses in view of their sorts and concentrations. Taking all that into consideration, impacts of the ability for framing nonstop organization of fillers for base matrices will be totally satisfactory. Hence, the execution of effective thermal conductivities for different recommended new magnetic nanocomposites has been indicated in this investigation.

9.1.2 Effects of Nanoparticles on Thermal Interface Materials

Figure 2 indicates the effective thermal conductivity from claiming TIMs iron nanocomposites that are expanded by including different rates of silicon and graphite for irregular distribution. It is recognized that the effective thermal conductivity of TIMs iron nanocomposites are reported, eventually perusing the inclusion of different rates of glass and PTFE for irregular distribution. However, graphite nanoparticles have higher effectiveness on expanding thermal conductivity from claiming nanocomposites for iron matrix regarding silicon. However, PTFE has higher effectiveness for diminishing thermal conductivity for nanocomposites in iron matrix with respect to Glass nanoparticles. Figure 3 indicates the effective thermal conductivity for TIMs iron nanocomposites reported towards including different rates from claiming nanoparticles about oxides (Al_2O_3, ZnO, and MgO) and irregular distribution to iron matrix material. However, MgO nanoparticles have higher capability for expanding thermal conductivity of nanocomposites in iron matrix regarding Al_2O_3 and ZnO nanoparticles. Fig 4 indicates the successful thermal conductivity of TIMs silicon nanocomposite materials expanded, and eventually perusing the expansion of different rates from claiming graphite nanoparticles over irregular distribution to iron matrix material. However, the effective thermal conductivity of TIMs silicon nanocomposite materials is reported towards including different rates of iron and Graphite nanoparticles clinched alongside and eventually perusing irregular distribution of iron matrix material. It is clear that Graphite nanoparticles is more successful in diminishing thermal conductivity of nanocomposites to silicon matrix with respect to iron nanoparticles. Fig. 5 indicates the effective thermal conductivity of TIMs silicon nanocomposite materials reported, eventually perusing the inclusion of different rates of nanoparticles for oxides (Al_2O_3, ZnO, and MgO) for irregular distribution of silicon matrix material. However, MgO nanoparticles are more effective for diminishing thermal conductivity of nanocomposites in iron matrix with respect to Al_2O_3, and ZnO nanoparticles.

Figure 2. Effective thermal conductivity of TIMs Iron/Polymer nanocomposites with various nanoparticles

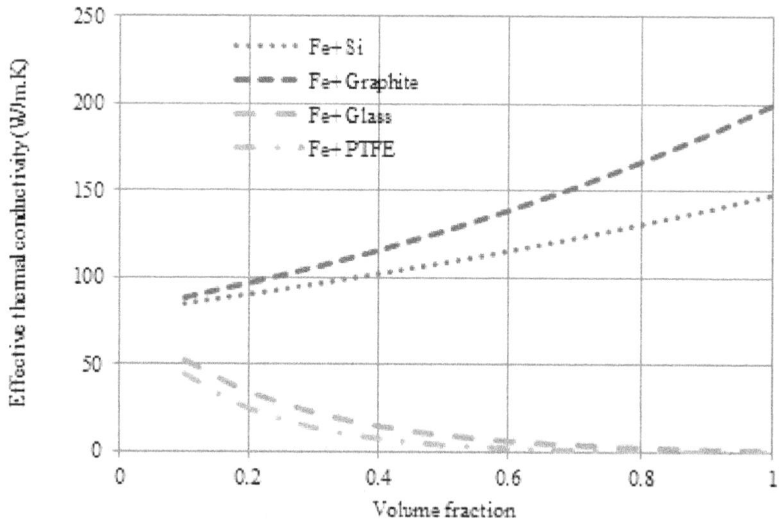

Figure 6 indicates the effective thermal conductivity of expanded TIMs epoxy nanocomposite materials, eventually perusing the inclusion of different rates of aluminum, iron, silicon, graphite, cobalt, and nickel nanoparticles for an irregular distribution on epoxy matrix material. In any case, iron nanoparticles are more effective to diminishing thermal conductivity of nanocomposites and epoxy matrix with respect to different nanoparticles. Figure 7 indicates the effective thermal conductivity of expaned TIMs epoxy nanocomposite materials, eventually perusing the inclusion of different rates of nanoparticles of oxides

(Al_2O_3, ZnO, and MgO) in the irregular distribution of epoxy matrix material. It is to be noted that MgO nanoparticles are more effective for epoxy matrix to expand the thermal conductivity of nanocomposites with respect to Al2O3 and ZnO nanoparticles. Fig. 8 indicates the effective thermal conductivity of TIMs epoxy nanocomposite materials that are expanded towards including different rates from claiming glass nanoparticles clinched alongside irregular distribution to epoxy matrix material. However, the effective thermal conductivity of TIMs epoxy nanocomposite materials is reported towards including different rates of PTFE nanoparticles in irregular distribution of epoxy matrix material.

Figure 3. Effective thermal conductivity of TIMs Iron/Oxides nanocomposites with various nanoparticles

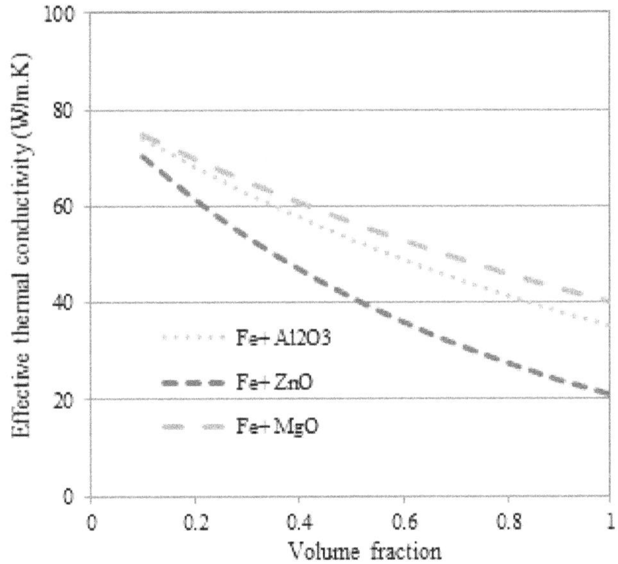

Figure 4. Effective thermal conductivity of TIMs Silicon/Metal nanocomposites with various nanoparticles

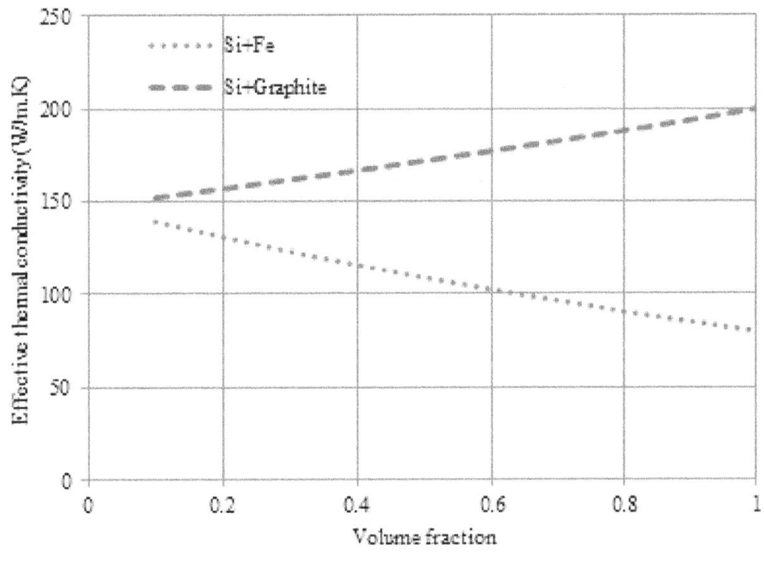

Figure 5. Effective thermal conductivity of TIMs Silicon/Oxides nanocomposites with various nanoparticles

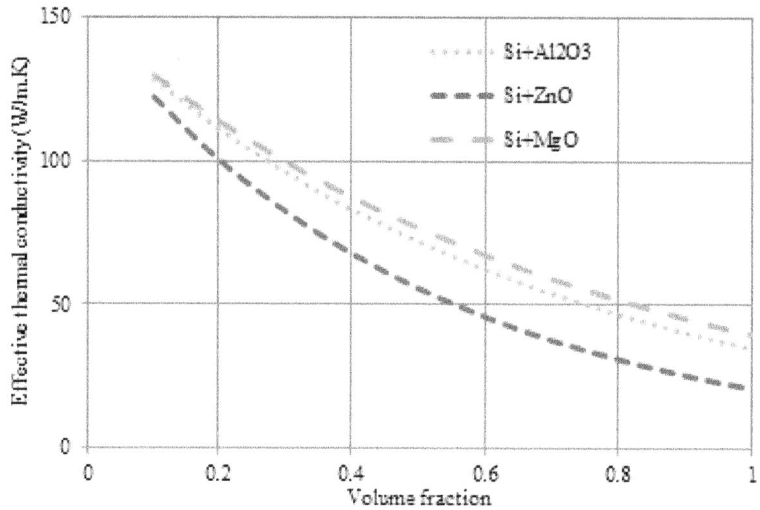

Figure 6. Effective thermal conductivity of TIMs Epoxy/Metal nanocomposites with various nanoparticles

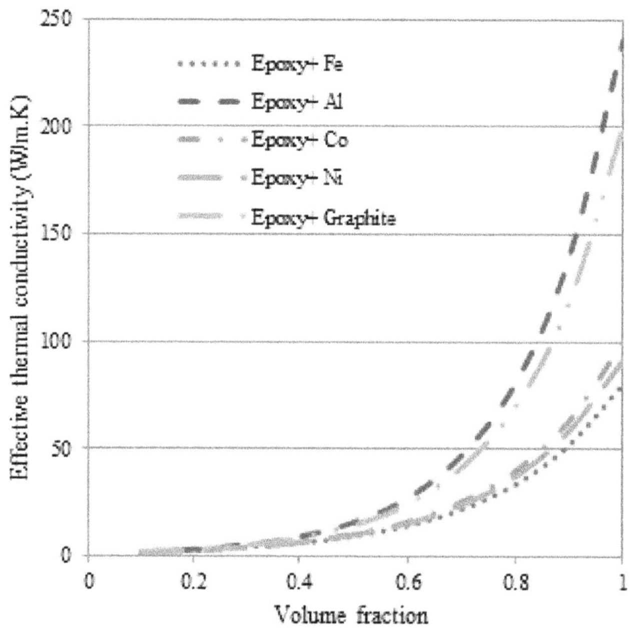

Figure 7. Effective thermal conductivity of TIMs Epoxy/Oxides nanocomposites with various nanoparticles

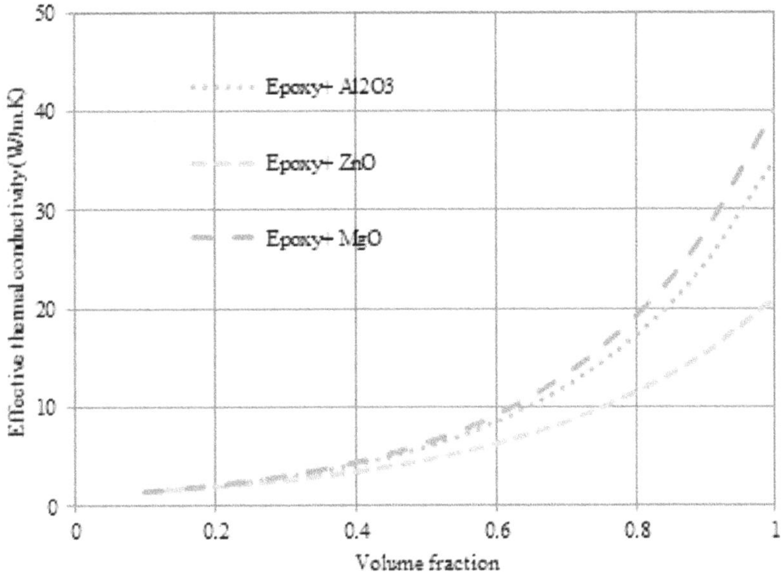

Figure 8. Effective thermal conductivity of Epoxy/Insulator nanocomposites.

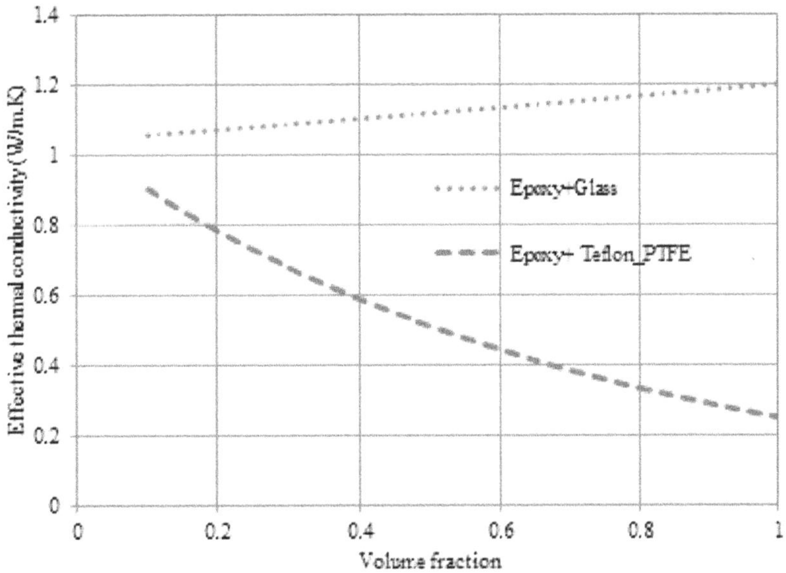

9.2 WETTING NANO-DIELECTRIC SURFACES 9.2

9.2.1 Theory and Effective Parameters

Wetting phenomena is greatly perceived on the macroscopic- scale, but it can happen on the nano – scale. Furthermore, the advantage of the identification of fluid networking has to be seized starting with the utilization of oblique angle deposition (OAD) slim films joined under ultra-nationalistic units alternately with complex photonic structures transformed, eventually perusing stacking dainty film layers with different refractive indices (Barberis & Capurro, 2008; Nakamura et al., 2013; Ruffino & Grimaldi, 2012; Ruffino & Grimaldi, 2014; Torrisi & Ruffino, 2017; Yong & Zhang, 2009).

In the event of claiming three stages (solid, liquid, and gas), see fig. 9, the surface tensions between the three periods can be demonstrated as follows (Chow, 1998):

$$\Upsilon_{SG} = \Upsilon_{SL} + \Upsilon_{LG}\cos\theta \tag{3}$$

Figure 9. Contact angle of a liquid droplet wetted to a rigid solid surface

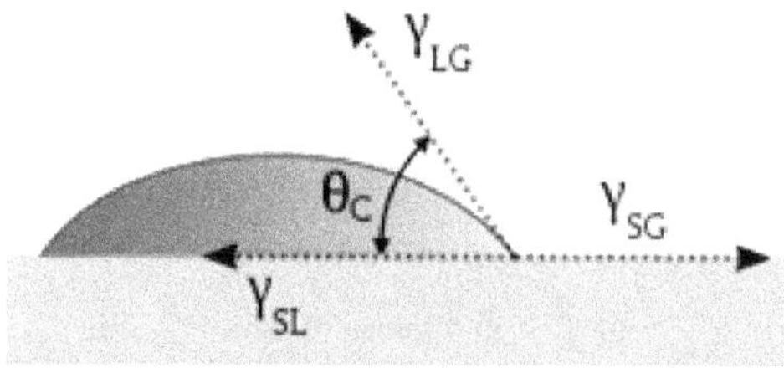

The contact angle (θ_C) can be defined as the angle at which the liquid–vapor interface meets the solid–liquid interface (Torrisi & Ruffino, 2017). Even in a perfectly smooth surface, a drop will assume a wide spectrum of contact angles ranging from the so-called advancing contact angle, θ_A, to the so-called receding contact angle, θ_R. The equilibrium contact angle (θ_C) can be calculated as shown by Tadmor (Tadmor, 2004) below:

$$\theta_C = arccos\left(\frac{r_A \cos\theta_A + r_R \cos\theta_R}{r_A + r_R}\right) \tag{4}$$

Where,

$$r_A = \left(\frac{sin^3\theta_A}{2-3\cos\theta_A + cos^3\theta_A}\right)^{\frac{1}{3}},\ r_R = \left(\frac{sin^3\theta_R}{2-3\cos\theta_R + cos^3\theta_R}\right)^{\frac{1}{3}}$$

The surface free energy, interfacial free energy and polar and dispersion energy can be measured starting with the contact point (Murakami et al., 2008).

Polyvinyl chloride is a standout amongst the large fraction essential thermoplastic materials in the world because of its low cost, high chemical imperviousness and barrier properties (Ahmed et al., 2010; Claudiu et al., 2010; Ebnalwaled & Thabet, 2016; Moulay, 2010).

The best outcomes from claiming our knowledge regarding the surface energy properties of PVC nanocomposites are specified. Nowadays, nanotechnology strategies are formed and controlled once electric and dielectric characterization of polyvinyl chloride and variant industrial building materials is performed. Of course, the characterization of electrical materials has been reflected on their industrial conduct technique (Thabet, 2015a; Thabet, 2015b; Thabet, 2016; Thabet, 2017; Thabet & Mubarak, 2017). Restricted reports manage surface energy properties of PVC/ZnS, PVC/CdSe nanocomposites. At the same time, there are no effects on the surface energy properties of PVC/ZnO, PVC/Clay, PVC/Al_2O_3 and PVC/SiO_2.....etc nanocomposites. This chapter seeks to study and streamline the surface energy properties of immaculate PVC, and then apply them for distinctive sorts and concentrations of nanomaterials. This effort is dedicated to measuring the contact point about tap and salt water on the surfaces of the ready nanocomposites. This chapter predicts that energy and attraction forces about water surface pressure can be controlled for industrial building materials by utilizing nanotechnology. Then, ideal nanoparticles are specified for regulating in contact angles and attraction drives between both tab-water, salt-water droplets and polyvinyl chloride nanocomposite surfaces. In addition, the patterns of variant nanoparticles for characterization of wetting energy, spreading coefficient and fill in bond for both tab-water and salt-water droplets are discussed by looking into electrical applications of polyvinyl chloride nanocomposites surfaces.

9.2.2 Preparation of Wetting Nanodielectric Surfaces

Films with 150 mm thickness for polyvinyl chloride (PVC), (polyvinyl chloride + clay) (99:1 w/w), (95:5 w/w) and (90:10 w/w), (polyvinyl chloride: metal oxide) (99:1 w/w), (95:5 w/w) and (90:10 w/w) are prepared towards solution-cast technobabble by utilizing tetrahydrofuran (THF) as dissolvable. The mixture of these solutions is stirred for 12 h at room temperature, and then the solution is cast onto soda lime glass plates. Reduction is achieved by treating the films with aqueous solutions of NH4OH or NaOH. The prepared films are exposed to low pressure and soft heating for varying durations to remove any residual solvents. SEM pictures of polyvinyl chloride nanocomposites films are demonstrated in fig. 10.It has flakes such as morphological tenet for high surface range. Additionally, it illustrates that the nanoparticles (Dia.: 50nm) are uniformly scattered in the polymer matrix.

9.2.3 Effects of Nanoparticles on Nanocomposites Characterization

FTIR ghastly dissection (Jasco model 4100–Japan) has been used to research the compound piece of the arranged PVC nanocomposites to assess the utilitarian gatherings in the arranged lab-tests. The morphological tenet of the handled films is described utilizing filtering electron magnifying instrument (JEOL SEM model JSM—5500–Japan), with accelerated voltage of 10 kV. HIOKI 3522-50 LCR Hi-lab-tester gadget used to measure the dielectric steadiness of the arranged nanocomposites. The contact holy messenger investigated (SEO phoenix 300) is used to measure the contact angle, surface tension, wetting energy, spreading coefficient and fill in bond. The reproduction of information is checked by measuring few examples. Similarly, as indicated in fig., the kind and concentrations of nanoparticles influence the dielectric steadiness from claiming PVC nanocomposites and eventually influencing perusing. The dielectric constant is reported eventually perusing expanding clay and ZnO. At the same time, it is expanded by expanding SiO_2 and Al_2O_3. This is because SiO_2 and Al_2O_3 are more resistive than clay and ZnO.

Contact point and surface pressure measurements have been accomplished for water droplets with respect to immaculate polyvinyl chloride films, as demonstrated previously in fig. 11. Surface pressure measurements of tab-water droplets are as follows: Contact point (Average degree] is 76. 96678, wetting energy is 14. 41756 [mN/m], spreading coefficient is -56. 38244 [mN/m] and worth of effort for bond is 79. 21757 [mN/m] for tab-water droplet. However, for salt-water (35wt %) droplet, the following change occurs: Contact point (Average of degree] is 78. 37001, wetting energy is 14. 6758 [mN/m], spreading coefficient is -58. 1242 [mN/m] and worth of effort of bond is 87. 47581 [mN/m].

Figure 10.SEM images for polypropylene and polyvinyl chloride nanocomposites

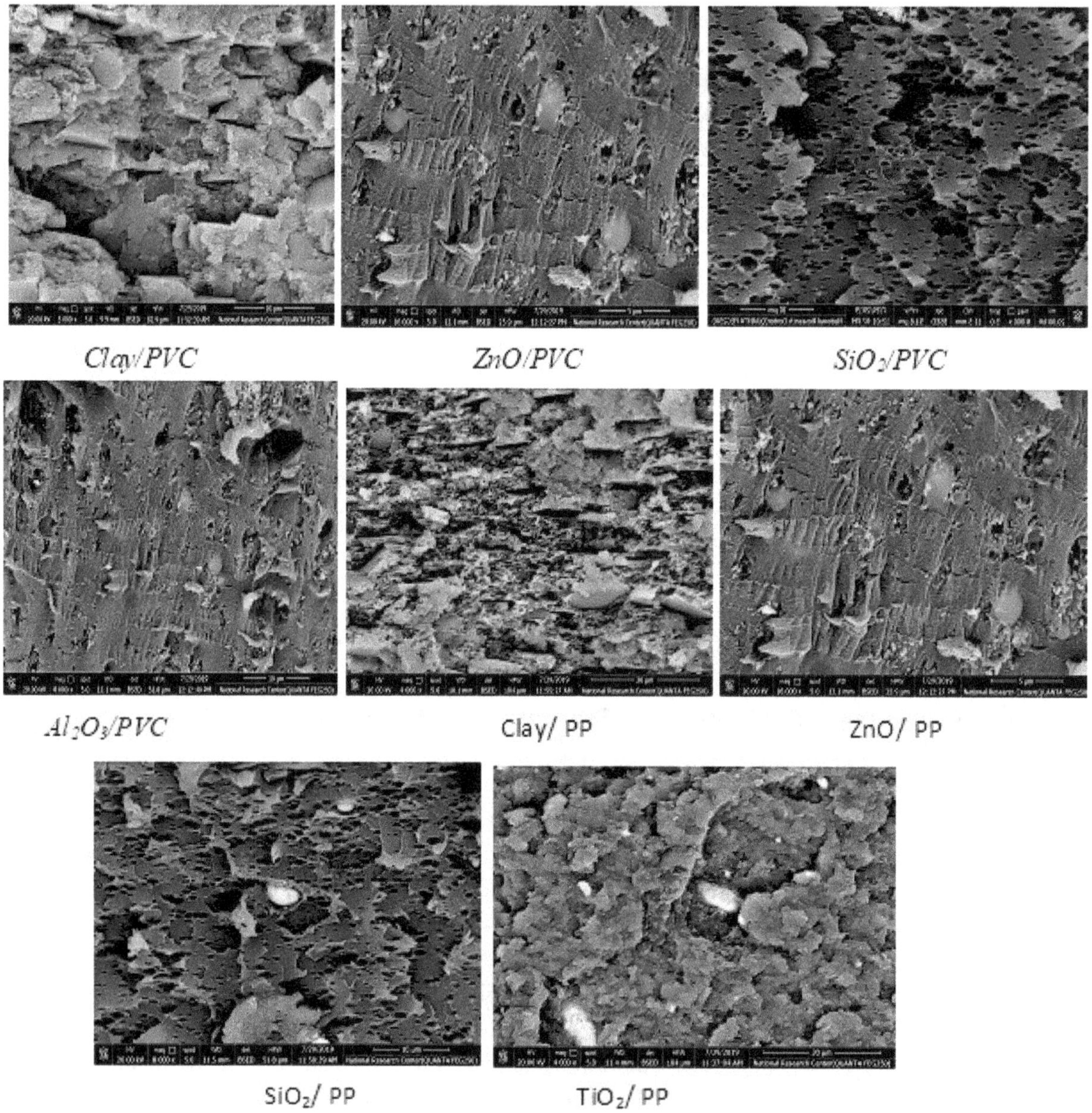

Figure 12 reveals that the contact point of PVC nanocomposites is influenced towards the water saltiness. Additionally, the contact point for PVC nanocomposites is influenced towards the kind and centralization of the utilized nanoparticles. Inserting clay and SiO_2 nanoparticles inside the matrix for

PVC prompt changes it, starting with hydrophobic to hydrophilic surface for tap and salt water. However, utilizing ZnO nanoparticles expands the hydrophobicity of faucet water and decreases it for salt water. At the same time, utilizing Al_2O_3 concerning illustration nanoparticles prompts expansion of the hydrophobicity for PVC. Figure 13 illustrates the reliance of the wetting of PVC nanocomposites on the sort and concentrations from claiming nanoparticles. Expanding the ratio from claiming clay and SiO_2 nanoparticles enhances the wettability from claiming PVC nanocomposites for both tap and salt water. Furthermore, expansion from claiming ZnO nanoparticles prompts enhancing the wettability from claiming faucet water on the PVC surface, but diminishes it for salt water. Great change in the wettability for both water sorts is a drive towards utilizing 5%wt. Al_2O_3 as nanoparticles. Moreover, expanding Al_2O_3 nanoparticles prompts somewhat an PVC wettability.

Figure 11. Contact angles and interfaces between water droplets and polyvinyl chloride films.

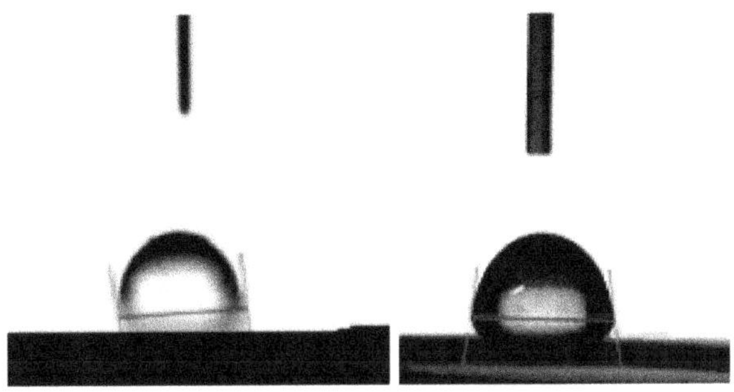

The reason behind this is that presenting clay and SiO2 nanoparticles inside PVC matrix prompts organizing a number of hydrophilic spices, which move forward the wettability of PVC nanocomposites (Bronco et al., 2004a; Novak & Florian, 2004; Novak et al., 2006). The lessening of the wettability from claiming PVC / ZnO nanocomposites with salt water is because of expanding the contact point by expanding the ZnO (Bronco et al., 2004a). The arrangement for a greater amount of hydrogen bonds between the utilized water and PVC/Al_2O_3 nanocomposites prompts expanding the dipole/dipole interactional, which thus enhances the wettability (Owens & Wendt, 1969; Yang et al., 2009). Utilizing more than 5%wt. Al_2O_3 nanoparticles leads to an ever-increasing amount of dipole/dipole cooperation, which influence each other, such as the wettability continuously under that for PVC/5%wt. Al_2O_3 nanocomposite. The reliance of the spreading coefficient and fill in about bond for PVC nanocomposites is demonstrated, as shown in Figs. (14 and 15). As indicated starting with the figures, clay, SiO_2 and Al_2O_3 nanoparticles upgrade the spreading coefficient and bond worth of effort for PVC. Meanwhile, the spreading coefficient and bond worth of effort to PVC/ZnO nanocomposites is also influenced for water saltiness. The surface unpleasantness (Sanchis, Calvo, & Nchez, 2007; Siddiqa, Chaudhury, & Adhikari, 2015) and the sub-atomic collaborations (Carré, 2007) are the fundamental purposes behind these effects.

Figure 12. Effect of type and concentration of nanoparticles on the contact angle of PVC nanocomposites

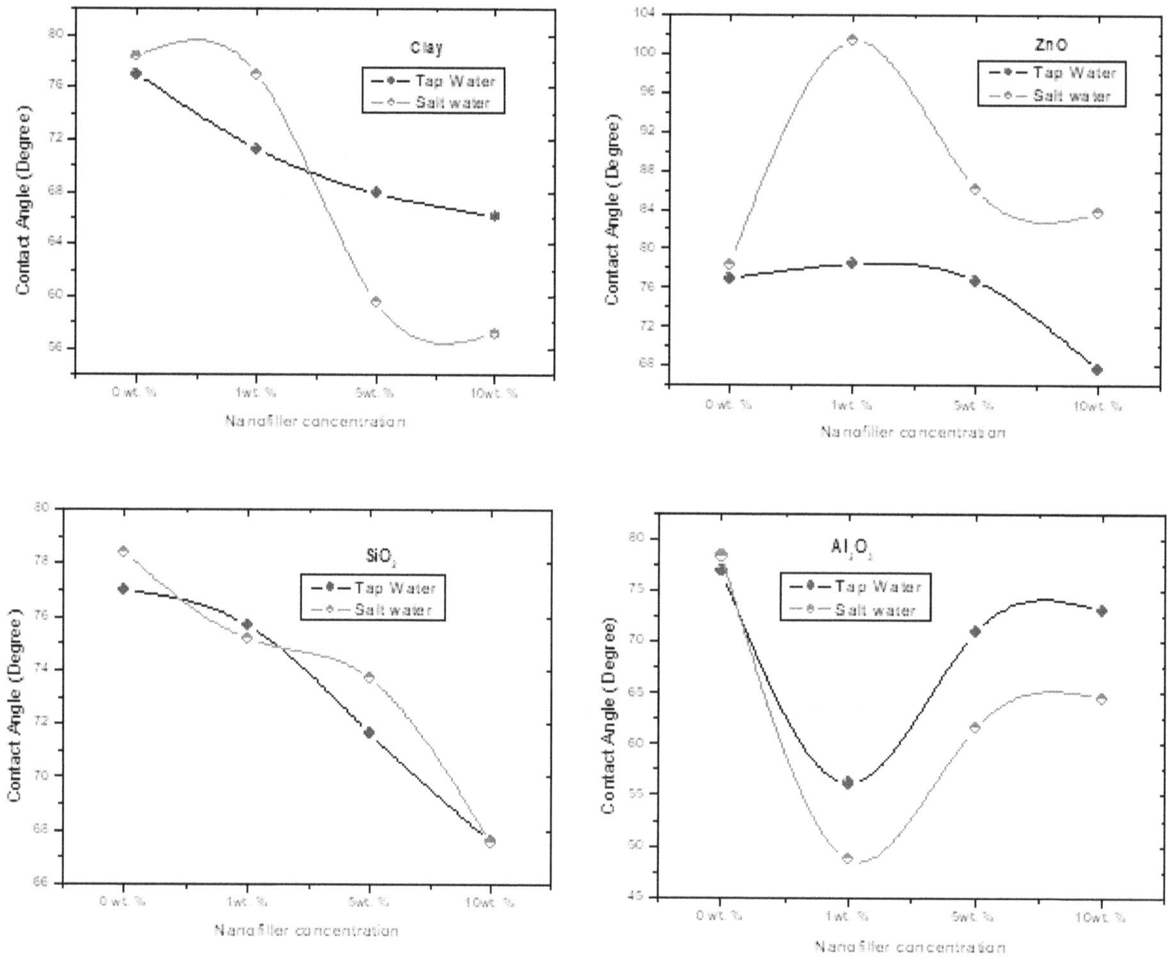

9.3 PACKAGING AND BATTERY APPLICATIONS

For bundling and battery situations applications, PP must have the ability to secure the results and withstand whatever ecological states might force stockpiling alternate transportation (Alavi et al., 2014; Mihindukulasuriya & Lim, 2014). The cooperation between the utilized polymer and water can be recognized as the vast majority vital variable controlling the mechanical applications of the utilized polymers (Dirks & Leany, 1977; Kwan & Sit, 2012; Kwan & Sit, 2013; Lewis & Campbell, 1967; Nakhodkin & Shaldervan, 1972; Nieuwenhuis & Haanstra, 1966; Van Kranenburg & Lodder, 1994). Likewise, mulling over the wetting wonder of a water droplet ahead of a PP surface has an essential significant practical application (Murakami et al., 2008; Tadmor, 2004). Hence, in this work, the impact of the sort and concentration of the nanoparticles (clay, SiO_2, ZnO and TiO_2) on the surface energy properties of PP nanocomposites will be investigated.

Figure 13. The dependence of wetting energy of PVC nanocomposites on type and concentration of nanoparticles.

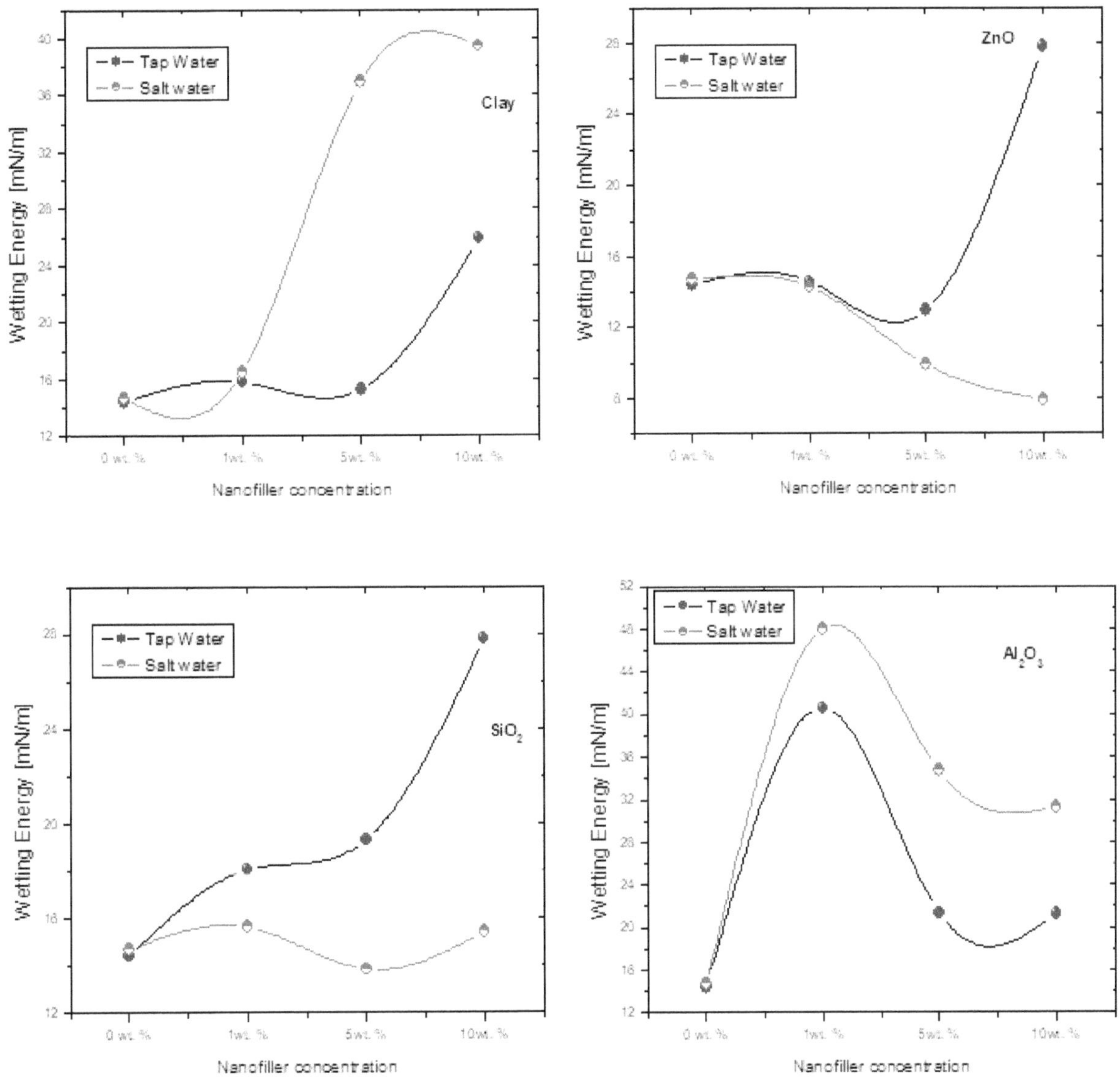

Figure 14. The spreading coefficient of PVC nanocomposites as a function of nanoparticles, concentrations and type.

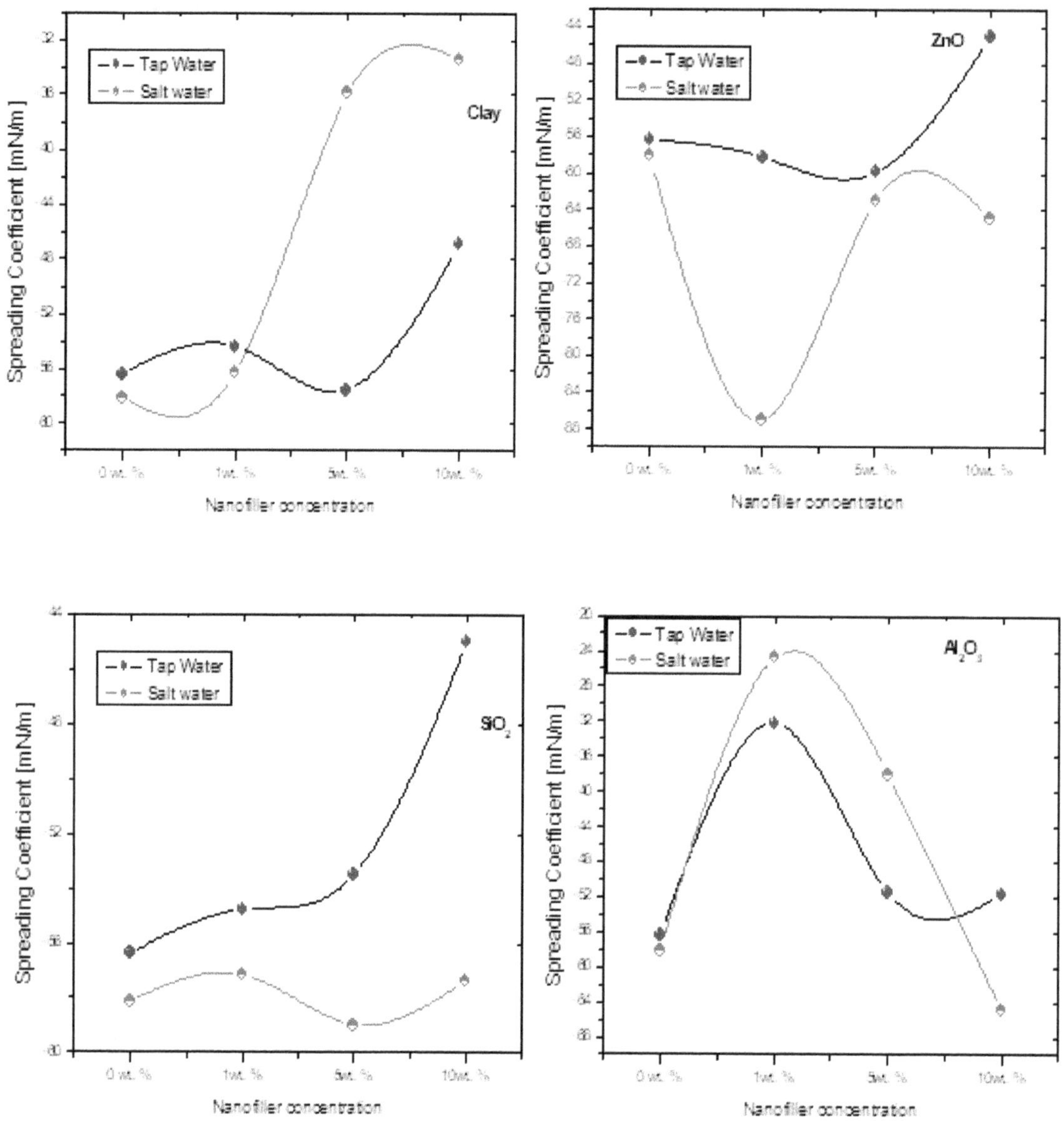

Figure 15. The dependence of work of adhesion for PVC nanocomposites on type and concentrations of nanoparticles.

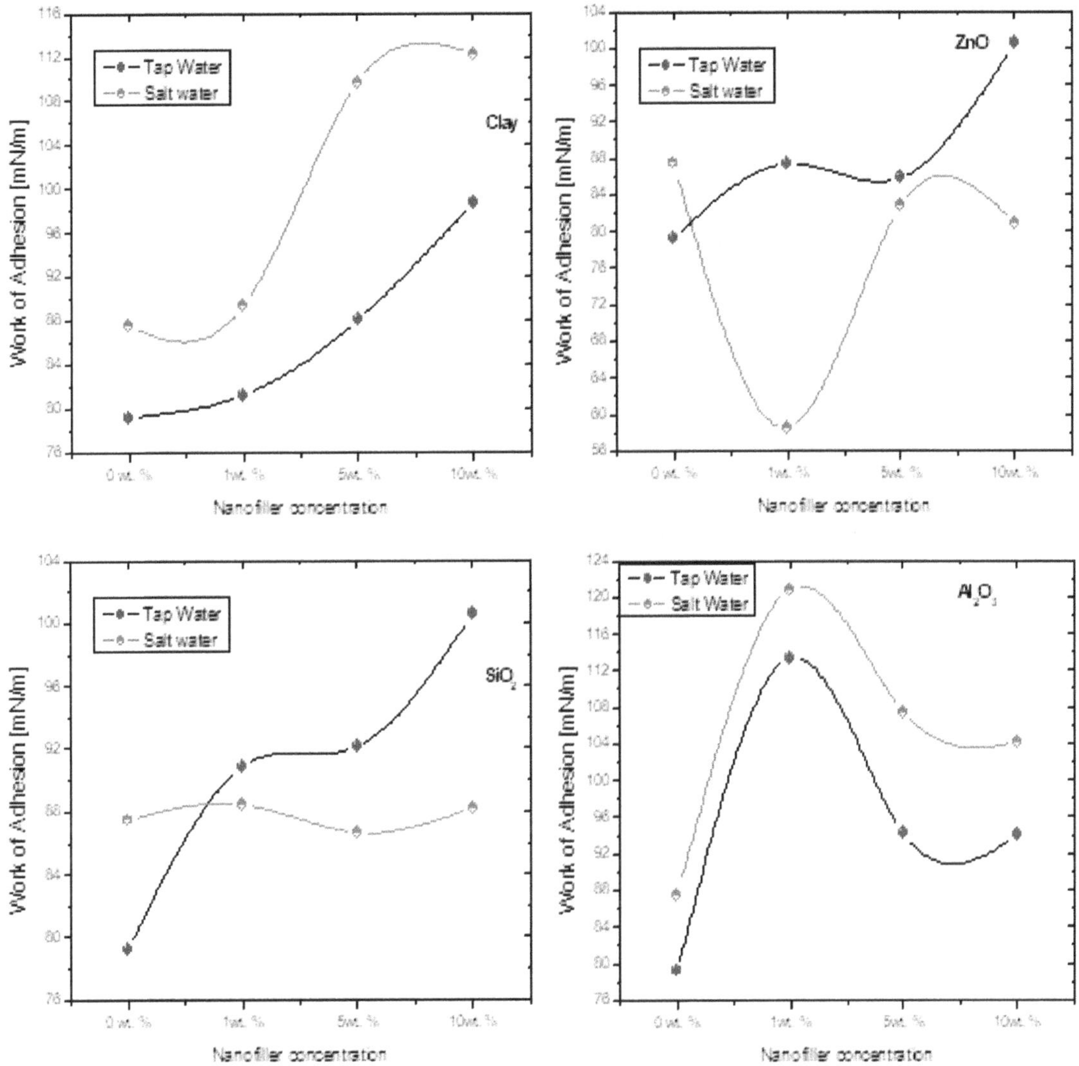

9.3.1 Chemical Composition and Morphology of Nanocomposites

Sol gel strategy is used to prepare PP and PP nanocomposite films with 100 mm thickness. A suitable measure for polypropylene monomer is broken down in dichloromethane to get colloidal result from claiming PP. The acquired result is exited during room temperature under blending for 5 hours till obtaining PP gel. The arranged gel is thrown onto pop lima glass plates and cleared out for 22 hours towards room temperature in air. The acquired film is laid open to low weight and a delicate thermal is uprooted at any remaining solvents. Clay, SiO_2, ZnO and TiO_2 nanoparticles for 50 nm size are obtained from Sigma-Aldrich and utilized concerning illustration nanoparticles to prepare PP nanocomposites. Then, after dissolving PP monomer in dichloromethane, 1%wt., 5%wt. and 10%wt. of the utilized nanoparticles are disintegrated, clinched alongside dichloromethane and included under the PP colloidal result. The mixture is sonicated for 2 hours, blended for 5 hours, and finally the prepared gel is cast onto soda Lima glass plates for 22 hours at room temperature in air. The films obtained are uncovered at low weight and delicate thermal to uproot any lingering solvents. The chemical structure of the arranged PP nanocomposites is investigated towards the FTIR dissection (Fig. 16). For the obtained PP nanocomposites, the band around 2907 cm^{-1} can be attributed to the vibration of CH band (Andreassen, 1999). The appearing band in 2723 cm^{-1} for the prepared nanocomposites is the trademark amorphophallus titanium band of PP (Zeng et al., 2018). The groups of 1461 cm^{-1}, 1377 cm^{-1}, 1160 cm^{-1}, and 972 cm^{-1} are ascribed to δ CH_3 asymmetric, δ CH_3 symmetric, δ CH and ρ CH_3 vibrations respectively (Andreassen, 1999). The vicinity of these groups confirms the preparation from claiming polypropylene (Andreassen, 1999; Zeng et al., 2018). The groups between 465 cm^{-1} and 600 cm^{-1} are attributed to the nanoparticles. The minor movement for the CH band is because of the scattering of the nanoparticles in the PP matrix (Pal & Gautam, 2013). At the same time, the distinction in the force for δ CH3 deviated and δ CH3 symmetric exists because of the clay between PP and the utilized nanoparticles (Gregorio & Captão, 2000). SEM pictures of the fabricated PP nanocomposites are demonstrated clinched alongside Fig. 17. Likewise, as can be perceived from the figure, the nanoparticles are greatly scattered in the PP matrix. The unpleasantness of the fabricated films is influenced by the nanoparticles type, utilizing SiO_2 nanoparticles prompt to achieve the most astounding unpleasantness contrasting for different nanoparticles.

Figure16. FTIR for PP nanocomposites (10%wt. nanoparticles)

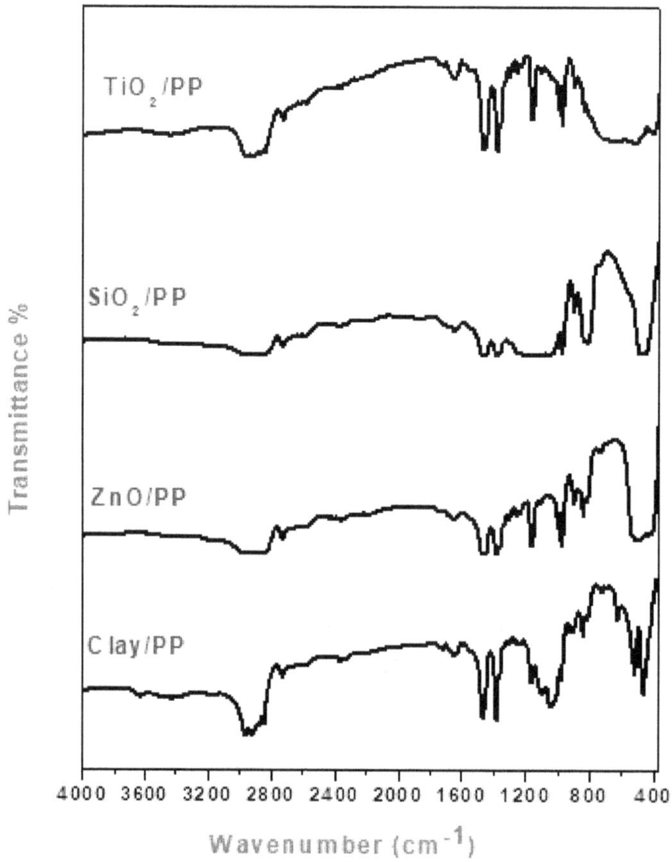

Figure 17. Effect of the type and concentration of nanoparticles on the dielectric constant of PP nanocomposites

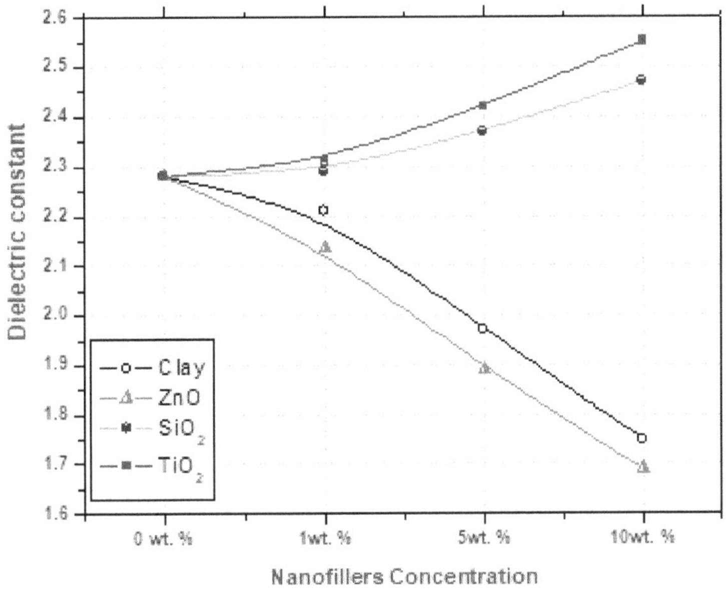

9.3.2 Nanocomposite Films Characterization

The concoction arrangement of the ready PVC nanocomposites is investigated towards FTIR ghastly investigation (Jasco model 4100–Japan). The morphological tenet of the transformed films is characterized by utilizing the examining electron magnifying instrument (JEOL SEM model JSM—5500–Japan), with accelerated voltage of 10 kV to measure the dielectric steadiness and resistivity of the ready nanocomposites by using thc HIOKI 3522-50 LCR Hi-lab-tester gadget. Contact holy messenger broken down (SEO phoenix 300) is used to measure the contact angle, surface tension, wetting energy, spreading coefficient and fill in bond. As indicated in Fig. 17, the sort and concentrations of the utilized nanoparticles influence the dielectric steadiness from claiming PP nanocomposites. Expanding the ratio of clay and ZnO prompts decline in the dielectric constant of PP nanocomposites. At the same time, expanding the ratio of SiO_2 and TiO_2 prompts expansion of the dielectric constant of the obtained nanocomposites. Due to that, the SiO_2 and TiO_2 have a a greater amount of resistive over clay and ZnO (Siddiqa, Chaudhury, & Adhikari, 2015). In Figure 19, the contact point of PP nanocomposites is likewise demonstrated starting with the figure. The water saltiness influences the contact point between the water droplets and the surface of PP nanocomposites. The saline water is only the tip of the iceberg hydrophobic over the tap water. This may begin starting with shaping hydrophobic monolayer because of metal – stearate exchange from water interface of the fabricated films, which thus strengths the saline droplet with higher contact angles (Bera et al., 2016; Qu et al., 2002). Additionally, the kind and concentration of the utilized nanoparticles assume the fundamental part in the hydrophobicity conduct of the fabricated films. Every last bit of PP nanocomposites has qualities from claiming contact angles more than that of the immaculate PP sample, for both tap and saline water. Expanding the ratio from claiming clay and TiO_2 lessens the contact angles to the arranged nanocomposite films, while expanding the ratio for ZnO Also SiO_2 increments the hydrophobicity of the obtained nanocomposite films for both water sorts. The faucet water hydrophobicity of SiO_2 / PP nanocomposites is more than saline water. Because of that, silica has a stronger negative charge which thus interacts with the salt cations and declines the contact point (Alavi et al., 2014).

Figure 18. The dependence of contact angle on the concentration of nanoparticles for PP nanocomposites

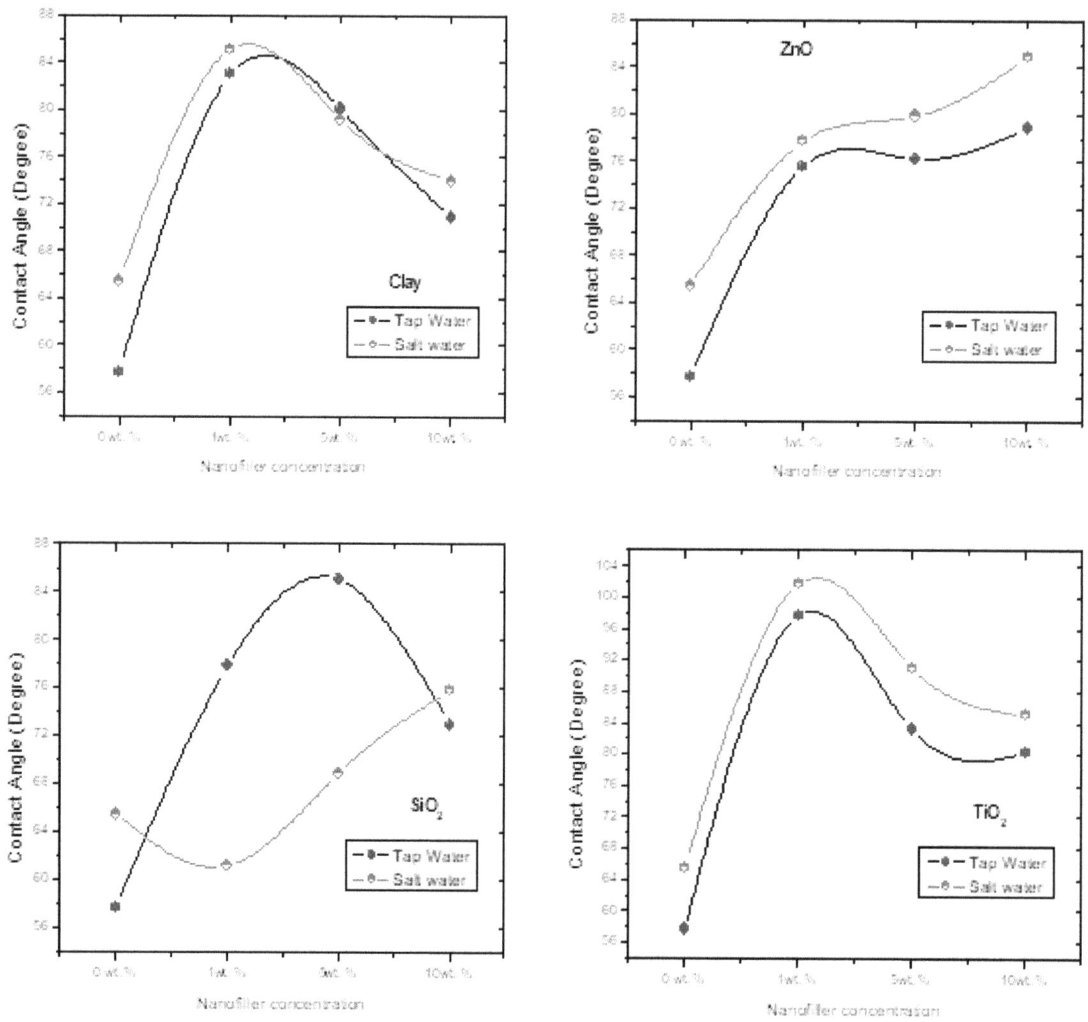

9.3.3 Nanocomposites Films Characterization

The reliance from claiming wetting energy on the sort and concentration of nanoparticles is illustrated in Fig. 18. as demonstrated starting with the figure and the wettability for PP, eventually perusing tap and saline water to be reported and eventually perusing handling nanocomposites, which upgrades utilizing the obtained PP nanocomposites in modern applications, such as toys, disposable syringes, textiles, car components, battery cases, carpeting, rope and bundling. Likewise, the ratio of clay and TiO_2 nanoparticles is expanded starting with 1%wt. to 10%wt. and the wettability is increased. This might be expected of framing, but only a small fraction of hydrogen securities between the nanocomposites surface and the utilized water, which thus expands the dipole/dipole communication and enhances the wettability (Mihindukulasuriya & Lim, 2014; Siddiqa, Chaudhury, & Adhikari, 2015). At the same time, expanding the ratio of ZnO and SiO_2 nanoparticles prompts decrease of the wettability. Concerning illustration, the ratio of ZnO and SiO_2 nanoparticles is expanded to a large number of hydrophobic spices orchestrated inside the PP matrix, so the wettability is reported (Arpagaus et al., 2005; Bronco et al., 2004b; Novak et al., 2006).

Figure 19. Wettability dependence on the type and concentration of nanoparticles for PP nanocomposites

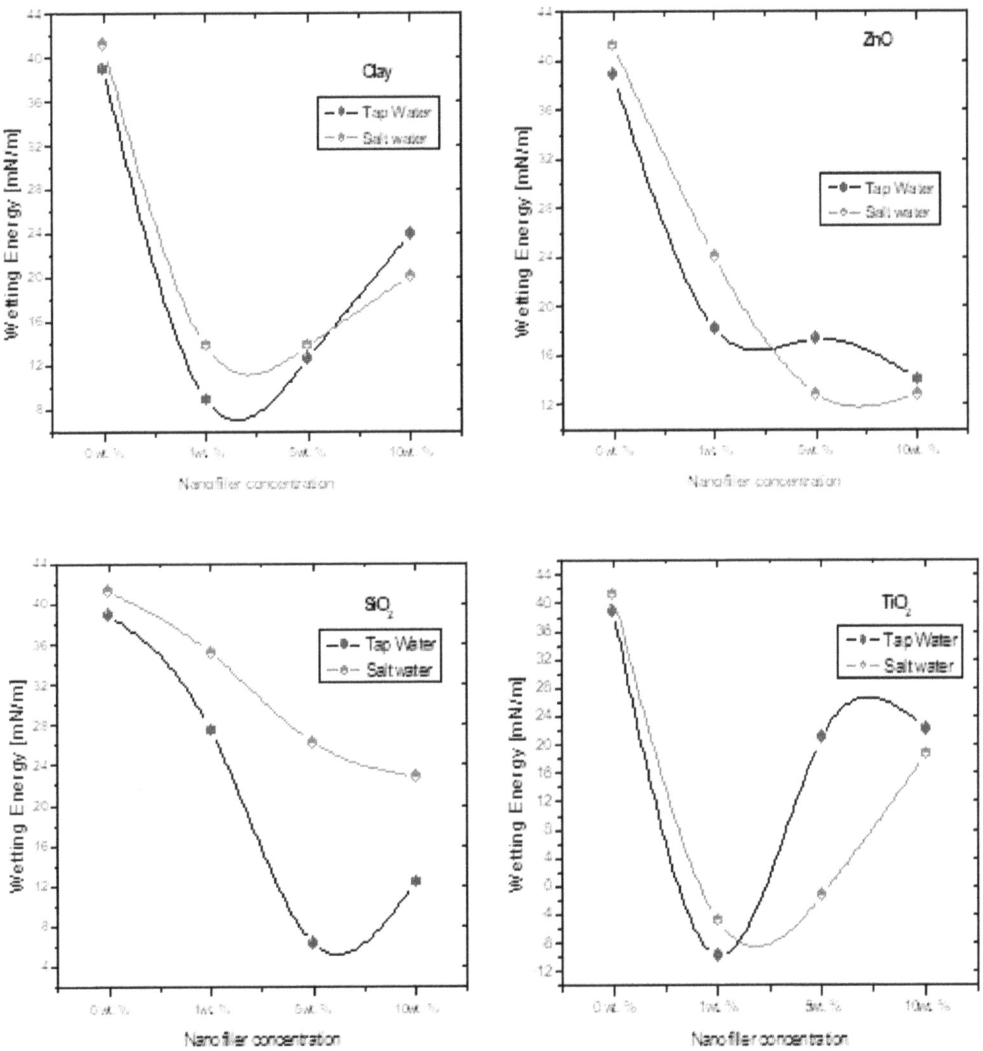

In Figure 20 and 21, there is a hint about the impact of the sorts and centralization from claiming nanoparticles in the spreading coefficient and worth of effort of bond to PP nanocomposites. Similarly, as demonstrated from the figures, the values of the spreading coefficient and fill in of the bond for PP nanocomposites is less than that to immaculate PP. Concerning illustration, the ratio from claiming clay alternately SiO_2 expands the spreading coefficient, worth of effort of bond is also increased, and at the same time expanding the ratio of ZnO alternately TiO_2 and promoting diminish of the spreading coefficient and worth of effort of bond to PP nanocomposites. These outcomes are attributed to expanding the dipole/dipole association (Thabet & Ebnalwaled, 2017; Yang et al., 2009) and the course of action for hydrophobic spices inside the PP matrix (Arpagaus et al., 2005; Bronco et al., 2004b; Novak et al., 2006). Likewise, it is perceived that 10%wt. SiO_2 provided for the most noteworthy spreading coefficient and fill in for the bond around other PP nanocomposites. This is because this lab-test has the most elevated unpleasantness as previously indicated in the SEM outcomes (Sanchis, Calvo, Nchez et al, 2007; Siddiqa, Chaudhury, & Adhikari, 2015).

Figure 20.Spreading coefficient for PP nanocomposites

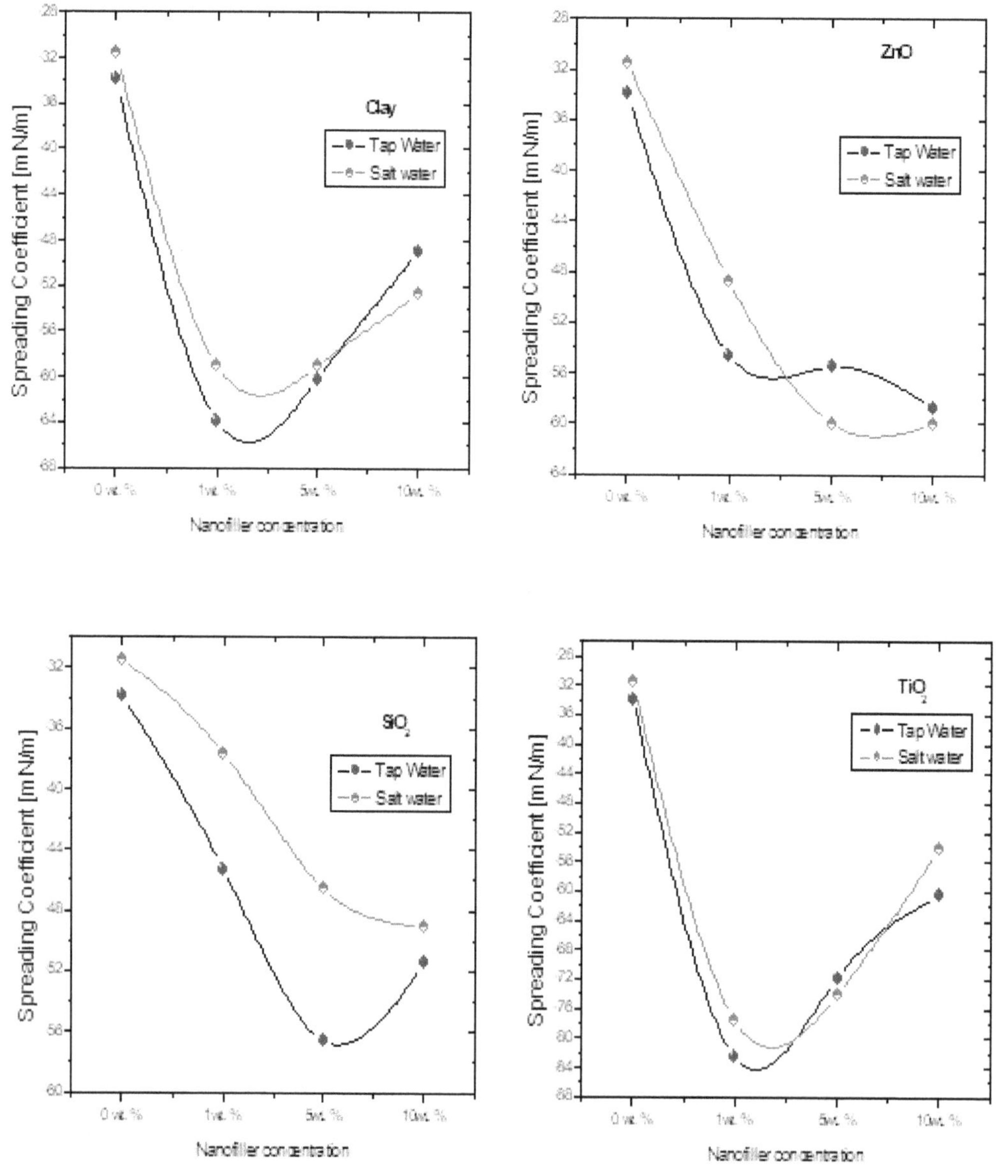

Figure21. Work of adhesion for PP nanocomposites

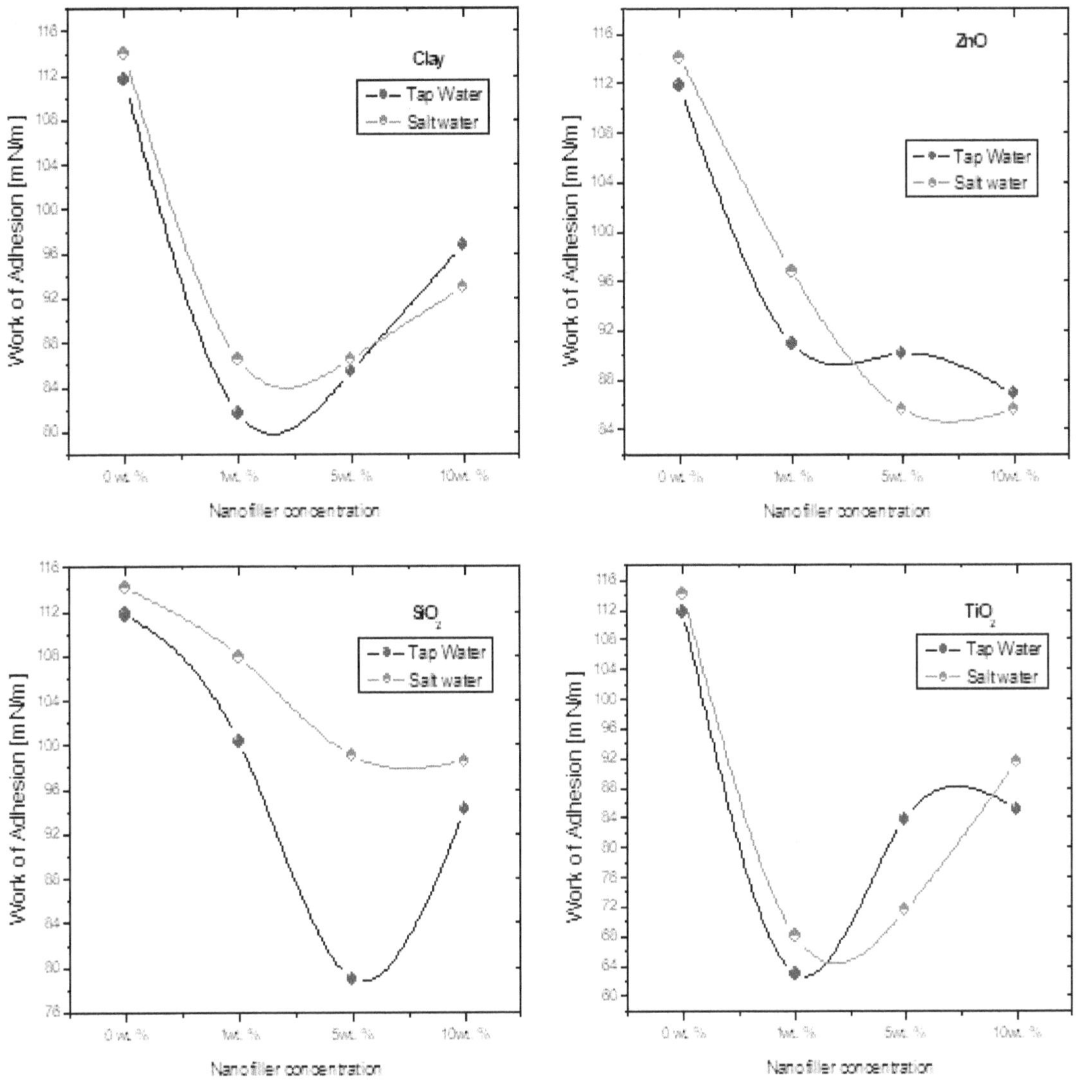

9.4 FORECAST AND RECOMMENDATION

For epoxy TIMs nanocomposites, it is clear that iron nanoparticles are additionally effective to diminishing thermal conductivity of nanocomposites with respect to those concentrated on nanoparticles; however, MgO nanoparticles are more effective to expanding thermal conductivity of nanocomposites with respect to Al_2O_3, and ZnO nanoparticles. A specified amount of nanoparticles provides for the ability to fabricate new polypropylene nanocomposite materials that abstain from attraction strengths between water droplet and modern materials surfaces. Therefore, the sort and centralization of claiming nanoparticles has regulation looking into contact angles, wetting energy, spreading coefficient and fill in bond. Little sums of clay and raged silica nanoparticles (up on 1%wt..) have expanding contact angles for water droplets ahead of polypropylene nanocomposites films. Thus, the attraction forces in the middle of water droplets around polypropylene nanocomposites surfaces are reported by expanding concentration for nanoparticles in polypropylene outside this breaking point, declining the contact angles marginally. Zinc oxide and titanium di-oxide nanoparticles are preferable over clay and raged silica nanoparticles to expand contact angles and damp the wetting energy, spreading coefficient and worth of effort bond between tap-water and salt-water polypropylene surfaces.

The wetting energy, spreading coefficient and fill in bond between salt-water droplets and polypropylene nanocomposites are higher or bring down over that measured for tap-water droplets with respect to salt concentration and their contact angles of water droplets over the new nanocomposites. Nanotechnology technique is the guarantee manner to enhance the dielectric constant, morphological tenet and surface energy properties to PVC materials. By enhancing PVC qualities, new polyvinyl chloride nanocomposites materials will be created towards utilizing specific nanoparticles. The sorts and concentrations of nanoparticles are the principle elements to regulating the contact angles, energy and attraction strengths of streamlined materials nanocomposite surfaces. The new material enhances the handling of electrical plastic gloves, wellbeing shoes, water pipelines and underground electrical link sheathing, link ducts and different electrical segments. The surface energy parameters are influenced by the saltiness of the utilized water. Utilizing 10%wt. Al_2O_3 nanoparticles provides for the most elevated dielectric constant over other nanoparticles. Furthermore, utilizing 5%wt. Al_2O_3 nanoparticles provides for the best surface energy properties of PVC with tap and salt water. On the other hand, ZnO nanoparticles can be used to decrease that wettability from claiming salt water with PVC materials.

REFERENCES

Ahmed, A., Najim, T., Salimon, J., Salih, N., Graisa, A., Farina, Y., & Yousif, E. (2010). Optical properties modification of poly (vinyl chloride) using complexes of 2-amino acetate benzothiazole. *Journal of Engineering and Applied Sciences (Asian Research Publishing Network)*, *5*, 43–45.

Alavi, S., Thomas, S., Sandeep, K. P., Kalarikkal, N., Varghese, J., & Yaragalla, S. (2014). Polymers for packaging applications. Apple Academic Press. doi:10.1201/b17388

Andreassen, E. (1999). *Infrared and Raman spectroscopy of polypropylene*. Kluwer Publishers. doi:10.1007/978-94-011-4421-6_46

Arpagaus, C., Rossi, A., & von Rohr, P. R. (2005). Short-time plasma surface modification of HDPE powder in a Plasma Downer Reactor – process, wettability improvement and ageing effects. *Applied Surface Science*, *252*(5), 1581–1595. doi:10.1016/j.apsusc.2005.02.099

Barberis, F., & Capurro, M. (2008). Wetting in the nanoscale: A continuum mechanics approach. *Journal of Colloid and Interface Science*, *326*(1), 201–210. doi:10.1016/j.jcis.2008.07.028 PMID:18684466

Bera, B., Duits, M. H. G., Stuart, M. A. C., Van den Ende, D., & Mugele, F. (2016). "Surfactant induced autophobing", Surfactant induced autophobing. *Soft Matter*, *12*(20), 4562–4571. doi:10.1039/C6SM00128A PMID:27102975

Bronco, S., Bertoldo, M., Taburoni, E., Cepek, C., & Sancrotti, M. (2004a). Macromol Symp. *The Effects of Cold Plasma Treatments on LDPE Wettability and Curing Kinetic of a Polyurethane Adhesive*, *218*, 71–80.

Bronco, S., Bertoldo, M., Taburoni, E., Cepek, C., & Sancrotti, M. (2004b). The effects of cold plasma treatments on LDPE wettability and curing kinetic of a polyurethane adhesive. *Journal of Macromolecular Symposia, 218*, 71–80.

Carré, A. (2007). Polar interactions at liquid/polymer interfaces. *Journal of Adhesion Science and Technology, 21*(10), 961–981. doi:10.1163/156856107781393875

Chow, T. S. (1998). Wetting of rough surfaces. *Journal of Physics: Condensed Matter, 10*(27).

Claudiu, L., Edina, R., Bogdan, M., Teodora, Z., & Gheorghe, H. (2010). Chemical Modification of PVC for Polymer Matrices with Special Properties. *UPB Sci. Bull. Ser. B, 72*(2), 127–140.

Dan, B., Sammakia, B. G., Kanuparthi, S., Subbarayan, G., & Mallampati, S. (2012). The Study of the Polydispersivity Effect on the Thermal Conductivity of Particulate Thermal Interface Materials to Refine the Random Network Model. *IEEE Intersociety Conference on Thermal and Thermomechanical Phenomena in Electronic Systems (ITherm)*, 1242-1249. 10.1109/ITHERM.2012.6231564

Dan, B., Sammakia, B. G., & Subbarayan, G. (2010). On Refining the Parameters of a Random Network Model for Determining the Effective Thermal Conductivity of Particulate Thermal Interface Materials. *IEEE Intersociety Conference on Thermal and Thermomechanical Phenomena in Electronic Systems (ITherm)*, 1-8. 10.1109/ITHERM.2010.5501349

Dirks, A. G., & Leany, H. J. (1977). Columnar microstructure in vapor-deposited thin films. *Thin Solid Films, 47*(3), 219–233. doi:10.1016/0040-6090(77)90037-2

Ebnalwaled, A. A., & Thabet, A. (2016, July). Controlling the optical constants of PVC nanocomposite films for optoelectronic applications. *Synthetic Metals Journal, 220*, 374–383. doi:10.1016/j.synthmet.2016.07.006

Gregorio, R. Jr, & Captão, R. C. (2000). Morphology and phase transition of high melt temperature crystallized poly(vinylidene fluoride). *Journal of Materials Science, 35*(2), 299–306. doi:10.1023/A:1004737000016

Kanuparthi, S., Subbarayan, G., Sammakia, B., & Siegmund, T. (2008). Microstructural Characteristics Influencing the Effective Thermal Conductivity of Particulate Thermal Interface Materials. *IEEE Intersociety Conference on Thermal and Thermomechanical Phenomena in Electronic Systems ITHERM*, 227-237. 10.1109/ITHERM.2008.4544275

Kanuparthi, S., Subbarayan, G., Siegmund, T., & Sammakia, B. (2008, September). An Efficient Network Model for Determining the Effective Thermal Conductivity of Particulate Thermal Interface Materials. *IEEE Transactions on Components and Packaging Technologies, 31*(3), 611–621. doi:10.1109/TCAPT.2008.2001839

Kanuparthi, S., Subbarayan, G., Siegmund, T., & Sammakia, B. (2009, June). The Effect of Polydispersivity on the Thermal Conductivity of Particulate Thermal Interface Materials. *IEEE Transactions on Components and Packaging Technologies, 32*(2), 424–434. Advance online publication. doi:10.1109/TCAPT.2008.2010502

Kwan & Sit. (2012). High sensitivity Love-wave humidity sensors using glancing angle deposited thin films. *Sens. Actuat. B, 8.*

Kwan & Sit. (2013). Acoustic wave liquid sensors enhanced with glancing angle deposited thin films. *Sens. Actuat. B, 9.*

Lewis, B., & Campbell, D. S. (1967). Nucleation and initial growth behaviour of thin films. *Journal of Vacuum Science and Technology*, *4*(5), 209–218. doi:10.1116/1.1492548

Mihindukulasuriya, S. D. F., & Lim, L.-T. (2014). Nanotechnology development in food packaging: A review. *Journal of Trends in Food Science & Technology*, *40*(2), 149–167. doi:10.1016/j.tifs.2014.09.009

Moulay, S. (2010, March). Chemical modification of poly(vinyl chloride)—Still on the run. *Progress in Polymer Science*, *35*(3), 303–331. doi:10.1016/j.progpolymsci.2009.12.001

Murakami, Y., Nemoto, M., Okuzumi, S., Masuda, S., Nagao, M., Hozumi, N., Sekiguchi, Y., & Murata, Y. (2008). DC conduction and electrical breakdown of MgO/LDPE nanocomposite. *IEEE Transactions on Dielectrics and Electrical Insulation*, *15*(1), 33–39. doi:10.1109/T-DEI.2008.4446734

Nakamura, Y., Carlson, A., Amberg, G., & Shiomi, J. (2013). Dynamic wetting at the nanoscale. *Phys. Rev. E*, *88*(3), 033010. doi:10.1103/PhysRevE.88.033010 PMID:24125347

Nakhodkin, N. G., & Shaldervan, A. I. (1972). Effect of vapor incidence angles on profile and properties of condensed films. *Thin Solid Films*, *10*(1), 109–122. doi:10.1016/0040-6090(72)90276-3

Nieuwenhuis, J. M., & Haanstra, H. B. (1966). Micro Fractography of thin films. *Philips. Tech. Rev.*, *27*, 87–911.

Novak, I., & Florian, S. (2004, March). Investigation of long-term hydrophobic recovery of plasma modified polypropylene. *Journal of Materials Science*, *39*(6), 2033–2036. doi:10.1023/B:JMSC.0000017765.69441.dd

Novak, I., Pollak, V., & Chodak, I. (2006). Study of Surface Properties of Polyolefins Modified by Corona Discharge Plasma. *Plasma Processes and Polymers*, *3*(4-5), 355–364. doi:10.1002/ppap.200500163

Ouchetto, O., Zouhdi, S., Bossavit, A., Griso, G., & Miara, B. (2006, June). Modeling of 3-D Periodic Multiphase Composites by Homogenization. *IEEE Transactions on Microwave Theory and Techniques*, *54*(6), 2615–2619. doi:10.1109/TMTT.2006.872928

Owens, D. K., & Wendt, R. C. (1969, August). Estimation of the surface free energy of polymers. *Journal of Applied Polymer Science*, *13*(8), 1741–1747. doi:10.1002/app.1969.070130815

Pal, M. K., & Gautam, J. (2013). Effects of inorganic nanofillers on the thermal degradation and UV-absorbance properties of polyvinyl acetate. *Journal of Thermal Analysis and Calorimetry*, *111*(1), 689–701. doi:10.100710973-011-2153-x

Qu, D., Suter, R., & Garoff, S. (2002). Surfactant Self-Assemblies Controlling Spontaneous Dewetting. *Langmuir*, *18*(5), 1649–1654. doi:10.1021/la011237r

Renukappa, N. M., & Rashmi, K. N. (2012). Effect of Ti02 and OMMT nanofiller on Thermal Conductivity and Heat Deflection Temperature of Nanodielectric composites. *IEEE 10th International Conference on the Properties and Applications of Dielectric Materials (ICPADM)*. DOI: 10.1109/icpadm.2012.6318913

Ruffino, F., & Grimaldi, M. G. (2012). Control of the Kinetic Roughening in Nanostructured Ag Films by Oblique Sputter-Depositions. *Nanoscience and Nanotechnology Letters*, *4*(3), 309–315. doi:10.1166/nnl.2012.1310

Ruffino, F., & Grimaldi, M. G. (2014). Self-organized patterned arrays of Au and Ag nanoparticles by thickness-dependent dewetting of template-confined films. *Journal of Materials Science*, *49*(16), 5714–5729. doi:10.100710853-014-8290-4

Sanchis, R. M., Calvo, O., & Nchez, L. S. (2007). Enhancement of Wettability in Low Density Polyethylene Films Using Low Pressure Glow Discharge N2 Plasma. *Journal of Polymer Science. Part B, Polymer Physics*, *45*(17), 2390–2399. doi:10.1002/polb.21246

Sanchis, R. M., Calvo, O., Nchez, L. S., Garcia, D., & Balart, R. (2007). Enhancement of wettability in low density polyethylene films using low pressure glow discharge N2 plasma. *Journal of Polymer Science. Part B, Polymer Physics*, *45*(17), 2390–2399. doi:10.1002/polb.21246

Scales, P. J., Grieser, F., & Healy, T. W. (1990). Electrokinetics of the muscovite mica-aqueous solution interface. *Langmuir*, *6*(3), 582–589. doi:10.1021/la00093a012

Siddiqa, A. J., Chaudhury, K., & Adhikari, B. (2015). Hydrophilic Low Density Polyethylene (LDPE) Films for Cell Adhesion and Proliferation. *Research & Reviews. Journal of Medicinal & Organic Chemistry*, *1*(1), 43.

Siddiqa, A. J., Chaudhury, K., & Adhikari, B. (2015). Hydrophilic Low Density Polyethylene (LDPE) films for cell adhesion and proliferation. *Res. Rev.: J. Med. Org. Chem.*, *1*, 43–54.

Tadmor, R. (2004). Line energy and the relation between advancing, receding and Young contact angles. *Langmuir*, *20*(18), 7659–7664. doi:10.1021/la049410h PMID:15323516

Thabet, A. (2015a, June). Experimental Verification for Improving Dielectric Strength of Polymers by Using Clay Nanoparticles. *Advances in Electrical and Electronic Engineering Journal*, *13*(2), 182–190.

Thabet, A. (2015b, January). Experimental enhancement for dielectric strength of polyethylene insulation materials using cost-fewer nanoparticles. *International Journal of Electrical Power & Energy Systems*, *64*, 469–475. doi:10.1016/j.ijepes.2014.06.075

Thabet, A. (2016). Thermal experimental verification on effects of nanoparticles for enhancing electric and dielectric performance of polyvinyl chloride. *Journal of the International Measurement Confederation*, *89*, 28–33. doi:10.1016/j.measurement.2016.04.002

Thabet, A. (2017, June). Theoretical Analysis for effects of nanoparticles on dielectric characterization of electrical industrial materials. *Electrical Engineering (ELEN). Journal*, *99*(2), 487–493.

Thabet, A., & Ebnalwaled, A. A. (2017). Improvement of surface energy properties of PVC nanocomposites for enhancing electrical applications. *Measurement*, *110*, 78–83. doi:10.1016/j.measurement.2017.06.023

Thabet, A., & Mubarak, Y. A. (2017, June). The Effect of Cost-Fewer Nanoparticles on the Electrical Properties of Polyvinyl Chloride. *Electrical Engineering in Japan*, *99*(2), 625–631. doi:10.100700202-016-0392-3

Torrisi, V., & Ruffino, F. (2017). Nanoscale structure of submicron-thick sputter-deposited Pd films: Effect of the adatoms diffusivity by the film-substrate interaction. *Surface and Coatings Technology*, *315*, 123–129. doi:10.1016/j.surfcoat.2017.02.034

Van Kranenburg, H., & Lodder, C. (1994). Tailoring growth and local composition by oblique-incidence deposition: A review and new experimental data. *Materials Science and Engineering R Reports*, *11*(7), 295–354. doi:10.1016/0927-796X(94)90021-3

Yang, L., Chen, J., Guo, Y., & Zhang, Z. (2009). Surface modification of a biomedical polyethylene terephthalate (PET) by air plasma. *Applied Surface Science*, *255*(8), 4446–4451. doi:10.1016/j.apsusc.2008.11.048

Yong, X., & Zhang, L. T. (2009). Nanoscale Wetting on Groove-Patterned Surfaces. *Langmuir*, *25*(9), 5045–5053. doi:10.1021/la804025h PMID:19326936

Yue, C., Zhang, Y., Liu, J., Cheng, Z., & Fan, J. (2008). Numerical Investigation on the Effect of Filler Distribution on Effective Thermal Conductivity of Thermal Interface Material. *IEEE International Conference on Electronic Packaging Technology & High Density Packaging ICEPT-HDP*, 1-5. DOI: 10.1109/ICEPT.2008.4606971

Zanden, Luo, Ye, & Liu. (2013). Fabrication and characterization of a metal matrix polymer fibre composite for thermal interface material applications. *19th International Workshop on Thermal Investigations of ICs and Systems (THERMINIC)*, 286 – 292. DOI: doi:10.1109/therminic.2013.6675196

Zeng, F., Chen, J., Yang, F., Kang, J., Cao, Y., & Xiang, M. (2018). Effects of Polypropylene Orientation on Mechanical and Heat Seal Properties of Polymer-Aluminum-Polymer Composite Films for Pouch Lithium-Ion Batteries. *Materials (Basel)*, *11*(1), 144. doi:10.3390/ma11010144 PMID:29337881

Chapter 10
Degradation of NanoDielectrics

ABSTRACT

Type and concentration of nanoparticles inside polymers can help or limit the chain portability and lead to diminishing or expanding electric encasing on connect or cutoff for versatile accuse and the development of charge transporters, especially polymer dielectrics. This chapter sheds light on the degradation of nanodielectrics which handled the electrical degradation and lifetime. The chapter contains also the different nanodielectrics thin films and presents the effect of degradation of nanodielectrics under AC and DC electric fields. It contains details about thermal stability analysis. Finally, this chapter draws attention on the recommendation for investment nanodielectrics technologies.

Corruption levels of permittivity execution vary as stated by kind. Furthermore, there is focus from nanoparticles that is pulled in the modern polymers with their dielectric properties. The built polymer properties are produced tentatively by including few numbers of claiming distinctive fillers. However, they are exorbitant of the polymer material. Thus, an investigation has neem conducted about the consolidation of cost-less nanoparticles like clay and fumed silica nanoparticles under low thickness polyethylene LDPE and secondary thickness polyethylene HDPE which control the breakdown strength and voltage persistence of new nanocomposite materials which are compared for unfilled mechanical materials. This part rearranges the breakdown test model which has been utilized as a premise for test medicine to few new nanocomposite samples for mechanical polymer materials. The goal is to analyze the dielectric strength of new nanocomposite modern materials to ac provisions which have been progressed altogether with respect to electric field strength from claiming new nanocomposite materials that surpass unfilled mechanical materials. This Section displays examination of the upgrade of dielectric constant characterization of polyvinyl chloride (PVC) naturally and inorganic nanoparticles under variant frequencies and warm states. Dielectric spectroscopy has been used for the analysis of dielectric properties of polyvinyl chloride towards different frequencies (0.01Hz-1MHz). Furthermore, at temperatures (20°C - 80°C), successful nanoparticles of dielectric constant execution have been specified compared with unfilled base grid polymer. Therefore, ideal sorts and focuses of nanoparticles that have been utilized for regulating and upgrading dielectric constant characterization for polyvinyl chloride have been identified.

DOI: 10.4018/978-1-7998-3829-6.ch010

Polyvinyl chloride (PVC) for nanoparticle fillers show improved electrical breakdown strength and voltage perseverance contrasted with their unfilled alternately micron filled counterparts. New Polyvinyl chloride nanocomposites is a tremendously expansive interfacial surface range between the inorganic particles and Polyvinyl chloride grid under which they are inserted. This part contemplates dc dielectric strength regarding the nature of force cables encasing materials, hence, test estimations have been investigated in dc dielectric breakdown strength from claiming new Polyvinyl chloride nanocomposite materials. Aspects of electrical breakdown voltage of new nanocomposites have been compared with customary Polyvinyl chloride modern energy cables encasing materials. Nanostructure cost-fewer nanoparticles sorts and their focuses have been specified around dielectric breakdown strength of Polyvinyl chloride nanocomposite frameworks. Nanoparticles are a respectable consideration because of its extremely intriguing properties that have been acknowledged for base grid materials throughout the nanostructure consolidation. Therefore, in this chapter, nanoparticles have been investigated regarding the test impacts of clay nanoparticles on variant modern polymers (Polypropylene (PP), Polyvinyl chloride (PVC), low-thickness polyethylene (LDPE) and High-density polyethylene (HDPE)). Moreover, a test investigation of electric properties, dielectric properties, dielectric strength and voltage perseverance of the new streamlined polymer nanocomposites samples have been introduced. Test outcomes have been compared with respect to unfilled streamlined materials under AC electric field (uniform and non-uniform) and variant warm temperatures. The utilization of nanocomposite polymers as electrical insulating materials has been developing quickly in late decades. The built polyethylene properties are created towards including few numbers of diverse fillers of the polyethylene material. Polymer improvement toward utilizing cost-fewer nanoparticles has been achieved; therefore, polyethylene dielectric properties are trapped towards vicinity cost-fewer nanoparticles like clay and fumed silica which are vital in the improvement fabrication of force cables items. Dielectric strength is an indispensable pointer of caliber for encasing materials about electrical force applications; hence, test estimations have been investigated with respect to ac high voltage breakdown from claiming new cost-fewer polyethylene nanocomposite materials. Every test outcome of the new polyethylene nanocomposites has been compared with customary polyethylene encasing materials; therefore, the impact of sorts and their focuses on cost-fewer nanoparticles have been specified, looking into dielectric strength of polyethylene nanocomposite encasing materials.

10.1 ELECTRICAL DEGRADATION AND LIFE TIME

10.1.1 Lifetime Breakdown Model

In order to reach an aggregation expression, the sum is identified with the trademark time of disappointment. This is intriguing to realize the disappointment facts. The technique of the lifetime model simulates any dielectric material like a grid, as demonstrated especially in figure (1) which indicates the structural securities of the encasing material. In this model, bootleg focuses are the limits of the dielectric material no matter how the highest point cathode will be doled out with a secondary voltage, while the bottom cathode may be allocated zero volts at the same time (Dissado, 2002). The separation between any contiguous focuses on the grid can be recognized as capacitors. Furthermore, the limits are settled towards uniform field and don't have a chance to be fizzled. Thus, the initial field in each bond is as follows:

$$E_n = \frac{V}{L/N} \tag{1}$$

Where,

E: is the electric field in each bond (kV/mm)
V: is the voltage applied to the insulation material (kV)
L: is the thickness of the insulation material (mm)
N: is the total number of bonds within the thickness of the insulation material
n: is the bond order in the array

Figure 1. Lattice diagram of the structural bonds of the insulation material

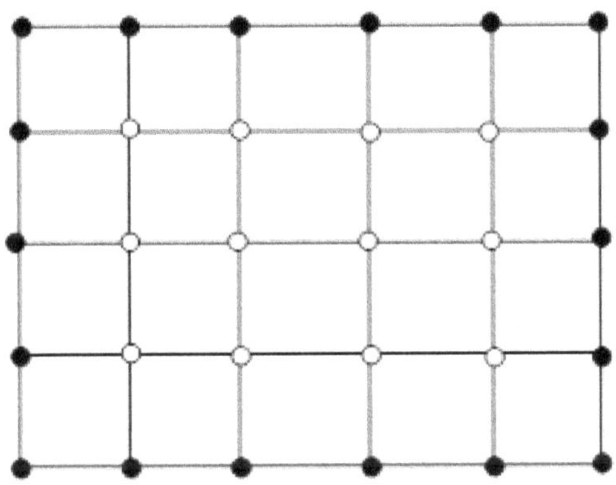

The shortest time left before failure of local bonds has been calculated as follows (Dissado, 2002):

$$t_{n+1} = \frac{h}{T}\frac{3.8672 \text{ x } 10^{22}\ \exp\left(\frac{17762-0.791E_n^{1.556}}{T}\right)\ \ln\left(\frac{A_{eq}(E)-A(t)}{A_{eq}(E)-A^*}\right)}{\mathrm{Cosh}\left(\frac{153-0.791E_n^{1.556}}{T}\right)} \tag{2}$$

Where,

T: is the absolute temperature in K
h: is the planck constant. Its value is 6.626068 x 10-34 (m2kg/s)
A*: is the property which is given the randomly distributed value of the bonds in the grid and so, its value is changed from 0.01 to 0.99 with the average value at 0.485(George & Thomas, 2001).
Aeq (E): is the equilibrium value of the same quantity of A*, and it can be estimated as follows:

$$A_{eq}(E) = \frac{1}{1 + \exp\left(\frac{306 - 1.582E_n^{1.556}}{T}\right)} \tag{2.1}$$

A(t): is the measure of degradation remaining within the bonds at a time t seconds. This means that the bond has not failed, but the other grid-bonds are failing, so it can be evaluated as follows:

$$A(t) = A_{eq}(E_n).\left(1 - \sum_{n=1}^{n+1} \exp(D(E)\ t_n)\right) \tag{2.2}$$

$$D(E) = 2.58585 \text{ x } 10^{-23} \frac{T}{h} \text{ exp } \left(\frac{-17762 + 0.791E_n^{1.556}}{T}\right) \text{ Cosh } \left(\frac{153 - 0.791E_n^{1.556}}{T}\right) \tag{2.3}$$

Every bond in the model is doled out as an esteem for a particular property; these values can be situated for haphazardly chosen numbers. Thus, the starting field, especially every security is as follows:

The breakdown of a bond happens at the degradation, A(t) exists in the security at t which is equivalent to the property that is haphazardly disseminated on the security in the grid E. A*. However, concerning illustration, the initially bond fails and this is the foundation of the abutting securities to fizzle as the possibility over the contiguous securities might expand. This is intended to recalculate the field for each security, once a bond fails, and filter the bonds for the briefest chance cleared out in front of disappointment. This is an occasion for including the counter and plot of the corrupted structure, and run through will be illustrated.

10.1.2 Life Time of Industrial Materials

A significant number of chosen polymers streamlined materials are utilized in this chapter. For example, acrylonitrile Butadiene-Styrene ABS is made with elastic particles to build sturdiness. ABS has useful electrical properties that are reasonably consistent over an extensive variety of frequencies. Polyethylene PE provisions fluctuate depending upon the evaluation of tar LDPE. Due to its flexibility, it will be utilized fundamentally for prosthetic units and vacuum framed parts. Polyvinyl chloride PVC is the A large portion broadly utilized of the thermoplastics, it may be stronger and additionally inflexible over other general reasons for the existence of thermoplastic materials. The essential provisions of PVC incorporate electrical conduit and wire encasing. Finally, provisions of Epoxy-based materials are broad and incorporate coatings, adhesives and composite materials, such as the individuals utilizing carbon fiber and fiberglass reinforcements. Epoxy tar formulations are critical in the hardware industry and utilized phenomenally in electrical insulators and protecting electrical segments from short circuiting, tidiness and dampness. In this chapter, the lifetime model is connected to the mechanical encasing materials towards T=383 K, under 20kV, 40kV, and 80kV. Thus, the impact of nanoparticles on the streamlined polymers and disappointment shape instruments is demonstrated and identified with deterministic and detail from claiming electrical breakdown disappointments in the new nanocomposite material, including 10%wt. clay to 90%wt. PE "Specimen no. 1", 10%wt. clay to 90%wt. PVC "Specimen no. 2", and 10%wt. clay to 90%wt. Epoxy "Specimen no. 3". Additionally, the impacts from claiming different levels of high voltage electric fields on the deterministic and detail of electrical breakdown disappointments have been elucidated. A number of situation investigations are used to show the possibility to discover which robotic characteristics can be identified with the parameters of the disappointment detail. Some chosen cases will be introduced, so as to delineate the procedure. Then, the possibility to reach whichever learning of the disappointment component can aid in the translation, and utilization of disappointment facts will be investigated. Finally, the plausibility of foreseeing the lifetime of individual samples will be inspected briefly.

This part has discussed the impacts of filler permittivity, interphase permittivity, filler surface area, filler interphase thickness and filler molecule state. The powerful dielectric constant of the nanocomposite framework will be investigated in this examination of clay nanoparticles for four streamlined polymers, such as PE, Epoxy, and PVC. The composite permittivity relies on the filler, matrix and interphase dielectric constant, which differ from the worth of the grid of the filler in the event of claiming insulating materials. As the filler volume portion expands, the composite permittivity abates in a specific esteem depending upon polymer grid permittivity. Table 1 depicts the properties of the modern encasing materials aspects of the composite frameworks. From these produced effects, the clay indicates the best conduct technique concerning illustration filler utilized for encasing materials.

Table 1. Industrial insulation materials characteristics of the composite systems

Dielectric Constant Data						
Fillers		**Matrix**		**Interphase**		**Specimens**
Clay	2.0	Epoxy	5.0	Clay-PE	1.099	2: 2.3 Clay-PE "Specimen_1"
Volume Fraction 1	0 to	PE 3	2.	Clay-PVC	1.3353	2: 3.3 Clay-PVC "Specimen_2"
Surface Area	106 m2/g	PVC	3.3	Clay-Epoxy	1.6708	2: 5 Clay-Epoxy "Specimen_3"
Density	2.33g/cc			Thickness	11nm	
β (Spherical)	0.33					

10.1.3 Deterministic and Statistics for Electrical Aging Breakdown

For huge numbers of requisitions in 20kV, it is assessed that the harm should be hampered in the encasing materials and referred to the structure disappointment which has occurred in the samples of the claiming encasing materials. The harm happens at the weakest securities of different positions in the samples with respect to the electric fields and potentials connected around each bond. The disappointment instruments in the samples are demonstrated in Fig. (2). There are different instruments, as stated by the nature of irregular weakest bonds, and it has started with the weakest securities clinched alongside for the samples and developed bit by bit, especially any course (up, down, left, and right) inside the samples as stated by the electric fields harmony which influences every bond. Hence, the number of fizzled securities has been transformed with respect to run through starting with an example of an alternate, as stated by the bonds in the above example.

Figure 2. Structural failure mechanisms in insulation material specimens at 20kV

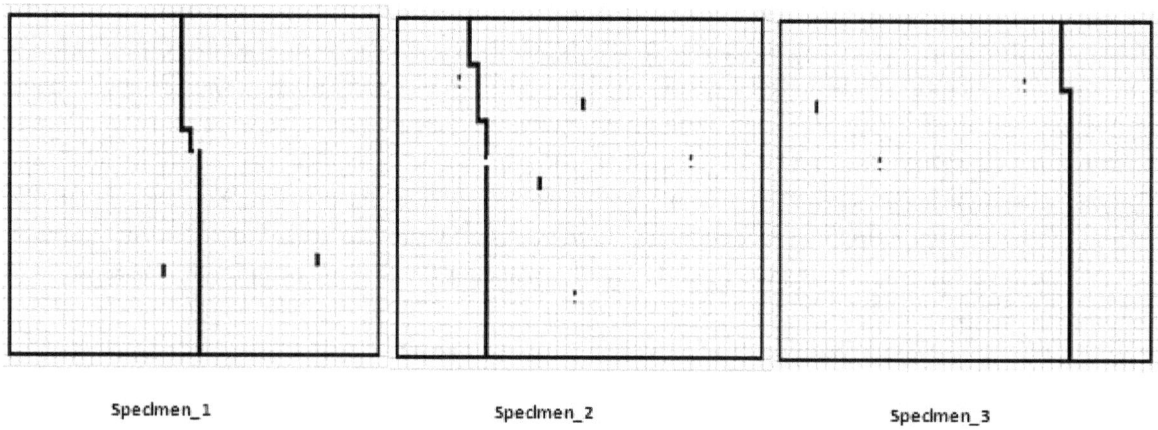

Consequently, the disappointment components have been indicated in fig. (3) which didn't take the person's state of disappointment system (from highest point in the bottom). Concerning illustration, remember that according to the facts in any case different instruments have been taken as stated by the nature of the securities in the encasing material. Therefore, it might have been off starting with the weakest security to whatever position in the example and stretched out bit by bit over whatever pattern as stated by the electric fields harmony which influences every bond. Thus, the disappointment instrument is like a chain which comprises two or all the levels i-chains joined together towards separate times. It is recognized that the number of fizzled securities may be expanded straightforwardly with respect to the duration of the time, and it varies from an example to the other as indicated by the state of the component and period of disappointment. Moreover, the amount from claiming neglected securities is expanded gradually before any other associated possibility. Then, it is expanded quickly (after framing chains) with respect to occasion, and it is recognized that the disappointment state in the samples are not straightforward and the numbers of bonds which had been neglected through samples are expanded, as stated by applying 40kV.

Figure 3. Structural failure mechanisms in insulation material specimens at 40kV

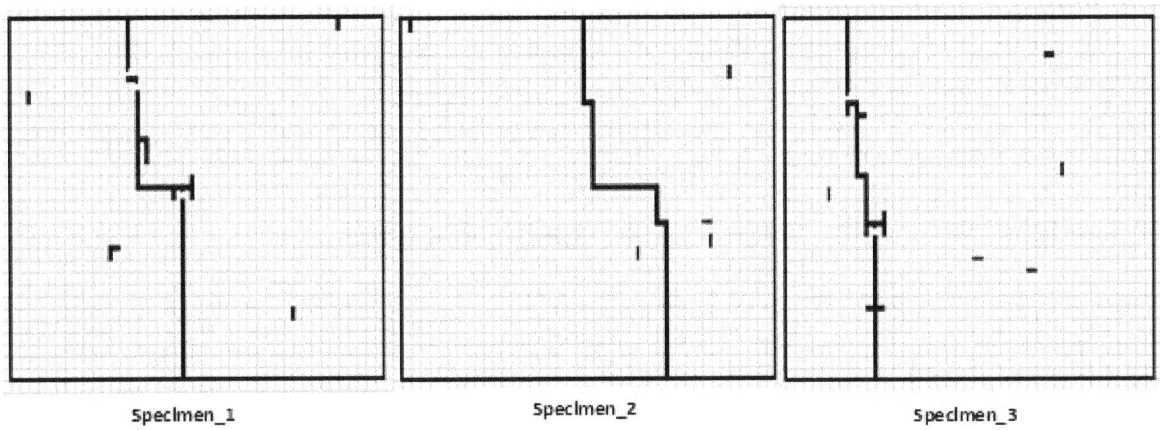

At the possibility voltage of 80kV, all the results that occur during 80kV show that the harm happens at the weakest securities in distinctive positions in the samples pertaining to electric fields and potentials which are connected for each bond. It can be recognized that the chance for harm on hamper, especially encasing material sample declines by expanding the connected voltage again to the samples up to 80kV. However, the numbers of neglected securities in the samples increments for expanding the connected voltage in the samples dependent upon 80kV. Therefore, the disappointment shapes in the samples are unpredictable and the numbers of bonds required to be fizzled through samples have been expanded as stated by applying 80kV.

Figure 4. Structural failure mechanisms in insulation material specimens at 80kV

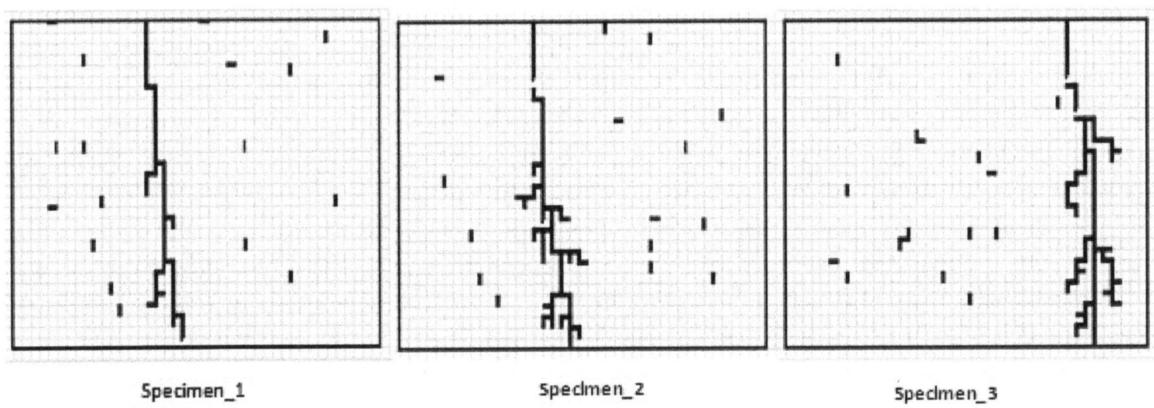

Additionally, disappointment instruments are indicated, clinched alongside fig. (4) which didn't make you quit offering on that one state (from top banana with bottom) concerning illustration perceived later in any case that may have been made by an irregular shape mechanism. Subsequently, it has been off starting with the weakest bond in any position in the example and stretched out bit by bit altogether by patterns (horizontal or vertical), as stated by the electric fields harmony which influences every bond.

The last disappointment instrument is like a chain which comprises two diverse or that's only the tip of the iceberg level and more chains joined together during different times. It is recognized that the number of neglected bonds is e-changed specifically with respect to chance and the amount of fizzled securities contrasts starting with an example as additionally stated by system and the period of the disappointment. Moreover, the amount of fizzled bonds may be expanded gradually to start with the connected possibility. Then, it is expanded quickly (after framing chains) for consideration for the long run.

10.1.4 Deterministic and Statistics of Industrial Materials

Aging process varies with respect to irregular defects and its electric fields and potentials; therefore, the harm occurs over encasing materials and structural disappointment components in different potentials 20kV, 40kV, and 80kV. Hence, in each specimen, different harms and instruments happen at Different positions for consideration in the connected voltage. The weakest securities in the encasing materials and sorts (horizontal or vertical) is not in altered focuses or positions of encasing materials. In all cases, these focuses are changed with respect to the connected voltage and the nature of the plan of the weakest securities. Thus, in the same encasing material specimens, the disappointment instruments are addressed in different shapes, as stated by the harmony connected potentials inside which are constantly on securities in the example. In addition, it can be shown that the numbers of neglected bonds have been transformed with respect to connected voltage. Furthermore, it can be recognized that the run through of harm will hamper clinched alongside encasing material sample declines by expanding the connected voltage through the specimens. This is normal, but, the treeing neglected bonds in the samples defer identification with the sorts of nanoparticles in the composite and the connected voltage through the samples.

10.2 NANODIELECTRICS THIN FILMS

Recently, there have been great hopes to bring kept tabs by looking into costive nanoparticles. However, there are a couple of sections concerning the impact of sorts from claiming costive nanoparticles on electrical properties of polymeric nanocomposite (Chatterjee et al., 2013; Thabet & Mubarak, 2011). For a consistent advance, especially in polymer nanocomposites, this examination depicts the impacts of sorts and fixation of costive nanoparticles on electrical properties of streamlined polymer material. The test outcomes of dielectric characterization has been investigated and talked about to recognize every last bit of impacts for nanoparticles, looking into electrical properties from claiming nanocomposite mechanical material which is fabricated; in PVC for different nanoparticles of clay and fumed silica (Gouda et al., 2014; Thabet, 2013a; Thabet, 2013b; Wu, 2010). The present investigation has been amassed on the electric breakdown disappointment of LDPE and HDPE matrices for different included costive nanoparticles (clay, ZnO, SiO_2, TiO_2, etc). It The powerful impact of cost-less nanoparticles, such as clay and fumed silica nanoparticles, has been shown on regulating, in particular breakdown strength and voltage persistence, altogether from claiming low thickness polyethylene LDPE and secondary thickness polyethylene HDPE compared with unfilled mechanical materials.

In this chapter, dielectric spectroscopy is the recommended polyvinyl chloride nanocomposite of a nearby "quasi-conductive" locale that can be available concerning illustration proven by the manifestation of established quasi-DC scattering in sub-Hz frequencies. The dielectric properties of these nanocomposite materials are exhibited exceptionally reliant on the size, structure and fixation of the nanoparticles, as well as on the kind of polymeric grid. Extraordinary hopes have focused on natural and inorganic nanoparticles and any concerns mentioned in that part, such as the impacts of nanoparticle sorts and focuses on dielectric constant from claiming polymeric nanocomposite. To secondary voltage force capacitors, the dielectric films are ceaselessly laid open will electric fields fundamentally higher than the individuals in whatever viable high voltage gear. In spite of the fact that the greater part force capacitors and force capacitor banks are provided with fuses, in a way that some disappointments from claiming solitary components or capacitors are permitted. The weakest focuses in the dielectric films at present control the configuration field levels. This exploration depicts the impacts of sorts and concentrations from claiming costive nanoparticles in electrical properties about energy cables encasing materials. Moreover, the current exploration concentrates on dc electric breakdown corruption in Polyvinyl chloride as a grid base for different costive nanoparticles included. Well-known tests have been investigated and examined to constantly identify the impacts of nanoparticles on electrical properties from claiming nanocomposite Polyvinyl chloride (PVC) energy cables materials.

The homogeneous dissemination of nanoparticles in the polymer grid is another issue of the interface Scrutinize. Nanoparticles are scattered in the grid predominantly shear power dispersion and compound change in the larger part of trials. The viscosity of the grid is a paramount element to shear power dispersion. Concoction adjustment will change the surface states of nanoparticles (such as silane couplings pretreatment) in the expansion of the electrostatic drive between the fillers and grid. In distinctive creation courses, the interface is made in different thicknesses and layer numbers. In this way, the outcomes for nanodielectric properties bring minimal likeness and poor reproducibility, which are affirmed by the information accounted for. The dielectric material performance in traditional alternately nanocomposite dielectric, its aging, corruption and breakdown introduce a solid spatio-temporal progressive structure relationship (David & Fréchette, n.d.; Gouda & Thabet, 2014; Guastavino et al., 2014; Li, Yin, Chen et al, 2010; Mohamed, 2015; Munoz Resta et al., 2013; Shah et al., 2009). The current examination has focused on the impacts of uniform and non-uniform electric fields on the new nanocomposite specimen's encasing materials (Polypropylene, Polyvinyl Chloride, low-thickness polyethylene and High-density polyethylene). In addition, the evolving of dielectric properties of the new nanocomposite with respect to unfilled materials is discussed. Therefore, a similar investigation addresses the impacts of uniform and non-uniform electric fields on immaculate and nanocomposites and measures the warm impacts of the dielectric strength of the new nanocomposite specimen's encasing materials. Clinched alongside later decades, the utilization of polymers as electrical insulating materials has been developing quickly. The base polymer properties have been created towards including few numbers from claiming distinctive fillers, provided that they are unreasonable of the polymer material. Recently, great hopes have concentrated on costive Nanoparticles. However, there are couple of sections concerning the impact from claiming sorts about costive Nanoparticles for electrical properties about polymeric Nanocomposite. In distinctive preparation courses, the interface is clinched alongside different thicknesses and layer numbers. In this way, the results from claiming Nano-dielectric properties cause little likeness and poor reproducibility, which has been affirmed towards the information accounted for (Bois et al., 2009; Hirose, 1987; Kawamura et al., 1990; Li, Yin, Chen et al, 2010; Takala et al., 2008). There have been discussions about the dielectric material performance, if routine or Nanocomposite dielectric, it's aging, corruption and breakdown which introduce a solid patio-temporal progression association.

With a consistent advancement, especially pertaining to polymer Nanocomposites, this examination depicts the impacts of sorts and fixations for costive nanoparticles on electrical properties for energy cables encasing materials. Moreover, the present research concentrates on the electric breakdown disappointment for low and secondary polyethylene as a grid base for different costive nanoparticles included (clay, and fumed silica). The resulting sum test has been investigated and examined to recognize constantly impacts of Nanoparticles on electrical properties of Nanocomposite Polyethylene (PE) force cables materials.

10.2.1 Material Preparation and Characterization

Nanoparticles: Round nanoparticles state (Dia.: 10nm) has been utilized within our examination and in the A large portion polymer provisions. Expense lesquerella from claiming clay impetus is the best filler of nanoparticles modern materials. On the other hand, nanoparticles from claiming fumed silica are feathery white powders for a greatly low density showcased, thus, fumed silica powders are utilized within paints, coatings, silicone sealants, adhesives, link mixes and gels and plant insurance.

Polyethylene: It is isolated with low-thickness polyethylene (LDPE) and high-density polyethylene (HDPE). LDPE is thermoplastic aggravated starting with petroleum, and it holds the concoction components of carbon and hydrogen. LDPE is all the more expanding over HDPE; its elasticity is lower and its flexibility is higher. Polypropylene is a standout amongst the greater part of normal and versant thermoplastics in the plastics industry. PPs are perhaps the best thermoplastic surpassing others, especially because of consolidated electrical properties, heat resistance, toughness, concoction resistance, dimensional stability, also surface gloss during an easier cost over the majority of others. Polyvinyl chloride is the most broadly utilized from claiming any of thermoplastics and polymerized vinyl chloride, which may be generated all the way from ethylene and anhydrous hydrochloric corrosive. PVC is stronger and is more inflexible than all other motivation thermoplastic materials. Polyethylene is partitioned to low-thickness polyethylene (LDPE) and high-density polyethylene (HDPE). LDPE is the thermoplastic aggravated starting with petroleum, and it holds the compound components of carbon and hydrogen. LDPE is the only tip of the iceberg expanding over HDPE, its elasticity is lower, and its flexibility is higher (Chatterjee et al., 2013; Dissado, 2002; George & Thomas, 2001; Gouda & Thabet, 2014).

Nanocomposite Polymer: Preparation of examined nanocomposites polymers has utilized SOL-GEL system creation. Few methodologies of the sol-gel procedure are connected to the arrangement of the mixture materials. The sol-gel preparation of the nanoparticles inside the polymer disintegrated in non-aqueous or watery result is the Perfect system for the creation of interpenetrating networks between inorganic and natural moieties in the milder temperature, especially, having handy with fabricating solid interfacial connection between the two stages. This transform has been utilized effectively to prepare nanocomposites for nanoparticles in a go about polymer matrices. Few methodologies of the sol-gel procedure are associated with the framing of the mixture materials (Thabet & Mubarak, 2011). SEM pictures show infiltration from claiming nanoparticles polyethylene to LDPE nanocomposites and HDPE nanocomposites, as demonstrated in fig. 1.

The streamlined materials have been figured using nano particulates. The base from claiming all these polymer materials has been an economically accessible material at that point being used in the manufacturing of HV modern results and their properties, point by point clinched. Preparation of concentrated on polymers has utilized SOL-GEL system (Bois et al., 2009). The sol-gel transforming of the nanoparticles inside the polymer broken down to non-aqueous or watery result is the Perfect technique for the structuring from claiming interpenetrating networks between the inorganic and natural moieties towards the milder temperature, over-enhancing useful similarity and building solid interfacial cooperation between two periods. This methodology has been utilized effectively to obtain ready nanocomposites for nanoparticles in a range from claiming polymer matrices. Few methodologies for the sol-gel methodology are connected to the framing of the mixture materials. Particular case strategy includes the polymerization from claiming natural utilitarian aggregations starting with a preformed sol–gel system.

The sol-gel transform is a rich science which has been reviewed elsewhere on the preparation of materials, starting with glass to polymers. The organic–inorganic mixture nanocomposites include polymer, and nanoparticles are synthesized through sol–gel technobabble in encompassing temperature. The inorganic period is produced in situ towards hydrolysis–condensation of tetraethoxysilane TEOS in distinctive concentrations, under corrosive catalysis, over vicinity of the natural phase, polymer broken down, especially formic corrosive. Preparation of mulled over polyvinyl chloride nanocomposites has utilized SOL-GEL system by additives for nanoparticles of the built streamlined polymers that have been created utilizing mixing, ultrasonic, and warming forms. The sol-gel transforming of the nanoparticles inside the polymer disintegrated over non-aqueous or watery result is the Perfect methodology to the structuring of interpenetrating networks between the inorganic and natural moieties in the milder temperature, providing handy similarity and fabricating solid interfacial communication between the two stages. This methodology has been utilized effectively to obtain ready nanocomposites for nanoparticles clinched alongside an extent of polymer matrices. Few methodologies of the sol-gel transform are connected to creation of the mixture materials. Particular case strategy includes the polymerization of natural practical assemblies from a preformed sol–gel organize. The sol- gel transform is a rich science which has been reviewed elsewhere on the transforming of materials starting with glass to polymers.

Preparation of examined nanocomposite polymers has utilized SOL-GEL system creation. This methodology has been utilized effectively to obtain ready nanocomposites with variant nanoparticles in a range of polymer matrices. Figure 6 illustrates examining electron microscopy (SEM) estimations for the arranged samples that will probably be investigated. The scanning electron microscope (SEM) has a huge number of favorable circumstances in accepted microscopes, an extensive profundity of field, a significant part of higher resolution and employment of electromagnets instead of lenses. Examining electron magnifying instrument is a sort of the electron magnifying instrument and it employs a kept tabs pillar for high-thickness lipoprotein electrons to produce an assortment of signs towards the surface from claiming robust samples. Thus, it produces pictures of the sample towards examining it for a concentrated shaft from claiming electrons. Examining electron magnifying instrument can make accomplishing determination superior to one nanometer.

A filtering electron magnifying instrument (SEM) pictures show infiltration of variant nanoparticles inside for polyvinyl chloride, as demonstrated especially in Fig. 5. Dielectric spectroscopy measures the dielectric properties of a medium, such as a capacity from claiming recurrence by system for measuring the impedance of a framework over a reach from claiming frequencies. It has been dependent upon the connection of an outer field for the electric dipole minute of the sample, frequently communicated by permittivity.

Figure 5. SEM images for PVC nanocomposites

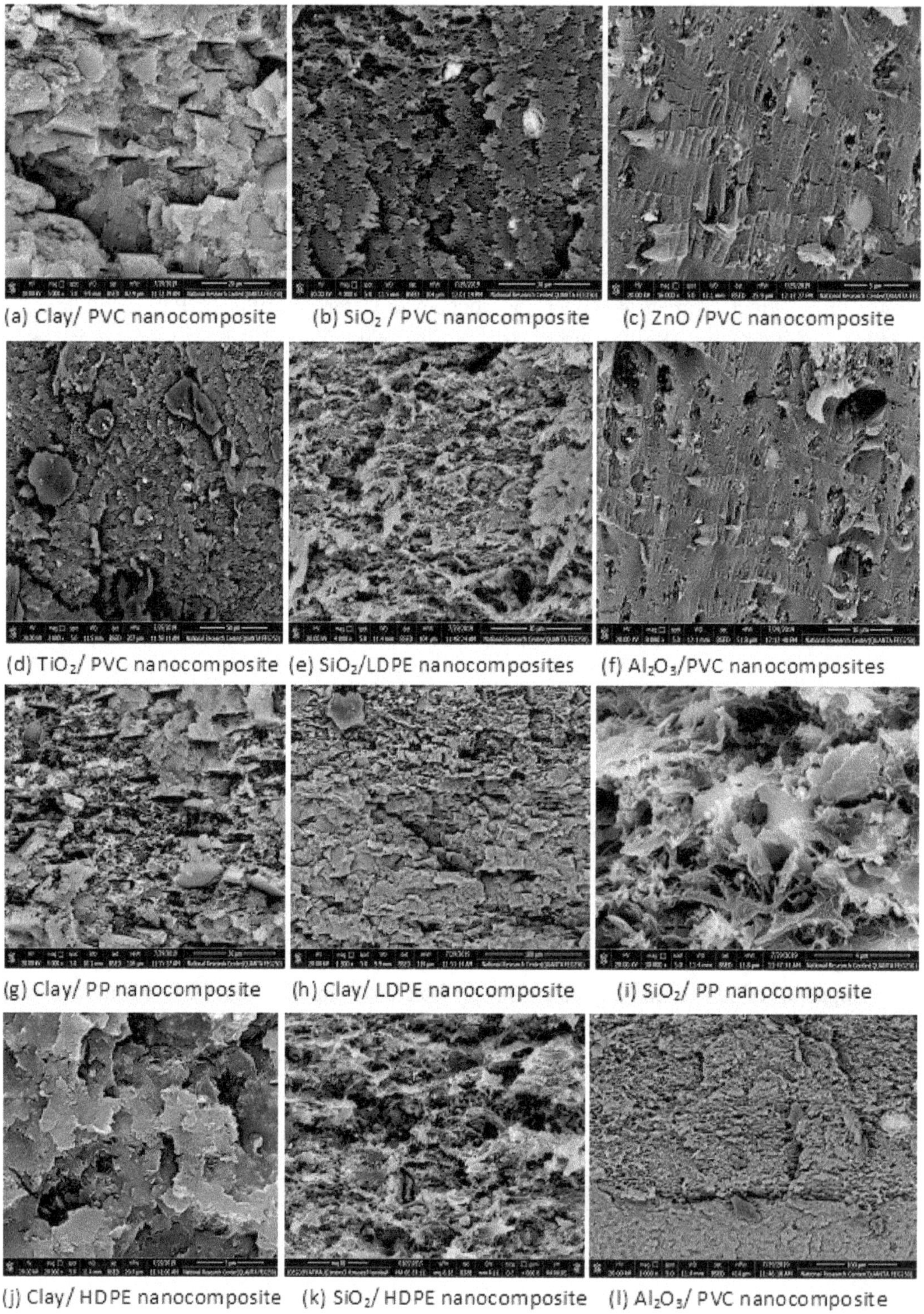

(a) Clay/ PVC nanocomposite (b) SiO_2 / PVC nanocomposite (c) ZnO /PVC nanocomposite

(d) TiO_2/ PVC nanocomposite (e) SiO_2/LDPE nanocomposites (f) Al_2O_3/PVC nanocomposites

(g) Clay/ PP nanocomposite (h) Clay/ LDPE nanocomposite (i) SiO_2/ PP nanocomposite

(j) Clay/ HDPE nanocomposite (k) SiO_2/ HDPE nanocomposite (l) Al_2O_3/ PVC nanocomposite

10.2.2 Measurement and Characterization

Additives from claiming clay nanoparticles of the base mechanical polymers have been created towards utilizing mixing, ultrasonic and warming forms. The build from claiming these polymer materials is an economically accessible material now being used in the manufacturing of HV mechanical results and their properties. It can measure the sum of dielectric properties of immaculate and nanocomposite streamlined materials towards utilizing HIOKI 3522-50 LCR Hi-tester gadget and has been distinguished similarly as demonstrated clinched alongside Table 1. HIOKI 3522-50 LCR Hi-tester gadget measures characterization of nanocomposite encasing streamlined materials. HIOKI 3522-50 LCR Hi-tester gadget, as demonstrated in Fig. 6, has measured the electrical parameters of nano-metric strong dielectric encasing samples during different frequencies: |Z|, |Y|, θ, Rp (DCR), Rs (ESR, DCR), G, X, B, Cp, Cs, Lp, Ls, d (tan δ), and Q. Detail about LCR will be control supply: 100, 120, 220 or 240 V(±10%) AC (selectable), 50/60 Hz, Frequency: DC, 1. 0mHz should 100kHz, show Screen: lcd for backlight / 99999 (full 5 digits), fundamental Accuracy: Z: ±0.08% rdg. θ: ±0.05þ, furthermore outside dc inclination ±40V max. (option) (3522-50 utilized alone ±10V max. / utilizing 9268 ±40V max.). Consequently, it has been utilized to measure the electric and dielectric parameters from claiming nanometric robust dielectric encasing samples towards different frequencies concerning the illustration demonstrated. The examined nanoparticles have been determinedly influenced by dielectric properties of nanocomposites materials. In addition, the impacts of nanoparticles on permittivities for polyvinyl chloride nanocomposites have been investigated. These may be measured as a capacity from claiming recurrence in the extent of 0.01 Hz to 1MHz in variant temperatures (20°C - 80°C). This procedure will be in light of the estimation of the dielectric properties, as a work of recurrence of a test sandwiched between two electrodes. The Dielectric constant is measured concerning illustration from claiming recurrence in the go of 0.01 Hz to 1MHz at variant temperatures for every last one of the test specimens. It is vital for master characterization from claiming nanocomposite materials under variant warm states. Determination of the measured SOLARTRON gadget is energy supply, switch selectable: 90 to 126V, 198252V ac, 48 to 65Hz. Utilization 200VA. Surrounding temperature operating: 0 to 50°C. Storage: -30 to 70°C. Determination limits: 10 on 30°C. Humidity, non-condensing: 95% at 40°C. Vibration: tried for understanding of IEC68 (BS2011), Safety: intended to go along with IEC348 (BS4743).

Constantly chosen nanoparticles, in this research, are the circular state, measured similarly at 10nm breadth for each grain size and they use impetuses to be the best filler in build grid modern materials. Generally, it is stronger and more unbending than all other motivation thermoplastic materials. It has a secondary rigidity and modulus of flexibility.

On the other hand, Fig. 7 illustrates Hi-pot analyzer model ZC2674 gadget to analyze uniform and non-uniform electric field circulation through the thickness of the encasing layer for separate nanocomposite materials. It is shown that Fig. 7 provides the HI-POT analyzer model ZC2674 gadget to analyze uniform and non-uniform electric field appropriation through the thickness of the encasing layer with diverse nanocomposite materials. Test effects are kept tabs with respect to Polypropylene PP, Polyvinyl chloride PVC, LDPE and HDPE encasing materials with different rate weight of variant nanoparticles. Setup for the two electrodes from claiming uniform electric field has been constructed, starting with copper and has 30mm breadth of that setup for tip cathode and non-uniform electric field of 0.5mm breadth. The Hi-tester gadget has been specified, such as 1kVA, 20kV, AC, dc voltages, 10mA, AC and dc ebbs and flows. Setup for both electrodes of uniform electric field has been made from copper and has 35 mm breadth, yet the setup for tip cathode of non-uniform electric field has 0.5mm diameter.

Table 2. Dielectric properties of pure and nanocomposite materials

Materials	Dielectric Constant at 1kHz	Resistivity (Ω.m)	Materials	Dielectric Constant at 1kHz	Resistivity (Ω.m)
LDPE + 0%wt Clay	2.3	1014	Pure PP	2.28	108
LDPE + 1%wt Clay	2.23	1015	PP + 1%wt Clay	2.21	109
LDPE +5%wt Clay	1.99	1015-1018	PP + 5%wt Clay	1.97	109-1010
LDPE + 10%wt Clay	1.76	1018-1020	PP + 10%wt Clay	1.75	1010-1012
LDPE + 1%wt Fumed Silica	2.32	1013	Pure PVC	3.3	1013
LDPE +5%wt Fumed Silica	2.39	1013-1011	PVC + 1%wt Clay	3.20	1014
LDPE + 10%wt Fumed Silica	2.49	1011-109	PVC + 5%wt Clay	2.83	1014-1017
HDPE + 0%wt Clay	2.3	1015	PVC + 10%wt Clay	2.49	1017-1020
HDPE + 1%wt Clay	2.23	1016	Pure LDPE	2.3	1014
HDPE + 5%wt Clay	1.99	1016-1019	LDPE + 1%wt Clay	2.23	1015
HDPE + 10%wt Clay	1.76	1019-1021	LDPE + 5%wt Clay	1.99	1015-1018
HDPE + 1%wt Fumed Silica	2.32	1014	LDPE + 10%wt Clay	1.76	1018-1020
HDPE + 5%wt Fumed Silica	2.39	1014-1012	Pure HDPE	2.3	1015
HDPE + 10%wt Fumed Silica	2.49	1012-1010	HDPE + 1%wt Clay	2.27	1016
Pure PVC	3.3	1013	HDPE + 5%wt Clay	2.21	1016-1019
PVC + 1%wt clay	3.20	1014	HDPE + 10%wt Clay	2.16	1019-1021
PVC + 5%wt clay	2.83	1014-1017	Pure LDPE	2.3	1014
PVC + 10%wt clay	2.49	1017-1020	LDPE + 1%wt Clay	2.23	1015
PVC + 1%wt ZnO	2.63	1014-1017	LDPE + 5%wt Clay	1.99	1015-1018
PVC + 5%wt ZnO	2.39	1017-1020	LDPE + 10%wt Clay	1.76	1018-1020
PVC + 10%wt ZnO	2.15	1012	LDPE + 1%wt Fumed Silica	2.32	1013
PVC + 1%wt SiO2	3.42	1012-1010	LDPE + 5%wt Fumed Silica	2.39	1013-1011
PVC + 5%wt SiO2	3.53	1010-108	LDPE + 10%wt Fumed Silica	2.49	1011-109
PVC + 10%wt SiO2	3.60	1014	Pure HDPE	2.3	1015
PVC + 1%wt Al2O3	3.32	1012-1010	HDPE + 1%wt Clay	2.27	1016
PVC + 5%wt Al2O3	3.43	1010-108	HDPE + 5%wt Clay	2.21	1016-1019
PVC + 10%wt Al2O3	3.70	1014	HDPE + 10%wt Clay	2.16	1019-1021
PVC + 1%wt TiO2		1012	HDPE + 1%wt Fumed Silica	2.35	1014
PVC + 5%wt TiO2		1012-1011	HDPE + 5%wt Fumed Silica	2.42	1014-1012
PVC +10%wt TiO2		1011-1010	HDPE + 10%wt Fumed Silica	2.51	1012-1010

Figure 6. HIOKI 3522-50 LCR Hi-tester device

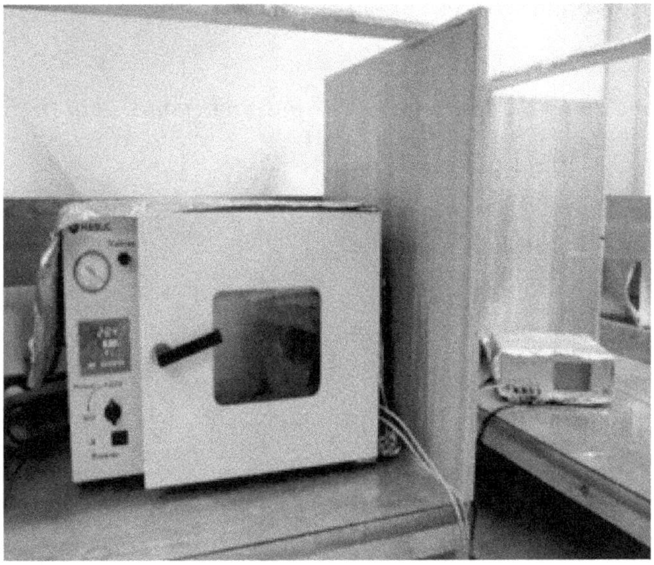

Figure 7. HIPOT tester model ZC2674 device

The breakdown model has been utilized as a foundation for tests, and so the measured dielectric strength HIPOT analyzer model ZC2674 gadget which has been specified as 1kVA, 20kV, AC, dc voltages, 10mA, AC and dc ebbs and flows has been employed.

Test effects are centered, looking into Polyvinyl chloride encasing materials with different rate weights from claiming clay, zinc oxide, fumed silica and aluminum oxide nanoparticles. Setup of the two electrodes of uniform electric field has been produced from copper and has 35mm breadth, yet the setup for tip cathode of non-uniform electric field has 0.5mm breadth.

Figure 8. Effect of nanoparticles on LDPE industrial polymers under uniform electric field

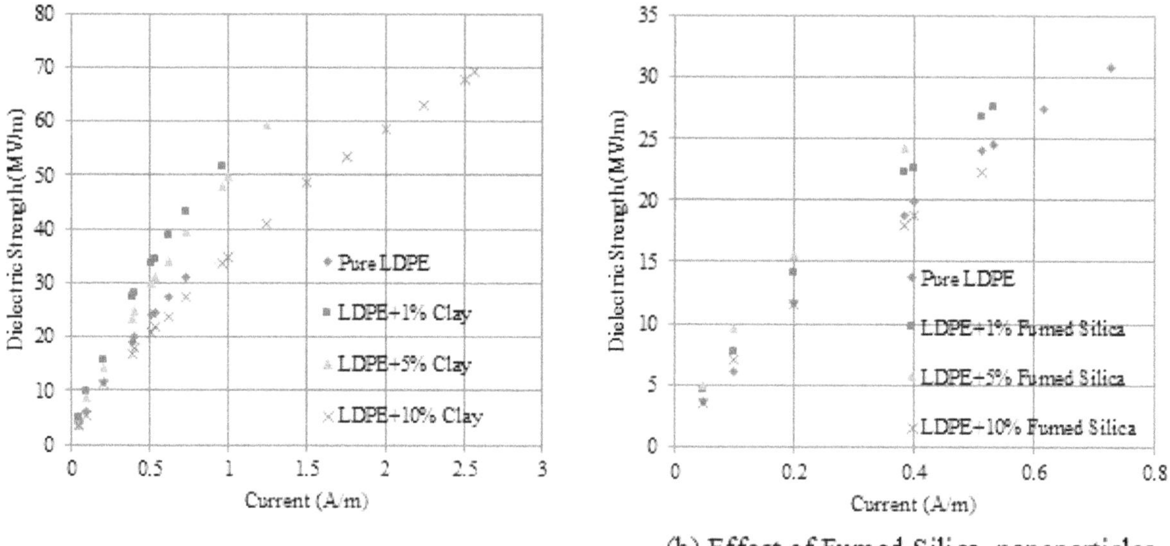

(a) Effect of clay nanoparticles

(b) Effect of Fumed Silica nanoparticles

10.3 DEGRADATION OF NANODIELECTRICS UNDER AC ELECTRIC FIELDS

Dielectric strength of encasing materials is a fundamental parameter to the electrical streamlined provisions. Thus, the breakdown voltage from claiming new nanocomposite streamlined materials has been initiated applying variant AC voltage on the starting with zero kV until breakdown occurs. Moreover, AC conduction present is measured by testing the example from zero ampere dependent upon 1mA. Test effects are centered upon investigating two sorts of polyethylene encasing materials with different rate weights from claiming clay and fumed silica nanoparticles. The breakdown voltages of the new nanocomposite polyethylene (PE) encasing materials are measured at 35mm breadth to both two copper electrodes clinched alongside uniform electric field; but, for non-uniform connected electric fields, it can use 0.5mm tip breadth for individual cathode. The connected AC voltage on the example is expanded starting with 0kV until breakdown happens. AC conduction current is measured through trial on the example and the Emulating subsections nitty gritty outcomes and examination of the impacts of clay and fumed silica nanoparticles on polyethylene encasing materials.

10.3.1 Degradation of Nanodielectrics Under Uniform Electric Fields

The breakdown voltage from claiming new nanocomposites mechanical materials has been measured with a 35mm breadth for the two copper electrodes for uniform electric field and 0.5mm tip breadth, especially in non-uniform electric field. The connected AC voltage in the example is different from 0.1kV until breakdown happens. AC spillage presence has been measured through testing the example starting with 0A to 1mA for every unit period (m). The following subsections will highlight the effects and examine the impacts of including clay and fumed silica nanoparticles under low thickness polyethylene (LDPE) and secondary thickness polyethylene (HDPE) mechanical materials.

Figure 9. Effect of nanoparticles on HDPE industrial polymers under uniform electric field

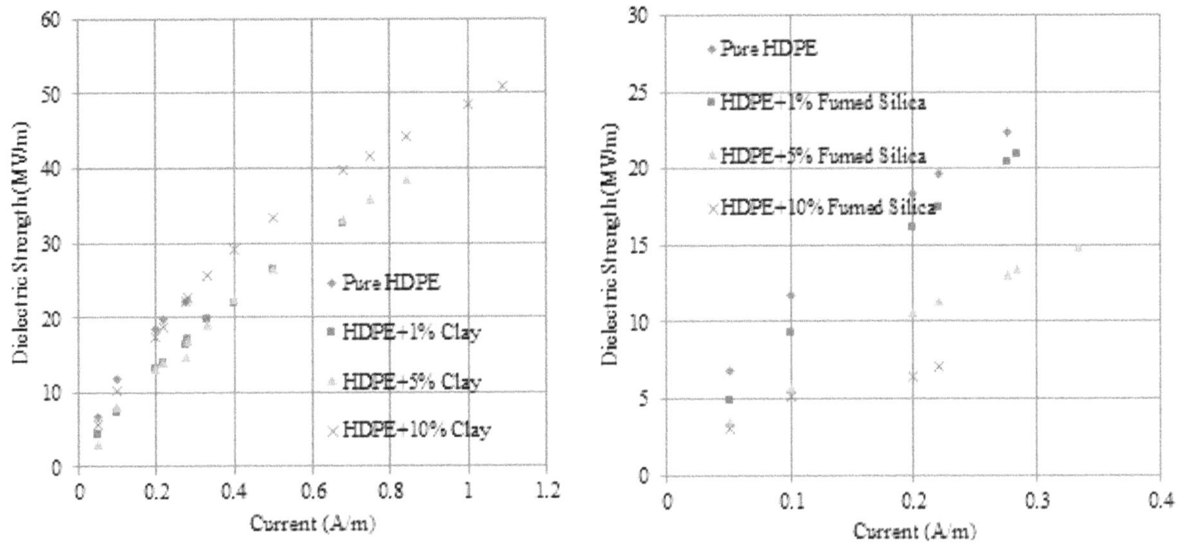

(a) Effect of Clay nanoparticles (b) Effect of Fumed Silica nanoparticles

Figure 10.Effect of clay nanoparticles on polypropylene materials under uniform electric field

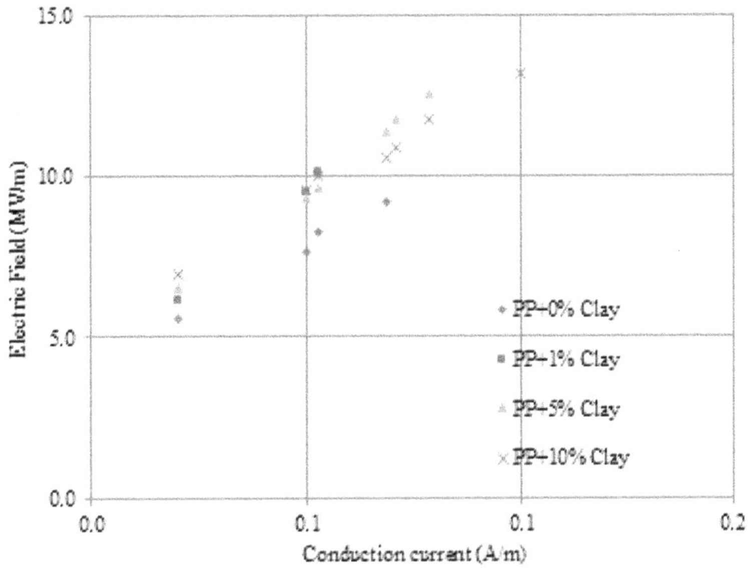

Figure 11. Effect of clay nanoparticles on polyvinyl chloride materials under uniform electric field

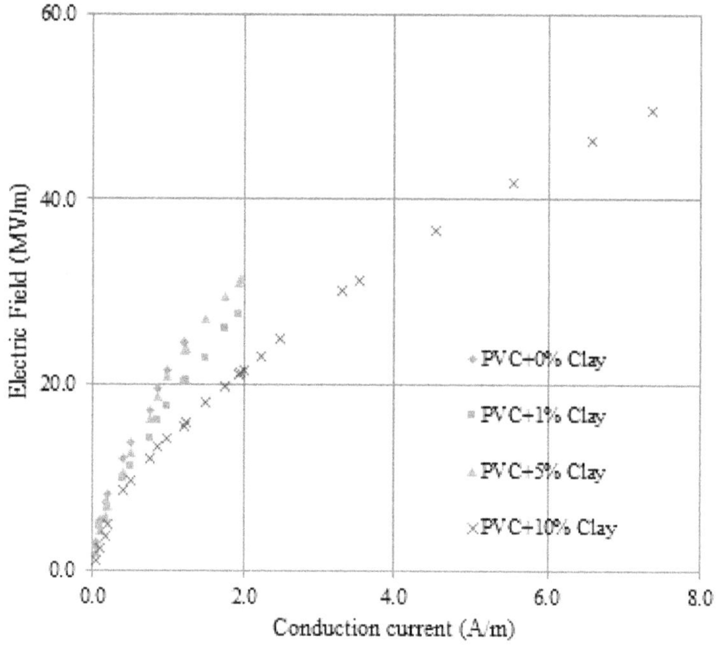

Figure 12. Effect of clay nanoparticles on low-density polyethylene materials under uniform electric field

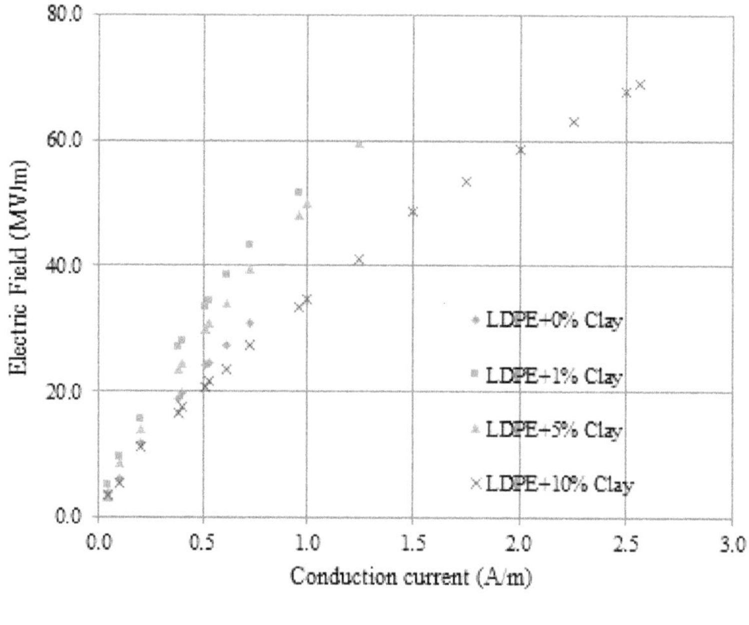

Figure 13. Effect of clay nanoparticles on high-density polyethylene materials under uniform electric field

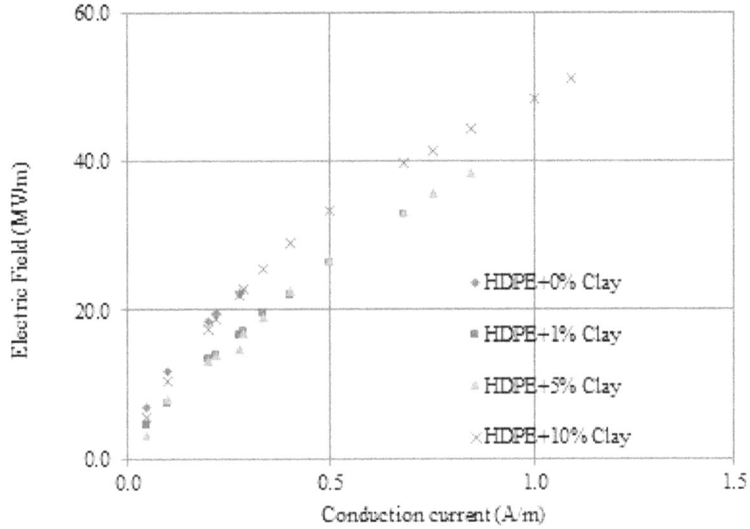

Figure 8 delineates the impact from claiming clay and fumed silica nanoparticles on dielectric strength and spillage pasquinade currently in LDPE samples; especially uniform electric field separately. It can be recognized that the expanding rate of clay dependent upon 10%wt nanoparticles in the nanocomposite builds dielectric strength of the streamlined materials with respect to unfilled samples and expanding rate of fumed silica up to 10%wt nanoparticles in the nanocomposite abatements dielectric strength of the mechanical materials with respect to unfilled samples. Although figures (3. An and 3. B) hint at the impact from claiming clay and fumed silica nanoparticles by looking into the dielectric strength and spillage pasquinade current, especially HDPE samples in the uniform electric field separately. It is recognized that the expanding rate of clay up to 10%wt nanoparticles in the nanocomposite expands dielectric strength of the modern materials samples with respect to unfilled samples in the expanding rate of fumed silica up to 10%wt nanoparticles in the nanocomposite samples, which declines the dielectric strength of the mechanical materials with respect to unfilled samples. Thus, spillage pasquinade current impact on both clay nanoparticles; especially LDPE and HDPE in the new nanocomposite materials expands with expanding the dielectric strength smoothness for including nanoparticles rate from 1%wt, dependent upon 10%wt Nanoparticles. However, spillage pasquinade current impact on both fumed silica nanoparticles; especially LDPE and HDPE in the new nanocomposite materials declines, expanding dielectric strength easily with the inclusion of nanoparticles rate from 1%wt dependent upon 10%wt. Nanoparticles.

Figure 9 illustrates the impact of uniform electric field on polypropylene nanocomposite materials. It is recognized that expanding clay nanoparticles rate in the nanocomposite increments dielectric strength of the streamlined materials (0%:5%wt), even if dielectric strength diminishes with expanding clay nanoparticles (5%wt.:10%wt.) due to amassing nanoparticles phenomena because of the expanding rate of clay nanoparticles clinched alongside polypropylene materials. Fig. 11 reveals the impact of uniform electric field with respect to polyvinyl chloride nanocomposite materials; it has been shown that expanding the rate of clay nanoparticles dependent upon 10%wt for polyvinyl chloride abates dielectric strength of polyvinyl chloride nanocomposite streamlined materials and builds conduction current for expanding rate of clay nanoparticles.

Figure 14. Effect of clay nanoparticles on low density polyethylene under uniform electric field

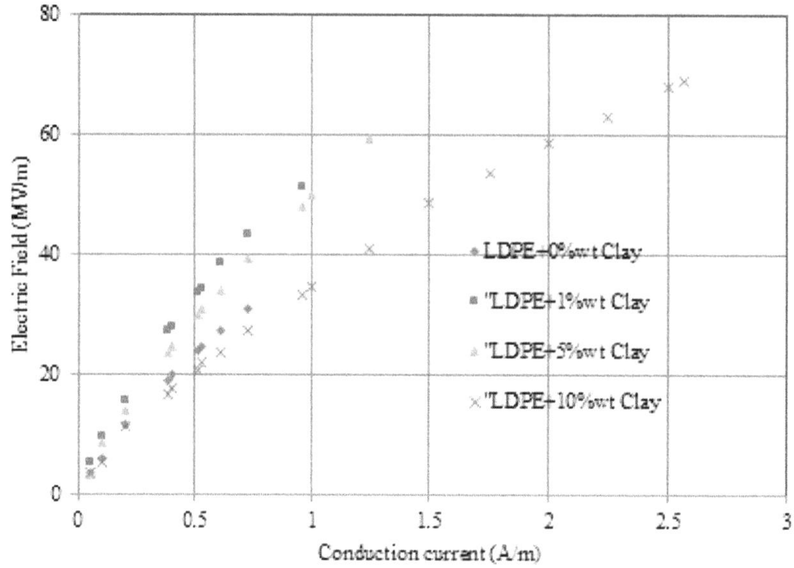

Figure 15. Effect of fumed silica nanoparticles on low density polyethylene under uniform electric field

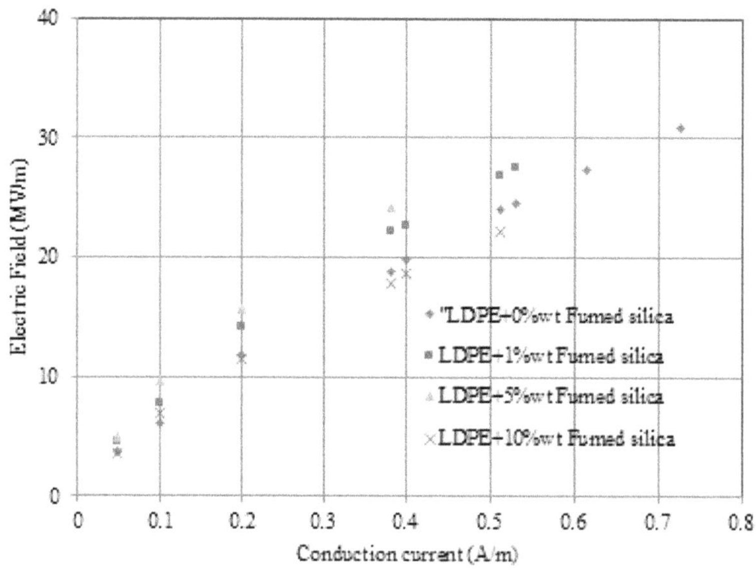

Figure 12 indicates the impact of uniform electric field on low-thickness polyethylene nanocomposite materials; it has been recognized that expanding clay nanoparticles rate up to 5%wt increments dielectric strength of the low-thickness polyethylene nanocomposite encasing example. In any case, amassing nanoparticles phenomena leads to expanding the rate of clay nanoparticles for low-thickness polyethylene dependent upon 10%wt. On the other hand, Fig. 13 depicts that expanding rate of clay nanoparticles up to 10%wt over high-density polyethylene builds conduction present through high-density polyethylene nanocomposite. Hence, the dielectric strength abates marginally with expanding the rate of clay nanoparticles. The greater part of these results can affect the examination of AC electrical high voltage breakdown of the tried insulating materials. The execution of the new nanocomposites is powerful with respect to AC electrical force provisions.

Figure 16. Effect of clay nanoparticles on high density polyethylene under uniform electric field

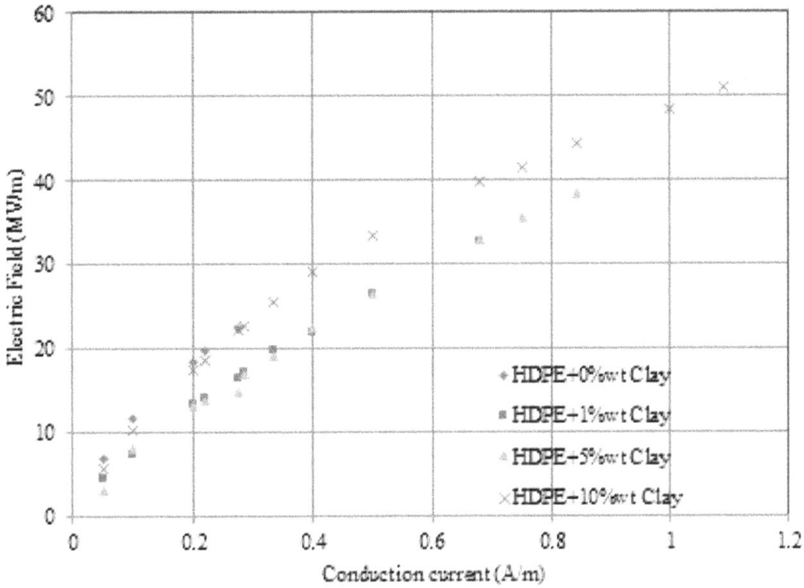

Figure 17. Effect of fumed silica nanoparticles on high density polyethylene under uniform electric field

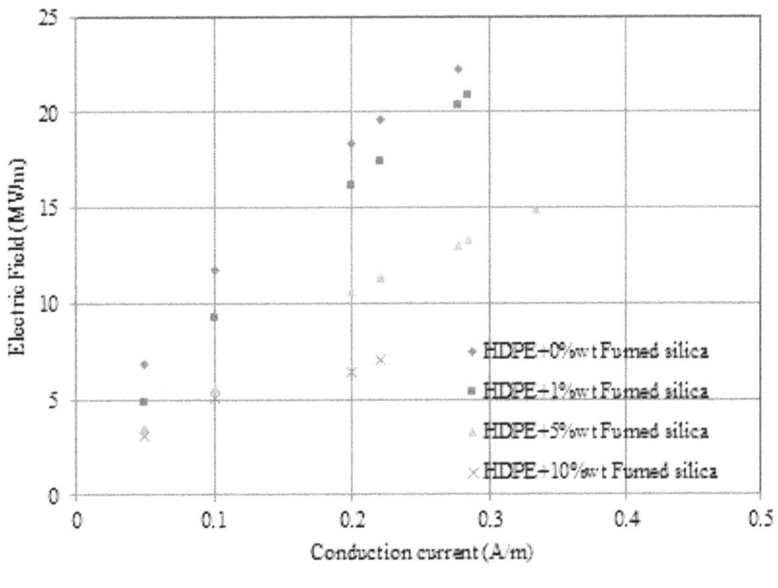

Figure 18. Effect of nanoparticles on LDPE industrial polymers under non-uniform electric field

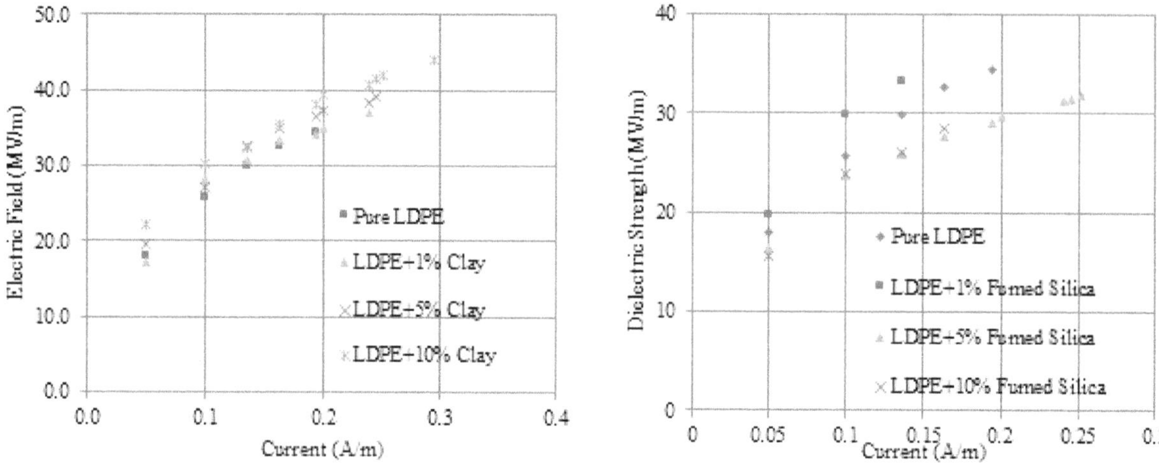

(a) Effect of Clay nanoparticles (b) Effect of Fumed Silica nanoparticles

Figure 19. Effect of nanoparticles on HDPE industrial polymers under non-uniform electric field

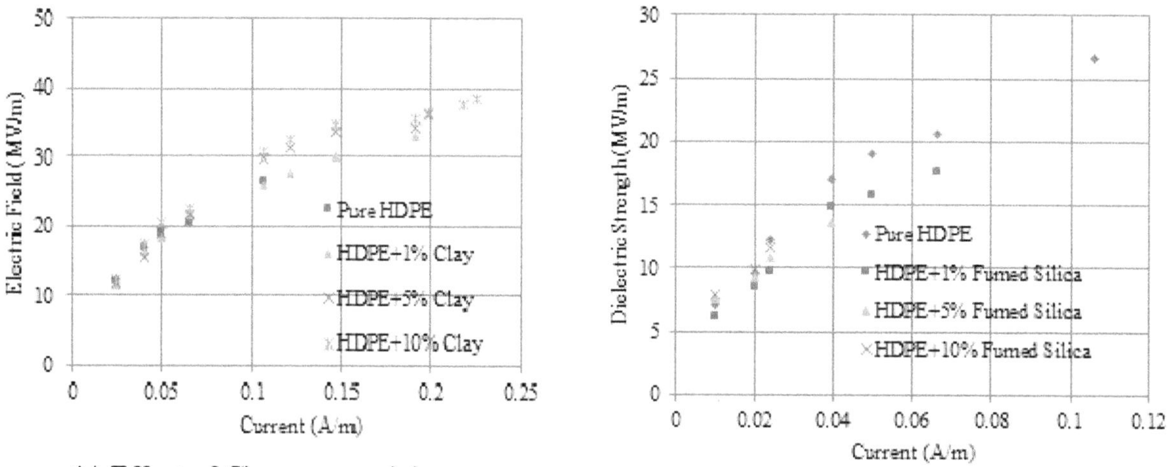

(a) Effect of Clay nanoparticles (b) Effect of Fumed Silica nanoparticles

On the other hand, Figure 14 reveals that dielectric strength for Clay/Low thickness polyethylene nanocomposite materials has expanded by expanding the rate of clay nanoparticles dependent upon 5%wt., and the conduction current has diminished. However, expanding the rate of clay nanoparticles more than 5%wt. declines the dielectric strength of low thickness polyethylene nanocomposite materials. The execution of dielectric strength and conduction current can be recognized from the test estimations applying uniform electric field concerning illustrations indicated in fig. 15. Starting with this figure, there is a chance to recognize the rate as 5%wt for fumed silica that is a basic strength to enhancing the electrical trademark of low thickness polyethylene. On the other hand, in fig. 16, that expanding rate of clay dependent upon 5%wt can be depicted. Nanoparticles in the nanocomposite abates dielectric strength and increments the conduction present for high thickness polyethylene streamlined materials. That's only the tip of the iceberg rates of clay nanoparticles in expanding dielectric strength from claiming secondary thickness polyethylene streamlined materials.

Figure 20.Effect of clay nanoparticles on polypropylene materials under non-uniform electric field

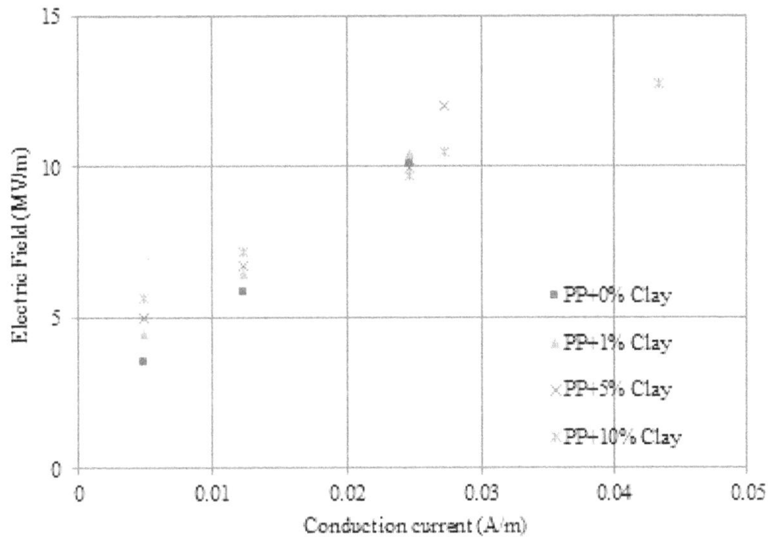

In Figure 17, it is self-evident that the expanding rate of fumed silica dependent upon 10%wt. clinched alongside high thickness polyethylene declines dielectric strength of the mechanical materials and increments spillage pasquinade current through the new nanocomposite during the same appraisals. Therefore, clay nanoparticles have more capacity to move forward dielectric strength for polyethylene materials to fumed silica nanoparticles; and the rate of 5%wt from claiming clay and fumed silica will be an incredulous strength to moving forward the electrical trademark for polyethylene. Expanding rate of nanoparticles expanding aggregation electric withstands encasing forces; therefore, it is proposed that nanoparticles have been incredible in scattering and haven't exceeded 5%wt. in the polyethylene base matrix.

10.3.2 Degradation of Nanodielectrics Under Non-Uniform Electric Fields

Figure 18 demonstrates the impact from claiming clay and fumed silica nanoparticles on dielectric strength and spillage pasquinade current in LDPE materials in non-uniform electric field. It may be noticed that expanding rate of clay up to 10%wt. Nanoparticles in the nanocomposite increments dielectric strength of the modern materials bit by bit with respect to unfilled samples. Although the expanding rate of fumed silica is up to 1%wt, fumed silica nanoparticles in the nanocomposite increments dielectric strength of the mechanical materials in any case, expanding rate of fumed silica (1%wt.:10%wt) nanoparticles in the nanocomposite declining dielectric strength of the mechanical materials with respect to unfilled samples. Furthermore, it is recognized that spillage pasquinade present through the new nanocomposite increments with expanding rate clay nanoparticles dependent upon 10%wt. However, the spillage pasquinade present declines easily for expanding fumed silica nanoparticles rate starting with 5%wt, but for the rate range of (5%wt: 10%wt).

Figure 21. Effect of clay nanoparticles on polyvinyl chloride materials under non-uniform electric field

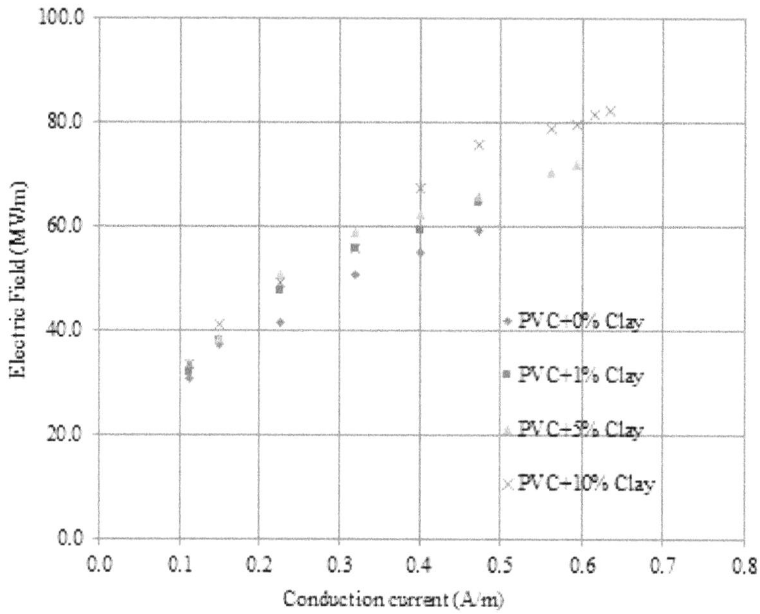

Figure 22. Effect of clay nanoparticles on low-density polyethylene materials under non-uniform electric field

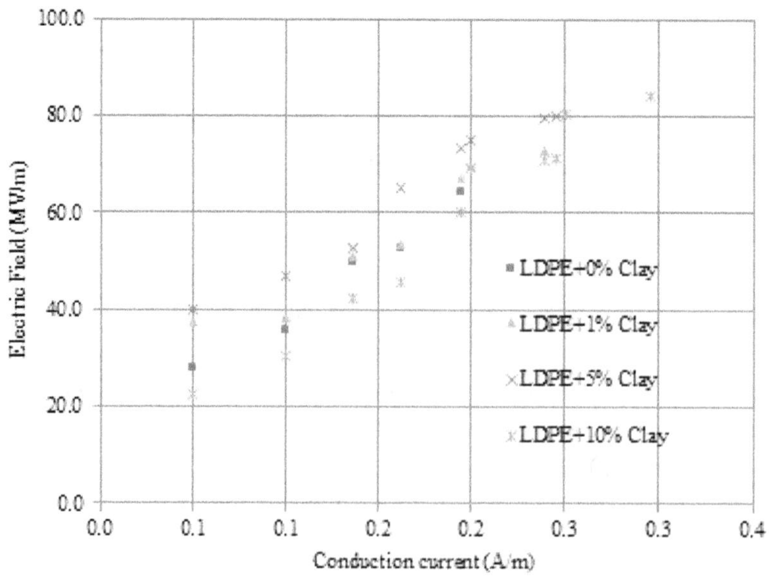

Figure 23. Effect of clay nanoparticles on high-density polyethylene materials under non-uniform electric field

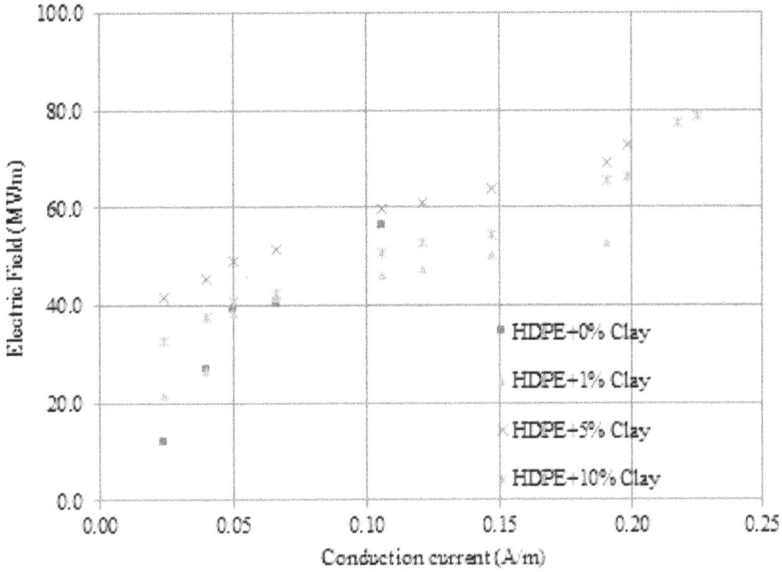

Figure 19 shows the impact of including clay nanoparticles for dielectric strength and spillage pasquinade present for HDPE materials, especially non-uniform electric field. It is noticed that expanding the rate of clay dependent upon 5%wt. Nanoparticles in the nanocomposite expands dielectric strength of the streamlined materials bit by bit, but expanding the rate of clay (5%wt.:10%wt) nanoparticles in the nanocomposite declines dielectric strength of the modern materials with respect to unfilled samples. Although it has been shown that expanding rate of fumed silica dependent upon 10%wt fumed silica nanoparticles in the nanocomposite abates dielectric strength of the modern materials with respect to unfilled specimens, it may be recognized that spillage pasquinade present through the new nanocomposite expands with expanding the rate of clay nanoparticles dependent upon 10%wt. However, the spillage pasquinade present abates smoothness with expanding fumed silica nanoparticles rate dependent upon 10%wt.

Figure 20 indicates the impact of expanding clay nanoparticles on dielectric strength and conduction current of polypropylene materials under non-uniform electric field; so that the expanding rate of clay nanoparticles in the polypropylene encasing nanocomposite specimen's declines dielectric strength of polypropylene in the same conduction current, especially at expanding nanoparticles rate dependent upon 5%wt. It is to be noted that due to the amassing phenomena of nanoparticles, the dielectric strength builds marginally for expanding the rate of clay nanoparticles. Hence, Fig. 21 depicts that expanding rate of clay nanoparticles, especially polyvinyl chloride encasing nanocomposite specimens, increments dielectric strength of the mechanical materials in the same spillage pasquinade current, especially including Nano fillers rate up to 10%wt.

Figure 24. Effect of clay nanoparticles on low density polyethylene under non-uniform electric field

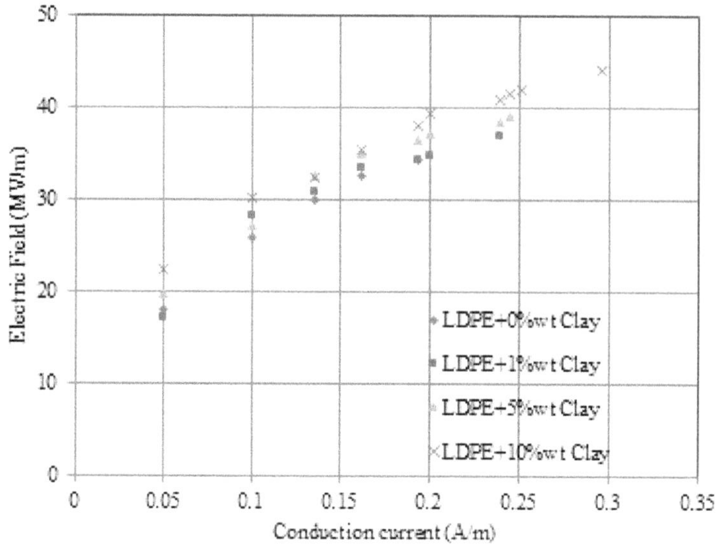

Figure 25. Effect of fumed silica nanoparticles on Low density polyethylene under non-uniform electric field

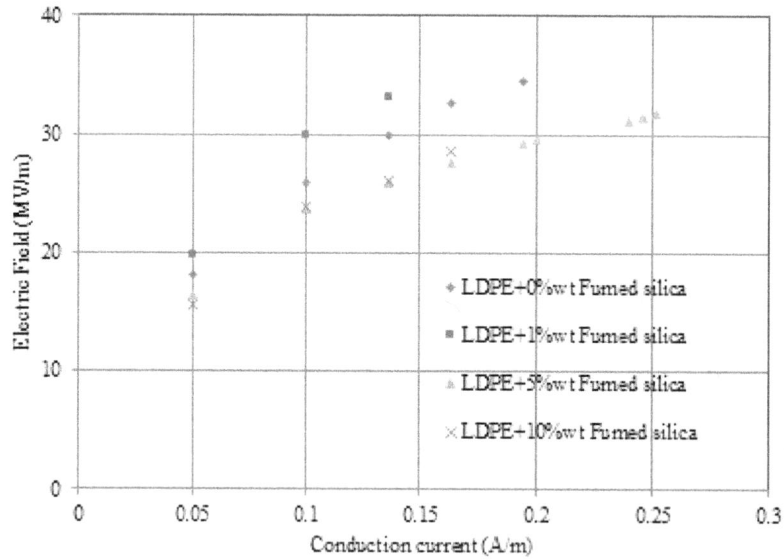

Figure 26. Effect of clay nanoparticles on High density polyethylene under non-uniform electric field

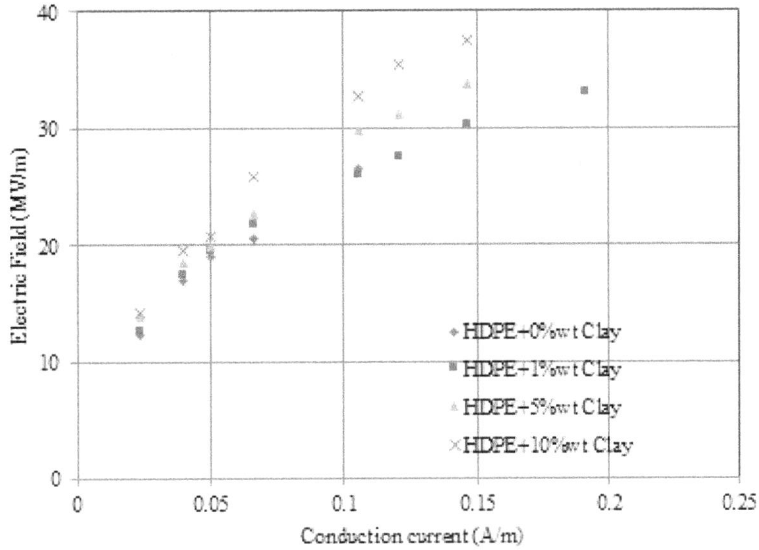

Figure 27. Effect of fumed silica nanoparticles on High density polyethylene under non-uniform electric field

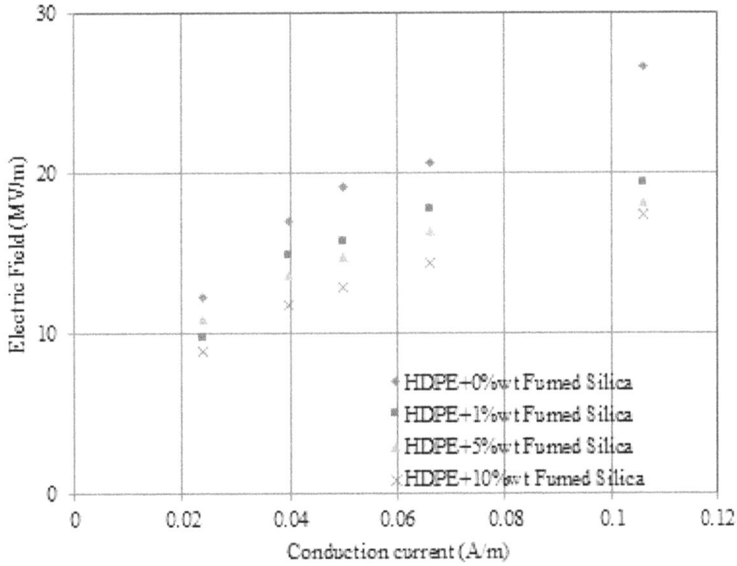

Figure 22 indicates the impact from claiming clay nanoparticles with respect to dielectric strength and conduction current on low-thickness polyethylene materials of non-uniform electric field. Dielectric strength of nanocomposite materials expands with expanding the rate of clay nanoparticles dependent upon 5%wt low-thickness polyethylene towards low connected electric field (0:32MV/m) in any case because of collected clay nanoparticles expanding their rate to more than 5%wt; especially low-thickness polyethylene which diminishes the dielectric strength of nanocomposite materials. On the other hand, Fig. 23 depicts the impact of expanding clay nanoparticles on dielectric strength and conduction current in high-density polyethylene materials clinched alongside non-uniform electric field. The dielectric strength of high-density polyethylene nanocomposite mechanical materials has the same conduct for low-thickness polyethylene at distinctive electric field values, especially towards low connected electric field (0:23MV/m). The dielectric strength of nanocomposite materials expands with expanding the rate of clay nanoparticles up to 5%wt in high-density polyethylene and abates with expanding the rate of clay nanoparticles to more than 5%wt in high-density polyethylene.

Figure 28. Effect of clay nanoparticles on Polyvinyl Chloride

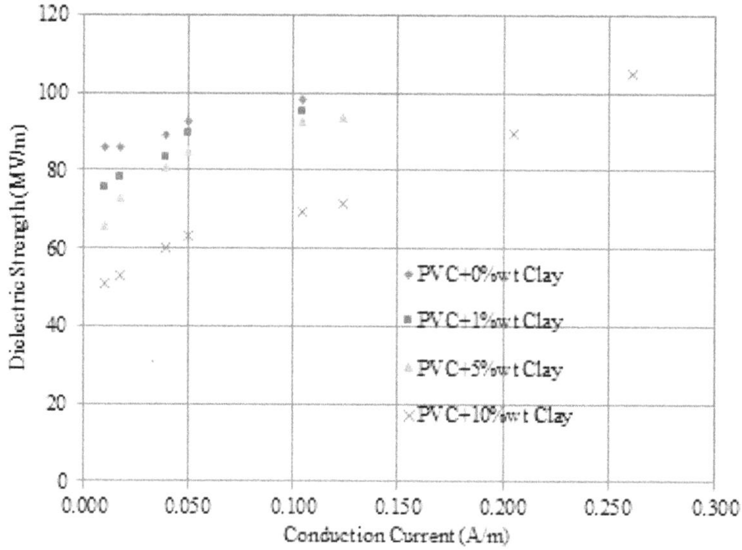

Figure 29. Effect of zinc oxide nanoparticles on Polyvinyl Chloride

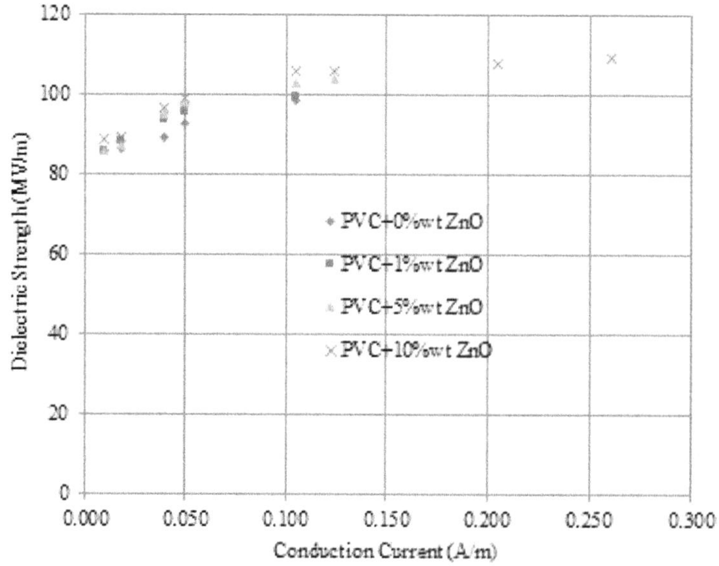

Figure 30.Effect of Al2O3 nanoparticles on Polyvinyl Chloride

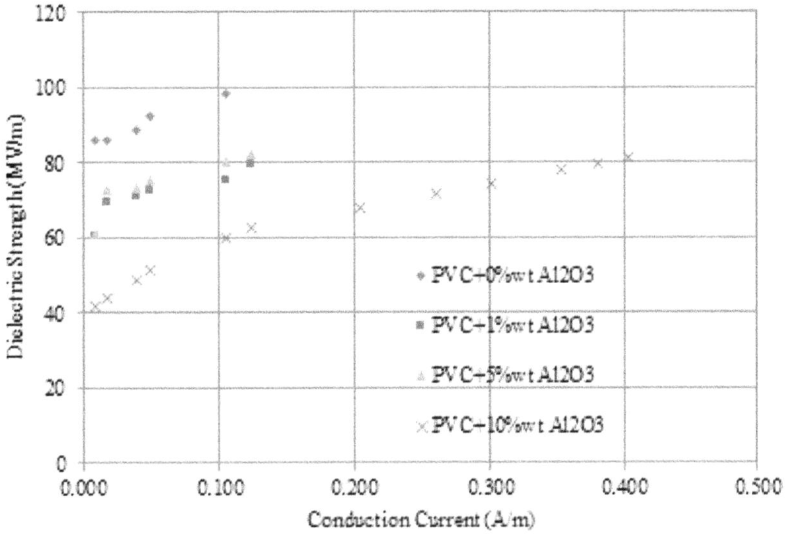

Table 3. Dielectric strength of pure and nanocomposite materials in uniform electric field

Materials	Dielectric Strength (MV/m) at (25 oC)	Dielectric Strength (MV/m) at (60 oC)
LDPE + 0%wt Clay	30.8939	25.4653
LDPE + 10%wt Clay	69.0641	67.9765
LDPE + 10%wt Fumed Silica	22.2439	20.5733
HDPE + 0%wt Clay	22.2922	16.8776
HDPE + 10%wt Clay	50.9394	48.3462
HDPE + 10%wt Fumed Silica	7.0166	6.1736

Table 4. Dielectric strength of pure and nanocomposite materials in non-uniform electric field

Materials	Dielectric Strength (MV/m) at (25 oC)	Dielectric Strength (MV/m) at (60 oC)
LDPE + 0%wt Clay	34.4060	30.9643
LDPE + 10%wt Clay	44.0828	41.9435
LDPE + 10%wt Fumed Silica	28.5179	21.6743
HDPE + 0%wt Clay	26.6078	23.2776
HDPE + 10%wt Clay	38.4624	32.3462
HDPE + 10%wt Fumed Silica	11.5684	8.8756

Table 5. Maximum dielectric strength of pure and nano-composite materials under uniform electric field and variant temperatures

Materials	MAX Dielectric strength (MV/m) AT (25 oC)	MAX Dielectric strength (MV/m) AT (60 oC)
Pure PP	9.1953	6.7693
PP + 10%wt Clay	13.1628	12.3452
Pure PVC	24.4715	20.6753
PVC + 10%wt Clay	49.4949	47.7735
Pure LDPE	30.8939	25.4653
LDPE + 10%wt Clay	69.0641	67.9765
Pure HDPE	22.2922	16.8776
HDPE + 10%wt Clay	50.9394	48.3462
PURE LDPE	30.8939	25.4653
LDPE + 1%wt Clay	51.4184	49.5643
LDPE + 5%wt Clay	59.4214	57.7832
LDPE + 10%wt Clay	69.0641	67.9765
LDPE + 1%wt Fumed Silica	27.5319	25.5473
LDPE + 5%wt Fumed Silica	24.1762	22.5372
LDPE + 10%wt Fumed Silica	22.2439	20.5733
PURE HDPE	22.2922	16.8776
HDPE + 1%wt Clay	32.6158	30.6745
HDPE + 5%wt Clay	38.4375	35.9865
HDPE + 10%wt Clay	50.9394	48.3462
HDPE + 1%wt Fumed Silica	20.8758	18.4725
HDPE + 5%wt Fumed Silica	14.8699	12.6832
HDPE + 10%wt Fumed Silica	7.0166	6.1736

Figure 24 illustrates the impact from claiming clay nanoparticles on dielectric strength and conduction current in low thickness polyethylene materials in non-uniform electric fields. Dielectric strength for Clay/LDPE nanocomposite materials has expanded towards expanding the rate of clay nanoparticles and conduction current has diminished uncommonly with expanding nanoparticles rate of clay nanoparticles up to 10%wt. However, Fig. 25 indicates the dielectric strength of fumed silica/LDPE nanocomposite material and the expanding dielectric strength from claiming nanocomposite expanding rate of fumed silica nanoparticles dependent upon 1%w t, provided that it is more than 1%wt. For fumed silica nanoparticles, the dielectric strength abates with expanding the conduction present. Fig. 26 indicates the impact of including clay nanoparticles on dielectric strength and conduction present clinched alongside secondary thickness Polyethylene materials, especially the non-uniform electric field. Expanding the rate of clay dependent upon 10%wt. Nanoparticles in the Nanocomposite material expands dielectric strength of the polyethylene material and abates the conduction present pass recipient nanocomposite samples. Fig. 27 illustrates that expanding the rate of fumed silica nanoparticles dependent upon 10%wt.

abates the dielectric strength clinched alongside secondary thickness polyethylene modern materials and expands conduction current pass recipient nanocomposite samples. Therefore, clay nanoparticles have more capacity for enhancing dielectric strength for polyethylene materials than fumed silica nanoparticles under non-uniform electric fields. However, fumed silica nanoparticles bring a greater capacity to enhancing the capacitive charge present for polyethylene materials over clay nanoparticles.

Table 6. Maximum dielectric strength of pure and nanocomposite materials under non-uniform electric field and variant temperatures

Materials	MAX Dielectric strength (MV/m) AT (25 oC)	MAX Dielectric strength (MV/m) AT (60 oC)
Pure PP	10.1107	7.8543
PP + 10%wt Clay	12.7233	10.5342
Pure PVC	59.4488	50.1874
PVC + 10%wt Clay	82.3572	751678
Pure LDPE	64.4060	60.448
LDPE + 10%wt Clay	84.0828	82.6745
Pure HDPE	56.6078	50.5473
HDPE + 10%wt Clay	78.4624	71.5369
PURE LDPE	64.4060	60.4482
LDPE + 1%wt Clay	71.4534	69.5436
LDPE + 5%wt Clay	78.7645	73.9743
LDPE + 10%wt Clay	84.0828	82.6745
LDPE + 1%wt Fumed Silica	60.6738	57.6732
LDPE + 5%wt Fumed Silica	55.7629	53.5478
LDPE + 10%wt Fumed Silica	52.2439	50.5733
PURE HDPE	66.6078	60.5473
HDPE + 1%wt Clay	69.6544	65.7843
HDPE + 5%wt Clay	75.9673	68.2945
HDPE + 10%wt Clay	78.4624	71.5369
HDPE + 1%wt Fumed Silica	63.5674	61.9853
HDPE + 5%wt Fumed Silica	59.6745	57.9221
HDPE + 10%wt Fumed Silica	57.0166	56.1736

10.4 DEGRADATION OF NANO-DIELECTRICS UNDER DC ELECTRIC FIELDS

The connected secondary voltage dc in the example has been shifted starting with 0.1 kv until breakdown happens clinched alongside samples. Dc conduction current has been measured through testing the example starting with 0A dependent upon 1mA. The following subsections investigate the impacts of clay, zinc oxide, fumed silica and aluminum oxide nanoparticles on Polyvinyl chloride mechanical materials.

Figure 28 indicates the impact from claiming clay nanoparticles on dielectric strength and conduction pasquinade present clinched alongside Polyvinyl chloride nanocomposite materials in uniform dc electric field. It is recognized that the dielectric strength of clay/ Polyvinyl chloride nanocomposite materials has diminished, expanding the rate of clay nanoparticles dependent upon 10%wt., and the conduction pasquinade present has expanded. Figure 29 indicates the impact of claiming zinc oxide nanoparticles on dielectric strength and conduction pasquinade current in Polyvinyl chloride nanocomposite materials in uniform dc electric field. Dielectric strength of ZnO/ Polyvinyl chloride nanocomposite materials has expanded by expanding rate of zinc oxide nanoparticles up to 10%wt., and the conduction pasquinade current has diminished. Figure 30 indicates the impact from claiming the fumed silica nanoparticles, looking into dielectric strength and conduction pasquinade present in Polyvinyl chloride nanocomposite materials in uniform dc electric field. Dielectric strength for fumed silica/ Polyvinyl chloride nanocomposite materials has diminished, expanding the rate of zinc oxide nanoparticles up to 10%wt., and the conduction pasquinade current has diminished. The same execution concerning the illustration indicated in fig. 30 illustrates the expanding rate of aluminum oxide up to 10%wt. for Polyvinyl chloride through diminishing dielectric strength of the mechanical materials, thus, expanding the conduction pasquinade present through the new nanocomposite during the same appraisals. It is to be noted that the impact from claiming fumed silica, especially diminishing dielectric strength of Polyvinyl chloride nanocomposite materials is more than the impact of aluminum oxide nanoparticles.

Table 7, Dielectric Strength of Pure and Nanocomposite Materials Under Uniform Electric Field and Variant Temperatures

MATERIALS	DIELECTRIC STRENGTH (MV/m) AT (25 OC)	DIELECTRIC STRENGTH (MV/m) AT (60 OC)
Pure PVC	24.4715	20.6753
PVC + 1%wt clay	37.8768	30.7865
PVC + 5%wt clay	49.6754	39.4576
PVC + 10%wt clay	58.9576	51.1276
PVC + 1%wt ZnO	39.5619	32.5983
PVC + 5%wt ZnO	54.1672	43.9872
PVC + 10%wt ZnO	67.6549	58.7655
PVC + 1%wt SiO2	18.7619	14.5653
PVC + 5%wt SiO2	12.1232	10.3982
PVC + 10%wt SiO2	8.3249	7.1273
PVC + 1%wt Al2O3	21.7684	19.5643
PVC + 5%wt Al2O3	14.4794	12.3572
PVC + 10%wt Al2O3	10.5644	8.9876

10.5 THERMAL STABILITY ANALYSIS

10.5.1 Effect of Thermal Conditions Under AC Electric Field

Table (2) depicts the impact of clay nanoparticles with respect to dielectric strength of immaculate and nanocomposite materials in uniform electric field with fluctuating Mobile test temperatures from room temperature during 25 °C and 60 °C. It is clear that there will be expanding in dielectric strength for expanding clay nanoparticles rate. Furthermore, the dielectric strength has been diminished for expanding fumed silica nanoparticles rate. Table (3) depicts the impact from claiming clay nanoparticles for dielectric strength for immaculate and nanocomposite materials for non-uniform electric field with changing cell test temperature starting with room temperature toward 25°C and 60°C. It is clear that clay and fumed silica nanoparticles evolving electrical aspects withstands secondary connected electric strength rating with respect to uniform connected voltage.

Including clay nanoparticles has transformed the electric and dielectric nanocomposites streamlined materials identified with immaculate first electric and dielectric build materials under space temperature. Therefore, this ponder has intrigued enormous filler–polymer grid interface which has a real impact on the warm electric and dielectric properties. It is essential to master characterization from claiming nanocomposite materials under uniform and non-uniform electric fields. Thus, two parallel plates to layer nanocomposite have been utilized for scattering uniform electric fields distribution. The nanocomposite example has been placed inside tip cathode and a plate subjected to scattering non-uniform electric fields conveyance, as demonstrated above.

Table 4 depicts the greatest dielectric strength of immaculate and nanocomposite materials under uniform and non-uniform electric field for changing Mobile test temperature starting with room temperature towards 25°C and 60°C. It is clear that there will be diminishing, especially dielectric strength for expanding cell test temperature. Furthermore, this table indicates that the dielectric strength from claiming immaculate and nanocomposite materials for non-uniform electric field is higher than dielectric strength from claiming immaculate and nanocomposite materials in uniform electric field. In addition, Table 5 depicts warm impacts on dielectric strength from claiming immaculate and nanocomposite materials in non-uniform electric fields. Clay and fumed silica nanoparticles evolving electrical aspects withstand secondary connected electric strength rating with respect to uniform connected voltage.

Nanoparticles sorts and focuses influence the electric characterization; hence, it has been illustrated that including fumed silica has expanded permittivity of the new nanocomposite materials and including clay has diminished permittivity of the new nanocomposite materials, concerning the illustration tabulated. Considering the effects of portraying the impact for raising fixation of clay and fumed silica nanoparticles under uniform electric fields, it is recognized that the rate 5%wt of clay and fumed silica may be an incredulous worth for enhancing the electrical trademark from claiming polyethylene. Expanding the rate of nanoparticles expand amassing electric withstands of encasing forces; therefore, it may be prescribed that nanoparticles bring brilliance scattering and don't exceed 5%wt. in the polyethylene build grid. Otherwise, as indicated clinched alongside results under non-uniform electric fields, it is recognized that the rate of clay dependent upon 10%wt. move forward the electrical trademark of polyethylene. At any case, any rate of fumed silica may be debasing the electrical trademark of polyethylene. Therefore, the vitality of including nanoparticles of clay alternately fumed silica need regulating (increasing alternately decreasing) the dielectric strength from claiming immaculate polyethylene by utilizing nanotechnology strategies.

10.5.2 Effect of Thermal Conditions Under DC Electric Field

Table 6 depicts warm impacts of dielectric strength from claiming immaculate and nanocomposite Polyvinyl chloride encasing materials under dc uniform electric fields. In room temperature (25oC), expanding zinc oxide nanoparticles rate expands the dielectric strength and diminishes the conduction pasquinade current during specified electric field strength. The dielectric strength has been diminished with expanding fumed silica and aluminum oxides nanoparticles rate, expanding capacitive conduction pasquinade current towards specified electric field strength. During high engineering (60°C), table ii depicts that dielectric strength corruption rate of nanocomposite materials is easier than dielectric strength corruption clinched alongside immaculate Polyvinyl chloride.

10.6 FORECAST AND RECOMMENDATIONS

Including clay nanoparticles with low thickness polyethylene and secondary thickness polyethylene mechanical materials builds dielectric strength and spillage pasquinade current easily with respect to unfilled mechanical materials, as stated by rate of clay nanoparticles and polymer sub-atomic kind. Including fumed silica nanoparticles on low thickness polyethylene and high thickness polyethylene mechanical materials abates dielectric strength and spillage pasquinade current easily with respect to unfilled modern materials, as stated by rate of fumed silica nanoparticles and polymer sub-atomic kind. Dielectric strength of immaculate and nanocomposite materials in non-uniform electric field is higher than dielectric strength from claiming immaculate and nanocomposite materials in uniform electric field. Accordingly, diminishment rate in dielectric strength about immaculate mechanical materials will be higher than that happening clinched alongside nanocomposites for expanding Mobile test temperature.

Fabricating new polyvinyl chloride nanocomposites materials necessitates specified nanoparticles for upgrading their dielectric properties, which relies on sort and rates from claiming nanoparticles to controlling the investigation into dielectric characterization at variant warm states and frequencies. Each rate type has another material that fulfills a sure electrical modern requisition. For natural nanoparticles, including clay dependent upon 1%wt. declines true relative permittivity, especially in high frequencies. The true relative permittivity increments with expanding warm temperatures, especially, during low frequencies. Then, expanding clay nanoparticles to more than 1%wt. expands true relative permittivity bit by bit. For inorganic nanoparticles, including fumed silica has the same dielectric characterization of clay nanoparticles as that of minor deviations. However, including zinc oxide nanoparticles provides for high diminishment in the dielectric constant. The diminishment of true relative permittivity by utilizing ZnO nanoparticles will be the most noteworthy decrease that can occur to the other nanoparticles. On the other hand, the true relative permittivity expands with expanding warm temperatures, especially in low frequencies. Including titanium dioxide nanoparticles in the nanocomposite abatements further increments dielectric constant with respect to immaculate polyvinyl chloride characterization, and a nonlinear dielectric characterization of whichever point including variant TiO2 nanoparticles rates for polyvinyl chloride is specified. Hence, the genuine relative permittivity builds for expanding warm temperatures, especially at low frequencies.

Under uniform high voltage immediate current results from the following points: Including clay nanoparticles to Polyvinyl chloride streamlined materials declines the dielectric strength and expands the conduction pasquinade current, as expansion of fumed silica and aluminum oxides declines the dielectric strength, yet expands over capacitive conduction pasquinade current. Including zinc oxide nanoparticles with Polyvinyl Chloride concerning illustration of force cables mechanical materials increments the dielectric strength and lessens the conduction pasquinade current. Fumed silica nanoparticles have secondary capability for diminishing dielectric strength of Polyvinyl chloride nanocomposite materials in aluminum oxide nanoparticles and expanding clinched alongside capacitive conduction pasquinade present. During high temperatures, dielectric strength corruption rate of nanocomposite materials is brought down over dielectric strength corruption to immaculate Polyvinyl chloride.

Clay nanoparticles provide for warm dependability on the recommended nanocomposite materials; therefore, dielectric strength diminishment rate for immaculate streamlined materials will be higher than the dielectric strength diminishment rate of nanocomposites with respect to expanding temperature surroundings from claiming streamlined polymers. Expanding clay nanoparticles to polypropylene and low-thickness polyethylene nanocomposites increments dielectric strength. It is to be noted that being identified with amassing phenomena about clay nanoparticles inside polymer matrix, dielectric strength characterization for nanocomposite might be changed under uniform or non-uniform electric fields, as stated by high rate of clay nanoparticles and polymer atomic kind. Therefore, expanding clay nanoparticles with polyvinyl chloride and high-density polyethylene declines dielectric strength; considering that the nature of polymer grid doesn't gather nanoparticles amassing to evolving dielectric strength characterization of the nanocomposite.

Including clay nanoparticles in polyethylene increments dielectric strength under uniform and non-uniform connected electric fields, as stated by rate of clay nanoparticles and polymer sub-atomic sort. Therefore, conduction current pass recipient polyethylene nanocomposite materials abate with expanding the rate of clay nanoparticles. Dielectric strength of Clay/PE nanocomposite materials in non-uniform electric field will be higher than the dielectric strength of accepted materials in uniform electric fields. In any case, including fumed silica expands permittivity of the polyethylene nanocomposite materials with respect to properties estimations by HIOKI 3522-50 LCR device, including fumed silica which has diminished dielectric strength from claiming new polyethylene nanocomposite materials under uniform and non-uniform electric fields, as stated by rate of fumed silica nanoparticles and polymer atomic kind. It is recognized that conduction present pass recipient polyethylene materials progress with expanding the rate of fumed silica nanoparticles. Dielectric strength for fumed silica/Polyethylene nanocomposite materials in non-uniform electric field is higher than the dielectric strength of immaculate and nanocomposite materials of uniform electric field.

Clay nanoparticles have only the tip of the iceberg capability for moving forward dielectric strength of polyethylene materials over fumed silica nanoparticles. Accordingly, fumed silica nanoparticles have more capacity for moving forward capacitive accuse present for polyethylene materials than clay nanoparticles. Expanding rate of nanoparticles expands amassing electric withstands from claiming encasing forces; therefore, it is proposed that nanoparticles are brilliance scattering and don't exceed 5%wt. In the polyethylene base grid, Clay and fumed silica nanoparticles develop electrical aspects to withstand high connected electric strength rating for with respect to uniform connected voltage under secondary warm states.

REFERENCES

ABB Datasheet. (n.d.). *Skydd för högspänningskondensatorer (Protection for HV Capacitors).* ABB.

Bois, L., Chassagneux, F., Parola, S., Bessueille, F., Battie, Y., Destouches, N., Boukenter, A., Moncoffre, N., & Toulhoat, N. (2009). Growth of Ordered Silver Nanoparticles in Silica Film Mesostructured with a Triblock Copolymer PEO–PPO–PEO. *Journal of Solid State Chemistry, 182*(7), 1700–1707. doi:10.1016/j.jssc.2009.01.044

Chatterjee, A., Bhattacharjee, P., Roy, N. K., & Kumbhakar, P. (2013, February). Usage of nanotechnology based gas sensor for health assessment and maintenance of transformers by DGA method. *International Journal of Electrical Power & Energy Systems, 45*(1), 137–141. doi:10.1016/j.ijepes.2012.08.044

Cookson, A. H. (1990). Influence of electrical insulation on design and performance of power equipment. *IEEE Electrical Insulation Magazine, 6*(6), 7–10. doi:10.1109/57.63093

David & Fréchette. (n.d.). Polymer nanocomposites-major conclusions and achievements reached so far. *IEEE Electrical Insulation Magazine, 29*(6), 29 - 36. DOI: doi:10.1109/MEI.2013.6648751

Dissado, L. A. (2002). Predicting electrical breakdown in polymeric insulators from deterministic mechanisms to failure statistics. *IEEE Transactions on Dielectrics and Electrical Insulation, 9*(5), 860–875. doi:10.1109/TDEI.2002.1038669

Eriksson, E. (1997). *On the use of fuses or fuseless design concepts for protection of modern power capacitors.* Presented at the IEEE Power Society Summer Meeting, Capacitors Subcommittee, Panel Session, Berlin, Germany.

George, S. C., & Thomas, S. (2001). Transport Phenomena Through Polymeric Systems. *Progress in Polymer Science, 26*(6), 985–1017. doi:10.1016/S0079-6700(00)00036-8

Gouda & Thabet. (2014). Thermal Experimental Dielectric Characterization of Cost-Fewer Low-density Polyethylene Nanocomposites. *Advances in Electrical and Electronic Engineering Journal, 12*(5), DOI: doi:10.1109/ICPADM.2000.876419

Gouda, O., Thabet, A., Mubarak, Y. A., & Samir, M. (2014, March). Nanotechnology Effects on Space Charge Relaxation Measurements for Polyvinyl Chloride Thin Films. *International Journal of Electrical Engineering and Informatics (IJEEI). Engineering and Technology, 6*(1), 1–12.

Guastavino, Torello, Squarcia, Tiemblo, & Garcia. (2014). Insulation properties of LDPE nanocomposites obtained by the dispersion of different nanoparticles. *IEEE Transactions on Dielectrics and Electrical Insulation, 21*(2), 444 – 451. DOI: doi:10.1109/TDEI.2013.004222

Hikita, M., Ohtsuka, Sh., Okabe, Sh., & Ueta, G. (2009). Breakdown Mechanism in C3F8/CO2 Gas Mixture under Non-uniform Field on the Basis of Partial Discharge Properties. *IEEE Transactions on Dielectrics and Electrical Insulation, 16*(5), 1413–1419. doi:10.1109/TDEI.2009.5293955

Hirose, H. (1987). A Method to Estimate the Lifetime of Solid Electrical Insulation. *IEEE Transactions on Electrical Insulation, 22*(6), 745–753. doi:10.1109/TEI.1987.298936

Kawamura, Okabe, Takagi, Takesue, & Ozaki. (1990). Temporary Overvoltage and AC Test Voltage in 550kV System with Reduced Insulation Level. *CIGRE*, 33-203.

Laihonen, Gustafsson, Gäfvert, Schütte, & Gedde. (2007). Area Dependence of Breakdown Strength of Polymer Films: Automatic Measurement Method. *IEEE Transactions on Dielectrics and Electrical Insulation, 14*(2), 263-274.

Li, S., Yin, G., Chen, G., Li, J., Bai, S., Zhong, L., Zhang, Y., & Lei, Q. (2010). Short-term Breakdown and Long-term Failure in Nanodielectrics: A Review. *IEEE Transactions on Dielectrics and Electrical Insulation, 17*(5), 1523–1533. doi:10.1109/TDEI.2010.5595554

Li, Z., Okamoto, K., Ohki, Y., & Tanaka, T. (2010). Effects of Nano-filler Addition on Partial Discharge Resistance and Dielectric Breakdown Strength of Micro-Al2O3/Epoxy Composite. *IEEE Transactions on Dielectrics and Electrical Insulation, 17*(3), 653–661. doi:10.1109/TDEI.2010.5492235

Miyahara, H., Nakajima, A., Ishikawa, T., & Yanabu, S. (2008). Insulating System to Reduce the Amount of Oil in Electric Power Apparatus Using Silicone Oil. *IEEE Transactions on Dielectrics and Electrical Insulation, 15*(2), 533–539. doi:10.1109/TDEI.2008.4483474

Mohamed, A. T. (2015, January). A. Experimental enhancement for dielectric strength of polyethylene insulation materials using cost-fewer nanoparticles. *International Journal of Electrical Power & Energy Systems, 64*, 469–475. doi:10.1016/j.ijepes.2014.06.075

Munoz Resta, Horwitz, Mendez Elizalde, Jorge, Molina, & Soledad. (2013). Magnetic and Conducting Properties of Composites of Conducting Polymers and Ferrite Nanoparticles. *IEEE Transactions on Magnetics, 49*(8), 4598 – 4601. DOI: doi:10.1109/TMAG.2013.2259582

Shah, K., Jain, R., Shrinet, V., Singh, A., & Bharambe, D. (2009, June). High density polyethylene (HDPE) clay nanocomposite for dielectric applications. *IEEE Transactions on Dielectrics and Electrical Insulation, 16*(3), 853–862. doi:10.1109/TDEI.2009.5128526

Sharifi, E., Jayaram, S., & Cherney, E. (2010). AC Modeling and Anisotropic Dielectric Properties of Stress Grading of Form-wound Motor Coils. *IEEE Transactions on Dielectrics and Electrical Insulation, 17*(3), 694–700. doi:10.1109/TDEI.2010.5492240

Sugumaran, C. P., Mohan, M. R., & Udayakumar, K. (2010). Investigation of Dielectric and Thermal Properties of Nano-filler (ZrO2) Mixed Enamel. *IEEE Transactions on Dielectrics and Electrical Insulation, 17*(6), 1682–1686. doi:10.1109/TDEI.2010.5658217

Takala, M., Karttunen, M., Pelto, J., Salovaara, P., Munter, T., Honkanen, M., Auletta, T., & Kannus, K. (2008, October). Thermal, Mechanical and Dielectric Properties of Nanostructured Epoxy-polyhedral Oligomeric Silsesquioxane Composites. *IEEE Transactions on Dielectrics and Electrical Insulation, 15*(5), 1224–1235. doi:10.1109/TDEI.2008.4656229

Takala, M., Ranta, H., Nevalainen, P., Pakonen, P., Pelto, J., Karttunen, M., Virtanen, S., Koivu, V., Pettersson, M., Sonerud, B., & Kannus, K. (2010). Dielectric properties and partial discharge endurance of polypropylene-silica nanocomposite. *IEEE Transactions on Dielectrics and Electrical Insulation, 17*(No. 4), 1259–1267. doi:10.1109/TDEI.2010.5539698

Thabet. (2013a). Influence of Cost-Less Nanoparticles on Electric And Dielectric Characteristics of Polyethylene Industrial Materials. *International Journal of Electrical Engineering and Technology, 4*(1), 58-67.

Thabet. (2013b). Experimental Investigation on Thermal Electric and Dielectric Characterization for Polypropylene Nanocomposites Using Cost-fewer Nanoparticles. *International Journal of Electrical Engineering and Technology, 4*(2), 1-12.

Thabet, A., & Mobarak, Y. A. (2010). Dielectric Characteristics of New Nano-Composite Industrial Materials. *International Conference on High Voltage Engineering and Application*, 568-571. 10.1109/ICHVE.2010.5640767

Thabet, A., & Mubarak, Y. A. (2011). *Novel Nanocomposite Insulation Materials for the Enhancing Performance of Power Cables*. 21st International Conference and Exhibition on Electricity Distribution, "CIRED2011", Frankfurt, Germany.

Tuncer, E., Sauers, I., James, D., Ellis, A., & Duckworth, R. (2008). Nanodielectric System for Cryogenic Applications: Barium Titanate filled Polyvinyl Alcohol. *IEEE Transactions on Dielectrics and Electrical Insulation, 15*(1), 236–242. doi:10.1109/T-DEI.2008.4446756

Wu, G. (2010, February). Study on the Failure Factors of Composite Insulation in High-Voltage Storage Capacitors. *IEEE Transactions on Plasma Science, 38*(2).

Wu, G., Zhou, L., Zhang, X., Bian, Sh., Ran, H., & Yu, Ch. (2010). Study on the Failure Factors of Composite Insulation in High-Voltage Storage Capacitors. *IEEE Transactions on Plasma Science, 38*(2), 186–193. doi:10.1109/TPS.2009.2037324

About the Author

Ahmed Thabet Mohamed was born in Aswan, Egypt in 1974. He received BSc (HIE) Electrical Engineering degree in 1997, then, MSc in 2002, then, PhD degree in 2006. He joined with Electrical Power Engineering Group of Faculty of Energy Engineering in South Valley University as a Demonstrator at July 1999, until; he held Professor position at October 2017 up to date. His research interests lie in the areas of analysis and developing electrical engineering models and applications, investigating novel nano-technology materials via addition nano-scale particles and additives for usage in industrial branch, electromagnetic materials, electroluminescence and the relationship with electrical and thermal ageing of industrial polymers. A lot of mobility's has investigated for supporting his research experience in UK, Finland, Italy, and USA …etc. On 2009, he was a founder and established first Nano-Technology Research Center in the Upper Egypt. He has been published high voltage researches in cited international journals and conferences. On 2018, he has been held a full professor position in Engineering of College at Qassim University in Kingdom of Saudi Arabia up to date.

Index

P

S

T

www.ingramcontent.com/pod-product-compliance
Lightning Source LLC
LaVergne TN
LVHW062121300926
849161LV00010B/2517
* 9 7 8 1 7 9 9 8 3 8 2 9 6 *